Quantum Reality, Relativ
and Closing the Epistemi

THE WESTERN ONTARIO SERIES
IN PHILOSOPHY OF SCIENCE

A SERIES OF BOOKS
IN PHILOSOPHY OF SCIENCE, METHODOLOGY, EPISTEMOLOGY,
LOGIC, HISTORY OF SCIENCE, AND RELATED FIELDS

Managing Editor

WILLIAM DEMOPOULOS

Department of Philosophy, University of Western Ontario, Canada

Managing Editor 1980–1997

ROBERT E. BUTTS

Late, Department of Philosophy, University of Western Ontario, Canada

Editorial Board

JOHN L. BELL, *University of Western Ontario*

JEFFREY BUB, *University of Maryland*

PETER CLARK, *St Andrews University*

DAVID DEVIDI, *University of Waterloo*

ROBERT DiSALLE, *University of Western Ontario*

MICHAEL FRIEDMAN, *Stanford University*

MICHAEL HALLETT, *McGill University*

WILLIAM HARPER, *University of Western Ontario*

CLIFFORD A. HOOKER, *University of Newcastle*

AUSONIO MARRAS, *University of Western Ontario*

JÜRGEN MITTELSTRASS, *Universität Konstanz*

WAYNE C. MYRVOLD, *University of Western Ontario*

THOMAS UEBEL, *University of Manchester*

ITAMAR PITOWSKY, *Hebrew University*

VOLUME 73

Wayne C. Myrvold · Joy Christian

Quantum Reality, Relativistic Causality, and Closing the Epistemic Circle

Essays in Honour of Abner Shimony

Springer

Wayne C. Myrvold
Department of Philosophy
University of Western Ontario
London, ON N6A 3K7
Canada
wmyrvold@uwo.ca

Joy Christian
Wolfson College
University of Oxford
Oxford, OX2 6UD
United Kingdom
joy.christian@wolfson.ox.ac.uk

ISBN: 978-90-481-8072-1   e-ISBN: 978-1-4020-9107-0

© Springer Science+Business Media B.V. 2009
Softcover reprint of the hardcover 1st edition 2009
No part of this work may be reproduced, stored in a retrieval system, or transmitted in any form or by
any means, electronic, mechanical, photocopying, microfilming, recording or otherwise, without written
permission from the Publisher, with the exception of any material supplied specifically for the purpose
of being entered and executed on a computer system, for exclusive use by the purchaser of the work.

Printed on acid-free paper

9 8 7 6 5 4 3 2 1

springer.com

Abner Shimony and Manana Sikic

Acknowledgments

This book collects papers stemming from a conference held in honour of Abner Shimony at the Perimeter Institute for Theoretical Physics in Waterloo, Ontario, Canada, July 18–21, 2006, together with some contributions from people who would have liked to have been at the conference, but were unable to attend.

We would like to thank Jeremy Butterfield for initiating the idea of a conference in honour of Abner Shimony. We are also grateful to Howard Burton and Lucien Hardy for providing the venue for the conference at the Perimeter Institute, which also provided generous financial support for the conference. Additional financial support was provided by contributions from the Institute for Quantum Computing and the University of Western Ontario. We thank the staff of the Perimeter Institute, in particular the conference coordinator Kate Gillespie, for their dedicated efforts, which made the conference smoothly. We are grateful to William Demopoulos for his support for this volume, to Lucy Fleet of Springer for her editorial assistance and patience, to David Malament for his assistance with the editing of Howard Stein's paper, and to Paul Tappenden for the superb job he did in transcribing the "Bistro Banter" dialogue.

Finally, we thank the contributors of this volume for their enthusiasm in joining us to honour our mentor and friend, Abner Shimony.

The editors.

Contents

Part I Introduction

Passion at a Distance   3
Don Howard

Part II Philosophy, Methodology and History

**Balancing Necessity and Fallibilism: Charles Sanders Peirce
on the Status of Mathematics and its Intersection with the Inquiry
into Nature** 15
Ronald Anderson

Newton's Methodology 43
William Harper

Whitehead's Philosophy and Quantum Mechanics (QM): *A Tribute
to Abner Shimony* 63
Shimon Malin

Bohr and the Photon 69
John Stachel

Part III Bell's Theorem and Nonlocality
A. Theory

**Extending the Concept of an "Element of Reality" to Work
with Inefficient Detectors** 87
Daniel M. Greenberger

**A General Proof of Nonlocality without Inequalities
for Bipartite States** 95
GianCarlo Ghirardi and Luca Marinatto

On the Separability of Physical Systems 105
Jon P. Jarrett

Bell Inequalities: Many Questions, a Few Answers 125
Nicolas Gisin

B. Experiment

**Do Experimental Violations of Bell Inequalities Require a Nonlocal
Interpretation of Quantum Mechanics? II: Analysis à la Bell** 141
Edward S. Fry, Xinmei Qu, and Marlan O. Scully

The Physics of $2 \neq 1 + 1$ 157
Yanhua Shih

**Part IV Probability, Uncertainty, and Stochastic Modifications
of Quantum Mechanics**

**Interpretations of Probability in Quantum Mechanics:
A Case of "Experimental Metaphysics"** 211
Geoffrey Hellman

**"No Information Without Disturbance": Quantum Limitations
of Measurement** 229
Paul Busch

How Stands Collapse II 257
Philip Pearle

**Is There a Relation Between the Breakdown of the Superposition
Principle and an Indeterminacy in the Structure of the Einsteinian
Space-Time?** 293
Andor Frenkel

Indistinguishability or Stochastic Dependence? 311
D. Costantini and U. Garibaldi

Part V Relativity

Plane Geometry in Spacetime 327
N. David Mermin

The Transient *nows* 349
Steven F. Savitt

Quantum in Gravity? 363
Michael Horne

Contents   xi

A Proposed Test of the Local Causality of Spacetime 369
Adrian Kent

**Quantum Gravity Computers: On the Theory of Computation
with Indefinite Causal Structure** 379
Lucien Hardy

**"Definability," "Conventionality," and Simultaneity
in Einstein–Minkowski Space-Time** 403
Howard Stein

Part VI Concluding Words

Bistro Banter: *A Dialogue with Abner Shimony and Lee Smolin* 445

Unfinished Work: A Bequest 479
Abner Shimony

Bibliography of Abner Shimony 493

Index 507

Part I
Introduction

Passion at a Distance

Don Howard

In 1984, Abner Shimony invented the expression, "passion at a distance," to characterize the distinctive relationship of two entangled quantum mechanical systems [1]. It is neither the local causality of pushes, pulls, and central forces familiar from classical mechanics and electrodynamics, nor the non-local causality of instantaneous or just superluminal action at a distance that would spell trouble for relativity theory. This mode of connection of entangled systems has them feeling one another's presence and properties enough to ensure the strong correlations revealed in the Bell experiments, correlations that undergird everything from superfluidity and superconductivity to quantum computing and quantum teleportation, but not in a way that permits direct control of one by manipulation of the other. Intended to echo Aristotle's distinguishing of "potentiality" from "actuality" as different senses of "being," Shimony's "passion at a distance" is all about tendency and propensity, not the concreteness whose misplacement in realm of the physical was lamented by Alfred North Whitehead.

No metaphor is better suited, however, to describe as well the feelings of Abner's students, colleagues, and friends for his presence in their lives and for the character that he brings to his own work, to his work with others, and to the world whose betterment has been his abiding aim. However great the distances later grow, world lines once intersecting Abner's remain forever entangled with his in this passionate way. Here are lives and careers with outcomes that will never again be independent, however much the outward parameters might differ and change.

There is passion, too, in all that Abner does, whether campaigning against nuclear weapons, helping a graduate student over a rough patch in dissertation research, or closing the last loophole in a no-go theorem. There is passionate sympathy and passionate intensity. These are passions driven in equal measure by high moral and intellectual principle and by simple joy in living.

D. Howard
Department of Philosophy, and Program in History and Philosophy of Science, University of Notre Dame, Notre Dame, IN 46556
e-mail: dhoward1@nd.edu

W.C. Myrvold and J. Christian (eds.), *Quantum Reality, Relativistic Causality, and Closing the Epistemic Circle,* The Western Ontario Series in Philosophy of Science 73,
© Springer Science+Business Media B.V. 2009

The main object of Abner's intellectual passion has been, of course, the philosophical foundations of physics. Within that larger arena, his having done the work that made possible the experimental tests of the Bell inequalities, which, in turn, made plain the ubiquity and deep importance of quantum entanglement, will be the achievement for which Abner is best remembered, and rightly so. No revolution in physics will as much transform either our understanding of nature or the range and subtlety of our mastery of it. No thinker has played a more crucial role in facilitating that revolution than has Abner. No wonder, then, that questions descending from this early and important work of Abner's dominate the present volume of papers honoring his legacy, as well as an earlier pair of volumes with a similar aim [2].

But Abner's contributions to philosophy, physics, and intellectual life more generally extend well beyond the work that made his reputation. Can one call to mind another philosopher-physicist of our era who has been a presence also in fields as diverse as naturalistic epistemology, inductive logic, and children's literature? Let's start with this last and work our way back to philosophy of physics.

Tibaldo and the Hole in the Calendar is a delightful story [3]. Young Tibaldo Bondi, who lives in Bologna, suffers the misfortune of his twelfth birthday's being lost when Pope Gregory XIII eliminated ten days in October, 1582, as part of his reform of the calendar. The aggrieved Tibaldo protests all the way to the Pope himself, who is persuaded to decree the celebration of all of the otherwise missing festivals, saints' days, and holidays. Beautifully illustrated by Abner's son, Jonathan, *Tibaldo* is a good read but also teaches the young reader a lot about Renaissance science, culture, and history. The science might be far removed from quantum mechanics, but the youthful enthusiasm for science is the same as that which animates all of Abner's work (yes, there is more than a little of Abner in the character of Tibaldo, not just the quick, wide-ranging mind, but also the moral character that will not rest until a wrong is righted). And the broad range of *Tibaldo* reminds one of Abner's many other enthusiasms, music and art prominent among them.

At some seeming remove from quantum physics is also Abner's long interest in naturalistic epistemology. The larger context here is a rich and old American tradition of embedding epistemology in a broadly Darwinian, evolutionary framework, a tradition dating back to James Mark Baldwin, William James, and John Dewey, among others, a tradition wherein cognition is to be regarded as an adaptive trait of the human organism. But an important local influence was Abner's meeting and learning from the work of the psychologist Donald Campbell, the person who invented the term "evolutionary epistemology" [4], and gave to the project so known a theoretical articulation not found in the broad-brush musings of that earlier generation.

Campbell's evolutionary epistemology, and Abner's, differs importantly from that of, say, Karl Popper, in that, for Popper, Darwinian evolution is mainly a metaphor or model for a diachronic structure of theory change, whereas, for Campbell and Shimony, evolution is more than just a model, the evolutionary psychology of cognition being continuous with the evolutionary biology of the whole human organism. One might, for that reason, think to compare Abner's naturalism with that of W.V.O. Quine, who celebrates a debt to Dewey and, like Dewey, draws anti-foundationalist and (some think) anti-realist conclusions from naturalistic premises. The evolutionary version of that argument would ask whether the relationship of an adaptive trait to its environment bears any noteworthy similarity or analogy to a semantic relationship between theory and world or proposition and fact. Crudely put, the question is whether utility in the guise of fitness has anything to do with truth. But Abner's naturalism is not Quine's. Abner is to Quine somewhat as the critical realist Roy Wood Sellars was to Dewey, arguing that a thoroughgoing evolutionary naturalism implies a robust realism and a strong notion of truth, it being precisely the presumptive truth of evolutionary theory that grounds one's naturalism and precisely the dependable, truth-tracking capacity of human cognition, especially in its highly refined scientific form, that renders it adaptive. And by emphasizing *truth-tracking* as the distinguishing adaptive feature of human cognition, Abner's naturalism can ground a normative epistemology – unlike Campbell's and Quine's mere descriptivism – wherein the biology and psychology of cognition give us tools for the critical assessment of scientific practice and epistemic practice more generally.

Abner's epistemological naturalism found an early and engaging expression in his 1971 paper, "Perception from an Evolutionary Point of View" [5]. But his most compelling presentation of the view came a decade later, in the paper "Integral Epistemology" [6], the significance of which is obvious from its placement as the lead essay in the 1993 collection of Abner's philosophical papers, *Search for a Naturalistic World View* [7]. That collection's title makes clear that naturalism is, indeed, integral to Abner's larger philosophical project, scientific perspectives on knowledge being necessary to achieve the theoretical "closure" that Abner prizes as a pre-eminent virtue of a comprehensive, philosophical world view [8].

More than anywhere else, it is in Abner's extensive work on inductive logic and scientific inference that his pursuit of a critical, normative, epistemological naturalism is on display. This, too, is a very old interest, dating back at least to his widely-discussed 1965 lecture and the resulting, vastly expanded 1970 paper, "Scientific Inference" [9]. The view that Shimony defends, "tempered personalism," is to be situated within the larger arena of Bayesian approaches to the formal, probabilistic modeling of scientific inference. It distinguishes itself within that crowded field mainly by its trying to avoid the problem of arbitrary priors by restricting attention to the scientists' seriously proposed hypotheses and by probability assignments that reflect not just guesses or subjective whims but the accumulated insights of science. This is a normative project, influenced by Harold Jeffreys, yielding not just rational criteria for the comparative preferment of hypotheses in specific instances, but also a pragmatic justification of induction in general. It is, at the same time, an exercise in philosophical naturalism precisely by the restriction to seriously considered hypotheses and the constraining of the priors on the basis of extant science. Scientific inference is thus justified, in part, on the basis of earlier scientific achievements. Critics of naturalism see this as vicious circularity, with science seeking self-justification. Champions of such a weakly normative naturalism, such as Abner, see the circularity as virtuous theoretical closure, science, as it were, pulling itself up by its bootstraps.

Abner's interest in epistemic probability and the probabilistic modeling of scientific inference is something that he shared with, among others, his teacher, Rudolf Carnap, and his longtime friend, Richard Jeffrey. As was true of Carnap, Abner's interest in the role of probability in scientific inference is continuous with a still broader interest in the role of probability and statistics in science, from statistical thermodynamics to evolution and population genetics. The concept of information and its physical avatar, entropy, is one especially important place where questions about epistemic probability and questions about physical probability intersect. No surprise, then, that this, too, has long engaged Abner's attention, whether it is a matter of his trying to make sense of Leo Szilard's 1929 "Demon" paper [10], which gave birth to the modern habit of connecting information and entropy, or assaying the contemporary project of providing an information-theoretic axiomatic foundation for quantum mechanics. One simultaneously professional and personal expression of this specific interest was Abner's bringing to public view two important, previously unpublished papers by Carnap on entropy [11].

Mention of probability in physics brings us full circle, of course, back to the suite of issues in the foundations of quantum mechanics where Abner made his name. The name thus made will be forever tied to experimental tests of the Bell inequalities, more on which in a moment. But there is a background to Abner's early work on Bell's theorem that provides a needed context for understanding that work. It is relevant to recall here that Abner began his career in the 1960s as one of a still small community of philosopher-scientists with two Ph.D.s. Abner's first was in philosophy, from Yale (also Abner's undergraduate alma mater), in 1953; his second was in physics, from Princeton, in 1962. At Yale, Abner got a heady dose of Whitehead from Paul Weiss and Robert Calhoun. That exposure to process metaphysics provided at least a vocabulary for describing how, in the quantum world, the actual is the realization of potentialities or propensities, which have equal claim to being part of our physical ontology, as well as for appreciating how, in a world where elementary particles might be endowed with "proto-mental characteristics," consciousness can play a role in physics, and physics a role in consciousness [12]. It was at Yale that Abner also got his first introduction to Peirce – Weiss again being the intermediary – and thus his introduction to a kind of pragmatism and evolutionary naturalism that does not eschew metaphysics but insists on its respecting the judgments of science.

At Princeton, Abner had the good fortune to study with physicists such as Eugene Wigner and John Archibald Wheeler, who encouraged both an interest in foundational questions and speculative approaches to physics, as long as the latter were subjected to the same critical standards in play elsewhere in science. This was the same Princeton physics environment in which, under Wheeler's direction, Hugh Everett had in 1957 produced the "many-worlds" or "relative-state" interpretation of quantum mechanics [13], and where, just five years earlier and immediately before his exile, David Bohm had developed his version of a non-local hidden variables theory [14].

So when Abner arrived at MIT in 1959 to begin his teaching career in the Humanities department, his had been a truly unusual intellectual formation. He was more deeply grounded in physics than any young philosopher since perhaps Moritz Schlick, Hans Reichenbach, or Rudolf Carnap, but, of course, with an importantly different philosophical pedigree, and more thoroughly steeped in philosophy than any young physicist since, perhaps, Albert Einstein, Erwin Schrödinger, or Werner Heisenberg. Here was a mind in which were married uncompromising technical rigor and a boldly speculative metaphysical imagination.

Abner's first major paper on the philosophical foundations of physics, "Role of the Observer in Quantum Theory," appeared in 1963 [15]. It has aged well and rewards the reader still today. In it, Abner assesses two views on quantum measurement, the von Neumann and London and Bauer picture wherein wave-packet collapse occurs when the results of measurement are registered in consciousness, and Bohr's complementarity interpretation, which relates the change of state upon measurement to the role of the experimental arrangement in the description of quantum phenomena. Abner judges each a failure, the former for want of corroborating psychological evidence and its making problematic the notion of intersubjective agreement about the results of observation, the latter for its vagueness and its making problematic the construction of a unified physical ontology because of its supposed fundamental distinction between a macroscopic classical realm and the quantum microworld. What is needed to solve the measurement problem, Abner concludes, is a reformulation of quantum mechanics itself.

Precisely here is to be found the origin of so many of Abner's more specific interests and so much of his own later work. Research into non-linear variants of the Schrödinger equation is one such. Better to treat wave packet collapse as a stochastic process or as induced by some physical factor, such as the gravitational potential, than as resulting from subjective mentality's intrusion into the objective physical realm, though the latter is not to be ruled out on *a priori* grounds. Better still if collapse were not needed in the first place. Thus Abner's decades-long interest in hidden variables theories.

In all such areas, the approach was basically the same. First, make sure that the space of theoretical possibilities is filled, which means encouraging sometimes highly heterodox theoretical variants. Second, seek a clear and natural way to parameterize and parse the resulting possibility space. Third, look for no-go theorems that might rule out large swaths of that space. And, fourth, within the space that remains, where logic and mathematics alone can't do the work, design and do experiments to kill off the rest of the unfit.

Thus it was that Abner came to co-write in 1969 his most important single paper, with John Clauser, Michael Horne, and Richard Holt, "Proposed Experiment to Test Local Hidden-Variable Theories" [16]. The story of this paper's genesis and aftermath has now been told several times and need not be told again in this brief resume of Abner's life and work [17]. Suffice it to record here that, the scrupulously honest and modest arraying of authorship notwithstanding, Abner was an engine in this effort to recast Bell's theorem in a form amenable to laboratory test and to him goes the full measure of credit for all that followed by way of proving the now virtually undeniable presence in our world of the deepest mystery of the quantum realm, the entanglement evinced by the joint state of previously interacting systems in orthodox quantum mechanics or the corresponding quantum potential in the Bohmian variant. Ever since, Abner has been the guiding spirit, generous collaborator, and gentle critic for those who carry on the work of refining and extending Bell's original insight.

We could dawdle over any of a number of Abner's more specific engagements in this later work. Most noteworthy, however, have been his reflections on Jon Jarrett's rederivation of Bell's theorem. In 1984, in a University of Chicago dissertation written under the supervision of Abner's old and close friend, Howard Stein, Jarrett showed that the original Bell locality condition is, in fact, a conjunction of two logically independent conditions, which Jarrett dubbed "locality" and "completeness" [18]. The potential gain from Jarrett's work was immediately obvious, for by so decomposing the Bell locality condition one could hope to identify more narrowly the reason or reasons why orthodox quantum mechanics, Bohm-type hidden variables theories, and nature itself, as revealed in the Bell experiments, agreed in violating the Bell locality condition. Especially tantalizing was the suggestion that the Jarrett "locality" condition captured more precisely than did the original Bell "locality" condition whatever might be the relevant constraint from the side of relativity theory. But otherwise the physical significance of the two Jarrett conditions was not transparent.

Enter Abner. In two deep and reflective papers [1, 19], he pushed and prodded, reworked and revised. The result was his now-classic recrafting of the Jarrett conditions as what are known as "parameter independence" and "outcome independence." The first – a cousin of relativistic locality constraints or the principle of microcausality in quantum field theory – asserts the stochastic independence of measurement outcomes in one wing of a Bell experiment from the choice of a parameter to measure in the other wing. The second, which one suspects is deeply related to quantum nonseparability or entanglement, asserts the stochastic independence of measurement outcomes in one wing from measurement outcomes in the other wing.

There are various reasons why Abner's gloss on Jarrett is important. High on Abner's own list would be that clearer insight into the precise nature of quantum mechanical non-locality makes possible our employing Bell's theorem and experimental demonstrations of violations of the Bell inequality as levers by means of which to uncover what Abner has long suspected is the really deep lesson, namely, that quantum mechanics points to the need for profound changes in our understanding of space-time structure. Nothing so simple as just discretizing space-time structure is envisioned here. No, for Abner the question has always been the Whiteheadean or Aristotelian one about the actualization of potentialities. That's the whole point of turning the Jarrett decomposition theorem explicitly into a theorem about the two different ways in which measurement *outcomes* (which are actualizations) can fail to be independent of physical events and processes – measurement outcomes and parameter settings – in space-like separated regions of space-time. The question, then, is this: What kind of structure in space-time is necessary to sustain the mode(s) of *dependence* between space-like separated events revealed in the Bell experiments? The possibilities are many, from novel topologies to Whiteheadean "prehension," and no possibility is to be ignored simply for being unfashionable.

Einstein thought that the arrow of implication ran in the other direction. Einstein thought that the more or less *a priori* necessity of what he called the "separation principle," a principle enshrined in general relativistic space-time structure and in any field theory built on such a framework, entailed that quantum mechanics could not be the right framework for fundamental physics [20]. Abner tends to think otherwise. A future physics will have to render judgment.

That physics will have to be the judge – not intuition, not *a priori* reason, and certainly not dogma – brings us to one final point that Abner has always stressed. The questions at issue here are not just physical, they are metaphysical questions, questions about the fundamental nature of reality. They are the really big questions, if you will. But their answers are to be sought not by some mode of inquiry or insight beyond physics (or biology or psychology). The answers are to be sought within physics, by doing more physics. The kind of inquiry intended here is what Shimony famously characterized as "experimental metaphysics." The basic idea is simply that one asks what is the manifest ontology of one's current best physics. The idea is controversial. Many metaphysicians are ill at ease with a metaphysics that is contingent and corrigible. Many physicists (and some philosophers) are still shy about pressing physics' claims too far beyond the bounds of the observable. But Abner is a bit unusual here, too, combining a kind of metaphysical temperament – the big questions are the ones worth asking – with the scientist's cautious skepticism. The resulting attitude toward metaphysics is as much a part of Abner's philosophical naturalism as is attitude toward epistemology. Here, too, the circle closes.

The professional world has long recognized Abner's achievements and their import for both the physics and the philosophy that have been his life. He is a Fellow of the American Academy of Arts and Sciences and a Fellow of the American Physical Society. He has been a visiting professor at Paris, Mount Holyoke, Geneva, and the ETH (Zurich). In 1996, Abner's collected philosophical papers [7] were awarded the Lakatos Prize in the Philosophy of Science. He was honored by election as President of the Philosophy of Science Association for 1995 and 1996. And, as mentioned above, this collection of papers is the second such by which his students, friends, and colleagues have sought to show their respect and affection.

One might guess, however, that Abner would prefer that we think well of him less for the honors that have come his way, and more for the service that he has done for others. Such service has been a constant in his life. His students know this well. They know the Abner who put so much energy into every dissertation he directed that he allowed himself only two Ph.D. students at a time, one each in physics and philosophy. Abner's colleagues know it well, too. His service to them has come in many forms. Consider only the evidence of the volumes he has edited, or with which he has otherwise assisted, to honor and showcase the works of those with whom he has enjoyed the closest intellectual and personal relationships, including Laszlo Tisza [21], Eugene Wigner [22], Daniel Greenberger [23], Howard Stein [24], John Stachel [25], and Martin Eger [26].

Let us celebrate, finally, Abner the person. Born in 1928 in Columbus, Ohio, and raised in Memphis, Tennessee. The son of Morris (Moshe) and Sara Altman Shimony, and the stepson of Dora Farber Shimony. Husband of the noted Wellesley anthropologist, Annemarie Anrod Shimony, with whom he had two sons, Ethan and Jonathan [27]. After Annemarie's untimely demise in 1995, he became in 1998 the husband of Helen Claire Walker, his close high school friend, who died in 2002. He is now married to Manana Sikic, to whom he was wed in 2005 by his former doctoral student, the late Fr. Ronald Anderson, who performed a miraculously moving ceremony combining Catholic and Jewish elements. All this and more is Abner. Army signalman and longtime Professor of both Philosophy and Physics at Boston University, where he has been a prominent intellectual and moral presence. Lover of literature, art, and music. Mentor, friend, and champion of peace.

References

1. Shimony, A. (1985). "Controllable and Uncontrollable Non-locality." In Kamefuchi, S. et al., eds. *Foundations of Quantum Mechanics in the Light of New Technology* (Tokyo: The Physical Society of Japan), 225–230; reprinted in [7], vol. 2, 130–139.
2. Cohen, R.S.; Horne, M.; and Stachel, J. eds. (1997). *Quantum Mechanical Studies for Abner Shimony*. Vol. 1, *Experimental Metaphysics*. Vol. 2, *Potentiality, Entanglement, and Passion-at-a-Distance* (Dordrecht: Kluwer).
3. Shimony, A. (1997). *Tibaldo and the Hole in the Calendar* (New York: Springer-Verlag).
4. Campbell, Donald (1974). "Evolutionary Epistemology," in Schilpp, P.A. ed., *The Philosophy of Karl R. Popper*. La Salle, Il: Open Court, 412–63.
5. Shimony, A. (1971). "Perception from an Evolutionary Point of View." *Journal of Philosophy* 68, 571–583; reprinted in [7], vol. 1, 79–91.
6. Shimony, A. (1981). "Integral Epistemology." In Brewer, M. and Collins, B., eds. *Scientific Inquiry and the Social Sciences* (San Francisco: Jossey-Bass), 98–123; reprinted in [7], vol. 1, 3–20.
7. Shimony, A. (1993). *Search for a Naturalistic World View*, 2 vols. (Cambridge: Cambridge University Press).
8. Another noteworthy expression of Abner's long interest in naturalism is the collection that he co-edited with Debra Nails, *Naturalistic Epistemology: A Symposium of Two Decades* (Dordrecht: Reidel, 1987), which celebrates and is mainly built around the various perspectives on naturalism long prominent in discussions at Abner's principal professional home, Boston University.
9. Shimony, A. (1970). "Scientific Inference." In Colodny, R., ed. *The Nature and Function of Scientific Theories* (Pittsburgh: University of Pittsburgh Press), 79-172; reprinted in [7], vol. 1, 183–273.
10. Szilard, L. (1929). "Über die Entropieverminderung in einem thermodynamischen System bei Eingriffen intelligenter Wesen." *Zeitschrift für Physik* 53, 840–856.
11. Carnap, R. (1978). *Two Essays on Entropy*. A. Shimony, ed. (Berkeley: University of California Press).
12. See Shimony, A. (1965). "Quantum Physics and the Philosophy of Whitehead." In Black, M., ed. *Philosophy in America* (London: Allen and Unwin), 240–261; reprinted in [7], vol. 2, 291–309.
13. Everett, H. (1957). "'Relative State' Formulation of Quantum Mechanics." *Reviews of Modern Physics* 29, 454–462.
14. Bohm, D. (1952). "A Suggested Interpretation of the Quantum Theory in Terms of 'Hidden' Variables." *Physical Review* 85, 166–179, 180–193.
15. Shimony, A. (1963). "Role of the Observer in Quantum Theory." *American Journal of Physics* 31, 755–773; reprinted in [7], vol. 2, 3–33.

Passion at a Distance 11

16. Clauser, J.F.; Horne, M.A.; Shimony, A.; and Holt, R.A. (1969). "Proposed Experiment to Test Local Hidden-Variable Theories." *Physical Review Letters* 23, 880–884.
17. Best is the brand-new telling in: Gilder, L. (2008). *The Age of Entanglement: When Quantum Physics Was Reborn* (New York: Knopf).
18. Jarrett, J. (1984). "On the Physical Significance of the Locality Conditions in the Bell Arguments." *NOÛS* 18, 569–589. See also Jarrett, J. (1989). "Bell's Theorem: A Guide to the Implications." In Cushing, J. and McMullin, E., eds. *Philosophical Consequences of Quantum Theory: Reflections on Bell's Theorem* (Notre Dame, IN: University of Notre Dame Press), 60–79.
19. Shimony, A. (1986). "Events and Processes in the Quantum World." In Penrose, R. and Isham, C.J., eds. *Quantum Concepts in Space and Time* (Oxford: Oxford University Press), 182–203; reprinted in [7] vol. 2, 140–162.
20. See Don Howard (1985), "Einstein on Locality and Separability," *Studies in History and Philosophy of Science* **16**, 171–201.
21. Shimony, A. and Feshbach, H., eds. (1982). *Physics as Natural Philosophy: Essays in Honor of Laszlo Tisza on His Seventy-fifth Birthday* (Cambridge, MA: MIT Press).
22. Shimony, A., ed. (1997). *The Collected Works of Eugene Paul Wigner*. Part A, Vol. 3, Part 2, *Foundations of Quantum Mechanics* (Berlin and Heidelberg: Springer-Verlag).
23. Horne, M.A.; Shimony, A.; and Zeilinger, A., eds. (2000). Festschrift for Daniel Greenberger. *Foundations of Physics* 29, nos. 3 & 4.
24. Malament, D., ed. (2002). *Reading Natural Philosophy: Essays in the History and Philosophy of Science and Mathematics* (Chicago: Open Court). This is a *Festschrift* for Howard Stein, with an introduction by Shimony.
25. Ashtekar, A.; Cohen, R.S.; Howard, D.; Renn, J.; Sarkar, S.; and Shimony, A. (2003). *Revisiting the Foundations of Relativistic Physics: Festschrift in Honor of John Stachel* (Dordrecht: Kluwer).
26. Shimony, A., ed. (2006). *Science, Understanding, and Justice: The Philosophical Essays of Martin Eger* (Chicago: Open Court).
27. See Abner's forthcoming edition, with an introduction, of Annemarie's *Iroquois Portraits* (Syracuse, NY: Syracuse University Press).

Part II
Philosophy, Methodology and History

Balancing Necessity and Fallibilism: Charles Sanders Peirce on the Status of Mathematics and its Intersection with the Inquiry into Nature

Ronald Anderson[*]

Abstract An interest in Charles Sanders Peirce and pragmatist thought in general emerged in the United States in the middle of last century to exert a powerful influence on a generation of American philosophers educated in the 1940s and 1950s, including Abner Shimony, whose thought is the occasion for this paper. Those threads in Peirce's work related to developing a scientifically informed worldview and metaphysics were the natural influences on Abner and this paper will begin by briefly reviewing a number of these threads and their influences in his writings. This sets the scene for the main project of the paper, an earlier historical project on a related aspect of Peirce's thought—his understanding of mathematics and its place in the description of nature. Mathematics was a foundational discipline for Peirce, one with qualities of necessity and certainty, features that stand in interesting contrast and tension to Peirce's view of an evolving nature which is governed by chance and our knowledge of which is always fallible and thus open to revision. Exploring these issues reveals deep background beliefs structuring Peirce's thought. The paper concludes in the contemporary realm with the speculation that due to the scientific developments of the 20th century, aspects of Peirce's work that formed a vision for

---

R. Anderson
Department of Philosophy, Boston College, Chestnut Hill, MA 02467, USA

[*]In Memory of Fr. Ronald Anderson, SJ.

It is with great sadness we note here that Fr. Ronald Anderson, SJ passed away soon after completing this essay in honour of his former mentor, Abner Shimony. Ron, who received a doctoral degree in particle physics in 1980 from the University of Melbourne, did his second doctorate in philosophy under Abner at Boston University; this was awarded in 1991.

It is likely that this essay will be the last work by Ron to appear in print. Readers should be aware that he did not have an opportunity to proof-read the final copy. The editors have had to edit a number of sentences in the text for grammar and punctuation, but the essay as printed here is essentially as he left it, and the voice in which it speaks is Ron's own. *The editors.*

W.C. Myrvold and J. Christian (eds.), *Quantum Reality, Relativistic Causality, and Closing the Epistemic Circle,* The Western Ontario Series in Philosophy of Science 73,
© Springer Science+Business Media B.V. 2009

a scientific metaphysics for earlier generations may be less relevant now. Nevertheless, the naturalistic spirit and orientation of Peirce's work remains compelling and productive.

1 Peirce Scholarship and Peirce as a Resource for Forming a Scientifically Informed Metaphysics

An intricate and complex yet engaging and enticing task awaits the contemporary student seeking to enter the worlds and thought of Charles Sanders Peirce (1838–1914). One reason for this is simply: Peirce's vast output of texts. Estimates are given of published texts of around 10,000 printed pages with unpublished manuscripts of around 80,000 sheets.[1] Other reasons are the often technical nature of Peirce's writings, and their wide range. The topics addressed by Peirce include (in intersecting combinations) logic, semiotics, purely mathematical and scientific topics, the philosophical significance of these topics, and a sustained concern for building a comprehensive philosophical system informed by mathematics and science.[2] A further complexity arises from Peirce's thought evolving during his life, generating a subtle layered effect in his work that needs navigating carefully.

In addition, an equally intricate task awaits in tracing through scholarship on Peirce, a body of writings that matches the richness of Peirce's own writings. From the 1940s, when studies of his thought emerged (see [3] and [4] as early examples) to the present, the student is confronted with a range of academic styles and concerns reflecting the changing philosophical landscape of the 20th century.[3] During this period those reading Peirce as an historical project in classical American philosophy are together with those primarily concerned with locating resources within Peirce for informing present projects. Moreover, for a generation formed in and after the historiographic and sociological revolution in the study of science of the past few decades, it is instinctive to locate thinking deeply within its context, to attend to unique scientific practices of a time and culture, and to eschew seeking any abiding general essence to science. Thus for this generation factors such as the radical difference in present science (and mathematics) from that of Peirce's world lead to an unease with the project of earlier generations of Peirce scholars, of locating resources in Peirce for building a contemporary naturalized form of metaphysics or epistemology. Instead contextual projects suggest themselves, with force given the dense nature of Peirce's writings, with the attendant dangers of getting caught in projects that resist closure.

For exploring Peirce's life, Joseph Brent's biography [8] provides a perceptive and sympathetic reading, one that traces Peirce's upbringing in an academically and

---

[1] Ketner and Putnam, introduction to [1, p. 8].

[2] For a careful comment on the last mentioned of these projects in Peirce, see [2].

[3] For a general history in the 20th century of pragmatist thought relevant to the issues here, see [5–7].

socially privileged world in Cambridge, Massachusetts, through to his final years, isolated in Milford, Pennsylvania, at a property acquired in 1888.

Brent brings out vividly the complexity of Peirce's personal life and the degree it was marked with loss and tragedy through health problems, psychological struggles, financial difficulties, and the absence of a steady institutional context in which to work. For Brent:

> The beauty of the past arises from its permanence, from the impossibility of changing what was done. It is this forgiving permanence, suffusing even folly and tragedy with melancholy beauty, that transform the brilliant, bitter, humiliating, and above all tragic life of the American philosopher Charles Sanders Peirce into an odyssey of spirit which is at once fascinating, saddening, and compelling [8, p. 1].

In the spirit of Brent's assessment (and as Brent recognizes in this passage), the invariable experience of reading Peirce is to grow in admiration for his persistent creativity in generating ideas and texts in spite of the tragedy surrounding his life. Also on this point, Ketner and Putnam remark: "At times he must have written without stop: perhaps this explains at least in part his difficult nature" [1, p. 8]. The compelling quality is evident in the way Peirce is able to captivate and inspire others, reflected in the steady positive assessments by those who have encountered his thought. For example, his friend and supporter William James, writing to the President of Harvard in 1895 to recommend Peirce for teaching a course, remarked:

> He is the best man by far in America for such a course, and one of the best men living. The better graduates would flock to hear him—his name is one of mysterious greatness for them now—and he would leave a wave of influence, tradition, gossip, etc. that wouldn't die away for many years (quoted in [8, p. 243]).

And, as a long time scholar of Peirce's thought, Max Fisch, assessed the place of Peirce in glowing terms:

> Who is the most original and the most versatile intellect that the Americas have so far produced?" The answer "Charles S. Peirce" is uncontested, because any second would be so far behind as not to be worth nominating (quoted in [8, p. 2]).

The introductions to the two volume collection of Peirce's philosophical writings by Houser and Kloesel [9] and Houser [10] provide a concise entry to Peirce's thought, and the texts by Goudge [11], Murphey [12] and Hookway [13], and the more recent collection of studies edited by Houser et al. [14], together give a secure and comprehensive view of Peirce's thought and texts.

On venturing into Peirce's writings the now canonical multi-volume collection of his papers by Paul Weiss and Charles Hartshorne, joined later by Arthur W. Burks [15] is still the standard source and is now available on-line.[4] In addition, a chronological edition of his writings has been emerging [16]. Peirce has captivated many a student, not only by the power and breath of his ideas but also by his direct and immediate style as he develops ideas in an exploratory manner. His sustained use of vivid metaphors and tropes adds color, irreducible complexity, and subtlety to his

---

[4] References to the *Collected Papers* will be designated by the customary letters *CP x.y*, where *x* is the volume and *y* the paragraph number.

thinking. Indeed, for Peirce metaphor is intrinsic to philosophy, as we can see in a passage sounding distinctly contemporary, given recent studies of the intrinsic and important place of metaphor in writing and cognition:

> Metaphysics has been said contemptuously to be a fabric of metaphors. But not only metaphysics, but the logical and phaneroscopical [phenomenological] concepts need to be clothed in such garments. For a pure idea without metaphor or other significant clothing is like an onion without a peel [10, p. 392].

Given the power of Peirce's texts in these ways, the use of quotations from Peirce in the following is a perspicuous way to bring out the points I wish to make.

One of the centers for Peirce studies mid-century was Yale's philosophy department. Peirce scholars such as Paul Weiss and Rulon Wells had arrived in 1945 and John E. Smith in 1952. Abner's undergraduate years there in the 1940s, with teachers such as Weiss, and subsequent PhD studies in the early 1950s, were in the ambiance of these studies. Also, his Master's degree at Chicago in between undergraduate and PhD studies had another Peirce scholar, Charles Hartshorne, as director, who was familiar as well with the thought of Alfred North Whitehead. Later in the 1950s, when Richard Rorty and Richard Bernstein graduated from the department, it has been characterized as a "hotbed of pragmatist activity" [17, p. 97]. Rorty's PhD thesis on the history of concept of Potentiality for example, ended with an explicit pragmatist conclusion: "our descriptions of logical empiricism's difficulties ... suggest that we need to strive for the sort of rapprochement between formal logic, semiotics, and traditional epistemology which is found in the work of Peirce." (quoted in [17, p. 96]). When speaking of his earlier years, the captivating quality of Peirce's thought is evident in Abner's recollections:

> I read Peirce avidly and assented to almost everything that I understood of his semiotics, phenomenology, scientific methodology, pragmatism, critical common-sensism, and evolutionary metaphysics. Peirce's mixture of logical toughness, immersion in the history and practice of the natural sciences, and metaphysical speculation was inspiring to me then and continues to be so [18, vol. I, p. x].[5]

> I read lots of Peirce's papers, and I loved Peirce. I love Peirce to this day, and I think my point of view is closer to Peirce than to anyone else ([19, p. 15]).

As an illustration of the resources from Peirce's writings for forming a scientifically informed metaphysics, in the rest of this section I will outline a number of themes from Peirce's thought that figure in Abner's writings. This part of the paper will also serve to bring into relief aspects of the nature of Peirce's inquiry into the "facts of nature" (*CP* 2.750) as well as respect the occasion for the paper.

One central characteristic of Peirce's thought is his "fallibilism." For Peirce, human inquiry is such that "... people cannot attain absolute certainty concerning questions of fact" (*CP* 1.149). Peirce's notion is woven into other themes such as

---

[5] Susan Haack [20] provides a similar list, attesting to the power of Peirce's thought to a generation: "Over time, it has been Peirce's work that has come to influence me the most: his formal fluency and logical innovations, of course, but also his distrust of easy dichotomies, his idea of the growth of meaning, his attractively naturalistic theory of inquiry, his constructive reconception of metaphysics and its role—not to mention his penchant for neologisms."

Balancing Necessity and Fallibilism 19

his critique of Descartes' foundational project of grounding knowledge, his position that chance is woven deeply and intrinsically into the universe, and his perspective on the evolutionary nature of the universe (e.g. *CP* 1.173 and 1.152). Peirce's fallibilism goes with a belief that although we have no assurance of the correctness about our knowledge of nature, over time, with inquiry, we converge closer to truth about reality.[6] For 20th century pragmatism, this quality continued as a spirit of anti-foundationalism and a rejection of secure absolutes that to Dewey figured in much of Western philosophy.[7] Moreover, as Bernstein observes: "It was Peirce who initially argued that fallibilism is essential for understanding the distinctive character of modern experimental science" [21].

For Abner this feature of Peirce's thought was associated with the use of probability theory and "... certainly prepared me for the point of view that probability is essential in our epistemology, and that judgments of very high probability in favor of one conjecture and against another are quite compatible with his overall fallibilism" [19, p. 18].

Peirce also has the elements of what is known as the propensity interpretation of probability in a notion of "would be," referred to by one commentator as a "watershed" separating the middle from the final years of his intellectual life [10, p. xx] and characterized by Abner succinctly:

> It is not surprising that two of the most eminent advocates of the frequency interpretation, Peirce (1932) [15, vol. II] and Karl Popper (1957) [73], abandoned the frequency interpretation in favor of a different ontic interpretation, or propensity. The propensity interpretation ascribes an ontological status to the tendencies of propensities of the various possible outcomes of a singular chance event, such as the toss of a coin or the decay of a nucleus [18, vol. II, p. 237 and associated discussion].

There is an interesting lineage of Peirce's idea to Abner's profound and striking development of the idea that quantum states prior to measurement can be characterized by a notion of "objective indeterminacy," a notion in continuity with Heisenberg's idea of potentiality. When asked on the origin of this idea in the AIP interview, he noted: "I was ripe for it. Because of my advocacy of Peirce's would-be analysis of probability I was ripe to accept Heisenberg's analysis of the wave function in terms of potentiality" [19, p. 7].

Related to his fallibilism is Peirce's abiding concern with the process of human knowing, particularly that associated with the sciences. He observed in 1897:

> From the moment when I could think at all, until now, about forty years, I have been diligently and incessantly occupied with the study of methods of inquiry. ..... I have paid most attention to the methods of the most exact sciences, have intimately communed with some of the greatest minds of our time in physical science (*CP* 1.3).

---

[6] See for example, Hookway [13, p. 73f] for a standard presentation of Peirce on these matters.

[7] As representative of that perspective: "Pragmatism which arose in the first instance through Peirce's canonical critique of Descartes, has always been a very pluralist movement centered on a concern to continue the discussion of knowledge on a non-foundationalist basis..." [6, p. 467]. See also [22].

In an essay of Abner's in 1981, developing an naturalist epistemology where scientific investigations are drawn on, Peirce is nearby: "Among classical philosophers, Peirce seems to come closest to the integral epistemology which I envisage" [18, vol. I, p. 5] and in a later comment, that Peirce had "… the makings of a balanced epistemology. … between dogmatism … and excessive skepticism … [and that] he also really anticipated so much of the epistemology of the latter half of the 20th century" [19, p. 18]. Further, on particularities of Peirce's characterization of methods of scientific inquiry:

> To summarize, I find at least four methodological ideas of great value in Peirce's paper on scientific inference: that the scientific method achieves its successes by submission to reality, that a hopeful attitude towards hypotheses proposed by human beings is indispensable to rational investigation of the unknown, that a usable criterion of fair sample involves subjective and ethical considerations, and that it is rational to make certain weak assumptions about the fairness of the data in order to permit inquiry to proceed [18, vol. I, pp. 234–235].

Two further aspects of Peirce's thought that one finds mention of in Abner's writings relate to Peirce's understanding of evolution and the notion that the laws of nature themselves are emerging and evolving features of the universe. Peirce expressed various doubts as to Darwin's account of the manner of evolution, although not doubting that evolution had taken place (see e.g., [11, p. 227f]), and Abner, on arguing for the non-existence of the principle of natural selection, sees an affinity with Peirce on this point:

> Peirce seems to subsume the theory of natural selection under the theory of probability. …
> I believe that the my thesis of the non-existence of a principle of natural selection fits the main current of his thought. It is honorable to be an epigone of Peirce [18, vol. II, p. 245].

A more radical idea may be found in Peirce—that the fundamental laws themselves have an evolutionary explanation (*CP* 6.33), a speculation the nature of which to Abner reminds us of the "continuity of modern physics with metaphysics" [18, vol. I, p. 29]. In general Peirce posits the universe as evolutionary on its deepest level:

> The evolutionary process is, therefore, not a mere evolution of the existing universe, but rather a process by which the very Platonic forms themselves have become or are becoming developed (*CP* 6.194).

A sympathetic assessment of the idea in a range of thinkers (yet critical of Peirce) is given in [23], and Paul Davies has drawn on the idea in a number of general publications, e.g., most recently, [24].

The final idea of Peirce I wish to mention is that of proto-mentality or mentalism, referred to more generally as a position of panpsychism. The idea is found rather widely late 19th to the middle of the 20th century in the writings of figures such as James, Royce, Bergson, Teilhard de Chardin, Whitehead, and Hartshorne (for a impressive history of panpsychism, see [25]). For panpsychism, at a lower level of matter there is a dimension of mind or mentality throughout the universe, one that gets concentrated on higher levels such as in human consciousness. The idea is in conflict with that of contemporary notions of emergence, when consciousness can

Balancing Necessity and Fallibilism   21

be seen as naturally emergent property, arising from the complexity of a pre-mental neurological matter. Debates on this topic continues, although as we increasingly understand how the brain generates the nature of consciousness, a "naturalist" perspective (and on that, in accord with the spirit of Peirce) would now seem to align with the notion of emergence. Peirce presented the idea in a series of papers (1891–1893) where he argued for a monism of mind and matter and a "dual aspect" theory of mind:

> The one intelligible theory of the universe is that of objective idealism, that matter is effete mind, inveterate habits becoming physical laws (*CP* 6.25) and ... what we call matter is not completely dead, but is merely mind hidebound with habits (*CP* 6.128).

> But all mind is directly or indirectly connected with all matter, and acts in a more or less regular way; so that all mind more or less partakes of the nature of matter. ... Viewing a thing from the outside, considering its relations of action and reaction with other things, it appears as matter. Viewing it from the inside, looking at its immediate character as feeling, it appears as consciousness (*CP* 6.268).

Abner mentions in his AIP interview how the idea was an attractive and important one for him, both religiously and intellectually, in his early encounter with Peirce as well as Whitehead, observing, in what constitutes the main reason for the notion, that if we evolved then our mental faculties are production of evolution, not just our bodies—and if they are, then "there must be something mental-like from which the faculties evolve" [19, p. 10]. More recently as well Abner has suggested that a naturalist "physicalism" can be a component in an epistemological naturalism when combined with a mentalism of a sort that "would have a fundamental status in nature, either coordinate with physical reality or yet more fundamental" [26, p. 306].

2 Peirce on Mathematics: Necessary and Hypothetical

While Peirce has been known for fields such as his studies on scientific as well as his work on logic and as one of the founders of American Pragmatism, the new wave of Peirce scholars from the 1960s onwards have drawn out and emphasized the central place of mathematics in his thought.[8] Benjamin Peirce, his father, was a well known mathematician in 19th century America and a powerful charismatic teacher at Harvard during Peirce's youth. He played an important role in Peirce's early education in a variety of ways. Peirce directly refers to the important influence of his father in his early education and in particular that "... without appearing to be so, he [Benjamin Peirce] was extremely attentive to my training when I was a child, and especially insisted upon my being taught mathematics according to his

---

[8] The work of Carolyn Eisele stands out here, both in numerous studies on Peirce's mathematics and scientific philosophy [28] as well as in editing the four volume *The New Elements of Mathematics* [27] containing Peirce's mathematical writings. Also, studies by Buchler [3], Goudge [11], Hookway [13], Joswick [29], Levy [30], Cooke [31] and Campos [32] have drawn out the importance and significance of aspects of Peirce's thought on mathematics.

directions ..." (quoted in [27, vol. 4, p. v]). Brent's biography of Peirce emphasizes the weighty legacy of his father on his life, that his father had "draped on his shoulders the crushing mantle of genius" and engaged him in an exacting and intellectual training, the effects of which "were to aggravate his neurological pathologies, to nourish his arrogance, and to set his ambition afire" [8, p. 16].

Peirce first of all makes mathematics central in the priority he gives it in a classification of the disciplines. Moreover, it is a discipline in need of no other disciplines.[9] In one of his disciplinary mappings of the sciences, and the "architectonic character" of philosophy, Peirce observed:

> .... mathematics meddles with every other science without exception. There is no science whatever to which is not attached an application of mathematics. This is not true of any other science, since pure mathematics has not, as a part of it, any application of any other science, inasmuch as every other science is limited to finding out what is positively true, either as an individual fact, as a class, or as a law; while pure mathematics has no interest in whether a proposition is existentially true or not. In particular, mathematics has such a close intimacy with one of the classes of philosophy, that is, with logic, that no small acumen is required to find the joint between them (*CP* 1.245).

> .... It might, indeed, very easily be supposed that even pure mathematics itself would have need of one department of philosophy; that is to say, of logic. Yet a little reflection would show, what the history of science confirms, that that is not true. Logic will, indeed, like every other science, have its mathematical parts (*CP* 1.247).

> .... But mathematics is the only science which can be said to stand in no need of philosophy, excepting, of course, some branches of philosophy itself. It so happens that at this very moment the dependence of physics upon philosophy is illustrated by several questions now on the tapis (*CP* 1.249).

Rather strikingly Peirce gives mathematics a central role in developing a philosophy, as in a letter of 1894:

> My special business is to bring mathematical exactitude,—I mean *modern* mathematical exactitude, into philosophy,—and to apply the ideas of mathematics in philosophy (quoted in [27, vol. 4, p. x]).

Moreover in the development of thought itself, mathematics was the "earliest field of inquiry" as mathematics is the "most abstract of all the sciences" and the first questions asked are "naturally the most general and abstract ones" (*CP* 1.52–53).

The relationship between logic and mathematics forms an entangled thread in Peirce's thought. In various passages Peirce stressed the independence of mathematics from logic:

> I will not admit that the mathematician stands in any need of logic. The mathematician must reason, of course; but he needs no theory of reasoning, because no difficulties arise in mathematics which require a theory of reasoning for their resolution. The metaphysician *does* require a theory of reasoning; because in his science such difficulties *do* arise. All the special sciences (especially the nomological sciences) repose, more or less, on metaphysics, and therefore, at least indirectly, and some of them directly too, require a theory of logic. But pure mathematics can postpone such a theory [27, vol. 4, p. 98].

---

[9] For more on the manner in which mathematics is foundational in Peirce see [33] and [13, Chapter 6].

Balancing Necessity and Fallibilism 23

> It does not seem to me that mathematics depends in any way upon logic. It reasons, of course. But if the mathematician ever hesitates or errs in his reasoning, logic cannot come to his aid (*CP* 4. 228, 1902).

> Logic can be of no avail to mathematics; but mathematics lays the foundation on which logic builds ... (*CP* 4.250).

He will also characterize mathematics as an activity of reasoning that is direct and intuitive. Logic, on the other hand, is a study of reasoning [34]. Referred to by Dipert [35, p. 46] as a "reverse-logicism," Peirce's priority of mathematics is a persistent strain in his writings. Yet, as commentators have noted, in other places Peirce comments on mathematics' dependency on logic.[10] And Peirce will note, when referring to Dedekind's work on numbers of 1888 that the "boundary between some parts of logic and pure mathematics ... is almost evanescent" (*CP* 2.215).

One can see resonances in Peirce of an analogous distinction made in Whately's *Elements of Logic* [36]—a widely used logic book in the 19th century.[11] On several occasions Peirce noted that Whately's text, which he had first read as a youth, was of considerable influence on him, reflecting in a latter to Lady Welby in 1908 that "... from the day when at the age of 12 or 13 I took up, in my elder brother's room a copy of Whately's Logic ... it has never been in my power to study anything— mathematics, ethics, metaphysics, gravitation, thermo-dynamics, optics, chemistry, comparative anatomy, astronomy, psychology, phonetics, economic, the history of science, whist, men and women, wine, metrology, except as a study of semeiotic" [37, p. 85]. Whately remarks that one can reason accurately prior to a study of logic, much as one can speak prior to the study of grammar he also likens logic to the "grammar of reasoning" [36, p. 11]. Analogously for Peirce, the ability to do mathematics is independent of a study of its methods of reasoning.

These foundational features of mathematics are woven into a number of other features Peirce ascribes to mathematics. First, stressing that he owes the idea to his father, Peirce often referred to mathematics as a science that draws "necessary conclusions":

> Of late decades, philosophical mathematicians have come to a pretty just understanding of the nature of their own pursuit. I do not know that anybody struck the true note before Benjamin Peirce, who, in 1870, declared mathematics to be "the science which draws necessary conclusions," adding that it must be defined "subjectively" and not "objectively" (*CP* 3.558).

> ... It was Benjamin Peirce, whose son I boast myself, that in 1870 first defined mathematics as "the science which draws necessary conclusions." This was a hard saying at the time; but today, students of the philosophy of mathematics generally acknowledge its substantial correctness (*CP* 4.229).

The phrase that Peirce quotes is the opening sentence in Benjamin Peirce's well known study, "Linear Associative Algebra" [38]. Peirce argues in a number of places against a traditional definition of mathematics as the science of quantity (e.g. *CP* 3.554). Peirce also knew Boole's work well and there are echoes in Peirce

---

[10] Comprehensive discussions of this topic may be found in [30] and [35].

[11] For a study of the influence of Whately's text on Peirce see [39].

of Boole's same questioning of the significance of mathematics as the science of quantity as in his essay of 1847 [40, p. 4]. Similar notions of mathematics occur in the "Preface" of *Analytical Society Memories*, by Charles Babbage and John Herschel, where the power of a symbolic language for mathematical reasoning is celebrated and mathematics is characterize as examining "... the varied relations of necessary truth" [41, p. i]. Peirce also proposed a more general significance to mathematics in philosophy—all necessary *a priori* thinking is a form of mathematical thinking:

> Philosophy requires exact thought, and all exact thought is mathematical ... I can only say that I have been bred in the lap of the exact sciences and I know what mathematical exactitude is, that is as far as I can see the character of my philosophical training (quoted in [27, vol. 4, p. x]).

> All necessary reasoning is strictly speaking mathematical reasoning [,] that is to say, it is performed by observing something equivalent to a mathematical Diagram .... [1, p. 116].

Perice refers positively to an analogous definition by George Chrystal in the ninth edition of the Encyclopedia Britannica (1883).[12] Hints of such a position may also be found in his father's writings [42]. Mathematics with a definition as the science that draws necessary conclusions is such that, to his father it "belongs to every inquiry, moral as well as physical." ([43, p. 97] and see also [44, p. 377]).

As one would suspect, to give such a foundational role for mathematics requires a rich conception of mathematics, which is indeed the case for Peirce. In particular for Peirce mathematical reasoning involves diagrams and a form of interior observation:

> ... What then is the source of mathematical truth? For that has been one of the most vexed of questions. I intend to devote an early chapter of this book to it.1 I will merely state here that my conclusion agrees substantially with Lange's, that mathematical truth is derived from observation of creations of our own visual imagination, which we may set down on paper in form of diagrams (*CP* 2.77).

> ... In mathematical reasoning there is a sort of observation. For a geometrical diagram or array of algebraical symbols is constructed according to an abstractly stated precept, and between the parts of such diagram or array certain relations are observed to obtain, other than those which were expressed in the precept. These being abstractly stated, and being generalized, so as to apply to every diagram constructed according to the same precept, give the conclusion. (*CP* 2.216).

Peirce's references to observation and mathematics occur shortly after a famous British Association for the Advancement of Science address by J.J. Sylvester in 1868 where a role is given for observation in the practice of the mathematics [45]. Perice quotes a phrase from Gauss that Sylvester had used in his address: "... for as the great mathematician Gauss has declared—algebra is a science of the eye—only it is observation of artificial objects and of a highly recondite character" (*CP* 1.34).

---

[12] Chrystal [42] characterized mathematics as: "any conception which is definitely and completely determined by means of a finite number of specifications, say by assigning a finite number of elements, is a mathematical conception. ... As an example of a mathematical conception we may take "a triangle"; regarded without reference to its position in space, this is determined when three elements are specified, say its three sides ...".

Balancing Necessity and Fallibilism 25

Peirce also associates perceptual judgments with mathematical proof noting the "... compulsiveness of the perceptual judgment is precisely what constitutes the cogency of mathematical demonstration" (*CP* 7.659, 1903). In this way the "compulsory" feature of mathematics is grounded.

For Peirce, this underlies a role mathematics can play in philosophy as errors will be reduced "to a minimum" in philosophy by:

> ... treating the problems as mathematically as possible, that is, by constructing some sort of a diagram representing that which is supposed to be open to observation by every scientific intelligence, and thereupon mathematically,—that is, intuitively,—deducing the consequences of that hypothesis (quoted in [27, vol. 4, p. x]).

Other features of Peirce's notion of mathematics include taking mathematical reasoning as a form of experimenting with diagrams. A particularly bold statement of his position on this occurs in "Notes on Ampliative Reasoning" in 1902 that "Mathematical proof is probably accomplished by appeal to experiment upon images or other signs, just as inductive proof appeals to outward experiment" (*CP* 2.782). Such mathematical diagrams are "iconic" which leads to Peirce's rich and extensive work on semiotics that would take us to far a field to consider here (on this see [13, p. 189f]). That all thinking for Peirce involves signs is another way mathematics is linked deeply to general reasoning. Peirce in the following, on the practice of the reasoning, weaves all these threads together:

> ... he searches his heart, and in doing so makes what I term an abstractive observation. He makes in his imagination a sort of skeleton diagram, or outline sketch, of himself, considers what modifications the hypothetical state of things would require to be made in that picture, and then examines it, that is, observes what he has imagined, to see whether the same ardent desire is there to be discerned. By such a process, which is at bottom very much like mathematical reasoning, we can reach conclusions as to what would be true of signs in all cases, so long as the intelligence using them was scientific (*CP* 2.227).

In addition to characterizing mathematics as the discipline that draws necessary consequences, Peirce stressed (as in the last quotation), and increasingly as his thought developed, that mathematics is hypothetical. In particular, that "... all mathematicians now see clearly that mathematics is only busied about purely hypothetical questions" (*CP* 1.52). In this way mathematics is distinguished from an inquiry into nature:

> For all modern mathematicians agree with Plato and Aristotle that mathematics deals exclusively with hypothetical states of things, and asserts no matter of fact whatever; and further, that it is thus alone that the necessity of its conclusions is to be explained. This is the true essence of mathematics ... (*CP* 4.232, 1902).

> Mathematics is the study of what is true of hypothetical states of things. That is its essence and definition. (*CP* 4.333, 1902)

Peirce emphasizes in other places that hypotheses are creations of the mathematician and that this is the origin of the necessary nature of mathematics (*CP* 3.560, 8.110). This cluster of features then—mathematics as manipulating with and experimenting on diagrams, as observational, as working with hypothesis that are other than to do with facts about the world, as that which draws necessary consequences, as the discipline that is foundational and central in philosophy—together constitute Peirce's vision of mathematics. Ketner and Putnam go so far to remark that many of these features meant mathematics "was the inspirational source for the pragmatic maxim, the jewel in the methodological part of the semeiotic, and the distinctive feature of Peirce's thought" [1, p. 2].

That a significant feature of Peirce's characterization of mathematics is blended with actual practices of the mathematician provides further support for the place of mathematics in pragmatism. When commenting on the nature of mathematics Peirce often refers to the beliefs and practices of mathematicians, with attention frequently to historical contexts. The words "mathematician" and "mathematicians," for example, occur 202 times in the *Collected Papers*, and while less that "mathematics" (340) and "mathematical" (334) the number is significant. The usage accords with Peirce's pervading epistemological concern with the nature of human reasoning. Campos [32] has drawn attention to this dimension of mathematics for Peirce, noting Peirce's definitions of mathematics as necessary and hypothetical are "descriptions of mathematical activity" and observed in a comment that concisely sums up various points in this section:

> The practice of imagining hypothetical states of things and asking what would necessarily be true about them provides the context in which mathematical icons are conceived, created and recreated, so as to explore a myriad would-be worlds.

3 Balancing Mathematics and Inquiry into Nature

A long persistent thread in reflection on the empirical and natural sciences has been on the role of mathematics in such sciences.[13] As mathematics is a structured symbolic system with features of a natural language and long taken, as expressed by Galileo's famous trope, as the language of the book of nature, the issue in the broadest sense is one of the relationship of a language to reality, on the junction of "word" and "thing", an issue that has haunted modern philosophy. Locke's clear and direct separation of words, things and ideas in *An Essay Concerning Understanding* leaves the unsettling question of their relationship, and forms a textual monument to this question that has haunted modernity:

> We should have a great many fewer disputes in the world if only words were taken for what they are, the signs of our ideas only, and not for things themselves [46, vol. III, p. 10].

As applied to mathematics, the question appears as a semantic one of how the symbols and notation of mathematics embody mathematical concepts and refer either to mathematical objects or to features of empirical objects, such as properties and the laws of nature.[14]

---

[13] For the manner in which this topic can be addressed in tracing the history of physics, see [47,48].

[14] As an aside, Benjamin Peirce's study of Algebra of 1870 in various places uses the textual image for mathematics; mathematics as a language and with a grammar [43, p. 98, 105].

Balancing Necessity and Fallibilism 27

During the 19th century the question was sharpened as mathematics was increasingly seen as abstract and as a discipline separate from the sciences of nature. The development of abstract symbolic algebra (separate from arithmetic algebra) by Peacocke, Hamilton, and De Morgan in the early part of the century was part of this development in mathematics while the development of non-Euclidean geometry in the latter part was another.[15] Herschel's influential *Preliminary Discourse* [49] is representative of these moves and draws a sharp distinction between the abstract sciences of mathematics and the natural sciences concerned with causality and laws of nature. In Whately's logic, too, the text mentioned earlier, there is a persistent emphasis on how a proper understanding of Logic requires recognizing that logical matters to do with reasoning are distinct from "the observations and experiments essential to the study of nature" [36, p. 9; see also, p. 25, 338].

This stress on the unique features of mathematics brings into clear relief the question of relationship of mathematics and the natural sciences in a discipline such as mathematical physics. This multi-sided question can be posed generally as one about probing the nature of the meeting point of the abstract, necessary and symbolic with the concrete, contingent and empirical. This question will set the agenda for tracing Perice's texts on this topic.

By stressing in various places that mathematics has a distinct identity, different from the natural sciences, Peirce is part of these movements within 19th century mathematics. His emphasis on the hypothetical nature of mathematics is one such place where this occurs: "Mathematics is engaged solely in tracing out the consequences of hypotheses. As such, she never at all considers whether or not anything be existentially true, or not" (*CP* 1.247). And in some striking passages:

> The mathematician lives in another world from the rest of us, in a world of pure forms. Here he is domiciled and spends part of his time, but he is a mere sojourner; this is not the world that he knows or that he cares for. If you tell him that something in the world of mathematical forms corresponds to something in the real world, be cautious not to speak as if such a correspondence could impart any value to the mathematical object, or he may consider you impertinent. Of what consequence is that reality to him? [16, vol. 6, p. 258].

> There is no essential difference between pure and applied mathematics. The mathematician does not, as such, inquire into facts. He only develops ideal hypotheses. These hypotheses are all more or less suggested by observation and all depart from or transcend, more or less, what observation fully warrants. But if the hypotheses are developed with a view to ideal interests, it is pure mathematics. If they are made crabbed and one sided in the interest of truth it is applied mathematics. [27, vol. 2, p. vi].

Both of these passages, and the second one in particular, bear a resemblance to his father's almost Pythagorean vision of a fusion of mathematics and nature. For example, in a series of lectures published shortly after his death his father writes, with vivid metaphors:

> But in the frozen cave of geometry, the thoughts which may trickle in from the actual world are crystallized into glittering, passionless, and unsympathizing stalactites; and the

---

[15] For an exploration of this topic see [50].

mathematical sage cares not whence they came,—whether they fell as dew from the quiet sky, or as rain from the clouds driven by the wind. Whatever their origin, they are ideal truth [43, p. 167].

And for his father, on the ready application of mathematics to the study of nature, the mathematics of quaternions to which the mathematician was led from imaginary numbers has become "the true algebra of space" that "clearly elucidates some of the darkest intricacies of mechanical and physical philosophy" (Ibid. p. 29).

In these passages and in his son's writings in particular there are hints of Cantor's view of "pure mathematics" as "free mathematics," presented in the *Grundlagen* of 1883. Such a mathematics is in opposition to that constrained by the empirical world, or "crabbed" in Peirce's phrase quoted above. As well Boole, in the text referred to above, remarks that mathematics considers operations in themselves, "independently of the diverse objects to which they can be applied" [40].

The spirit here is in accord with another characteristic of mathematics that Peirce stresses, viz., generalization, and this too is outlined in a context that places it in opposition to applied mathematics:

> Another characteristic of mathematical thought is that it can have no success where it cannot generalize. One cannot, for example, deny that chess is mathematics, after a fashion; but, owing to the exceptions which everywhere confront the mathematician in this field—such as the limits of the board; the single steps of king, knight, and pawn; the finite number of squares; the peculiar mode of capture by pawns; the queening of pawns; castling—there results a mathematics whose wings are effectually clipped, and which can only run along the ground (*CP* 4.236).

Interestingly Peirce then will often identify aspects of mathematics by placing them in contrast to science. The creative and free nature of forming hypotheses in mathematics, the necessary features of mathematics, and the pursuit of generalization in mathematics all stand in apparent contrast to the practices of the natural sciences.

The frequency with which Peirce places the intersection of mathematics and study of nature in this way is striking. It is a particular way of doing mathematics:

> The truths of mathematics are truths about ideas merely .... Thomson and Tait (Natural Philosophy §438) wisely remark that it is "utterly impossible to submit to mathematical reasoning the exact conditions of any physical question." A practical problem arises, and the physicist endeavors to find a soluble mathematical problem that resembles the practical one as closely as it may. ... The mathematics begins when the equations or other purely ideal conditions are given. "Applied Mathematics" is simply the study of an idea which has been constructed to look more or less like nature [27, vol. 4, p. xv].

Peirce continues this passage to mention that geometry is an example of "applied mathematics." The mathematician, will use a "space imagination" to form "icons of relations which have no particular connection with space." These are diagrams visually imagined of a space. But at the same time "space is a matter of real experience" (Ibid. p. xv). Elsewhere too, Peirce dwells on geometry's dual nature: non-Euclidean geometry is securely established in abstract mathematics, yet "geometry, while in its main outlines, it must ever remain within the borders of philosophy, since it depends and must depend upon the scrutinizing of everyday experience, yet at certain special points it stretches over into the domain of physics" (*CP* 1.249). Only measurements

Balancing Necessity and Fallibilism 29

will tell the nature of the geometry of actual space. Peirce also intriguingly specu-
lates on the existence of higher dimension, a topic of sustained interest in the latter
part of the 19th century: "Thus, space, as far as we can see, has three dimensions;
but are we quite sure that the corpuscles into which atoms are now minced have not
room enough to wiggle a little in a fourth?" (*CP* 1.249). With practice a mathematician at home in universal geometry can adjust to a space of four dimensions: "Give
a higher geometer sixty days to accustom himself to a four-dimensional space, and
he would be ever so much more at home there than he ever can be in this perverse
world" [51, vol. 3. p. 182].

The overall context for this seeming opposition between mathematics as the
abstract hypothetical study and mathematics as practiced in the midst of the investigation into nature is one where Peirce is often addressing the practices of the mathematician and the practices of the scientist. It is here I propose we have a clue to a
pervasive feature of Peirce's thought: that the apparently more systematic issue such
as that posed above of the relationship of mathematics as a formal system to natural
science and its objects, Western philosophy's old haunting issue of representation
of thought to reality, appears invisible to Peirce. Instead it appears as steadily posed
instead in terms of activities.[16] This is illustrated nicely in a passage where Peirce
directly addresses the use of mathematics for physics:

> The complex plane is one of the meeting-grounds of mathematicians and physicists, and
> the latter are now quite at home in the presence of that coy handmaiden, the complex vari-
> able; indeed, the well-known transformation scene in which she and her image play such
> a prominent part, is now an important feature in the solution of some practical problems
> [27, vol. 3, p. 145].

Also mathematics is useful for the work of the physicist as, "First, it enables him
to solve his own problems instead of employing a mathematician …. Secondly, it
supplies him with fundamental conceptions and methods of thinking without which
he never can rise from the ranks of the army of science" [27, vol. 3, p. 121].

While posing the issue of the meeting places of mathematics and the natural sciences in terms of the practices of both disciplines is a persistent feature of Peirce's
thought, there's a deeper more systematic question: how does mathematics' necessary and certain nature fits with Peirce's Fallibilism? The issue has been directly
addressed by Haack [52] and Cooke [31] and to both there are unresolved tensions in
Peirce's writings on this topic. For Haack the puzzle is that Peirce seems able to hold
that our mathematical beliefs could be mistaken while still holding to a position that
mathematical truths are necessary [52, p. 37]. Indeed Peirce in places stresses how
mistakes can be made in doing mathematics and it is clear it is an uneasy problem
for him (see, e.g., *CP* 1.149, *CP* 4.237). For Haack the tension resides in Peirce's
failure to specify fully what is meant by fallibilism (a point other commentators
have remarked on) and with a more elaborate specification, there are ways in which
it could coexist for Peirce with mathematics necessary nature.

---

[16] There is an intriguing link between Peirce on this point and J.J. Sylvester and others in the British
context such as James Clerk Maxwell that awaits further exploration. Maxwell, for example, in
an British Association address in 1870, soon after Sylvester's address considers the relationship
between mathematics and physics largely in terms of the activities of those in both disciplines.

Cooke argues that Peirce "can and should hold a position of fallibilism within mathematics, and that this position is more consistent with his overall pragmatic theory of inquiry and general commitment to the growth of knowledge" [31, p. 159]. In particular, for Peirce to hold for a type of theoretical infallibilism for mathematics would be deeply incompatible with his rejection of the separation of a science's intelligibility from its human knowers. Yet for Cooke Peirce could consistently allow error in the practice of the mathematician who for Peirce experiments with hypothetical truths via diagrams, and could be brought about by allowing a different form of fallibilism from that associated with investigating empirical features about the world. This would be a particular type of "internal fallibilism" such to allow for the obvious way mathematicians can make errors in doing mathematics, and further, recognizing such doubt in this realm for Cooke allows a general conclusion that it allows inquiring into new areas in mathematics—consequently discovering new relations and new systems [31, p. 174]. Such a position accords with Peirce remarks when commenting as indicated earlier on how deduction (or "analytical reasoning") involves perception and experimentation:

> Deduction is really a matter of perception and of experimentation, just as induction and hypothetic inference are; only, the perception and experimentation are concerned with imaginary objects instead of with real ones. The operations of perception and of experimentation are subject to error, and therefore it is only in a Pickwickian sense that mathematical reasoning can be said to be perfectly certain. It is so only under the condition that no error creeps into it; yet, after all, it is susceptible of attaining a practical certainty. (*CP* 6.595)

There is another deep issue here related to that to do with of foundations of knowledge. As those in the later pragmatist tradition of American thought have emphasized, Peirce's fallibilism can be seen as a form of anti-foundationalism, one that is not an either or sort where the opposite to foundationalism is a relativism (e.g. see [21, 53]). Moreover, we are now in the wake of a long sustained consideration in the 20th century of the pursuit of foundations in mathematics (see, e.g., [54, 55]). Peirce's famous critique of Descartes' grounding of the edifice of knowledge on an indubitable inner intuition is the basis of his anti-foundationalism (*CP* 5.264).

Also Peirce's metaphors have an anti-foundationalist flavor. Peirce will indeed use the metaphor of architecture, positively remarking when treating the classification of science and the "architectonic of philosophy" that the "... universally and justly lauded parallel which Kant draws between a philosophical doctrine and a piece of architecture has excellencies which the beginner in philosophy might easily overlook" (*CP* 1.176).[17] However, for Peirce the metaphor functions more as a way to comment on the texture and structure of a philosophical system: "that is why philosophy ought to be deliberate and planned out; and that is why, though pitchforking articles into a volume is a favorite and easy method of bookmaking it is not the one which Mr. Peirce has deemed to be the most appropriate to the exposition of the principles of philosophy ...." (*CP* 1.179). The architecture metaphor for

---

[17] Also, when characterizing philosophical systems Peirce will invoke the architectural metaphor: "There is a synchronism between the different periods of medieval architecture, and the different periods of logic. The great dispute between the Nominalists and Realists took place while men were building the round-arched churches ..." (*CP* 4.27).

Balancing Necessity and Fallibilism 31

knowledge therefore is not taken, as commonly taken in the philosophical tradition, to describe the building knowledge built on firm foundations.

Peirce also has various other powerful metaphors for knowledge which argue against knowledge being grounded on foundations, one being his famous metaphor of knowledge as on a bog and another, that of a bottomless lake.[18]

> The only end of science, as such, is to learn the lesson that the universe has to teach it. In Induction it simply surrenders itself to the force of facts. But it finds ... that this is not enough. It is driven in desperation to call upon its inward sympathy with nature, its instinct for aid, just as we find Galileo at the dawn of modern science making his appeal to il lume naturale. But in so far as it does this, the solid ground of fact fails it. It feels from that moment that its position is only provisional. It must then find confirmations or else shift its footing. Even if it does find confirmations, they are only partial. It still is not standing upon the bedrock of fact. It is walking upon a bog, and can only say, this ground seems to hold for the present. Here I will stay till it begins to give way. (*CP* 5.589)

> Consciousness is like a bottomless lake in which ideas are suspended at different depths. Indeed, these ideas themselves constitute the very medium of consciousness itself. Percepts alone are uncovered by the medium. We must imagine that there is a continual fall of rain upon the lake; which images the constant inflow of percepts in experience. All ideas other than percepts are more or less deep, and we may conceive that there is a force of gravitation, so that the deeper ideas are, the more work will be required to bring them to the surface. (*CP* 7:533)

Then there is Peirce's powerful and famous metaphor of knowledge as constituted by the fibers of a cable given when criticizing Descartes (adapted, as Haack [2] notes, from Thomas Reid):

> Philosophy ought to imitate the successful sciences in its methods, so far as to proceed only from tangible premises which can be subjected to careful scrutiny, and to trust rather to the multitude and variety of its arguments than to the conclusiveness of any one. Its reasoning should not form a chain which is no stronger than its weakest link, but a cable whose fibers may be ever so slender, provided they are sufficiently numerous and intimately connected. (*CP* 5.265)

All these metaphors, which capture the spirit of Peirce's understanding of the inquiry into nature, are at odds with the spirit of mathematics. A chain metaphor in particular, one in opposition with that of a cable of fibers, has a long association with the deductive structure of mathematics in figures such as Descartes and Hume and in early 19th century writings on mathematics in the British context. Yet, as we have seen above, mathematics for Peirce has a foundational place in philosophy, it is acritical in that it stands in need of no other discipline to proceed and there is a necessary quality to its deductions. Moreover these qualities are often outlined

---

[18] Both Thagard [56] and Abrams [57] address Peirce's use of these metaphors. One may speculate too on the influence of Peirce's cultural context. As commentators on the anti-foundationalist dimension of pragmatist though have remarked, the world of the America following the civil war was one to encourage the development of "... a more flexible, open experimental way of thinking that would avoid all forms of absolutism and ideologies that result in intolerance" [21]. And in more general terms the expansionist spirit of a new country, with vast territory arguably lent itself to such thinking rather than the trend of European philosophy to search for secure foundations.

in contrast with the nature of the other sciences. There's a complex and apparent tension then, one that invites further consideration on the nature of mathematics.

A dimension of mathematics that mutes the foundationalist image is the role mathematicians play in the creation of hypotheses. Here, as Peirce stresses, they are not constrained by the nature of the world, and in this process lies a creative freedom for the mathematician. Thus a natural way to think of mathematics as foundational by virtue of its axioms and starting points is not immediately to the foreground in Peirce. However, and balancing this, Peirce is careful to note the process of hypothesis creation is not an arbitrary one. Peirce has hints in place of Platonist conception of mathematics, a potential foundation for mathematics. When addressing the issue that one would expect with arbitrary hypothesis creation, namely that "different mathematicians to shoot out in every direction into the boundless void of arbitrariness" Peirce remarks that this does not happen and this phenomena:

> ... is not an isolated one; it characterizes the mathematics of our times, as is, indeed, well known. All this crowd of creators of forms for which the real world affords no parallel, each man arbitrarily following his own sweet will, are, as we now begin to discern, gradually uncovering one great cosmos of forms, a world of potential being. The pure mathematician himself feels that this is so ... if you enjoy the good fortune of talking with a number of mathematicians of a high order, you will find that the typical pure mathematician is a sort of Platonist. Only, he is [a] Platonist who corrects the Heraclitan error that the eternal is not continuous. The eternal is for him a world, a cosmos, in which the universe of actual existence is nothing but an arbitrary locus. The end that pure mathematics is pursuing is to discover that real potential world. (*CP* 1.646)

Peirce here makes the commonplace observation that the practicing mathematician is a Platonist, and there's a hint of convergence of mathematics to a given form that parallels Peirce's notion of scientific investigators converging in time to truth about nature. Peirce's Platonist phrases can take lyrical form:

> That passage of the mathematician, Plato, strikes a sympathetic chord in every mathematicians' breast when he says that these heavens and earth we gaze upon are but the walls and floor of a dismal cavern which shut out from our direct view the glories of the world of forms beyond [16, vol. 6, p. 258].[19]

He leaves open however what this could mean for particular mathematical systems. It would take us to far a field to pursue the idea, but in giving a place to observation in mathematics, and experimentation on diagrams as part of mathematical reasoning, Peirce could be read as grounding a form of mathematical Platonism in a naturalist manner.[20] In general for Peirce the most likely source of inspiration for the mathematician's practices come for situations in the world, not a Platonic world of the beyond.

---

[19] As a further example of Peirce's balance of this mathematical world of the beyond with the study of nature, Peirce continues: "Yet, what would steam-engines, electric cables, turbine wheels, life-insurance and a thousand things be but for the hints which mathematicians have vouchsafed?" (*Ibid.* vol. 6, 258).

[20] Abner interestingly explores Gödel's Platonism in this manner, noting Gödel's fondness of "comparing intuition of mathematical objects with sensory perception of physical objects of ordinary experience" [26, p. 301].

Balancing Necessity and Fallibilism 33

The hypothetical nature of mathematics nevertheless dominates Peirce's account of mathematics, despite the hints of a grounding in a Platonic realm of potential form. Peirce also resists a Kantian move of grounding the axioms and starting points of mathematics in a metaphysical or otherwise foundation. In particular, Peirce denies any dependence of mathematics on space, time, or any form of "intuition" (*CP* 3.556).

Here Peirce's view on mathematical truth and certainty has interesting resonances with the Scottish mathematician Stewart (1753–1828). Stewart stressed that the starting points of mathematics are assumed: "we have in view [...] not to ascertain truths with respect to the actual existences, but to trace the logical filiation of consequences which follow from an assumed hypothesis. If from this hypothesis we reason with correctness, nothing [...] can be wanting to complete the evidence of the result; as this result only asserts a necessary connexion between the supposition and the conclusion" [58, vol. II, p. 114]. Stewart's view was opposed by the Cambridge philosopher William Whewell, who sought to ground mathematical truths in broader metaphysical foundations of a Kantian nature.

Peirce's stress on the hypothetic nature of mathematics goes along as well with the spirit of characterizing logical inference in a hypothetical manner: "To say that an inference is correct is to say that if the premises are true the conclusion is also true; or that every possible state of things in which the premises should be true would be included among the possible state of things in which the conclusion would be true." (*CP* 2.710) It is also in accord with his support of a "Philonian" interpretation of conditional statements such "If A then B" as being true if A is either an empty class or A is untrue (for a discussion of this point see, Ketner and Putnam in [1]). What matters essentially is the structure of inference or mathematical or logical deduction, not its grounding in initial axioms or premises. Peirce's account has also later been associated with a position of "If-Thenism" or "deductivism" where truth as understood in this manner of connection and deducibility within a system (see [59], Chapter 10 for a modern discussion of this position).

In places when considering scientific investigations, Peirce sees that as hypothetical as well:

> Nothing is vital for science: nothing can be. ... The scientific man is not in the least wedded to his conclusions. He risks nothing upon them. He stands ready to abandon one or all as soon as experience opposes them. Some of them, I grant, he is in the habit of *calling established truths*; but that merely means propositions to which no competent man today demurs. It seems probable that any given proposition of that sort will remain for a long time upon the list of propositions to be admitted. Still, it may be refuted tomorrow; and if so, the scientific man will be glad to have got rid of an error. There is thus no proposition at all in science which answers to the conception of belief ([1], Lecture 1).

Here then an activity of science shares a feature of mathematics.

By dwelling on the hypothetical nature of mathematics (and science), and deductive relations Peirce on these issues appears as an early exemplification of the structuralism that was to flourish in the 20th century. Bourbaki's text, for example, *Elements of the History of Mathematics*, notes on that history that it would be

"... be tempting to say that the modern notion of "structure" is attained in substance around 1900; in fact it will need still another thirty years of apprenticeship before it appears in all its glory" [60, p. 21].

There remains the clear foundational nature of the "necessary" nature of inferences of the mathematician, and exploring this leads to a key distinction Peirce makes in mathematical reasoning. One type, "corollarial," involves immediate deductions in a straightforward way from axioms. They need not involve the iconic diagrams directly. The other type is "theorematic" reasoning, which involves a more active creation of strategies and experimentation with diagrams to achieve a result (*CP* 2.267 and *CP* 4.613 and for a discussion of this distinction see [29, 30]). And example of the latter would be a supplementary construction needed in a proof to bring about the conclusion. The significance of such reasoning to Peirce had been overlooked in the tradition, and Peirce's remarks here are part of his attention to activities of the mathematician.

As we have seen, Peirce will ground the necessary nature of mathematics in various ways. A further way for Peirce is in the intuition—to imply a type of mathematical intuitionism. Goudge perceptively remarks that while Peirce can be read this way it is "entirely out of harmony with his naturalism" [11, p. 259]. It is not, though, grounded in a psychological form of intuition, as, for Peirce, "the mathematician clothes his thought in mental diagrams, which exhibit regularities and analogies of abstract forms almost quite free from the feelings that would accompany real perceptions" [51, vol. 3, p. 258]. Among recent commentators on this point, Joswick [29] takes the semiotic dimension of Peirce's mathematics as providing of seeing how Peirce grounds mathematical necessity. Of the threefold types of signs for Peirce—symbols, icons and indexes—it is only an icon that can bring out the inferential nature of mathematics, as it exhibits the form of an object and thus presents the relationships in the object. For Joswick,

> The icon is the essential mathematical sign because by "direct observation of it other truths concerning its object can be discovered" (2.280). Through the direct examination of an icon necessary connections in the object can be seen and unexpected relations revealed. "The whole of inference," Peirce contends, "consists in *observation*, namely in the observation of icons" (7.557) [29, p. 111].

Such a position is in accord with Peirce's notion that all necessary reasoning involves the use of diagrams, stated strongly in manuscript notes of 1896: "All valid necessary reasoning is in fact thus diagrammatic" (*CP* 1.54). What appears as significant is that again for Peirce a foundational dimension is significantly grounded in the very activity of the mathematician, not in a formal independent feature of a mathematical knowledge or mathematical objects. In this way it parallels the quality of fallibilism that attends the inquiry into the facts of nature.

In tracing in Peirce's thought the qualities of fallibilism and necessity that attend the natural sciences and mathematics respectively, one can see a subtle overlap of both realms. Yet there is a persistent tension. In a recent essay, Cooke, on this very topic, remarks that on a "pragmatic level" as to "how it is practiced" as indicated here, mathematics is like the empirical sciences, even though Peirce "so frequently holds that mathematics and science must be conceived as separate" [61].

Balancing Necessity and Fallibilism   35

A further point where the apparent contradictory qualities appear in balance in Peirce is on the topic of abstraction in mathematics. For Peirce the abstract is an important feature of mathematics: "Another characteristic of mathematical thought is the extraordinary use it makes of abstractions" (*CP* 4.234) and "... it may be said that mathematical reasoning (which is the only deductive reasoning, if not absolutely, at least eminently) almost entirely turns on the consideration of abstractions as if they were objects" (*CP* 3.509). Yet for Peirce the use of abstractions are woven into everyday life as well as mathematics and science. In a rich play of metaphor Peirce weaves together these contexts:

> These examples exhibit the great rolling billows of abstraction in the ocean of mathematical thought; but when we come to a minute examination of it, we shall find, in every department, incessant ripples of the same form of thought, of which the examples I have mentioned give no hint (*CP* 4.235).

The point here is similar to an observation of Whitehead, that, as mathematics increasingly entered into ever greater extremes of abstract thought, it became at the same time increasingly relevant for the analysis of particular concrete facts [62, p. 47], and to Dewey's remark that the very power of mathematics in physics arises from its free and abstract nature [63, p. 412].

The final point I wish to address the question originally posed on how mathematics relates to nature: what is that meeting place of mathematics and nature? Here two commentators on Peirce can provide a way to focus two threads in Peirce's thought.

The first arises from Peirce discussion of how maps, as icons and diagrams represent (*CP* 5.329 and *CP* 8.122). To Hookway [13], this example, plus a consideration of how for Peirce a color sample may be taken to represent color schemes of a house, provide a way to understand what Peirce would take to be the applicability of mathematics to nature. Maps represent and require interpretation, and in a similar way mathematical systems represent when interpreted and applied to "state of affairs" of the same form as the relational structure of the mathematical system [13, p. 191]. Various phrases in Peirce support such a perspective, for example, as quoted above, to Peirce for a practical problem "... the physicist endeavors to find a soluble mathematical problem that resembles the practical one as closely as it may." In such a way then Peirce can be seen as using the old metaphor of representation theory: mathematics mirrors and maps a reality other than it. The perspective is surely one of the dominant ways mathematics is seen to function in a scientific theory.

Another thread though places the issue of the union between mathematics and the natural sciences in an activity associated with mathematics. For Peirce, mathematics with its observational nature and manner of experimenting on diagrams, as well as its hypothetical nature shares similar practices to those of the natural science. Plus as we have seen, any necessary type of thinking for Peirce is mathematical. Such a position has been argued recently by Daniel Campos:

> I would claim that for Peirce the most important application of mathematics does not consist in the deployment of this or that particular mathematicaltheory to solve this or that practical

problem, but in the overall deployment of necessary reasoning to investigate problems in, say, phenomenology, aesthetics, ethics, logic, and the practical, physical and practical sciences [64, p. 73].

The abiding focus in Peirce is on the practices of the mathematician and scientist, plus the pervasive and central feature he gives to mathematics makes this a compelling perspective. It is one that sidesteps the long standing issue of how one realm of human endeavor, the mathematical and the resultant mathematical structures and theories, can represent a different realm that is implicated in notions of representation. It may be as well that in the background is Peirce's evocative expression that dealing with matters of representation entails further representations in an unending manner:

> The meaning of a representation can be nothing but a representation. In fact, it is nothing but the representation itself conceived as stripped of irrelevant clothing. But this clothing never can be completely stripped off; it is only changed for something more diaphanous. So there is an infinite regression here. Finally, the interpretant is nothing but another representation to which the torch of truth is handed along; and as representation, it has its interpretant again. Lo, another infinite series. (*CP* 1.399)

Moreover in a modern guise the position is similar to Hacking's proposal that traditional questions of realism when placed in the form of exploring classical issues to do with "representation" are intractable and a better perspective is obtained by exploring the instrumentality of our engagement with the world [65]. Hacking links his position directly to pragmatism: "The final arbitrator in philosophy is not what we think, but what we do" (Ibid. p. 31).

Further support from this position I'd suggest, although indirect, is related to Perice's panpsychism, his ascribing of a mental dimension to matter, and to a closely help belief that there is a natural mapping between mind and matter. That latter glides into a residue of idealism present in Peirce's writings, as, e.g., quoted above: "objective idealism, that matter is effete mind, inveterate habits becoming physical laws" (*CP* 6.25).

Here there is a blurring of traditional boundaries, not so much between the activity of the knower doing mathematics and the one involved in investigating nature, but between the knower doing mathematics and the realm which is the subject of that investigation, nature.

Peirce's father is the likely influence here.[21] Peirce in 1889, in a dictionary entry on the topic of ideal-realism described his father's position as "the opinion that nature and the mind have such a community as to impart to our guesses a tendency toward the truth, while at the same time they require the confirmation of empirical evidence" (quoted in [2, p. xxv]). In various places in Benjamin Peirce's writings hints of such a fusion of mind and matter emerge such as from a textbook written when teaching at Harvard: "Every portion of the material universe is pervaded by the same laws of mechanical action, which are incorporated into the very constitutions of the human mind" [44, p. 30; 66, p. 495]. Then later, that the "identity between the laws of mind and matter" suggests their common origin, one that if it is "conceded

---

[21] For a discussion of the influences of Benjamin Peirce on Charles, see [67].

Balancing Necessity and Fallibilism 37

to reside in the decree of a Creator," ceases to be mystery (Ibid. p. 31). To suggest alternatively that consciousness was "evoked out of the unconscious" would fail to give an adequate cause for it. And in an address to the American Association for the Advancement of Science in 1853, noted that the sciences and geometry in particular show "the world to which we have been allotted is peculiarly adapted to our minds, and admirable fitted to promote our intellectual progress" [68, p. 12]. A striking poetic Pythagorean fusion of matter and mathematics occurs in the following:

> The highest researches undertaken by the mathematicians of each successive age have been especially transcendental ... but the time has ever arrived ... when the progress of observation has justified the prophetic inspiration of the geometers, and identified their curious speculations with the actual workings of Nature. [44, p. 29]

> Long before ... observation had begun to penetrate the veil under which nature has hidden her mysteries, the restless mind sought some principle of power strong enough and of sufficient variety to collect and bind together all parts of the found. This seems to be found, where one might least expect it, in abstract numbers. Everywhere the exactest numerical proportion was seen to constitute the spiritual element of the highest beauty. (Benjamin Peirce, quoted in [69, p. 101])

His father refers to his position as one of "ideality" and will write that "the whole domain of physical science is equally permeated with ideality" [43, p. 17].

Peirce was immersed in this world view from his earliest years and given the influence of his father overall in his life, this would account for beliefs that mathematics may be applied to nature and that the worlds of nature and mathematics cohere together. Moreover I would claim, they form background assumptions and beliefs in Peirce, a haunting presence from his father's world. They are invisible to him in the sense they are not to the foreground to be subject to philosophical investigation.

In addition, Peirce's ready and powerful use of metaphor is such to allow a background belief to persist, carried subtly in images beyond full explication. Against this backdrop, the tensions of the two fields of inquiry, mathematics and the natural sciences, as focused abstractly above, can remain invisible. This supplements the unity in the knower due to the overlapping similarities in the practices of the knower of both fields. The complexity here is in need of further elaboration and contextualization, but its presence is a pointer to the deep currents guiding Peirce's thought. If correct, there is exemplified what I would propose is a general lesson: that pressing the status of mathematics in a system of thought and its relationship to the study of nature is a sure path to the depth structures, often silent ones, that constitute that system of thought.

4 Concluding Reflections

Together, the topics of this paper leave us with the question of what resources from Peirce and his understanding of mathematics and its place in the natural science can we use to inspire and inform contemporary projects. In some ways Peirce sounds rather modern (and postmodern). The foundationalist projects of 20th century foundations of mathematics have receded. Long reflections on the implications of Gödel's incompleteness results have taught us that foundations in grounding deductive thought tend to recede and elude us. Also, naturalist movements in the philosophy of mathematics, which see similarities of mathematics with the empirical sciences have taken hold and have undertaken to explore the practices and activities of the mathematician (see, for example [70, 71]). On both of these points, Peirce appears as a fellow traveler who initiated new paths.

Other parts of Peirce's world now appear dated. The complexity of neurological structure as revealed by contemporary cognitive sciences have made projects of understanding consciousness possible in new ways, such as an emergent phenomena of (pre-mental) matter. John Searle, expresses this vision powerfully, if polemically:

> Some traditional philosophical problems, though unfortunately not very many, can eventually receive a scientific solution. This actually happened with the problem of what constitutes life. We cannot now today recover the passions with which mechanists and vitalists debated whether a "mechanical" account of life could be given. The point is not so much that the mechanists won and the vitalists lost, but that we got a much richer conception of the mechanisms. I think we are in a similar situation today with the problem of consciousness. It will, I predict, eventually receive a scientific solution. But like other scientific solutions in biology, it will have to give us a causal account. It will have to explain how brain processes cause conscious experiences, and this may well require a much richer conception of brain functioning than we now have [72].

In continuity with this perspective Peirce's (and Whitehead's) panpsychism, that placed a mental dimension on lower levels of matter, now, through the advances of science, appears superfluous. Also studies on the practices of mathematics with the resources of contemporary projects in the sociology of science and naturalist accounts of reasoning have surpassed what Peirce achieved. Both of these developments mean that Peirce's blend and balance of mathematics and the natural sciences that I've suggested are tied into deeply held beliefs on the unity of mind and matter inspired by his father, and grounded in commonality of practices, are similarly dated.

In addition, 20th century physics, with its new understandings of the nature of chance in nature arising from quantum theory have supplanted Peirce's worlds. Overall our emerging theories on the structure of matter and space and time from decades of particle physics and the more recent string theory and loop quantum loop gravity have revealed a complexity and richness of matter unknown in Peirce's time, and thus dated various of the themes mentioned in section I above. And again, Whitehead's elaborate metaphysics of the event appears as from an earlier time in physics, prior to our present micro-theory of fundamental reality (even if presently incomplete) that's of such a nature to supplant many features of Whitehead's metaphysics. As a lighthearted observation, the complexity and details of string theory then can be seen to rival and surpass the difficulties previous generations had in working though the elaborate structures of Whitehead's *Process and Reality*.

Still questions to do with the nature of mathematics tend to persist and the vigor and complexity of Peirce's thought on mathematics and the activity of the mathematician are such that the very exercise to enter into Peirce's texts and those of the

Peirce scholarship on this topic remains valuable. The exercise is valuable historically in order to understand a key part in American intellectual history and how that unfolded in 20th century thought, and its present configuration. Here though projects still await on contextualizing Peirce's thought in more complete ways than some of those touched on above. The exercise is also of value to develop a set of skills to explore analogous issues on the contemporary landscape. Moreover, as the work of both Peirce and Abner witness to: a naturalist vision of using the resources of the natural sciences to pursue the deep questions associated with our philosophical tradition remains productive. And something else, very rewarding, remains for all who encounter the writings of Peirce: the inspiring example of what it means to live the life of a scholar, on how, with persistence and single mindedness, to explore ideas in spite of personal struggles and setbacks and at the same time to write, steadily, persistently, and relentlessly.

Acknowledgements Various of the ideas in this paper where discussed with Abner Shimony for which I'm grateful and that is part of a larger gratitude to Abner for discussions, encouragement and inspiration over the years on all manner of issues to do with understanding the nature of science. The help of Alison Walsh, Elizabeth Cooke, David Skrbina and Daniel Campos for sources and ideas for the paper is gratefully acknowledge. I am also grateful to Joy Christian and Wayne Myrvold for the opportunity to be part of the conference honoring Abner's thought, from which this paper emerged, and to the American Institute of Physics for permission to quote from the recorded interviews with Abner made in 2002.

References

1. Peirce, Charles S. (1992). *Reasoning and the logic of things: the Cambridge conferences lectures of 1898*. Cambridge, MA, Harvard University Press. Edited by Kenneth Laine Ketner; with an introduction by Kenneth Laine Ketner and Hilary Putnam.
2. Haack, Susan (1996). "Between Scientism and Conversationalism." *Philosophy and Literature* **20**(2): 455–474.
3. Buchler (1939), Justus (1939). *Charles Peirce's Empiricism*. New York: Harcourt Brace and Company.
4. Nagel, Ernest (1940). "Charles S. Peirce, Pioneer of Modern Empiricism." *Philosophy of Science* **7**: 69–80.
5. Kloppenberg, James T. (1996). "Pragmatism: An Old Name for Some New Ways of Thinking?" *Journal of American History* **83**: 100–138.
6. Rockmore, T. (2004). "On the Structure of Twentieth-Century Philosophy." *Metaphilosophy* **35**: 466–478.
7. Bernstein, Richard J. (1992). "The Resurgence of Pragmatism." *Social Research* **59**: 813–840.
8. Brent, Joseph (1993). *Charles Sanders Peirce: a life*. Bloomington, IN, Indiana University Press.
9. Peirce, Charles S. (1992). *The essential Peirce: selected philosophical writings*, Vol. 1. Bloomington, IN, Indiana University Press. Edited with an introduction by Nathan Houser and Christian J. W. Kloesel.
10. Peirce, Charles S. (1998). *The essential Peirce: selected philosophical writings*, Vol. 2. Bloomington, IN, Indiana University Press. Edited by the Peirce Edition Project; with an introduction by Nathan Houser.
11. Goudge, Thomas A. (1950). *The thought of C. S. Peirce*. Toronto, University of Toronto Press.

12. Murphey, Murray G. (1961). *The development of Peirce's philosophy*. Cambridge, MA, Harvard University Press.
13. Hookway, Christopher (1985). *Peirce*. London/Boston, Routledge & Kegan Paul.
14. Houser, Nathan, Don D. Roberts, and James Van Evra (1997) (eds.). *Studies in the logic of Charles Sanders Peirce*. Bloomington, IN, Indiana University Press.
15. Peirce, Charles S. (1931–1958). *Collected papers of Charles Sanders Peirce*. Vols. 1–8 Cambridge. Originally published: Cambridge, MA, Harvard University Press. Edited by Paul Weiss and Charles Hartshorne and A. W. Burk.
16. Peirce, Charles S. (1982). *Writings of Charles S. Peirce: a chronological edition*. Bloomington, IN, Indiana University Press. Edited by Max Harold Fisch and Christian J. W. Kloesel.
17. Gross, Neil (2003). "Richard Rorty's Pragmatism: A Case Study in the Sociology of Ideas." *Theory & Society* **32**: 93–148.
18. Shimony, Abner (1993). *Search for a naturalistic world view*. 2 volumes, Cambridge [England]/New York, NY, Cambridge University Press.
19. Shimony, Abner (2002). Transcript of oral history interview of Abner Shimony by Joan Bromberg, September 9 and 10, 2002. Niels Bohr Library, American Institute of Physics, 147 pp.
20. Haack, Susan (2005). "Formal Philosophy? A Plea for Pluralism" in Vincent F. Hendricks and John Symons, eds., *Formal Philosophy*, Automatic Press, pp. 77–98.
21. Bernstein, Richard J. (2006). "Pragmatic Reflections on Tolerance." Available online at http://www.pucp.edu.pe/eventos/congresos/filosofia/programa_general/lunes/plenaria/BernsteinRichard.pdf
22. Rockmore, T. (2005). "On Classical and Neo-Analytic Forms of Pragmatism." *Metaphilosophy* **36**: 259–271.
23. Shimony, Abner (1999). "Can the Fundamental Laws of Nature Be the Result of Evolution?" in C. Pagonis and J. Butterfield, eds., *From physics to philosophy*. Cambridge/New York, Cambridge University Press, pp. 208–223.
24. Davies, Paul (2007). "The Universe's Weird Bio-Friendliness." *The Chronicle of Higher Education* **53**(31): B14.
25. Skrbina, David (2005). *Panpsychism in the West*. Cambridge, MA, MIT Press.
26. Shimony, Abner (2002). "Some Intellectual Obligations of Epistemological Naturalism" in D. B. Malement, ed., *Reading natural philosophy: essays in the history and philosophy of science and mathematics*. Chicago, IL, Open Court, pp. 297–313.
27. Peirce, Charles S. (1979). *The new elements of mathematics*. 4 volumes. The Hague, Mouton/Humanities Press. Edited by Carol Eisele.
28. Eisele, Carolyn (1979). *Studies in the scientific and mathematical philosophy of Charles S. Peirce: essays by Carolyn Eisele*. The Hague/New York, Mouton. Edited by R. M. Martin.
29. Joswick, Hugh (1988). "Peirce's Mathematical Model of Interpretation." *Transactions of the Charles S. Peirce Society* **24**: 107–121.
30. Levy, Stephen H. (1997). "Peirce's Theoremic/Corollarial Distinction and the Interconnections between Mathematics and Logic" in N. Houser, D. D. Roberts, and J. Van Evra, eds., *Studies in the logic of Charles Sanders Peirce*. Bloomington, IN, Indiana University Press, pp. 85–110.
31. Cooke, Elizabeth F. (2003). "Peirce, Fallibilism, and the Science of Mathematics." *Philosophia Mathematica* **11**: 158–175.
32. Campos, Daniel (2007). "Peirce on the Role of Poetic Creation in Mathematical Reasoning." *Transactions of the Charles S. Peirce Society* **43**: 470–489.
33. Eisele, Carolyn (1971). "The mathematical foundations of Peirce's Philosophy" in *Studies in the scientific and mathematical philosophy of Charles S. Peirce: essays by Carolyn Eisele*. The Hague/New York, Mouton. Edited by R. M. Martin.
34. Hull, Kathleen (1994). "Why Hanker after Logic? Mathematical Imagination, Creativity, and Perception in Peirce's Systematic Philosophy." *Transactions of the Charles S. Peirce Society* **30**: 271–296.

35. Dipert, Randall R. (1994). "Peirce's Underestimated Place in the History of Logic: A Response to Quine," in Kenneth L. Ketner, ed., *Peirce and Contemporary Thought: Philosophical Inquiries*, New York: Fordham University Press, 32–58.
36. Whately, Richard (1826). *Elements of logic: comprising the substance of the article in the Encyclopedia Metropolitana; with additions*, &c. London, J. Mawman.
37. Peirce, Charles S. and Victoria Welby (2001). *Semiotic and significs: the correspondence between Charles S. Peirce and Victoria Lady Welby*. Elsah, IL, Arisbe Associates. Edited by Charles S. Harwick and James Cook.
38. Peirce (1870) Peirce, Benjamin (1881). "Linear Associative Algebra." *American Journal of Mathematics* **4**: 97–229. This paper is based on a memoir read before the National Academy of Sciences in Washington, 1870. It was published posthumously with notes by Charles S. Peirce.
39. Seibert, Charles H. (2005). "Charles Peirce's Reading of Richard Whately's 'Elements of Logic'." *History and Philosophy of Logic* **26**: 1–31.
40. Boole, George (1847). *The mathematical analysis of logic: being an essay towards a calculus of deductive reasoning*. Cambridge, England, Macmillan Barclay & Macmillan.
41. Babbage, Charles and John Herschel (1813). "Preface." *Memoirs of the Analytical Society*: i–xxii.
42. Chrystal, George (1883). "Mathematics". *The ninth edition, Encyclopaedia Britannica*. Vol. 15. Edinburgh, Adam & Charles Black.
43. Peirce, Benjamin (1881). *Ideality in the physical sciences*. Boston, FL, Little Brown & Co.
44. Peirce, Benjamin (1880). "The impossible in mathematics", in Mrs. John T. Sargent, ed., *Sketches and reminiscences of the Radical Club of Chestnut St. Boston*. Boston, FL, James R. Osgood, pp. 376–379.
45. Sylvester, James J. (1868). "Presidential Address to the British Association." reproduced in *Sylvester's mathematical papers*, vol. 2, p. 654.
46. Locke, John (1690). *An Essay Concerning Human Understanding*. London.
47. Anderson, Ronald and Girish Joshi (1993). "Quaternions and the Heuristic Role of Mathematical Structures in Physics." *Physics Essays* **6**: 308–319.
48. Anderson, Ronald and Girish Joshi (2008), "Interpreting Mathematics in Physics: Charting the Applications of SU(2) in 20th Century Physics." *Chaos, Solitons & Fractals*, **36**: 397–404.
49. Herschel, John F. W. 1831. *Preliminary Discourse on the Study of Natural Philosophy*. London: Longman, Rees, Orme, Brown, and Green.
50. Gray, Jeremy J. (2004). "Anxiety and Abstraction in Nineteenth-Century Mathematics." *Science in Context* **17**: 23–47.
51. Peirce, Charles S. (1975). *Charles Sanders Peirce: contributions to the nation*. Lubbock, TX, Texas Tech Press. Compiled and annotated by Kenneth Laine Ketner and James Edward Cook.
52. Haack, Susan (1979). "Fallibilism and Necessity." *Synthese* **41**: 37–63.
53. Bernstein, Richard J. (1989). "Pragmatism, Pluralism and the Healing of Wounds." *Proceedings and Addresses of the American Philosophical Association* **63**: 5–18.
54. Mancosu, Paolo (1997). *From Brouwer to Hilbert: the debate on the foundations of mathematics in the 1920s*. Oxford: Oxford University Press.
55. Giaquinto, M. (2002). *The search for certainty: a philosophical account of foundations of mathematics*. Oxford/New York, Clarendon Press/Oxford University Press.
56. Thagard, Paul and Beam, Craig (2004). "Epistemological Metaphors and the Nature of Philosophy." *Metaphilosophy* **35**: 504–516.
57. Abrams, Jerold J. (2002). "Philosophy After the Mirror of Nature: Rorty, Dewey, and Peirce on Pragmatism and Metaphor." *Metaphor and Symbol* **3**: 227–242.
58. Steward, Dugald (1854–1860). *The collected works of Dugald Stewart*. Edited by Sir W. Hamilton: 11 vols., Edinburgh, T. Constable: Vols. 2–4, Elements of the Philosophy of the Human Mind, vol. II first published 1814.
59. Chihara, Charles S. (2004). *A structural account of mathematics*. Oxford/New York, Clarendon/Oxford University Press.
60. Bourbaki, Nicolas (1994). *Elements of the history of mathematics*. Berlin/New York, Springer-Verlag.

61. Cooke, Elizabeth F. (2007). "Peirce's General Theory of Inquiry and the Problem of Mathematics" (under review).
62. Whitehead, Alfred N. (1926). *Science and the modern world.* Cambridge: Cambridge University Press.
63. Dewey, John (1938), *The later works, 1925–1953*, Volume 12, Carbondale and Edwardsville, IL, Southern Illinois University Press, Edited by Jo Ann Boydston.
64. Campos, Daniel G. (2005). *The Discovery of Mathematical Probability Theory: A Case Study in the Logic of Mathematical Inquiry.* Ph.D. thesis, Pennsylvania State University, UMI AAT 3202479.
65. Hacking, Ian (1983). *Representing and intervening: introductory topics in the philosophy of natural science.* Cambridge/New York, Cambridge University Press.
66. Peirce, Benjamin (1855). *A system of analytic mechanics.* Boston, FL, Little Brown & Co.
67. Walsh, Alison (2000). *Relationships Between Logic and Mathematics in the Works of Benjamin and Charles S. Peirce*, Ph.D. thesis, Middlesex University.
68. Peirce, Benjamin (1853). Address of Professor Benjamin Peirce, president of the American Association for the year 1853, on retiring from the duties of president.
69. Emerson, Edward Waldo (1918). *The early years of the Saturday club, 1855–1870.* Boston, FL/New York, Houghton Mifflin Co.
70. Goodman, Nicolas (1991). "Modernizing the Philosophy of Mathematics." *Synthese* **88**: 119–126.
71. van Kerkhove, Bart (2006). "Mathematical Naturalism: Origins, Guises and Prospects." *Foundations of Science* **11**: 5–39.
72. Searle, John R. (2006). "Minding the Brain, Review of 'Seeing Red: A Study in Consciousness' by Nicholas Humphrey." *The New York Review of Books* **53**.
73. Popper, Karl R. (1957). "The Propensity Interpretation of the Calculus of Probability, and the Quantum Theory," in S. Körner, ed., *Observation and Interpretation.* London: Butterworths Scientific Publications, pp. 65–70.

Newton's Methodology

William Harper

Abstract Newton's methodology is richer than the hypothetico-deductive model of scientific inference that was the focus of many philosophers of science in the last century. These enrichments focus on theory-mediated measurements of theoretical parameters by phenomena. It is argued that this richer methodology of Newton's informs a pre-relativity response to the Mercury perihelion problem, endorses the transition from Newton's theory to Einstein's, and continues to inform the testing frameworks for relativistic gravity theories today. On this rich methodology of Newton's, science is very informative about the world, without any commitment to progress toward an ideal limit of a final theory of everything.

Newton's scientific methodology is much richer than the models of scientific inference that have been studied by philosophers of science. I will be explaining several salient features that make this richer methodology more informative about the world than, even, quite sophisticated Bayesian models of scientific inference of the sort Abner Shimony has developed in his classic papers [23, 24]. Abner, Wayne Myrvold and I have begun a program of joint research designed to enrich the Bayesian model with resources to accommodate Newton's richer methodology. This paper will characterize some features that I shall argue ought to be accommodated in order to do justice to Newton's methodology. The job of how to enrich the Bayesian framework to do justice to these features will left to be addressed in future work.[1]

1 Newton's Methodology vs. Hypothetico-Deductive Model of Scientific Method

On the basic hypothetico-deductive (H-D) model of scientific method hypotheses are verified by the conclusions to be drawn from them and empirical success is accurate prediction.

W. Harper
Department of Philosophy, University of Western Ontario

[1] A beginning is made in [25].

W.C. Myrvold and J. Christian (eds.), *Quantum Reality, Relativistic Causality, and Closing the Epistemic Circle,* The Western Ontario Series in Philosophy of Science 73,
© Springer Science+Business Media B.V. 2009

There are various versions of this basic H-D model of scientific method. One version is that of Karl Popper. On his account, scientific inference would be limited to rejecting hypotheses that failed to make the right predictions of observable data. The more usual versions of this basic H-D model are inductive ones, according to which successful prediction would lead to increases in the epistemic probability assigned to the hypothesis being tested. The Bayesian learning model, on which agents update their epistemic probabilities by conditionalizing on the empirical outcomes of experiments, can be viewed as an extension of this more liberal inductive version of the H-D approach to scientific inference.

Newton's methodology adds features that go beyond all these versions of the hypothetico-deductive model of scientific inference and its usual Bayesian extensions. I will ague that these features that enrich Newton's methodology make it more informative about the world. They are important to the scientific practice in the investigation of gravity from Newton's time to our own; but, unfortunately, they have been largely ignored by philosophers of science. These important features are also not often mentioned in what scientists articulate to the public about their scientific practice. Even people developing and applying the testing frameworks for relativistic theories of gravity, which I will be arguing is a clear example of Newton's methodology at work, sometimes sound like they are just testing theories when they say what they are doing.

The first thing that I claim Newton's methodology adds to the basic hypothetico-deductive model is a new and stronger sort of empirical success. To realize this richer sort of empirical success a theory needs to do more than just accurately predict the phenomena it purports to explain. In addition, it needs to have those phenomena accurately measure the parameters which explain them. This will be illustrated in some detail with Newton's classic inferences from phenomena.

A second feature added is an appeal to theory-mediated measurements. In this methodology from Newton, one exploits, in so far as possible, theory-mediated measurements from phenomena so as to give empirical answers to theoretical questions. Many of the talks that we have seen at this conference are about experiments which are obviously doing this. There has been, however, a whole tradition in philosophy of science where "theory" is treated as a bad word.

A third feature added is provisional acceptance of theoretical propositions as guides to research. Without this, theory-mediated measurements do not even get off the ground. As we shall see, Newton starts by accepting his laws of motion, and uses propositions derived from them to make orbital motion phenomena afford information about centripetal forces.

All three of these features come together in a method of successive approximations that informs applications of universal gravity to motions of solar system bodies. On this method deviations from the model developed so far count as new theory-mediated phenomena to be exploited as carrying information to aid in developing a more accurate successor.

I am not claiming that there are no other significant ways Newton's methodology goes beyond the basic H-D model. I will argue, and I hope to convince some of

Newton's Methodology   45

you, that these three ways in which Newton's methodology is richer are important, because they really do make it a more informative methodology for investigating the world.

2 Newton's Classic Inferences from Phenomena

I am going to discuss Newton's classic inferences from phenomena. We will see how each of these illustrates the feature of having the phenomenon measure a theoretical parameter. The proposition inferred is a certain value of that parameter—the one that's measured by the cited phenomenon. The first of these classic inferences goes from Kepler's area law for an orbit to the centripetal direction of the force deflecting a body into that orbit

Kepler's area law $\Rightarrow$ centripetal direction

The second is from Kepler's harmonic law for a system of orbits about a common center, e. g. the planets about the Sun, to the inverse square variation with respect distance from that common center of the centripetal accelerations induced by the centripetal force maintaining bodies in those orbits.[2]

Kepler's harmonic law $\Rightarrow$ inverse-square variation

Newton also infers the inverse-square variation of the centripetal force maintaining a single body in its orbit from the absence of orbital precession.

Absence of precession $\Rightarrow$ inverse-square variation

I will be focusing on the systematic dependencies that back up these inferences, which Newton proves from his laws of motion as assumed premises. We shall see that these systematic dependencies turn these inferences into measurements by the cited phenomena.

Kepler's area law phenomenon is that the rate at which areas are swept out by radii drawn from the center is constant. According to the first Proposition of the *Principia*, a body that is deflected from inertial motion by a force that is directed toward an inertial center moves in a plane, and the rate at which areas are swept out by radii to that center is constant. This gives

Centripetal direction of force $\Rightarrow$ areal rate is constant

So, if you were doing a scientific inference according to the hypothetico-deductive method, that would be it. Proposition 1 shows that a centripetal force

---

[2] Newton defines such a centripetal force as a capacity, a field of attraction, surrounding the center. A fundamental characteristic of the inverse-square centripetal forces toward the Sun and planets, which Newton infers from orbits, is that these inverse-square varying induced centripetal acceleration components would be equal for any bodies at any equal distances from the center. Howard Stein has called these "fields of acceleration" (see [26–29]). Newton counts such a centripetal force as the common cause of the several motive forces corresponding to the centripetal accelerations times the masses of bodies being attracted toward that center. See [3], 405–408.

would predict the area law phenomenon. Therefore, the hypothesis that the force maintaining the body in its orbit is directed toward this center is confirmed by the fit of this area law phenomenon to the data.

Newton's inference is backed by things that make it far more compelling than any such hypothetico-deductive confirmation. According to the second Proposition of the *Principia*, if you have a body that is moving in a plane relative to an inertial center and the rate at which areas are being swept out by radii to that center is constant then the force is directed toward that center. This gives

Areal rate is constant $\Rightarrow$ centripetal direction of force

This is the proposition Newton cites in support the inference. Notice how you have theoretical background assumptions that are getting the centripetal direction from the area law phenomenon. It is even better than this. According to Corollary 2 of Proposition 2

Areal rate is increasing $\Rightarrow$ force is directed off-center forward
and, in the absence of resistance,
Areal rate is decreasing $\Rightarrow$ force is directed off-center backward.

These implications, which are coming from the laws of motion, support counterfactuals.[3] They are systematic dependencies that afford a very powerful way that Kepler's area law carries the information that the force maintaining a body in an orbit is directed toward the center about which the rate at which it sweeps out areas is constant.

Consider next Kepler's harmonic law for a system of several orbits about a common center. He discovered that the ratio of the square of the period to the cube of the mean distance of its orbit is the same constant value for all six of the primary planets known in his day. That is, the periods are proportional to the 3/2 power of the mean distances. One way to illustrate the fit of this harmonic law to cited periods and mean distances is to plot the logarithms of the periods against the logarithms of the distances. Such a plot, for the data cited by Newton for Mercury, Venus, Earth, Mars, Jupiter and Saturn, is shown in Fig. 1, Newton cites periods agreed to by astronomers and mean distances from Kepler and also from the French astronomer Boulliau.

To have the periods be some power or other of the mean distances is to have these log periods plotted against log distances be fit by some straight line. To have the harmonic law hold is to have these fit by a straight line of slope 3/2. Notice the good fit of this harmonic law phenomenon to the data Newton cites. The line in the diagram is one with a 3/2 slope, not a fit to the data.

---

[3] According to Newton's Laws of Motion, if at a given instant the rate at which a body is sweeping out areas by radii to a given inertial center were increasing (decreasing) then the total force deflecting that body from inertial motion at that instant would be directed off center in a forward (backward) direction.

The counterfactual-supporting nature of these dependencies make Newton's inferences immune to the counter examples based on constructing "unnatural" material conditionals that led to the demise of Clark Glymour's boot-strap confirmation as a serious candidate for explicating scientific inference. See [30].

Newton's Methodology

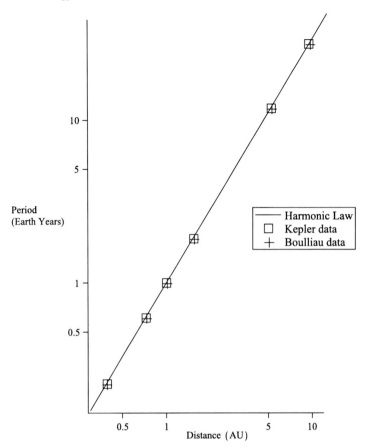

Fig. 1 Log-log plot of planetary periods against distances, from data cited in Newton's *Principia*

This diagram also illustrates another interesting thing about Newton's inferences from phenomena. Phenomena are not just data. They are patterns exhibited in sets of data. In this case we have the data originally cited by Kepler as well as additional data obtained later by Boulliau.

This harmonic law phenomenon measures the inverse- square relation among the centripetal accelerations exhibited by these orbits at their respective mean-distances.[4]

---

[4] What is counted as the mean-distance of a Keplerian orbit with force toward a focus is a distance equal to the length of the semi-major axis (half the major axis) of the ellipse.

Newton proves proposition 4 and its corollaries for concentric uniform motion circular orbits, but the results extend exactly to the corresponding relations among the centripetal accelerations of con-focal elliptical orbits at their respective mean- distances from that focus.

At its mean distance from the focus, toward which the force maintaining a body in such a Keplerian elliptical orbit is directed, the acceleration toward that focus equals the centripetal acceleration of a uniform motion concentric circular orbit with radius equal to the mean-distance of the ellipse and period equal to that of the elliptical orbit. Note that

According to Corollary 7 of Proposition 4 of book 1,[5]

$$t \propto R^s \Leftrightarrow F \propto R^{1-2s}$$

To have the periods be as some power s of the mean-distances is equivalent to having the centripetal force count as one producing accelerations that vary as the $1-2s$ power of those distances. Newton's sixth corollary

$$t \propto R^{3/2} \Leftrightarrow F \propto R^{-2}$$

is a special case that follows immediately from the more general seventh corollary.

Notice, however, that the more general relation established in corollary seven backs up this equivalence with additional systematic dependencies. We have not only the above equivalence, but also

$$s > 3/2 \Leftrightarrow 1 - 2s < -2$$

as well as

$$s < 3/2 \Leftrightarrow 1 - 2s > -2$$

So, in addition to having $s = 3/2$ equivalent to having the accelerations induced by the force be as the inverse-square of the distances, we also have $s > 3/2$ just in case the induced accelerations would fall off faster than the inverse-square, while $s < 3/2$ would be to have those induced accelerations fall off less fast than the inverse-square of these mean-distances. Alternative values of s would carry information about alternative values of a power-law relating the accelerations produced by the centripetal force to the mean distances of the orbits it maintains. These systematic dependencies make the harmonic law phenomenon for a system of such orbits measure the inverse-square power law relating the accelerations produced at their mean-distances on bodies being maintained in those orbits.

In the opening proposition of his argument for universal gravitation Newton argues for an inverse-square centripetal force toward Jupiter from the area law motion and harmonic law relation of the orbits of its satellites. By the second edition of the *Principia* he had enough data on such orbits of Saturn's moons to add an inference to an inverse-square centripetal force toward Saturn. In the next proposition, Newton infers an inverse-square centripetal force toward the sun from the area law and harmonic law phenomena for the orbits of the primary planets.

$$acc_r = -(4\pi^2 a^3/t^2)(1/r^2)$$

where acc_r is the centripetal acceleration corresponding to distance (e.g. radius vector) r from the focus in an elliptical orbit of semimajor axis a and period t with the force toward that focus (See French 1971, 588). Now set $r = a$. This yields

$$acc_r = -(4\pi^2 r/t^2),$$

which equals the centripetal acceleration of a concentric circular uniform motion orbit of radius r and period t about that center.

[5] This is a special case, for power-law forces, of a more general relation established in Corollary 2 of Proposition 4, according to which $F \propto R/t^2$.

Newton's Methodology   49

For these primary planets Newton also infers the inverse-square from the absence of orbital precession. In a stable elliptical orbit a planet would traverse against the fixed stars exactly 360 degrees in order to go from and return to the aphelion (the most distant point from the sun) of its orbit.[6] The orbit would count as precessing with p degrees of precession per revolution just in case the planet would traverse against the fixed stars $360 + p$, rather than just 360 degrees, to return to its aphelion. In the time it takes the planet to get back to the same point in its ellipse the axis of that ellipse will have traversed p degrees against the stars. The orbit is precessing forward if p is positive and is precessing backward (in the opposite direction from the motion of the planet in the ellipse) if p is negative. A stable orbit has zero precession.

Here is Corollary 1 of Proposition 45 book 1 of Newton's *Principia*, which applies to the orbit of a body acted upon by a centripetal force F that is proportional to the power n of the distance R from the center toward which that force is directed.[7]

$$p° \text{ precession/revolution} \Leftrightarrow n = [360/(360 + p)]^2 - 3$$

Having $p = 0$ is equivalent to having $n = -2$, so that zero orbital precession is equivalent to having the centripetal force F maintaining a body in that orbit vary inversely as the square of the distance R from the center toward which it is directed. As in the case of the harmonic law inference, the basic equivalence is backed up by systematic dependencies. Having p be positive is equivalent to having n be less than -2, while having p be negative is equivalent to having n greater than -2. Forward precession corresponds to having the centripetal force fall off faster than the inverse-square of the distance, while backward precession corresponds to having the centripetal force fall off less fast than the inverse-square of the distance. Zero precession would exactly measure the inverse-square variation with distance of the centripetal force maintaining a body in such an orbit.

Here, from his *System of the World*, is a comment by Newton on the absence of precession in the orbits of the planets about the Sun.

> But now, after innumerable revolutions, hardly any such motion has been perceived in the orbits of the circumsolar planets. Some astronomers affirm there is no such motion; others reckon it no greater than what may easily arise from causes hereafter to be assigned, which is of no moment to the present question [1, p. 561].

Newton can apply the foregoing one-body results to systems, like the orbits of the planets about the sun, where there are forces of interaction between planets which perturb what would be motion under the inverse-square centripetal force toward the sun alone. If all the orbital precession of a planet is accounted for by perturbations, the zero left-over precession can be counted as measuring the inverse-square variation of the basic centripetal force that maintains that planet it in its orbit about the sun.

---

[6] In Newton's time it was common to refer to orbital precession as motion of the aphelion. We now refer to it as motion of the perihelion (the opposite end of the major axis), which is the closest point in the orbit to the sun.

[7] This is a special case, for a power law force, of a more general theorem, Newton's Proposition 45, relating rate of orbital precession to the dependency of a centripetal force on distance.

3 Successive Approximations and Newton's 4th Rule as a Characterization of Acceptance in Science

In August 1684 Edmund Halley, who would become the Astronomer Royal in 1720, visited Newton in Cambridge. According to a much retold story, Halley's visit convinced Newton of the importance of a calculation in which Newton had connected the ellipse with an orbit produced by an inverse-square force. By November Newton had sent Halley a small but revolutionary treatise *De Motu*. An extraordinarily intense and productive effort by Newton over the next few years transformed this small treatise into his *Principia*.

One of the intermediate versions of the *De Motu* text has the scholium we are about to quote. What probably led Newton to this was his realization that the center of mass of a group of interacting bodies is not disturbed by the interactions among those bodies. So, if you are going to find anything that could count as a center relative to which you can fix what can be counted as true motions among the solar system bodies it would be the center of the mass.

> By reason of the deviation of the Sun from the center of gravity, the centripetal force does not always tend to that immobile center, and hence the planets neither move exactly in ellipses nor revolve twice in the same orbit. There are as many orbits of a planet as it has revolutions, as in the motion of the Moon, and the orbit of any one planet depends on the combined motion of all the planets, not to mention the action of all these on each other. But to consider simultaneously all these causes of motion and to define these motions by exact laws admitting of easy calculation exceeds, if I am not mistaken, the force of any human mind. [2, p. 253]

This would have been written some time in 1684. Instead of giving up, Newton embarked upon the prodigious additional effort that resulted in his *Principia*, which was published in 1687. I like to think of this effort as one devoted to developing resources for dealing with this daunting complexity problem by successive approximations.

Newton's center of mass resolution of the two chief world systems problem in his application of his theory of universal gravity to solar system motions showed that neither the sun nor the earth could be considered as a center which fixes what are to count as the true motions. But, while the center of the Earth is hopelessly deviant, the center of the Sun could be counted upon as a good enough approximation from which to begin. Elliptical orbits with the sun at a focus could be counted as an initial model of the orbital motions of the planets corresponding to the basic inverse-square gravity toward the Sun. Deviations could then be treated as theory-mediated phenomena to be exploited as carrying information about interactions to be taken into account in constructing more accurate successor models. As is well known, this was the enterprise that, in the hands of such successors as Laplace, developed into the science of celestial mechanics. Newton's fourth rule for doing natural philosophy is a general statement of his methodology. It actually got published only in the third edition of the *Principia*.

Newton's Methodology   51

Rule 4. In experimental philosophy, propositions gathered from phenomena by induction should be considered either exactly or very nearly true notwithstanding any contrary hypotheses, until yet other phenomena make such propositions either more exact or liable to exceptions.

This rule should be followed so that arguments based on induction may not be nullified by hypotheses. [3, p. 796]

We shall see that this rule characterizes the important role of acceptance of propositions as guides to research in the enterprise of using empirical deviations to find successively better approximations.

Newton's characterization of his enterprise as "experimental philosophy" is an explicit contrast to the mechanical philosophy of his continental critics. The aim of the mechanical philosophy was to render motion phenomena intelligible by giving some hypothesis about how it **could** be caused by contact action between bodies. The initial reaction to the *Principia*, especially on the continent by such figures as Huygens and Leibniz, who were in their various ways committed this mechanical philosophy, was that it didn't look like there could be any reasonable hypothesis that could recover Newton's gravitational interactions at a distance by contact action between bodies. For example, Newton's applications of the third law of motion to construe the equal and opposite reaction to the attraction of a planet toward the Sun to be an attraction of the Sun toward that planet were objected to.[8] According to the mechanical philosophy the natural application of the third law would be to hypothesized vortical particles pushing the planet toward the Sun.

Newton's comment tells us that Rule 4 should be followed so that arguments based on induction may not be nullified by hypotheses. Consider an appeal to this rule to prevent his argument from being undercut by the contrary hypothesis that Jupiter is maintained in its orbit by vortical particles pushing it toward the Sun. In order to assess such an application of this Rule we will need to better understand what are to be counted as propositions gathered from phenomena by induction and how these differ from what are to be dismissed as mere contrary hypotheses.

Newton tells us "propositions gathered from phenomena by induction are to be considered either exactly or very nearly true ... until yet other phenomena make such propositions either more exact or liable to exceptions." One thing we see here is *acceptance subject to correction* rather than just assigning and adjusting probabilities. Acceptance of theoretical propositions is really central to Newton's methodology. A second thing is that the accepted propositions can be *accepted as approximations* even if they are not exactly true.

---

[8] Huygens [42, p. 159] objected to Newton's gravity as a mutual attraction between whole bodies, as well as to Newton's universal gravity as a mutual attraction between all the small parts of bodies, in comments on Newton's argument that he added to his own *Discourse on the Cause of Gravity* after reading Newton's *Principia*. Huygens published his *Discourse* on gravity as an addition to his *Treatise on Light* in 1690.

Leibniz developed his theory of harmonic vortices specifically to give a vortex theoretic alternative to Newton's inverse-square centripetal forces. See Aiton 1995, 10–13.

In his classic paper on scientific inference, Abner Shimony[9] discusses commitment to a theory h as belief that:

i. Within the domain of current experimentation, h yields almost the same observational predictions as the true theory.
ii. The concepts of the true theory are generalizations or more complete realizations of those of h.
iii. Among the currently formulated theories competing with h, there is none that better satisfies conditions (i) and (ii).

This is very much in line with Newton's provisional acceptance of propositions as approximations appropriate to guide further research.

According to Newton's Rule 4, the acceptance of propositions he counts as "gathered from phenomena by induction" is not to be undercut by mere contrary hypotheses. I want to say more about what ought to be counted interesting clear cases where mere hypotheses can be distinguished from an alternative explanation to be taken seriously. Consider the following strong formulation of an ideal of empirical success that informs Newton's methodology.

Newton's Strong Ideal of Empirical Success: Convergent accurate measurement of parameters by the phenomena to be explained.

By convergent I mean agreeing measurements of the same parameter by different phenomena. Where the propositions to be counted as gathered from phenomena by induction realize this strong ideal of empirical success we can characterize what should be dismissed as mere contrary hypotheses by contrast.

Mere Contrary Hypothesis: An alternative that does not realize this ideal of empirical success sufficiently well to count as a serious rival.

On this interpretation of Newton's Rule 4, the alternative hypothesis that Jupiter is maintained in its orbit by vortex particles pushing it toward the sun could be dismissed as a mere contrary hypothesis, unless it were backed up by measurements of vortical parameters that could rival the agreeing measurements of the relative masses of solar system bodies afforded by Newton's treatment of gravity as an interaction.

The dismissal of contrary hypotheses in Rule 4 focuses the engine of change on empirical phenomena. Newton tells us to continue to consider a proposition gathered from phenomena by induction as either exactly or very nearly true until yet other phenomena make such propositions either more exact or liable to exceptions. Start from the basic Keplerian orbits corresponding to the inverse-square gravitation toward the Sun. Empirically established deviations from motion in accord with these orbits are examples of yet other phenomena that make propositions asserting motion in accord with them liable to exceptions. But, to the extent that these exceptions can be accounted for as perturbations due to additional interactions, they can be used to make the propositions characterizing the details of the gravitational model of the solar system motions more exact. This enterprise of treating such deviations as further phenomena carrying information to help find more accurate corrected models was

---

[9] See [23, p. 199].

Newton's Methodology   53

very impressively realized in the hands of Newton's successors. They succeeded in
generating increasingly precise perturbation-corrected phenomena that would accurately fit large open-ended bodies of increasingly precise data. This extraordinary
predictive success was backed up by the increasingly accurate agreeing measurements of the relative masses of the interacting solar system bodies afforded from the
perturbation corrections to the phenomena. These agreeing measurements of such
parameters count as an equally extraordinary realization of Newton's stronger ideal
of empirical success.

4 The Mercury Precession Problem[10]

Van Fraassen [4, p. 265–266] has claimed that Newton's methodology is too con-
servative to do justice to radical theory change. Thomas Kuhn has gone so far as
to challenge the very idea that there can be non-question begging standards of theory assessment that apply across scientific revolutions (see quotation below from
Kuhn [5, p. 94]). I want to argue that Einstein's solution to the Mercury precession problem gave good grounds to expect that Newton's standard of empirical success would favor general relativity over Newton's own theory. I will also argue that
Newton's rich methodology applies to the empirical assessment of a pre-relativity
proposal to alter the inverse-square law to accommodate the extra precession and
that this same rich methodology applies to the empirical assessment of a proposed
alternative to general relativity that also involved the precession of Mercury. This
last episode illustrates that Newton's rich methodology is very much exemplified
in the formulation and application of testing frameworks for assessing alternative
relativistic gravity theories today.

(a) The Classic Problem: Hall's Hypothesis and Brown's Measurement

The actual precession of Mercury's orbit is approximately 574 s per century. About
531 of those are due to Newtonian perturbations. In addition you have the famous 43 s per century that are not explainable by Newtonian gravitation (see, e.g.,
[6, p. 181; 7, p. 91]). This value of about 43 s per century was arrived at by Simon
Newcomb in 1882 when he revised and corrected an earlier estimate from Le Verrier
[8, p. 473]. It has been holding up ever since.

Hall [9] proposed to account for this anomalous extra precession by revising the
inverse-square law.

> Applying BERTRAND's formula to the case of *Mercury* I find, taking NEWCOMB's value of
> the motion, or 43″, that the perihelion would move as the observations indicate by taking

$$n = -2.00000016$$

---

[10] A slightly different version of this material was given as a separate paper at PSA 2006 and is
published in [36]. The author is grateful for permission to use this material.

The formula Hall appeals to is equivalent to Newton's (see [10]). The alternative value −2.00000016 for the power of distance is just what Newton's formula would take to be the value measured by the 43 s per century precession left over after perturbations had been accounted for.[11]

In 1903 Ernest Brown had refined the complex theory of our moon's orbit with sufficient precision to empirically constrain departures δ from the inverse-square to less than 0.00000004. Here, from the *Monthly Notices of the Royal Astronomical Society* is Brown's statement of his result and its implication for Hall's hypothesis.

> If the new theoretical values of the motions of the Moon's perigee and node are correct, the greatest difference between theory and observation is only $0''.3$, making $\delta < .00000004$. Such a value for δ us quite insufficient to explain the outstanding deviation in the motion of the perihelion of *Mercury*. It appears, then, that this assumption must be abandoned for the present, or replaced by some other law of variation which will not violate the conditions existing at the distance of the Moon [11].

Notice Brown has used the motion of our moon to measure the inverse-square out to at least seven decimal places. It is this measurement bound on differences from the inverse-square that rules out Hall's hypothesis. This measurement bound from data limiting extra precession of our moon's orbit is taken as a measurement bound limiting deviations from the inverse-square for gravitation generally.

(b) Einstein and General Relativity: An Answer to Kuhn's Challenge on Criteria Across Revolutions

Here, from Einstein's 1915 Berlin Academy paper on Mercury's perihelion is his statement of the fact that his theory does account for the residual precession of Mercury's orbit.

> The calculation yields, for the planet Mercury, a perihelion advance of $43''$ per century, while the astronomers assign $45'' \pm 5''$ per century as the unexplained difference between observations and the Newtonian theory [12].

This paper was written 2 weeks before he had the full field equation for general relativity. He got the result from constraints on what the field equation would have to be like (see [13]). Within 2 weeks of this encouraging positive result Einstein had gone on to develop his full field equation for General Relativity.

Einstein's reaction to accounting for this anomalous precession was quite strong. There are stories about how Einstein did not take much interest in the light bending experiments. Some have taken this to indicate that Einstein's great confidence in General Relativity was based almost completely on its theoretical virtues alone. The following quotations from Pais' biography makes clear that his successful account of Mercury's precession was very important to him.

---

[11] Here is the calculation for Newton's formula from corollary 1 of proposition 45 bk 1. The period of Mercury is approximately 0.24 Julian years. 43 seconds of precession per century is therefore 0.000029 degrees per revolution. On Newton's formula the corresponding power law for the centripetal force is as the $(360/360.000029)^2 - 3 = -2.00000016$ power of distance.

Newton's Methodology

The first result was that his theory 'explains…quantitatively…the secular rotation of the orbit of Mercury, discovered by Le Verrier, … without the need of any special hypotheses.' This discovery was, I believe, by far the strongest emotional experience in Einstein's scientific life, perhaps in all his life [14, p. 253].

Here is more.

Nature had spoken to him. He had to be right. 'For a few days, I was beside myself with joyous excitement'. Later, he told Fokker that his discovery had given him palpitations of the heart. What he told de Haas is even more profoundly significant: when he saw that his calculations agreed with the unexplained astronomical observations, he had the feeling that something actually snapped in him. (*Ibid.*)

Now I want to claim that this attitude of Einstein's was appropriate to the fact that with this result he could be confident that his theory of General Relativity would beat Newton's theory on Newton's own standard of empirical success.

I am not saying that Einstein would have seen these facts in just this way, but it is important that he had good grounds for counting his theory as better than Newton's without any question begging appeal to some new standard that would favor it. One essential part of this is that he knew the Newtonian limit of General Relativity could recover the 531 s per century of Mercury's precession that had been successfully accounted for by Newtonian perturbations. Without this recovery of the Newtonian perturbations, his accounting for those 43 s would not have counted as a solution of the Mercury precession problem. The Newtonian limit recovers all the empirical successes of Newton's theory including all the agreeing measurements of parameters such as the relative masses of the Sun and planets. In addition to overcoming the Mercury precession anomaly Einstein's solution to that problem also affords a new agreeing measurement of the mass of the sun. In the years since 1915 many post-Newtonian corrections to solar system motion phenomena have afforded additional empirical successes that clearly favor Einstein's theory over Newton's without relying on any question begging appeal to new standards.

This raises deep problems for Thomas Kuhn's famous thesis that pre and post revolutionary theories are incommensurable paradigms.

Like the choice between competing political institutions, that between competing paradigms proves to be a choice between incompatible modes of community life. Because it has this character, the choice is not and cannot be determined merely by the evaluative procedures characteristic of normal science, for these depend in part upon a particular paradigm, and that paradigm is at issue. When paradigms enter, as they must, into a debate about paradigm choice, their role is necessarily circular. Each group uses its own paradigm to argue in that paradigm's defense [5, p. 94].

The fact that Newton's standard of empirical success clearly favors Einstein's theory over his own should count very heavily against Kuhn's doctrine that all revolutionary theory changes are to be understood as transitions between incommensurable paradigms.

(c) Mercury's Precession and a Challenge to General Relativity

One proposed alternative to General Relativity was the Brans–Dicke theory, which was motivated in part by Mach's objections to Newton's bucket experiment (see

56   W. Harper

[34]). Where Newton had used the rise of water against the sides of a spinning bucket to argue that absolute rotation can be empirically distinguished by dynamical effects, Mach challenged the assumption that such a dynamical effect would obtain if there were no background of fixed stars. In 1961 C. Brans and R. H. Dicke argued that General Relativity shares with Newton's theory an objectionable commitment to absolute rotation (see [15, p. 78]). A key feature of their alternative theory can be represented in the Parametrized Post Newtonian (PPN) Formalism for comparing alternative metrical gravity theories by the assignment of

$$\gamma = (1 + \omega)/(2 + \omega)$$

where γ is the PPN parameter representing the amount of space curvature per unit mass and ω is an additional parameter to be interpreted as representing contributions of distant masses to local curvature. In General Relativity the PPN parameter γ is fixed at the value 1. Therefore, the larger the value assigned to its additional parameter ω, the closer will be the Brans–Dicke theory assignment to agreement with General Relativity.

In 1967, Dicke and H. Mark Goldenberg reported measurements of solar oblateness which suggested that about 4 s per century of Mercury's precession would be accounted for by the quadrupole moment generated by a rapidly rotating inner core of the Sun [16–18]. This would make a version of the Brans–Dicke theory with ω set at 5, which would give 39 rather than 43 extra seconds per century, do better than General Relativity at accounting for Mercury's orbital precession.

In 1964 Irwin Shapiro proposed radar time delay as a test of general relativity. On a relativistic gravity theory there is round-trip time delay for radar ranging to planets, due to the gravitational potential of the sun along the path of the radiation, when that path passes close to the sun [19, 20]. This round-trip time delay measures γ, the parameter representing space curvature at issue above.[12]

In the positive detection of Shapiro's time delay, the exhibited pattern counts as a phenomenon (Fig. 2).

> The quantity $\Delta\tau$ is not an observable but is indicative of the magnitude and behavior of the measurable effect as predicted by general relativity [20, p. 1132].

In the 1979 Viking experiment with Mars, this time delay phenomenon was established with sufficient precision to measure $\gamma = 1 \pm 0.002$. Such measurements are precise enough to rule out any versions of the Brans–Dicke theory with $\omega < 500$ [21]. It was some time later, perhaps into the late 1980s, before further investigation established that effects of solar rotation were not great enough to undercut GR's account of Mercury's perihelion motion (see [6]).

---

[12] Shapiro's round trip time delay $\Delta\tau$ for radar ranging to planets measures γ:

$$\Delta\tau = (2r_0/c)((1+\gamma)/2) \ln((r_e + r_p + R)/(r_e + r_p - R))$$

where $r_0 = 2GM_S/c^2$, M_S is the solar mass, r_e and r_p are respectively the distances of the earth and the target planet from the sun, and R is the distance of the target planet from the earth. See [21].

Newton's Methodology 57

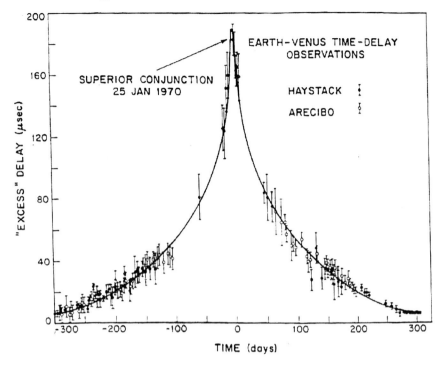

Fig. 2 Radar time delay (courtesy of Irwin Shapiro).

The development and applications of testing frameworks for relativistic gravitation theories is very much an illustration of Newton's methodology. The many successful tests of GR measure parameters that constrain alternative theories to approximate GR for scales and field strengths similar to those explored by the phenomena exhibited in those tests (see [6]).

5 Modest Newton: Progress Without Appeal to a Final Theory

Now I want to end with two pictures. This will be a comment on Newton inspired by a wonderful presentation that Lucien Hardy gave at a public debate on interpretations of quantum mechanics (Fig. 3). This was a debate put on by the Perimeter Institute that also had Antony Valentini talking about Bohm type interpretations and Wayne Myrvold talking about collapse interpretations. It ended with Lucien's presentation which included two depictions of Newton.

The first was the famous Blake depiction of Newton (Fig. 3). You can see that this Newton is a very impressive figure. He also can be seen as at least aspiring to the idea that his theory, represented by the diagram on which his attention is so sharply focused, is giving him a God like access to the world. He looks like he thinks of

Fig. 3 William Blake, *Newton* (1795)

the theory as giving the whole story about the world. That is, he takes his theory to afford him something very like a God's eye view of the whole world from the outside.

Lucien suggested that this Blake depiction of Newton represented the extremely optimistic view of the prospects for science that he had before struggling with the problems of Quantum Mechanics. He then contrasted the Blake depiction with a more modest depiction of Newton, done in 1870 and printed in the *The Illustrated London News* (Fig. 4). Newton is in the world and he's finding out more about it; but there is no pretension to anything like a God-like view from the outside. Lucien told us that Quantum Mechanics had driven him from the God-like prospects for science corresponding to the view of the Newton in Blake's depiction to the more modest aspirations corresponding to the depiction of Newton investigating light and colors.

Now I want to suggest that this more modest aspiration for the prospects of science was Newton's own view. Not just for his theory of light and colors but also for his theory of gravity. My evidence for this is that in one of his queries in the *Optiks* he comments that one of the features of his theory of gravity is that you would expect stability-threatening perturbations to the planets and suggested that one could use this as an argument for God [22, p. 402]. You would need to have God to interfere from time to time to keep the solar system stable.

The important thing for us today is that Newton did not see the prospect of a final theory of everything as a commitment needed to underwrite confidence in his methodology for using empirical deviations from the accepted theoretical model as theory mediated phenomena carrying information to be exploited in finding a

Fig. 4 "Newton Investigating Light," from *The Illustrated London News*, June 4, 1870

more accurate successor model. Newton's rich empirical methodology for generating deeper understanding by finding successively more accurate theoretical models need not be undercut by doubts about the conception scientific progress as progress toward the ideal limit of a theory of everything.

References

1. Cajori, F. (trans.) (1962). *Sir Isaac Newton's Mathematical Principles of Natural Philosophy and his System of the World*. Los Angeles: University of California Press.
2. Wilson, C. A. (1989). "The Newtonian achievement in astronomy," in [41, 233–274].
3. Cohen, I. B., and A. Whitman (trans.) (1999). *Isaac Newton The Principia, Mathematical Principles of Natural Philosophy: A New Translation*. Los Angeles: University of California Press, 1999.
4. van Fraassen, B. C. (1985). "Empiricism in the Philosophy of Science," in [35, 245–308]
5. Kuhn, T. S. (1970). *The Structure of Scientific Revolutions*, 2nd edition. Chicago: University of Chicago Press.
6. Will, C. M (1993). *Theory and Experiment in Gravitational Physics*, 2nd revised edition. Cambridge: Cambridge University Press.

7. Will, C. M. (1986). *Was Einstein Right? Putting General Relativity to the Test*. New York: Basic Books.
8. Newcomb, S. (1882). "Discussion and Results of Observations on Transits of Mercury from 1677 to 1881." *Astronomical Papers Prepared for the Use of the American Ephemeris and Nautical Almanac* **I** 473.
9. Hall, A. (1894). "A suggestion in the theory of Mercury," *The Astronomical Journal* **14**, 35–46.
10. Valluri, S. R., C. Wilson, and W. L. Harper (1997). "Newton's apsidal precession theorem and eccentric orbits," *Journal for the History of Astronomy* **28**, 13–27.
11. Brown, E. W. (1903). "On the Verification of the Newtonian Law," *Monthly Notices of the Royal Astronomical Society* **14**, 396–397.
12. Einstein A. ([1915] 1979). "Explanation of the perihelion motion of Mercury by means of the general theory of relativity," *Prussian Academy Proceedings* **11**, 831–839. B. Doyle (Trans.) in [38, 822–825].
13. Earman, J. and M. Janssen (1993). "Einstein's Explanation of the Motion of Mercury's Perihelion," in [40, 129–172].
14. Pais, A. (1982) *Subtle is the Lord: The Science and the Life of Albert Einstein*. New York: Oxford University Press.
15. Dicke, R. H. (1965). *The Theoretical Significance of Experimental Relativity*. New York: Gordon and Breach Science Publishers.
16. Dicke, R. H., and H. M. Goldenberg (1967). *Physical Review Letters* **18**, 313.
17. Dicke, R. H., and H. M. Goldenberg (1974). "The oblateness of the Sun," *Astrophysics Journal Supplement Series* No. 241, **27**, 131–182.
18. Dicke, R. H. (1974). "The oblateness of the Sun and relativity," *Science* **184**, 419–429.
19. Shapiro, I. (1964). "Fourth test of general relativity," *Physical Review Letters* **13**, 789–791.
20. Shapiro, I. I., M. E. Ash, R. P. Ingalls, W. B. Smith, D. B. Campbell, R. B. Dyce, R. F. Jurgens, and G. H. Pettengill (1971). "Fourth test of general relativity: new radar result," *Physical Review Letters* **26**, 1132–1135.
21. Reasenberg, R. D., I. I Shapiro, P. E. MacNeil, R. B. Goldstein, J. C. Breidenthal, J. P. Brenkle, D. L. Cain, T. M. Kaufman, T. A. Komarek, and A. I. Zygielbaum (1979). "Viking relativity experiment: Verification of signal retardation by solar gravity," *The Astrophysical Journal* **234**, L219–221.
22. Newton, I. ([1704] 1952) *Optics: Or a Treatise of the Reflexions, Refractions, Inflexions and Colours of Light*. (Based on the 4th edition of 1730.) New York: Dover Publications.
23. Shimony, A. (1970). "Scientific Inference," in R. G. Colodny, ed., *The Nature and Function of Scientific Theories*. Pittsburgh: Pittsburg University Press. Reprinted in Shimony (1993), 183–273.
24. Shimony, A. (1993). "Reconsiderations on Inductive Inference," in *Search for a Naturalistic World View, Vol. I: Scientific Method and Epistemology*. Cambridge: Cambridge University Press, 274–300.
25. Harper, W. L. (2007). "Acceptance and Scientific Inference," in W. Harper and G. Wheeler, eds. (2007), *Probability and Inference: Essays in Honour of Henry E. Kyburg, Jr.* London: Kings College Press, 33–52.
26. Stein, H. (1970). "On the Notion of Field in Newton, Maxwell, and Beyond," in R. H. Stuewer, ed., *Historical and Philosophical Perspectives of Science*, Minneapolis: University of Minnesota Press, 264–287.
27. Stein, H (1977). "Some Philosophical Prehistory of General Relativity," in J. Earman, C. Glymour, and J. Stachel, eds., *Minnesota Studies in the Philosophy of Science*, Vol. 8, Minneapolis: University of Minnesota Press, 3–49.
28. Stein, H (1991). "'From the Phenomena of Motions to the Forces of Nature': Hypothesis or Deduction?" in *PSA 1990*, Vol. 2, 209–222.
29. Stein, H (2002). "Newton's Metaphysics," in I. B. Cohen and G. Smith, eds., *Cambridge Companion to Newton*, Cambridge: Cambridge University Press, 256–307.
30. Harper, W. L. (1998). "Measurement and Approximation: Newton's Inferences from Phenomena versus Glymour's Bootstrap Confirmation," in P. Weingartner, G. Schurz, and G. Dorn, eds., *The Role of Pragmatics in Contemporary Philosophy*, Vienna: Hölder-Pinchler-Tempsky, 265–287.

Newton's Methodology   61

31. French, A. P. (1971). *Newtonian Mechanics*. New York: W. W. Norton & Company.
32. Aiton, E. J. (1995) "The Vortex Theory in Competition with Newtonian Celestial Mechanics" in R. Taton and C. Wilson, eds., *The General History of Astronomy, Vol. 2, Planetary astronomy from the Renaissance to the rise of astrophysics, Part A: Tycho Brahe to Newton*. Cambridge: Cambridge University Press, 3–21.
33. Shimony, A. (1993). *Search for a Naturalistic World View, Vol. I: Scientific Method and Epistemology*. Cambridge: Cambridge University Press.
34. Brans, C., and R. H. Dicke, (1961). "Mach's principle and a relativistic theory of gravitation" *Physical Review* **124**, 925–935. Reprinted in Dicke, R. H. (1965). *The Theoretical Significance of Experimental Relativity*. New York: Gordon and Breach Science Publishers, 77–96.
35. Churchland P. M. and Hooker C. A., eds. (1985). *Images of Science: Essays on Realism and Empiricism with a reply from Bas C. Van Fraassen*. Chicago: University of Chicago Press.
36. Harper, W. L. (2007). "Newton's Methodology and Mercury's Perihelion before and after Einstein" *Philosophy of Science* **74**, 932–942.
37. Harper, W. L. and G. R. Wheeler, eds. (2007). *Probability and Inference: Essays in Honour of Henry E. Kyburg, Jr*. London: Kings College Press.
38. Lang, K. R. and O. Gingerich, eds. (1979). *A Source Book in Astronomy and Astrophysics, 1900–1975*. Cambridge: Harvard University Press.
39. Roseveare, N. T. (1982). *Mercury's Perihelion from Le Verrier to Einstein*. Oxford: Oxford University Press.
40. Earman, J., M. Janssen, and J. Norton, eds. (1993). *The Attraction of Gravitation: New Studies in the History of General Relativity*. Boston: Birkhäuser.
41. Taton, R., and C. Wilson, eds. (1989). *The General History of Astronomy, Vol. 2, Planetary astronomy from the Renaissance to the rise of astrophysics, Part A: Tycho Brahe to Newton*. Cambridge: Cambridge University Press.
42. Huygens, C. (1690). *Discourse on the Cause of Gravity*, manuscript translation by Karen Bailey with annotations by Karen Bailey and George Smith. Translation of *Discours de la Cause de la Pesanteur, in Oeuvres completes de Christian Huygens*, vol. 21 (La Haye: Nijhoff, 1944), pp. 462–71 and pp. 476ff.

Whitehead's Philosophy and Quantum Mechanics (QM)

A Tribute to Abner Shimony

Shimon Malin

Abstract This paper is a tribute to Abner Shimony and a continuation of my discussions with him. In the first part some of Whitehead's concepts, and, in particular, actual entities and atemporal processes, are introduced. These are shown to correspond to the objectivized aspects of the collapse of quantum states. Next we reconcile the entanglement of quantum states with the speed of light barrier for the transmission of information by modifying Whitehead's system: We suggest that events that take place far apart can be aspects if the same actual entity. We show that this takes care of Lovejoy's objection to Whitehead's system.

1 The Challenge

Last year, when Abner and I conducted a dialogue on Whitehead's philosophy and QM, in Vienna, on the occasion of Anton Zeilinger's 60th birthday conference, Abner assigned me the task of winning the argument and restoring his enthusiasm and love for Whitehead.

I tried and failed. But I haven't given up. I consider this talk a second chance to fulfill Abner's task.

Following J. M. Burgers, who established, back in the 1960s, the intriguing connection between Process philosophy and quantum mechanics, Abner has been one of the champions, if not *the* champion of this connection. It was primarily A. Lovejoy's book, *The Revolt Against Dualism* [1], that convinced him otherwise. The point that convinced him seems to be the vagueness in the term "experience," which made it so far from the usual meaning of the term that it seemed to him virtually meaningless.

Early on Abner showed that certain modifications of Whitehead's philosophy as originally formulated are indispensable. As we shall see, other modifications are

---

S. Malin
Department of Physics and Astronomy, Colgate University, Hamilton, NY 13346, USA
e-mail: shimon@sover.net

W.C. Myrvold and J. Christian (eds.), *Quantum Reality, Relativistic Causality, and Closing the Epistemic Circle,* The Western Ontario Series in Philosophy of Science 73,
© Springer Science+Business Media B.V. 2009

needed to deal with the entanglement issue. The question is, whether the modified version enriches or destroys the Whiteheadian paradigm.

2 Whitehead's Philosophy and QM

Since the very existence of the relationship of Whitehead's philosophy (also known as "Process philosophy") and QM may be news for some of us, I will begin by establishing this intriguing connection.

The connection is based on the Whiteheadian concept of "actual entities," also known as "actual occasions," "throbs of experience," or "pulses of experience," depending on the context. They are, according to Whitehead, the "atoms of reality." The enormous gulf between Whitehead and contemporary scientific thinking is revealed already at this stage: According to Whitehead this universe we live in is an alive universe, a universe of experiences, rather than a universe of mostly inanimate matter.

Actual entities are processes (hence the name "Process philosophy"). They are the processes of their own self creation. They have both an objective and subjective aspects. And—as soon as they are completed they die. Dying, they become immortal in the sense that the fact that they did exist cannot be erased. In saying that actual entities are processes of their own self creation, creativity is implied. Whitehead's universe is not a completely deterministic one. It is a universe characterized by the phrase "creative advance into novelty."

The time constraints do not allow me to elaborate further. I am mentioning these points just to whet your appetites. Being introduced to Whitehead's thinking is a challenging, difficult and rewarding experience.

But how is it related to QM?

3 Schrödinger's Principle of Objectivation

Before addressing the question of a possible correspondence between Process Philosophy and QM, we need to put one more ingredient into the soup. This ingredient is Schrödinger's "Principle of objectivation."

When I recently discussed with Abner the role of the great physicists as philosophers, he said that Schrödinger is "in a class by himself." I agree.

According to Schrödinger, science, as it is practiced now, is based on two principles. First, the belief that nature is comprehensible, and, second, the principle of objectivation.

> By this [i.e., by "the principle of objectivation"] I mean what is also frequently called the 'hypothesis of the real world' around us. I maintain that it amounts to a certain

Whitehead's Philosophy and Quantum Mechanics (QM)   65

simplification, which we adopt in order to master the infinitely intricate problem of nature. Without being aware of it and without being rigorously systematic about it, we exclude the Subject of Cognizance from the domain of nature that we endeavor to understand. We step with our own person back into the part of an onlooker who does not belong to the world, which by this very procedure becomes an objective world. [2]

4 The Correspondence Between Actual Entities and Collapse

I suggest that the QM process that corresponds to an actual entity is the collapse of a quantum state. Like an actual entity, a collapse is an atemporal process that is not completely deterministic. Hence there is room for creativity.

I used the term "atemporal process." What are these?

The *process of self-creation of an actual entity is not a process in time; it is, rather, an atemporal process leading to the momentary appearance of the completed actual entity in spacetime.* Quoting Whitehead: "[In the process of self-creation which is an actual entity] the genetic passage from phase to phase is not in physical time ... the genetic process is not the temporal succession ... Each phase in the genetic process presupposes the entire quantum." [3] For example:

1. The creation of time in Plato's *Timaeus* comes after many other acts of creation—all of these must be atemporal.
2. The Platonic "participation" of the Forms in sensible things is another example of atemporal processes.

Whitehead's thinking was Platonic, yet his precision was a mathematician's. Therefore his inclusion of atemporal processes in his system is significant.

Having digressed to discuss atemporal processes, let us return now to the correspondence between actual entities and the collapse.

The final result of the collapse is "an elementary quantum event" in spacetime. The final phase of an actual entity is, likewise, an event in spacetime. The correspondence works, except for one problem.

The problem with the correspondence of the collapse to an actual entity is that QM, as a part of our science, is subject to the principle of objectivation. The collapse, as it is now understood, has nothing subjective about it. An actual entity, however, has both a subjective aspect and an objective aspect.

The main conclusion I am driving to is this: **The collapse corresponds to an actual entity to the extent that our science would allow it to; i.e., it corresponds to the objective aspects of an actual entity**.

When one follows, point by point, the characteristics of actual entities, one is amazed to realize what extent one can think of collapse as an objectivized actual entity.

5 Bell's Correlations and Actual Entities

The relationship between Whitehead's thought and QM is a two-way street. QM adds credence to the Whiteheadian vision, and Whitehead's philosophy helps us understand the apparently weird aspects of QM. We will now embark on a reexamination of EPR and Bell's correlations from a Whiteheadian perspective.

The EPR/Bell correlations *seem* to show that *something* (call it "influence") travels faster-than-light, but this faster-than-light travel cannot be harnessed to transmit *signals*. How can we understand this strange state of affairs?

A modification of Process philosophy is needed. Not knowing about entanglement, Whitehead naturally assumed that an actual entity is spatially confined to a small region. Let's modify Process philosophy by dropping this requirement. Let's assume that one actual entity can end up occupying two or more distant locations.

How is this modification related to the EPR/Bell situation?

In an EPR/Bell experiment two events that take place at the same time seem to influence each other, regardless of the distance between them. In principle this distance can be astronomical. Even when the events take place very far apart, they seem to be "entangled," they "feel" each other.

Is it possible that such a connection takes place because *both events are a single creative act, a single "actual entity," arising out of a common field of potentialities?* A single act of transition from the potential to the actual that occurs in two places is not the result of the propagation of anything between these two places; hence the speed of light barrier does not apply.

This is why such creative acts cannot be utilized to transmit signals faster than light: To transmit a signal two creative acts are required. The transmission and reception of a signal is, precisely, the creation of a situation where one completed actual entity affects another. Such transmissions cannot propagate faster than light.

The distinction between influences and signals reflects the distinction between two events that are components of a single act of self-creation and two events that are connected, yet distinct creative acts.

A mundane analogy may help clarify this distinction: Think of a dancer in the act of performing. The single creative act we have been discussing corresponds to the dancer gracefully lifting her left leg and right arm in one harmonious movement. In contrast, the two distinct events correspond to the appearance of an itch on the dancer's left leg, and scratching with her right hand.

The graceful lifting of arm and leg is really a single movement; its two correlated components take place simultaneously. The two events of itching and scratching are separated in time, since one is a reaction to the other.

The material covered so far is a prelude to the big question that we are now ready for. I will put this way:

6 Is Lovejoy's Objection as Convincing as Abner Takes It to Be?

The main point of Lovejoy's objection, which Abner endorses, is the loose use of terms like "experience," or "mentality." Differently stated, the possible meaning of such terms, as Whitehead uses them, is so vague that one wonders whether they mean anything at all.

This is a serious objection. I believe, however, that by using these terms, Whitehead tries to indicate something that is not vague at all. The idea I believe he had in mind (and, obviously, this is *my* formulation) is this: If the basic units of the universe are dead, no amount of complexity will make them alive.

This negation of the physicalistic approach means that the presence of mentality at the human level tells us that some level of mentality, however insignificant in itself, must be present at all levels. Abner's term "protomentality" is appropriate in this context.

And here is another point. To the best of my knowledge, Lovejoy was not aware of either the correspondence of Process philosophy and QM or of the issue of entanglement. Thus he staked his position against Whitehead being blissfully ignorant of the two unexpected, major triumphs of Process philosophy in terms of what Whitehead called "elucidation of things observed." [4]

Because of the results of what Abner called "experimental metaphysics," we no longer have the luxury of blissful ignorance. One cannot ignore a metaphysical approach that corresponds so well to QM and gives an elegant explanation of the characteristics of entanglement.

7 Is There an Agreement, At Least in Part, Between Abner and Myself?

Let me begin by stating what I am trying to achieve and what I am not trying to achieve in this presentation. *I am not* trying to arrive at a metaphysical statement that is, in any sense, final. Rather, I accept Whitehead's statement, "There remains the final reflection, how shallow, puny and imperfect are efforts to sound the depth in the nature of things. In philosophical discussions, the merest hint of dogmatic certainty as to finality of statement is an exhibition of folly." [5]

What *I am* trying to arrive at is the metaphysical system that is most true in the following sense: (1) it feels true (2) it is non-dogmatic in the sense of being open to possible changes, and (3) it is in line with the findings of physics in general and quantum mechanics in particular. Does a modified version of Process Philosophy fit the bill?

(1) It does feel true. If it doesn't, than Abner has to explain howcome he was a Whiteheadian for so many decades.
(2) It is non-dogmatic. As we saw, it can be modified to great advantage and no major damage.

(3) It is in line with the finding of QM in general, and the EPR/Bell entanglement in particular.

So, to conclude, here is my $64,000 question to Abner: Given that we are not looking for a final metaphysical statement, would you agree that a modified Process philosophy is the best one we currently have?

References

1. A. Lovejoy, *The Revolt Against Dualism*, Open Court Publishing, Chicago, IL, 1955.
2. A. Schrödinger, *What is Life? With Mind and Matter and Autobiographical Sketches*, Cambridge University Press, UK, 1966, p. 118.
3. A.N. Whitehead, *Process and Reality*, Corrected edition, Free Press, New York, 1978, p. 283.
4. A.N. Whitehead, *Modes of Thought*, Free Press, New York, 1966, p. 152.
5. A.N. Whitehead, *Process and Reality*, Corrected edition, Free Press, New York, 1978, p. xiv.

Bohr and the Photon*

John Stachel

Abstract Contrary to legend, in his quantum theory of the hydrogen atom Bohr did not utilize the photon concept. In fact, he rejected the concept vehemently until the mid-1920s, when experiments forced a change in his outlook. Exchanges with Einstein during this period contributed to the development of Bohr's concept of complementarity and subsequently, he recognized the role of the photon concept in describing one of the complementary aspects of electromagnetic phenomena: energy and momentum exchanges with ponderable matter. Yet, in accord with his interpretation of the correspondence principle, he still denied equal status to the wave and particle pictures, stressing the primacy of the classical wave picture of light and of the classical particle picture of the electron. Curiously enough, Einstein agreed.

1 Introduction—The Textbook Story

I started to survey textbook discussions of the Bohr atom, to see how they present the relation between Bohr's work on the hydrogen atom and Einstein's light quantum hypothesis. The first book at which I looked is so perfect an example of what I expected to find that I stopped my search—lest further research invalidate my belief that the presentation in one of the best texts available, Arnold B. Arons' *Development of the Concepts of Physics*,[1] is typical of many others.

J. Stachel
Center for Einstein Studies, Boston University

*A talk given March 11, 1986 at "A Centenary Symposium: In Memory of Niels Bohr" of the Boston Colloquium for the Philosophy of Science. I have added some notes and more recent references, notably to Dresden 1987 [1], an excellent biography of Kramers, but have not updated the talk.

[1] Prof. Arons made major contributions to physics education (see his obituary in *Physics Today* [2]) and the fact that I singled out his book is not intended to denigrate his work. On the contrary, it highlights the need for care in using even the best textbook accounts of the history of modern physics.

W.C. Myrvold and J. Christian (eds.), *Quantum Reality, Relativistic Causality, and Closing the Epistemic Circle,* The Western Ontario Series in Philosophy of Science 73,
© Springer Science+Business Media B.V. 2009

70   J. Stachel

After quoting from the opening of Bohr's classic 1913 paper, Arons continues:

With this background of motivation, Bohr suggested a direct application of Einstein's photon hypothesis in the following manner:

(1) Abandon classical electrodynamics to the extent of assuming that at radii of atomic dimensions . . . electrons can revolve in stable orbits *without* continuously radiating energy
in the form of electromagnetic waves . . . .

(2) Invoking Einstein's heuristic model, assume that electromagnetic radiation is absorbed
or emitted in *transfer* of electrons from one orbit to another and that such absorption and
emission of energy by individual electrons is associated with absorption or emission of
individual photons or radiation quanta of energy hv–as suggested by Einstein's heuristic
explanation of the photoelectric effect [3, p. 856].

2 The Actual Story

The actual story is quite different, of course. It would be more correct to say (but still
not really accurate) that Bohr was an inveterate opponent of the photon concept until
1925, when the results of Bothe–Geiger and Compton–Simon experiments forced
him to incorporate the photon concept into his thinking.

My purpose here is to document in some detail the story of Bohr and the photon,
providing the nuances that are missing from straw-man story I have just told. I think
the full story is worth telling, because it provides much of the raw material for
a better understanding of how Bohr developed his views on correspondence and
complementarity; and, although I shall not elaborate much on this theme, how he
later understood and applied them.

3 Planck's Second Theory of Radiation

The starting point of Bohr's attempt [4] to explain the radiation spectrum of atoms
was not Einstein's light quantum hypothesis of 1905,[2] but (as first noted in [6])
Planck's 1910–1911 [7, 8] so-called "second theory" of radiation.[3] Once he had
been forced to admit an element of discontinuity into his theory of radiation, Planck
attempted to localize the discontinuous element, first in the act of absorption, and
then later in the act of emission:

---

[2] The main reference to Einstein's work in Bohr 1913 [4] reads: "The general importance of
Planck's theory for the discussion of the behavior of atomic systems was originally pointed out
by Einstein. The considerations of Einstein have been developed and applied especially by Stark,
Nernst and Sommerfeld" (translation cited from [5, p. 137]. The only other reference is to Einstein's photoelectric law.

[3] See [1, p. 31]: "In this controversy [between Einstein and Planck over the nature of radiation]
Bohr unquestionably sided with Planck." This book Dresden cites much additional evidence for
this assertion.

Bohr and the Photon

The discontinuity must enter somehow ... Therefore, I have located the discontinuity at the place where it can do the least harm, at the excitation of the oscillator; its decay can then occur continuously, with constant damping (Planck to Lorentz, January 1910, quoted from [9, p. 236]).

Planck soon switched from absorption to emission as the locus of discontinuity, deriving his radiation law in a way that

does not depart from the core of classical electrodynamics and electron theory more than is absolutely necessary in view of the undeniably irreconcilable differences with the quantum hypothesis (Planck 1912 [10], a talk to German Physical Society on 12 January, 1912, translation quoted from [11, p. 47]).

Planck was at pains to distance himself from what he called the "extreme attitude" of Einstein and a few others:

They tend to the view that even the electrodynamic processes in pure vacuum, even the light waves, do not propagate continuously but in discrete quanta ... of magnitude $h\nu$, where ν denotes the frequency ([7], translation quoted from [12, p. 123]).

4 Bohr's First Theory of the Atom

The existence of both absorption and emission spectra forced Bohr in 1913 to introduce complete symmetry between discontinuous processes of emission and absorption; but otherwise his attitude at this time seems to have paralleled that of Planck. He constantly refers to "Planck's theory of radiation," citing only Planck 1910–1912 [7, 8, 10].

The clearest statement of Bohr's views on radiation at this time is found in a 1914 letter to his mentor and friend the Swedish physicist Carl W. Oseen:

I wish I could once really learn your opinion of the assumptions on which I built. As far as I can see they need not be in conflict with the assumption of Maxwell's equations in empty space. Since I, contrary to Planck, assume that emission and absorption go perfectly together. To obtain mutual consistency it seems to me necessary to break much more sharply with the customary mechanics than Planck would, and not for example assume that the systems in the stationary states neither emit nor absorb (Bohr to Oseen, 3 March 1914, in [13, p. 555]).

In a slightly later letter to Oseen, there is a passage fraught with significance for Bohr's later viewpoint, when read with hindsight (the clearest sight of all):

In Göttingen I talked quite a bit with Debye about the general foundation for my considerations. For the most part, he took a friendly line but stated that if there should be any reality in this kind of considerations, there had to be in his opinion a general principle, which allowed one to understand the connection between the quantum theory and the usual electrodynamics. In this discussion I tried to say that the necessity of such a principle was perhaps not evident, and that the problem which classical mechanics and electrodynamics had tried to solve perhaps was very different from the one which the phenomena confronted us with, *that the possibility of a comprehensive picture should perhaps not be sought in the generality of the points of view, but rather in the strictest possible limitation of the applicability of the points of view* (Bohr to Oseen 28 September, 1914, in [13, p. 563], emphasis added).

5 The Correspondence Principle

As seems to have been first noted in Jammer 1966 [11, p. 50], Bohr also may have taken from Planck the beginnings of what he later elaborated into his correspondence principle. But it was probably Einstein's well-known work of 1916–1917 on a new derivation of Planck's black-body radiation law that most directly inspired Bohr's formulation, around 1918, of the Correspondence Principle, which thereafter played such a large role in his attempts to understand quantum phenomena. Speaking of Bohr 1918 [14], Oskar Klein—who joined Bohr in Copenhagen in that year, notes:

> [I]n the paper ... Bohr had made an important advance by means of the correspondence viewpoint in showing that—in spite of the abyss, whose depth he never ceased to emphasize, between the quantum-theoretical mode of description and that of classical physics—a detailed correspondence is exhibited between these two modes of description, so that their results coincide in the limit where Planck's quantum of action is very small compared with the actions to be described. In this work he had built on a new derivation by Einstein of Planck's radiation law, the very origin of quantum theory. Einstein obtained this result by formulating probability laws for the transitions of an atom from one stationary state to another. I well remember Bohr's great admiration for Einstein arising equally from this great scientist's contributions to statistical molecular theory, to quantum theory and to relativity theory.

> Bohr, however, could not reconcile himself to Einstein's concept of light quanta, which had been further elaborated in the work I have just mentioned. Bohr's objections came from his thorough familiarity with the wave theory of light, and, when these things were mentioned, he used to emphasize the fantastic accuracy and completeness of this theory in accounting for the many experiments on the propagation of light. Especially he underlined that the definition of the frequency of a light quantum, which determines its energy, is itself derived from the wave theory. Einstein, on the other hand, believed that a true theory of light must in some way combine wave and particle features, so that light energy is concentrated within small regions; and he looked for experiments, through which deviations from the superposition principle might be discovered. How deep a revision of our accustomed ideas Bohr was already prepared to accept, appeared in his remark that perhaps one would have to give up the rigorous validity of the energy principle. Many years later he returned to this idea, which, however, was refuted by direct experiments [15, p. 77].

The importance of the correspondence principle for Bohr's approach has been emphasized in [16]:

> This principle of correspondence became a very powerful method for treating specific problems in the theory of spectra, but Bohr saw its real significance as being more than that: it made it possible, he wrote, "in a certain sense to regard this theory [the quantum theory of spectra] as a natural generalization of our ordinary ideas of radiation." In his search for a new theory, the correspondence principle was one of the few sure guides; it gave Bohr a way of keeping in contact with the solid results of classical electro-magnetic theory, while seeking the quantum theory which would be its "natural generalization" [16, pp. 18–19].

It was indeed, his reliance on the correspondence principle that seems to have been a principal motive for Bohr's distrust of the photon concept and his related willingness to give up energy–momentum conservation to save the classical electrodynamic picture of radiation.

Bohr and the Photon 73

But before going in this question, let me document Bohr's continued adherence to the classical picture of radiation and rejection of the photon concept. The earliest evidence I know occurs in a 1919 draft of a letter to George Darwin, written in July 1919, but not sent. Bohr wrote:

Next as regards the wave theory of light I feel inclined to take the often proposed view that the fields in free space (or rather in gravitational fields) are governed by the classical electrodynamical laws and that all difficulties are concentrated on the interaction between the electromagnetic forces and matter. Here I feel on the other hand inclined to take the most radical or rather mystical views imaginable. On the quantum theory conservation of energy seems quite out of question and the frequency of the incident light would just seem to be the key to the lock which controls the starting of the interatomic process.

As regards the question of conservation of energy there is quite apart from its validity in the usual sense the problem of what there become[s] of the energy if this particle has to be abandoned and here we meet with a curious state of affair[s] in the quantum theory which does not make it so criminal as it looks at first sight to speak with such light heart of the fundamental difficulties touched upon above and still to attempt to be a serious worker in the present crippled field of physics. In fact quite independent of the mechanism of interaction of radiation and matter we can in the worlds of stationary states obtain a rational definition of the energy by means of the principle of mechanical transformability of stationary states. (Independent support of this transference potentials.)

Finally the often seen sentence that the electrons cannot know the final state of transition and adapt its [their] frequency to this beforehand is to me a misconception of the fundamental ideas. Why should the atom itself not as well as we know the stationary states [17, p. 16].

In a 1921 manuscript on "Applications of the Quantum Theory to Atomic Problems in General," Bohr wrote:

As well known, Einstein has several years ago, in connection with his considerations on the photoelectric effect, proposed the view, that quite apart from the problem of the mechanism of the emission and absorption of electromagnetic radiation from atomic system, already the propagation through space of this radiation should take place in a way widely different from that, corresponding to the classical electromagnetic theory. Thus according to this theory of light quanta, electromagnetic radiation from an atom should not spread as a system of spherical waves, but should be propagated in a definite direction as a concentrated entity, containing within a very small volume the energy hv. On one hand such a conception seems to offer the only simple possibility of accounting for the phenomena of photoelectric action, if we adhere to an unrestricted application of the notions of conservation of energy and momentum. On the other hand, it does not appear reconcilable with the phenomena of interference of light, which constitute our only means of analysing radiation in its harmonic constituents and determining the frequency and state of polarisation of each of these constituents. At this state of things it would appear, that the interesting arguments brought forward more recently by Einstein, and which are based on a consideration of the interchange of momentum between the atom and the radiation rather than supporting the theory of light quanta will seem to bring the legitimacy of a direct application of the theorems of conservation of energy and momentum to the radiation processes into doubt [18, pp. 412–413].

In Bohr's notes for the 1923 Second Silliman Lecture, he wrote:

Einstein also pointed out that energy should be emitted in quanta.... Einstein's lead from this view to the suggestion that the transmission of light does not take place by waves but is atomic in nature. This cannot however be considered as a serious theory of light

74 J. Stachel

transmission. Light is not only a flow of energy, but our description of radiation involves a large amount of physical experience involving optical apparatus including our eyes for the understanding of the working of which nothing seems satisfactory except wave theory of light. No significance for the quantity v without waves. This paradox an example of the present state of physics, which is a promising state. Inadequacy of present conception. In the sequel philosophical problems will not be treated but we shall see how quantum ideas will furnish a clue to the interpretation of the properties of matter [18, p. 587].

In his manuscript on "Problems of The Atomic Theory," written in 1923 or 1924, he wrote:

While the theory of light quanta undoubtedly is suited to stress essential features of the laws governing the exchange of energy and momentum in radiation processes, it is hardly compatible with a simple interpretation of the numerous optical phenomena which have been explained, in so many respects satisfactorily, with the aid of the classical electromagnetic theory of light. It is more probable that the chasm appearing between these so different conceptions of the nature of light is an evidence of unavoidable difficulties of giving a detailed description of atomic processes without departing essentially from the casual description in space and time that is characteristic of the classical mechanical description of nature.

This circumstance, however, does not constitute a hindrance to an account of the connection between the observable physical phenomena; in fact, it seems possible, without abandoning either the conception of the propagation of radiation in empty space, held by the classical electromagnetic theory, or the postulate of the stability of the stationary states, to obtain a basis suited to describe all known optical phenomena. According to this description, the interaction between radiation and atom is uniquely determined, as far as the continuous change of the radiation field is concerned, by the state of this field and the instantaneous state of the atom. On the other hand, every change of the atom is regarded as contingent on probability laws. In fact, such a change consists in a transition to another state and is considered to be of so short duration that it is without essential significance for the radiation field and, hence, can be described as discontinuous as far as the description of the optical phenomena is concerned. The introduction of probability laws in the description of the course of interaction between atoms and radiation is due to Einstein, who used such considerations in his derivation of Planck's law of heat radiation on the basis of the existence and stability of stationary states and relation (1) [$hv =$ the energy difference between the two atomic states]. As is well known, through this derivation Einstein proved at the same time that relation (2) [$hv/c =$ the momentum difference between the two atomic states] was necessary and believed in this way to have found a decisive support for the reality of light quanta. However, the theory of light quanta may be characterized as an endeavor to uphold the unlimited validity of the classical principles of the conservation of energy and momentum. On the other hand, in a description as that considered above, it is a principle feature that these principles lose their strict validity for atomic processes and appear only as statistical results of probability laws [18, pp. 571–572].

In Bohr 1924 [19], we read:

[T]he hypothesis under discussion [of light quanta] can in no wise be regarded as a satisfactory solution. As is well known, this hypothesis introduces insuperable difficulties, when applied to the explanation of the phenomena of interference, which constitute our chief means of investigating the nature of radiation. We can even maintain that the picture, which lies at the foundation of the hypothesis of light-quanta, excludes in principle of the possibility of a rational definition of the concept of a frequency v, which plays a principle part in this theory. The hypothesis of light-quanta, therefore, is not suitable for giving a picture of the processes, in which the whole of the phenomena can be arranged, which are considered

Bohr and the Photon

in the application of the quantum theory. The satisfactory manner in which the hypothesis reproduces certain aspects of the phenomena is rather suited for supporting the view, which has been advocated from various sides, that, in contrast to the description of natural phenomena in classical physics in which it is always a question only of statistical results of a great number of individual processes, a description of atomic processes in terms of space and time cannot be carried through in a manner free from contradiction by the use of conceptions borrowed from classical electrodynamics, which, up to this time, have been our only means of formulating the principles which form the basis of the actual applications of the quantum theory (cited from [18, p. 492]).

These quotations seem to me fully justify Martin Klein's conclusion, summarizing Bohr's attitude:

One conclusion could be drawn from all the difficulties: "A general description of the phenomena, in which the laws of the conservation of energy and momentum retain in detail their validity in their classical formulation, cannot be carried through." As a result, Bohr warned, "We must be prepared for the fact that deductions from these laws will not possess unlimited validity." It was not the conservation laws but rather the correspondence principle and Ehrenfest's adiabatic principle to which Bohr looked for guidance. They were "suited, in a higher degree, to point out new ways for further extensions of the quantum theory of atomic structure," and they offered "a hope in the future of a consistent theory, which at the same time reproduces the characteristic features of the quantum theory ... and, nevertheless, can be regarded as a rational generalization of classical electrodynamics" [16, p.22].

Note that Ehrenfest's adiabatic principle, which Bohr called "the principle of mechanical transformability of stationary states", is vital: it allows a "rational definition of the energy" in the absence of the conservation laws.

In 1923 Hendrik Kramers, who was working closely with Bohr at the time, together with Helge Holst wrote a popular book on quantum theory. In it, they took pains to dissociate Bohr from the photon concept.[4] After rehearsing the difficulties with the concept, they concluded:

The theory of light quanta may thus be compared with medicine which will cause the disease to vanish but kill the patient. When Einstein, who has made so many essential contributions in the field of the quantum theory, advocated these remarkable representations about the propagation of radiation energy he was naturally not blind to the great difficulties just indicated. His apprehension of the mysterious light in which the phenomena of interference appear on his theory is shown in the fact that in his considerations he introduces something which he calls a "ghost" field of radiation to help to account for the observed facts. But he has evidently wished to follow the paradoxical in the phenomena of radiation out to the end in the hope of making some advance in our knowledge.

This matter is introduced here because the Einstein light quanta have played an important part in discussions about the quantum theory, and some readers may have heard about them without being clear as to the real standing of the theory of light quanta. The fact must be emphasized that this theory in no way has sprung from the Bohr theory, to say nothing of its being a necessary consequence of it [20, p. 175].

---

[4] It is particularly ironic that in 1921 Bohr had dissuaded Kramers from publishing a theoretical prediction of what later came to be known as the Compton effect (see [1], Chapter 14, "The Curious Copenhagen Interlude," pp. 289–298).

76   J. Stachel

6 The Bohr–Kramers–Slater Interlude

As is well known,[5] Bohr attempted to synthesize his views on the non-conservation
of energy-momentum in individual processes and the validity of the classical picture of radiation for the free field in a new version of quantum theory by grafting
them onto John Slater's virtual oscillator model of the radiation process. Slater had
originally intended his model to include photons; indeed, his original model included something rather like Einstein's ghost field, which guided the photons (see
the quotation in the previous section from [20]), but Bohr persuaded Slater to jettison
the photons. The resulting Bohr–Kramers–Slater 1924 [21] theory was immediately
subjected to a barrage of theoretical criticism by Einstein and Pauli, in particular;
most crucially, it was unable to withstand the experimental criticism provided by
the results of the Bothe–Geiger and Compton–Simon experiments.

Bohr summed up his reaction to the failure of the Bohr–Kramers–Slater theory
in an "Addendum," added in July 1925, to his paper "On the Behavior of Atoms in
Collision," which had been written in March of that year before the results of the
Bothe–Geiger experiment were known:

> Since the above was written the question of the strict validity of the conservation laws has
> entered a new phase through the publication by *Bothe* and *Geiger* of the results of their
> important experiments on the scattering of X-rays ... The renunciation of the strict validity
> of the conservation laws, and consequently of a coupling between the individual transition
> processes, was occasioned by the fact that no space-time mechanism seemed conceivable
> that permitted such a coupling and at the same time achieved a sufficient connection with
> classical electrodynamics, which has been successful to such a great extent in describing
> optical phenomena. In this connection it must be emphasized that the question of a coupling
> or an independence of the individual observable atomic processes cannot be looked at as
> simply distinguishing between two well-defined conceptions of the propagation of light in
> empty space corresponding to either a corpuscular theory or a wave theory of light. Rather,
> the problem is to what extent the space-time pictures, by means of which the description
> of natural phenomena has hitherto been attempted, are applicable to atomic processes. In
> fact, the analysis of optical phenomena can hardly be formulated without the assumption
> that the radiative activity of individual atoms is influenced by the presence of other atoms
> in the sense to be expected in the picture of the wave propagation of light. In this respect,
> the analysis of these phenomena with the aid of the correspondence principle—as indicated
> in the paper by *Bohr, Kramers* and *Slater*—may touch upon something essential in this
> matter and may be suited to give hints for the further extension of this analysis. However,
> the hope of giving a general formulation of the laws of quantum theory in the manner
> attempted would have the ground cut from under it by the demonstration of a coupling
> between individual atomic processes, which forces upon us the picture of a corpuscular
> propagation of light corresponding to Einstein's theory of light quanta. In this state of affairs
> one must be prepared to find that the generalization of the classical electrodynamic theory
> that we are striving for will require a fundamental revolution in the concepts upon which
> the description of nature has been based until now [17, pp. 204–205].

It appears that, at this time, Bohr was willing to consider the possibility that some
totally novel set of concepts was needed to describe quantum phenomena—a possibility that he later vigorously denied.

---

[5] For a particularly good account (see [1], Part 2, Chapter 13, Section III, "The Bohr–Kramers–
Slater Theory," pp. 159–215).

7 Einstein's Experiment

In January 1926, Bohr wrote Slater about earlier conversations Bohr had with Einstein in 1925 in Leiden:

> Although of course we were wrong in Copenhagen as regards the question of the coupling of the quantum process—in which respect I have a bad conscience in persuading you to our view [Bohr is alluding to Slater's dropping of the photon from his model]—I believe that Einstein agrees with us in the general ideas, and that especially he has given up any hope of proving the correctness of the light quantum theory by establishing contradictions with the wave theory description of optical phenomena (Bohr to Slater, 28 January 1926 [17, pp. 68, 497]).

Bohr is alluding to the second of two attempts by Einstein [22, 23] to design a "crucial" optical experiment, the result of which would distinguish between the light quantum theory and the classical wave theory of light. In both cases, it became clear to Einstein [24, 25]—after considerable resistance—that his experiment actually did not predict a result that was not predicted by the classical theory.

Analysis of the failure of such attempts as Einstein's proposed experiments may well have been one of the important clues that led Bohr to formulate his complementarity interpretation of the new quantum mechanics of Born and Heisenberg, together with the new wave mechanics of de Broglie and Schrödinger, both developed within the two years following the failure of the Bohr–Kramers–Slater theory. At any rate, as noted by Jørgen Kalckar,[6] it was in a letter to Einstein[7] (which included the proofs of Heisenberg's "uncertainty principle" paper) that Bohr seems first to have sketched out the complementarity concept:

> It has of course long been recognized how intimately the difficulties of quantum theory are connected with the concepts, or rather the words that are used in the customary description of nature, and which all have their origin in the classical theories. These concepts leave us only with the choice between Charybdis and Scylla, according to whether we direct our attention towards the continuous or discontinuous aspect of the description [27, p. 21].

After describing Heisenberg's treatment of a wave packet of light, Bohr goes on:

> Through the new formulation we are presented with the possibility of bringing the requirement of conservation of energy into harmony with the consequences of the wave theory of light, since according to the character of the description, the different aspects of the problem never appear at the same time (*ibid.*, p. 22).

Bohr then analyzes Einstein's most recent experimental proposal. As Kalckar notes:

> In the paper cited, Einstein had shown from general arguments that light emitted from a moving atom must be expected to exhibit the same interference effects as radiation from a classical moving oscillator. On this basis he concluded that, in the experiment considered, there would be no effects associated with the light quanta, contrary to his earlier expectations.

---

[6] See [26, pp. 16, 21].

[7] Bohr to Einstein, 15 April 1927, German text in [27, pp. 418–421], translation in [27, pp. 21–24].

78   J. Stachel

First Bohr analyzes the experiment from the viewpoint of classical wave theory, showing that a certain range of uncertainty in the frequency of the diffracted light is to be expected classically. Then he analyzes it from the viewpoint of the light quantum theory, using conservation of energy for the individual light quanta. Bohr shows that the frequency range to be expected on the basis of the classical optical picture just corresponds to the range of energies expected for the light quanta, because of the different recoil energies associated with the beam of emitting atoms, depending on the range of possible directions of their emission. He concludes the discussion by stating:

> That one can observe not merely a statistical, but an individual energy balance is connected to the fact that as you [i.e., Einstein] indicate in your footnote, no possible 'light quantum description' can ever explicitly do justice to the geometrical relations of the 'ray path' (*ibid.*, p. 23).

The footnote by Einstein, to which Bohr alludes, reads:

> In particular, one may not assume that the quantum processes of emission, which is energetically determined by position, time, direction and energy, is also determined in its *geometrical* characteristics by these quantities [25, p. 337].

This discussion of Einstein's second experiment is the first example known to me, in which Bohr discusses what he would soon call the complementary nature of a description in terms of the conservation laws and one in terms of a space-time picture; an example in which he goes into great detail in discussing a particular physical situation—or rather, two complementary situations.

8 Bohr's Complementarity

Soon Bohr began to expound this concept of complementary descriptions, and to develop his concept of "quantum phenomenon," which (as I shall discuss a little later) ultimately came to encompass nothing short of a complete cycle of preparation, interaction and registration.[8] Sometimes his interpretation of the wave-particle duality is misunderstood to imply that Bohr considered the wave and particle aspects of ordinary matter on one hand, and of radiation on the other to have equal validity. But, even in his earliest discussions of complementarity, this was not the case. The reason for the distinction he makes lies in the correspondence argument, which played a crucial role in Bohr's thinking: In each individual case, the classical aspect of wave-particle duality has predominant significance. He summarized his views on the role of classical concepts in the introduction to the first collection of his essays [30]:

> [I]t would be a misconception to believe that the difficulties of the atomic theory may be evaded by eventually replacing the concepts of classical physics by new conceptual forms.

---

[8] For further discussion of this concept, stressing its relation to Feynman's approach to quantum mechanics, see [28, 29].

Bohr and the Photon

Indeed, as already emphasized, the recognition of the limitation of our forms of perception by no means implies that we can dispense with our customary ideas or their direct verbal expressions when reducing our sense impressions to order. No more is it likely that the fundamental concepts of the classical theories will ever become superfluous for the description of physical experience. The recognition of the indivisibility of the quantum of action, and the determination of its magnitude, not only depend on an analysis of measurements based on classical concepts, but it continues to be the application of these concepts alone that makes it possible to relate the symbolism of the quantum theory to the data of experience. At the same time, however, we must bear in mind that the possibility of an *unambiguous* use of these fundamental concepts solely depends upon the self-consistency of the classical theories from which they are derived and that, therefore, the limits imposed upon the application of these concepts are naturally determined by the extent to which we may, in our account of the phenomena, disregard the element which is foreign to classical theories and symbolized by the quantum of action [30, p. 16].

He drew the consequences of the correspondence argument for the differing nature of matter and radiation in a number of places. In his 1930 Faraday Lecture [31], he said:

The extreme fertility of wave pictures in accounting for the behavior of electrons must, however, not make us forget that there is no question of a complete analogy with ordinary wave propagation in material media or with non-substantial energy transmission in electromagnetic waves. Just as in the case of radiation quanta, often termed "photons," we have here to do with symbols helpful in the formulation of the probability laws governing the occurrence of the elementary processes which cannot be further analysed in terms of classical physical ideas. In this sense, phrases such as "the corpuscular nature of light" or "the wave nature of electrons" are ambiguous, since such concepts as corpuscle and wave are only well defined within the scope of classical physics, where, of course, light and electrons are electromagnetic waves and material corpuscles respectively (cited from [27, p. 394]).

One of his most detailed and clear discussion of this question is in a paper given in Cambridge on the occasion of the Maxwell Centenary in 1931 [32]:

When one hears physicists talk nowadays about 'electron waves' and 'photons', it might perhaps appear that we have completely left the ground on which Newton and Maxwell built; but we all agree, I think, that such concepts, however fruitful, can never be more than a convenient means of stating characteristic consequences of the quantum theory which cannot be visualized in the ordinary sense. It must not be forgotten that only the classical ideas of material particles and electromagnetic waves have a field of unambiguous application, whereas the concepts of photons and electron waves have not. Their applicability is essentially limited to cases in which, on account of the existence of the quantum of action, it is not possible to consider the phenomena observed as independent of the apparatus utilised for their observation. I would like to mention, as an example, the most conspicuous application of Maxwell's ideas, namely, the electromagnetic waves in wireless transmission. It is a purely formal matter to say that these waves consist of photons, since the conditions under which we control the emission and the reception of the radio waves preclude the possibility of determining the number of photons they should contain. In such a case we may say that all trace of the photon idea, which is essentially one of enumeration of elementary processes, has completely disappeared (cited from [27, pp. 359–360]).

A more careful and mathematically complete discussion of this question forms a (small) part of the content of the Bohr–Rosenfeld (1933) [33] paper. Let me just quote a few lines:

80   J. Stachel

[T]here are, in the quantum domain, the peculiar fluctuation phenomena which derive from
the basically statistical character of the formalism....

The fluctuations in question are intimately related to the impossibility, which is characteristic of the quantum theory of fields, of visualizing the concept of light quanta in terms
of classical concepts. In particular, they give expression to the mutual exclusiveness of an
accurate knowledge of the light quantum composition of an electromagnetic field and of
knowledge of the average value of any of its components in a well-defined space-time region (translation cited from [34, p. 365]).

[F]or a more detailed comparison of the measurement possibilities and the requirements
of the quantum-electromagnetic formalism one must also take into account the limitation
imposed on the classical mode of calculation by the quantum-theoretical features of any
field effect, symbolized by the concept of light quanta (*ibid.*, p. 382).

[I]t is a major result of the quantum theory of fields that all predictions concerning field averages which do not rest on true field measurements, but on the light quantum composition
of the field to be investigated or on the knowledge of classically described field sources,
must be of an essentially statistical nature (*ibid.*, p. 381).

In field measurements, this complementary feature of the description, essential for consistency, corresponds to the fact that the knowledge of the light quantum composition of the
field is lost through the field effects of the test body; and in fact, the more so, the greater the
desired accuracy of the measurement. Moreover, it will appear from the following discussion that any attempt to re-establish the knowledge of the light quantum composition of the
field through a subsequent measurement by means of any suitable device would at the same
time prevent any further utilization of the field measurement in question (*ibid.*, p. 388).

9 Bohr's "Phenomena"

In a talk at the 1938 Warsaw meeting of the International Institute for Intellectual
Cooperation [35], Bohr gave a particularly clear exposition of the way he ultimately
came to use the word "phenomenon":

The essential lesson of the analysis of measurements in quantum theory is thus the emphasis on the necessity, in the account of the phenomena, of taking the whole experimental
arrangement into consideration, in complete conformity with the fact that all unambiguous
interpretation of the quantum mechanical formalism involves the fixation of the external
conditions, defining the initial state of the atomic system concerned and the character of
the possible predictions as regards subsequent observable properties of that system. Any
measurement in quantum theory can in fact only refer either to a fixation of the initial state
or to the test of such predictions, and it is first the combination of measurements of both
kinds which constitutes a well-defined phenomenon [35, p. 20; 36, p. 312].

He proceeded as usual to give examples, but with his new use of "phenomenon"
made clearer:

Instructive examples of this situation are offered respectively by the interference effects of
electrons and by the Compton effect, which are equally paradoxical from the point of view
of classical physics. In the former case, the phenomenon is in fact only defined when the relative positions of all scattering bodies and photographic plates are known with an accuracy
excluding the possibility, by means of a control of momentum transfer, of discriminating between various imaginable paths of the electron, to an extent incompatible with the very idea
of interference. In the latter case, a control of the space-time co-ordination of the scattering

Bohr and the Photon   81

process irreconcilable with the definition of momentum and energy quantities is excluded in advance by any arrangement allowing a test of the momentum and energy conservation such as is implied in the specification of the phenomenon itself. In neither case is there indeed any question of a simple replacement of the classical particle picture of electrons and wave picture of light with the electron wave idea or the photon concept respectively; rather we have to do with individual phenomena, which cannot be analyzed on classical lines, and which exhibit the peculiar complementary relationship of superposition principle and conservation laws in quantum theory (*ibid.*, pp. 22–23).

10 Bohr and Einstein Agree!

I shall close by noting that, curiously enough and contrary to some simplified accounts of Einstein's views on light quanta, he and Bohr ended up not differing on the question of the distinction between the nature of electrons and photons. Although Einstein claimed never to have made sense of Bohr's concept of complementarity— or rather never to have been able to give it a precise meaning—at least once Einstein made a statement quite similar to Bohr's:

> I do not believe that the light-quanta have reality in the same immediate sense as the corpuscles of electricity [i.e., electrons]. Likewise I do not believe that the particle-waves have reality in the same sense as the particles themselves. The wave-character of particles and the particle-character of light will—in my opinion—be understood in a more indirect way, not as immediate physical reality. (Einstein to Paul Bonofield, September 18, 1939, translation quoted from [28, pp. 373–374]).

The difference, of course, is that where Bohr saw a solution to the problem of their nature, Einstein saw the beginning of the puzzle.

References

1. Dresden, Max (1987). *H. A. Kramers: Between Tradition and Revolution*. New York: Springer-Verlag.
2. McDermott, Lillian C., Wilson, Kenneth G., and Jossem, E. Leonard (2001). "Arnold Boris Arons", *Physics Today* **54**: 76–77.
3. Arons, Arnold Boris (1965). *The Development of Concepts of Physics from the Rationalization of Mechanics to the First Theory of Atomic Structure*. Reading, MA: Addison-Wesley.
4. Bohr, Niels (1913). "On the constitution of atoms and molecules", *Philosophical Magazine* **26**: 1–25. Reprinted in [13], 161–185.
5. ter Haar, D. (1967). *The Old Quantum Theory*. Oxford: Pergamon.
6. Hirosige, Tetu and Nisio, Sigeko (1964). "Formation of Bohr's theory of atomic constitution", *Japanese Studies in History of Science* **3**: 6–28.
7. Planck, Max (1910). "Zur Theorie der Wärmestrahlung", *Annalen der Physik* **31**: 758–768.
8. Planck, Max (1911). "Eine neue Strahlungshypothese", *Verhandlungen der Deutschen Physikalischen Gesellschaft* **13**: 138–148.
9. Kuhn, Thomas S. (1978). *Black-Body Theory and the Quantum Discontinuity, 1894–1912*. Oxford: Clarendon Press/New York: Oxford University Press.
10. Planck, Max (1912). "Über die Begründung des Gesetzes der Schwarzen Strahlung", *Annalen der Physik* **37**: 642–656.

82   J. Stachel

11. Jammer, Max (1966). *The Conceptual Development of Quantum Mechanics*. New York: McGraw-Hill.
12. Mehra, Jagdesh and Rechenberg, Helmut (1982). *The Historical Development of Quantum Theory*, Vol. 1, *The Quantum Theory of Planck, Einstein, Bohr and Sommerfeld: Its Foundation and the Rise of Its Difficulties* 1900–1925. New York/Heidelberg/Berlin: Springer-Verlag.
13. Bohr, Niels (1981). *Collected Works Vol. 2. Work on Atomic Physics (1912–1917)*, ed. Ulrich Hoyer. Amsterdam: North-Holland/Elsevier.
14. Bohr, Niels (1918). "The quantum theory of line spectra. Part I—On the general theory", *Det Kongelige Danske Videnskabernes Selskab. Skrifter. Naturvidenskabelig og Matematisk Afdeling* 8(4): 5–36. Reprinted in [18], 71–102.
15. Klein, Oskar (1967). "Glimpses of Niels Bohr as a Scientist and Thinker". In [37] 74–93.
16. Klein, Martin (1970). "The first phase of the Bohr–Einstein dialogue", *Historical Studies in the Physical Sciences*, **2**: 1–39.
17. Bohr, Niels (1984). *Collected Works Vol. 5. The Emergence of Quantum Mechanics (mainly 1924–1926)*, ed. Klaus Stolzenberg. Amsterdam: North-Holland/Elsevier.
18. Bohr, Niels (1976). *Collected Works Vol. 3. The Correspondence Principle (1918–1923)*, ed. J. Rud Nielsen. Amsterdam: North-Holland/Elsevier.
19. Bohr, Niels (1924). "The application of the quantum theory to atomic structure, Part I, The fundamental postulates", Chapter III, "On the formal nature of the quantum theory", *Proceedings of the Cambridge Philosophical Society* **22**, Supplement: 1–42.
20. Kramers, H. A. and Holst, Helge (1923). "The Atom and the Bohr Theory of Its Structure". New York: Knopf.
21. Bohr, Niels, Kramers, Hendrik A., and Slater, John C. (1924). "The quantum theory of radiation", *Philosophical Magazine* **47**: 785–802. Reprinted in Bohr 1984, 101–118.
22. Einstein, Albert (1921). "Über ein den Elementarprozess der Lichtemission betreffendes Experiment", *Sitzungsberichte der Preussischen Akademie der Wissenschaften (physik.-math. Klasse)* [n.v]: 882–883.
23. Einstein, Albert (1926). "Vorschlag zu einem die Natur des elementaren Strahlungs-Emissionsprozesses betreffenden Experiment", *Die Naturwissenschaften* **14**: 300–301.
24. Einstein, Albert (1922). "Theorie der Lichtfortpflanzung in dispergierenden Medien", *Sitzungsberichte der Preussischen Akademie der Wissenschaften (physik.-math. Klasse)* [n.v]: 18–22.
25. Einstein, Albert (1926). "Über die Interferenzeigenschaften des durch Kanalstrahlen emittierten Lichtes", *Sitzungsberichte der Preussischen Akademie der Wissenschaften (physik.-math. Klasse)* [n.v]: 334–340.
26. Kalckar, Jørgen (1985). Introduction. In [27], 7–51.
27. Bohr, Niels (1985). *Collected Works. Vol. 6. Foundations of Quantum Physics I. (1926–1932)*, ed. Jørgen Kalckar. Amsterdam: North-Holland/Elsevier.
28. Stachel, John (1986). "Einstein and the Quantum", in Robert S. Colodny, ed. *From Quarks to Quasars: Philosophical Problems of Modern Physics*. Pittsburgh: University of Pittsburgh Press, 367–402. Reprinted in John Stachel, *Einstein From 'B' to 'Z.'* Boston/Basel/Berlin: Birkhäuser, 2002, 367–402.
29. Stachel, John (1997). "Feynman Paths and Quantum Entanglement: Is There Any More to the Mystery?" In Robert S. Cohen, Michael Horne and John Stachel, eds., *Potentiality, Entanglement and Passion-at-a-Distance/Quantum Mechanical Studies for Abner Shimony, Volume Two*. Dordrecht/Boston/London: Kluwer Academic, 245–256.
30. Bohr, Niels (1934). *Atomic Theory and the Description of Nature*. Cambridge: Cambridge University Press.
31. Bohr, Niels (1932). "Chemistry and the quantum theory of atomic constitution", *The Journal of the Chemical Society*: 349–384. Reprinted in [27], 373–408.
32. Bohr, Niels (1931). "Maxwell and modern theoretical physics", *Nature* (Suppl.) **12**: 691–693. Reprinted in [27], 359–360.
33. Bohr, Niels and Rosenfeld, Leon (1933). "Zur Frage der Messbarkeit der elektromagnetischen Feldgrössen", *Det Kongelige Danske Videnskabernes Selskab. Skrifter. Naturvidenskabelig og Matematisk Afdeling* 12(8). Reprinted in [36], 57–121.

34. Rosenfeld, Leon (1979). *Selected Papers*, eds. Robert S. Cohen and John Stachel. Dordrecht/Boston/London: D. Reidel.
35. Bohr, Niels (1938). "The causality problem in atomic physics". In [38], 11–30. Reprinted in [36], pp. 303–322.
36. Bohr, Niels (1996). *Collected Works*. Vol. 7: *Foundations of Quantum Physics II (1933–1958)*, ed. Jørgen Kalckar. Amsterdam: North-Holland/Elsevier.
37. Rozental, Stefan, ed. (1967). "Niels Bohr:His Life and Work as Seen by His Friends and Contemporaries". Amsterdam: North-Holland.
38. International Institute of Intellectual Cooperation (1939). *New Theories in Physics/Conference Organized in Collaboration with the International Union of Physics and the Polish Intellectual Co-operation Committee/Warsaw, May 30th–June 3rd 1938*. Paris: International Institute of Intellectual Co-operation.

Part III
Bell's Theorem and Nonlocality
A. Theory

Extending the Concept of an "Element of Reality" to Work with Inefficient Detectors

Daniel M. Greenberger

Abstract In two previous papers, we have shown that one may perform a Greenberger–Horne–Zeilinger (GHZ) analysis on a two-particle state formed by "entanglement swapping", so the two particles have never met. We showed that even for perfect correlations, the quantum mechanical result is inconsistent for any local, deterministic, realistic theory, even with very inefficient detectors.

Here we discuss in more detail the assumption that one may extend the Einstein–Rosen–Podolsky (EPR) reality condition to the case where one has inefficient detectors.

1 Introduction

I would like to thank the organizers for inviting me here, especially since the purpose of the meeting is to honor Abner Shimony. I have known Abner for many years, and have had many discussions with him, on many topics. He is a very wise man, a very learned man, a wonderful, generous person, and a dear friend. My life would have been much poorer had I not known him, and I am very thankful that this blessing has come my way.

Mike Horne, Anton Zeilinger and I recently produced a Bell's Theorem [1] for two entangled particles that uses a Greenberger–Horne–Zeilinger (GHZ)-type argument [2]. The argument occurs in two papers (the second of which was also co-authored by Marek Zukowski) and applies to the case where the two particles have a perfect correlation, meaning that if one knows the outcome of a measurement on one of them, one can predict the outcome of a corresponding measurement on the other with absolute certainty, so that an Einstein–Podolsky–Rosen (EPR) element of reality [3] exists. Another feature of the argument is that it involves no inequalities, and discusses only perfectly correlated states.

D.M. Greenberger
Department of Physics, City College of New York, New York, NY 10031, USA
e-mail: greebgr@sci.ccny.cuny.edu

W.C. Myrvold and J. Christian (eds.), *Quantum Reality, Relativistic Causality, and Closing the Epistemic Circle,* The Western Ontario Series in Philosophy of Science 73,
© Springer Science+Business Media B.V. 2009

This argument used a two-particle entangled state that was produced by the method of "entanglement-swapping" [4]. In this method, two pairs of particles, each pair in a singlet state, are independently produced. Then one catches one particle of each pair simultaneously (which correlates them into what we call a "cross-entangled" state). This automatically correlates the other particles, which have never met, into an entangled state, the "entanglement-swapped" state. Because the particles have never met and have no shared history, there are many limitations present on the capability of a deterministic, realistic, local theory that attempts to model the behavior of such a state. The first of these papers (which we will call paper A) assumes detectors of 100% efficiency, and it had no need to exploit all the limitations inherent in the system in order to prove that any such realistic, deterministic, local theory is inconsistent.

However there is a natural extension of the idea of reality proposed by EPR that applies to inefficient detectors, and in the second paper (which we will call paper B) we showed that we can model such detectors in the type of experiment we are considering. Then, exploiting the EPR non-locality assumptions, we showed that the Bell functions that describe the outcome of our experiments must factor in such a way that the instructions to the system contained in the hidden variables cannot make use of the angular settings of the polarization rotators used in the experiment. It follows from this that the predictions of such local realistic models are self-contradictory, even when one has very inefficient detectors. This is a new type of result, that can be used to rule out such realistic theories, even when using detectors of low efficiency. We also do not need to assume any kind of random sampling hypothesis, and thus our result closes two of the important loopholes in this field [5]. The experiment we discuss uses the technology of experiments that have already been performed, and the Zeilinger group is actively planning to perform an experiment using two independent sources.

A recent paper by Broadbent and Méthot [6] argues that entanglement swapping experiments can be explained by local hidden variables. But it gives an example that is much simpler than our experiment, and their results do not apply to our experiment[1].

In this paper, we want to discuss in a little more detail than in paper B the justification for being able to extend the notion of "element of reality" to those cases when the detectors do not have 100% efficiency. We will review just enough of the original papers to give a flavor of what we are doing, in order to make our discussion relatively self-contained.

---

[1] There are several reasons why Ref. [6] does not apply to our analysis, Ref. [1]. However the biggest is that they leave out the four independent angles of rotation in our experiment, which rotate the polarization of each beam, in ways not known to the observers in the other beams, so that in our case their auxiliary experiments cannot be done as they describe. These rotations are crucial as our chief result is to be able to separate the angular and hidden variable knowledge. There is no counterpart to this in their experiment. (Even in the original EPR experiment, leaving out the angles leads to a trivial explanation by hidden variables.) As an afterthought they discuss the case when these variables are in continual long range contact, but this defeats the entire locality motivation of EPR.

2 A Quick Review of the Experiment

The Bell states of a two-particle system are a particular set of four orthogonal entangled states that form a complete set of states for the system. For a two-photon system they are

$$
\begin{aligned}
\left|\phi^+\right\rangle &= \tfrac{1}{\sqrt{2}}(|H_1\rangle|H_2\rangle + |V_1\rangle|V_2\rangle), \\
\left|\phi^-\right\rangle &= \tfrac{1}{\sqrt{2}}(|H_1\rangle|H_2\rangle - |V_1\rangle|V_2\rangle), \\
\left|\psi^+\right\rangle &= \tfrac{1}{\sqrt{2}}(|H_1\rangle|V_2\rangle + |V_1\rangle|H_2\rangle), \\
\left|\psi^-\right\rangle &= \tfrac{1}{\sqrt{2}}(|H_1\rangle|V_2\rangle - |V_1\rangle|H_2\rangle).
\end{aligned}
\tag{1}
$$

Here the subscripts 1,2 refer to two different momentum states for the different photons. The notations $|\phi^\pm\rangle, |\psi^\pm\rangle$ in Eq. (2) represent the conventional labeling of each of these states. With present technology, by making suitable unitary transformations between the four Bell states, one can detect any two of the four states.

Our experiment also involves rotating the polarizations of each of our photons. This is given by the equations

$$
\begin{aligned}
R(\varphi)|H\rangle &= |H\rangle\cos\varphi + |V\rangle\sin\varphi, \\
R(\varphi)|V\rangle &= |V\rangle\cos\varphi - |H\rangle\sin\varphi.
\end{aligned}
\tag{2}
$$

We apply this to an experiment that was recently performed by the Zeilinger group [4], in which they swapped the entanglement of two photons as mentioned in the introduction, and as depicted in Fig. 1. In this experiment, two independent pairs of photons are created, each in the photon equivalent of a singlet state, $\frac{1}{\sqrt{2}}(H_1V_2 - V_1H_2) = |\psi^-\rangle$, and the polarizations of the four photons are independently rotated, through the angles $\varphi_1, \varphi_2, \varphi_3,$ and φ_4, as in Eq. (2).

The initial state of the system produced by the two independent lasers and two independent down-conversions is a product of two singlet states, photons a and b produced by one laser, and c and d produced by the other

$$
|\psi_I\rangle = \tfrac{1}{2}(|H\rangle_a|V\rangle_b - |V\rangle_a|H\rangle_b)(|H\rangle_c|V\rangle_d - |V\rangle_c|H\rangle_d),
\tag{3}
$$

We call this the "Volkswagen state" (VW state), from the shape of Fig. 1. Next, the polarization of each of the photons a, b, c, and d, gets rotated through their respective angles, φ_i. The subsequent experiment combines photons b and c at a Bell state analyzer (BSA) and also detects photons a and d separately, as well as their polarizations. The state in Eq. (3), rewritten in terms of the Bell states of particles b and c, and the Bell states of particles a and d, after the polarization of each of the photons has been rotated, is (after a lot of algebra)

$$
\begin{aligned}
|\psi_I\rangle = \tfrac{1}{2}\big\{ &\left|\phi_{bc}^+\right\rangle \left[-\left|\phi_{ad}^+\right\rangle\cos\xi + \left|\psi_{ad}^-\right\rangle\sin\xi\right] \\
+ &\left|\psi_{bc}^-\right\rangle \left[-\left|\psi_{ad}^-\right\rangle\cos\xi - \left|\phi_{ad}^+\right\rangle\sin\xi\right] \\
+ &\left|\phi_{bc}^-\right\rangle \left[+\left|\phi_{ad}^-\right\rangle\cos\eta + \left|\psi_{ad}^+\right\rangle\sin\eta\right] \\
+ &\left|\psi_{bc}^+\right\rangle \left[+\left|\psi_{ad}^+\right\rangle\cos\eta - \left|\phi_{ad}^-\right\rangle\sin\eta\right]\big\},
\end{aligned}
$$

$$\xi = ((\varphi_1 - \varphi_2) + (\varphi_3 - \varphi_4)),$$
$$\eta = ((\varphi_1 - \varphi_2) - (\varphi_3 - \varphi_4)). \tag{4}$$

As they approach the BSA, the two particles b and c are in a superposition of all four Bell states, according to Eq. (4). With a fully functional BSA, all of the four Bell states would be detectable. Then Eq. (4) shows that, once the BSA result is registered, particles a and d are in general thrown into a superposition of the Bell states, and with a suitable choice of the φ_i, into a very specific Bell state. In what follows, we will assume a full Bell state detection of particles b and c has been made.

3 Analysis of the Arrangement in Terms of Elements of Reality

In paper A we showed that in the experiment shown in Fig. 1, where we considered only 100% efficient detectors, the EPR local reality conditions imply that one can establish the existence of the four functions $A(\varphi_1, \lambda_1), D(\varphi_4, \lambda_4), \kappa(\lambda_1, \lambda_4)$, and $F_{\kappa(\lambda_1,\lambda_4)}(\varphi_1, \varphi_4, \lambda_1, \lambda_4)$, each of which can take on the values ± 1. Here λ_1 refers collectively to any hidden variables whose values are set when particles a and b are created, which can determine a specific outcome of a measurement of the polarization of these particles when they are detected. These functions exist as a consequence of the perfect correlations that arise in special cases of our measurements. (A perfect correlation occurs when it is possible in an event for one to make a measurement on three of the particles, and thereby predict with 100% certainty some property

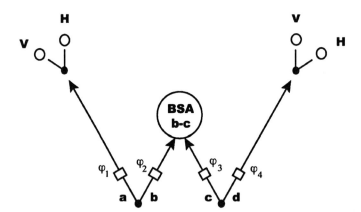

Fig. 1 Schematic diagram of the creation of the two-particle state. In this experiment there are two independent down-conversions, one creating the pair of photons **a–b**, and the other the pair **c–d**. Each of them undergoes a rotation through the angle φ_i, and particles **b** and **c** enter a Bell-state-analyzer (BSA), which will annihilate them while detecting which Bell state they were in. If the angles φ_i are set properly, as one of the perfect correlation cases, this process forces the particles a and d into a two-particle Bell state. In the experiment the Bell state of **a** and **d** is not determined, only their polarizations, but this is sufficient to rule out locally realistic, deterministic theories as an explanation of their observed properties

for the fourth particle, without interacting with that particle. In such a situation, the property was labeled an "element of reality" by EPR.) The function A specifically records the outcome of a polarization measurement of particle a. It can only depend on the settings of λ_1 and φ_1, and it can only take on the values $+1$, for horizontal polarization (H), and -1, for vertical polarization (V), so that one has $A(\varphi_1, \lambda_1) = \pm 1$. The hidden variable λ_4 plays a similar role for particles c and d, and the function D records the polarization of particle d, analogous to the role played by the function A.

Examination of Eq. (4) shows that whenever particles b and c occur in one of the Bell states $|\phi_{bc}^+\rangle$ or $|\psi_{bc}^-\rangle$, then particles a and d will also occur in one of the same two Bell states, regardless of the settings of the φ_i. This leads to the existence of an element of reality, described by a function κ, which can depend on both λ_1 and λ_4, but which is independent of the φ_i. These two Bell states determine the value $+1$ for κ. A similar argument holds for the occurrence of the other two Bell states, $|\phi_{bc}^-\rangle$ and $|\psi_{bc}^+\rangle$, also with the particles a and d in the same two Bell states, and when this occurs, it determines the value -1 for κ.

The function F records the results of the Bell state measurement for particles b and c. There are four separate outcomes, one for each of the Bell states, but it is more convenient for us to give the outcome equivalently as the product of the polarizations of particles b and c, so that $F_\kappa = +1$ for the outcomes HH or VV, and $F_\kappa = -1$ for the outcomes HV and VH. Thus the equivalency becomes

$$
\begin{aligned}
F_1 &= +1, \quad \textit{refers to} \quad \phi_{bc}^+, \\
F_1 &= -1, \quad\quad " \quad\quad\quad \psi_{bc}^-, \\
F_{-1} &= +1, \quad\quad " \quad\quad\quad \psi_{bc}^+, \\
F_{-1} &= -1, \quad\quad " \quad\quad\quad \phi_{bc}^-.
\end{aligned}
\tag{5}
$$

All of the perfect correlations in a classical deterministic, realistic, local description of this experiment, which are contained quantum mechanically in Eq. (4), for the case of 100% efficient detectors, are given by

$$
\begin{aligned}
A(\varphi_1, \lambda_1) F_{\kappa(\lambda_1, \lambda_4)}(\varphi_2, \varphi_3, \lambda_1, \lambda_4) D(\varphi_4, \lambda_4) &= 1, \quad \zeta_\kappa = 0, \pm\pi, \\
A(\varphi_1, \lambda_1) F_{\kappa(\lambda_1, \lambda_4)}(\varphi_2, \varphi_3, \lambda_1, \lambda_4) D(\varphi_4, \lambda_4) &= -1, \quad \zeta_\kappa = \pm\tfrac{\pi}{2}, \\
\zeta_\kappa = \varphi_1 - \varphi_2 + \kappa(\lambda_1, \lambda_4)(\varphi_3 - \varphi_4), \quad &(\zeta_+ = \xi, \zeta_- = \eta).
\end{aligned}
\tag{6}
$$

Equation (6) records the product of the polarizations of the four particles.

4 Extending the EPR Analysis to the Case of Inefficient Detectors

These results hold in the case where the detectors are 100% efficient, which means that the functions $A(\varphi_1, \lambda_1), D(\varphi_4, \lambda_4), F_\kappa(\varphi_2, \varphi_3, \lambda_1, \lambda_4)$, and $\kappa(\lambda_1, \lambda_4)$, exist according to the EPR postulates, and are equal to ± 1 for every value of their arguments, which in turn means that every one of the four photons that is generated in each event is counted at a detector. In paper A we showed that this situation, given by Eq. (16), is inconsistent. In paper B we assumed that this 100% efficiency is not necessarily the case, but rather that the particles may reach their detectors and not be counted.

This introduces a complication into the argument since the existence of the functions A, D, κ, and F, depends critically on the EPR postulates. However if the particle is not always counted, then one no longer has the one-to-one correspondence between predictability and reality needed to define an element of reality, and therefore completeness. Nonetheless, if we are considering a realistic, deterministic model, there is a natural extension of the EPR argument to cover this case. We discussed this case in paper B, but nonetheless, we would like to amplify on it here, as we do not believe that a deterministic GHZ-type analysis has been used before for inefficient detectors.

The standard classical EPR element of reality situation is often discussed in terms of the following example. I take two pieces of paper and draw a $\sqrt{}$ mark on one ($\equiv +1$) and an X mark on the other ($\equiv -1$). I put each piece in a separate envelope and hand them to two students and ask them to go to opposite ends of the room.

Now if one student opens his envelope and finds a $\sqrt{}$ mark, he immediately knows that the other student has an X. The element of reality argument asserts that although by opening his envelope he obtains knowledge of the other student's mark, in this act he in no way interacted with or affected the other student's result. Therefore the other student's mark, since it wasn't affected by the first student's action, must be a real, objective property, and so it must have been there before the other student looked at his mark. In fact it must have been there since the students were first given their envelopes. Its not being affected by the other student's distant action is what determines the mark to be what EPR called an *element of reality*.

Now consider the case where I again draw the marks, but I tell them I might not put the slips of paper into the envelopes I give them. It is still true that when the first student opens his envelope, if he has a $\sqrt{}$, he knows that if the other student has a slip of paper, it will contain an X. He does not know whether the other student has a slip or not, but it is also still true that he knows that the other student does not have a $\sqrt{}$. So this is an element of reality in the original sense. But it is still also true that *his action has no effect on whether the other student has a piece of paper or not*. So that information must also be an objective fact that was determined when the other student received his envelope. (In our experiment the evidence is even stronger, because by energy and momentum conservation, we know that the particle reached its detector, whether it gets measured or not. So *reaching the detector is an element of reality*.) Therefore the function A can take this into account, and may be extended to take on the value 0, to cover the deterministic fact that the envelope is empty, corresponding to inefficient detection in our case.

In our experiment, in any run where the conditions for a perfect correlation are met, if we successfully detect three of the particles in a given event then there are only two possibilities for the fourth particle. The first is that we detect it, in which case we can predict in advance what its polarization will be. If this happens to be particle a, say, we can say that in this case $A(\varphi_1, \lambda_1)$ exists, and has the value ± 1,

Extending the Concept of an "Element of Reality" to Work with Inefficient Detectors 93

which was determined when the particle was created. The second possibility is that it passes through its detector, but is not detected. But because it is not detected, it has no further effect on the experiment, and we can consistently assign to $A(\varphi_1, \lambda_1)$ the value A = 0.

In a deterministic theory, we can assume that this value was assigned to the particle when it was created. In other words this photon, with these particular values of λ_1 and φ_1, was destined at its creation not to be detected. The alternative is that the particle is not recorded simply because the detector is inefficient. It counts only a certain percentage of particles impinging upon it, independently of any state variables λ_i, and angles φ_i, that may determine the properties of the particles. This case is conceptually rather simple in that one may then merely consider those particles that are counted, knowing that one is counting a fair sampling of all the particles that impinge upon the counters. Then the outcome is independent of the properties of the interaction of the counters with the particles, except for random efficiency effects which will not prejudice any results one might obtain in the case of 100% efficiency and one can apply Bell-like theorems in this case. We will not be concerned with this case in what follows. We are concerned with a deterministic theory, for which no random sampling assumptions need be made. This case is more general than the stochastic case mentioned above, since a deterministic theory can be modelled to duplicate the results of a stochastic theory.

One may well question whether what we have left after extending the EPR theory to inefficient counters can truly be called an "element of reality". The answer is definitely "yes", because one must remember the motivation for introducing the term. *Since one can predict a property of the particle without in any way interacting with it, then according to EPR we cannot have affected this property, and so the property must have existed before we made the measurement.* Thus this is a true, objective property of the particle that it must have possessed since it was created, or at least since it last interacted with another particle, and hence the designation "element of reality". This argument still holds in our situation since, while we cannot predict whether it will be detected, we can predict this property precisely, *if* it is detected. Thus the particle must either possess this property beforehand, or it must be determined beforehand that it will not be detected. In either case, the existence of the property does not depend on the measurement, and so it is an objective element of reality.

Everything we have said about particle *a* also applies to particle *d*. So the functions A and D are to be considered as deterministic functions representing instructions to the particle not only to have a particular polarization if it is counted, but also to determine whether the particle is to be counted or not. Specifically, we will amend the definitions of the functions A and D in the inefficient case to read

$$A(\varphi_1, \lambda_1) = \pm 1, 0; \quad D(\varphi_4, \lambda_4) = \pm 1, 0, \tag{7}$$

In Eq. (7), no limits are placed on the functions, except that we will demand the consistency condition that the product of all the functions agrees with the results for perfect correlations whenever all four particles are actually detected, an important condition we discussed in papers A and B. The existence of these functions extends the concept of completeness to the case of inefficient counters.

The situation for particles b and c is similar, but a little more subtle. These particles are not counted separately, but as part of an entangled state. However, this does not affect our argument above, so we shall not comment on it here (although it is discussed in A and B). But it is also true in this case that for F, as well as for A and D,

$$F_\kappa(\varphi_2, \varphi_3, \lambda_1, \lambda_4) = \pm 1, 0. \tag{8}$$

Here, the ± 1 values represent the product of their polarizations.

5 Summary

We have shown that one can extend the concept of "element of reality" in the case of inefficient detectors. One must remember that the entire point of introducing the concept was because EPR objected to the idea that the reality of an object can be determined by actions one takes elsewhere that do not affect the object in question. Our extension of the concept does not interfere with this motivation, and in this sense it is a natural extension.

References

1. The paper referred to as paper A has been published, as D. M. Greenberger, M. A. Horne, and A. Zeilinger, *Phys. Rev. A* **78**, 022110 (2008). It is on the archive as quant-ph. ArXiv:05010201, ver. 2. The paper for inefficient counters, paper B, has an extra author, and has also been published, as D. M. Greenberger, M. A. Horne, A. Zeilinger, and M. Zukowski, *Phys. Rev. A* **78**, 022111 (2008). It is on the archive as quant-ph. ArXiv:05010207, ver. 2.
2. D. M. Greenberger, M. A. Horne, and A. Zeilinger, in *Bell's Theorem, Quantum Theory, and Conceptions of the Universe*, M. Kafatos, ed., Kluwer, Dordrecht, 1989, p. 69; D. M. Greenberger, M. A. Horne, A. Shimony, and A. Zeilinger, *Am. J. Phys.* **58**, 1131 (1990).
3. A. Einstein, B. Podolsky, and N. Rosen, Phys. Rev. **47**, 777 (1935). See also J. S. Bell, *Physics (NY)* **1**, 195 (1964). Both articles are reprinted in J. Wheeler and W. Zureck, *Quantum Measurement Theory*, Princeton University Press, Princeton, 1983. Bell's paper is also reprinted in the two collections of Bell's papers on the subject, J. S. Bell, *Speakables and Unspeakables in Quantum Mechanics*, Cambridge University Press, Cambridge (England), 1987; and *John S. Bell on the Foundations of Quantum Mechanics*, M. Bell, K. Gottfried, and M. Veltman, eds., World Scientific, Singapore, 2001.
4. T. Jennewein, G. Weihs, J-W Pan, and A. Zeilinger, *Phys. Rev. Lett.* **88**, 017903/1–4 (2002). The concept of entanglement swapping was first introduced by M. Zukowski, A. Zeilinger, M. A. Horne, and A. K. Ekert, *Phys. Rev. Lett.* **71**, 4287 (1993).
5. For a discussion of the low detection efficiency, random sampling, and other loopholes, see for example, M. Redhead, *Incompleteness, Nonlocality, and Realism*, Clarendon Press, Oxford, 1987; F. Selleri, *Quantum Paradoxes and Physical Reality*, Kluwer Academic, Dordrecht, 1990; A. Shimony, *An Exposition of Bell's Theorem*, in *Search for a Naturalistic World View, Vol. II* (Essays of A. Shimony), Cambridge University Press, Cambridge (England), 1993; and references within these.
6. A. Broadbent and A. A. Méthot, arXiv: quant-ph/0511047 v2.

A General Proof of Nonlocality without Inequalities for Bipartite States

GianCarlo Ghirardi and Luca Marinatto

Abstract We exhibit a general nonlocality argument without inequalities for bipartite (pure and mixed) states belonging to Hilbert space of arbitrary dimensionality. The argument, which makes use of simple set-theoretic manipulations, comprise, as its particular instances, the nonlocality proofs for bipartite states existing in the literature. Moreover, its relation with the Clauser-Horne inequality is investigated.

1 Introduction

It is a great pleasure to contribute to this volume honoring Abner Shimony with a paper dealing with some aspects of nonlocality, a topic to which he has given significant contributions.

Hardy's nonlocality without inequalities proof [1] is an ingenious argument, seemingly not involving Bell-like inequalities [2, 3], which demonstrates that there cannot exist a local hidden variable model reproducing the quantum mechanical predictions of almost any entangled bipartite pure state of two spin$-1/2$ particles. To this end, appropriate joint-probability distributions are considered which cannot be simultaneously accounted for by a local model in which the outcomes of single-particle measurements are predetermined by the knowledge of the hidden variables. Subsequently, refinements of the original proof [4] as well as generalizations to

---

G. Ghirardi and L. Marinatto
Department of Theoretical Physics of the University of Trieste, Italy
Istituto Nazionale di Fisica Nucleare, Sezione di Trieste, Italy
e-mail: ghirardi@ts.infn.it, marinatto@ts.infn.it

G. Ghirardi
International Centre for Theoretical Physics "Abdus Salam," Trieste, Italy
e-mail: ghirardi@ts.infn.it

W.C. Myrvold and J. Christian (eds.), *Quantum Reality, Relativistic Causality, and Closing the Epistemic Circle*, The Western Ontario Series in Philosophy of Science 73,
© Springer Science+Business Media B.V. 2009

multipartite entangled state vectors and to higher dimensional Hilbert spaces [5] have been presented. Other works have extended the original proof to the case of certain classes of nonseparable mixed states [6]. The relevant feature which is claimed to be common to all the mentioned proofs is that they do not involve explicitly Bell-like inequalities [2, 3], that is, inequalities involving linear combinations of correlation functions. In fact, in these proofs the inconsistency between quantum mechanical predictions and any conceivable hidden variable model is established either by resorting to counterfactual reasonings, as in Refs. [1, 4, 5, 7], or to set-theoretic manipulations, as in Ref. [6].

The aim of this paper is twofold: from one side to exhibit, by resorting to a recent approach [6], an extremely general scheme for a nonlocality proof without inequalities which is valid for bipartite (pure and mixed) states whose constituents belong to Hilbert spaces of arbitrary (finite) dimensions. The new aspect of the approach derives from its being based on the consideration of probability distributions which, contrary to what is usually the case in the literature [1,4,7], may all differ from zero. Our proof recovers well-known results, such as, for example, the original Hardy's proof [1] or those of Ref. [7], as particular instances of a more general scheme. Moreover, it is able to detect the nonlocal nature of some states which do not fall in the range of applicability of the existing proofs (as, for example, the maximal entangled states of two spin$-1/2$ particles).

The second and, in our opinion, more interesting point of our analysis is that it makes clear the relations between the alleged proofs *without inequalities* and the validity of the Clauser-Horne inequality [3] (CH in what follows). To be more precise, we will show that the Hardy-like conditions of nonlocality presented in the literature (the original one [1] as well as its generalizations [4, 6, 7]) are nothing more than particular instances of the violation of the Clauser-Horne inequality. As a consequence, the claimed absence of Bell-like inequalities in all Hardy-like nonlocality proofs is only seeming.

2 Generalized Hardy's Proof

Let us start by exhibiting a generalized version of Hardy's nonlocality proof, inspired by the techniques presented in Ref. [6]. For pedagogical reasons, we deal first with the simpler case of a bipartite quantum state $|\psi\rangle \in \mathbb{C}^2 \otimes \mathbb{C}^2$ and only subsequently we will extend our proof to cover the case of higher dimensional Hilbert spaces. To this end, let us suppose that there exist appropriate (dichotomic) spin-observables X_i and Y_i ($i = 1, 2$ being the particle index), such that the following joint probability distributions are satisfied:

$$P_\psi(X_1 = +1, X_2 = +1) = q_1, \tag{1}$$
$$P_\psi(Y_1 = +1, X_2 = -1) = q_2, \tag{2}$$
$$P_\psi(X_1 = -1, Y_2 = +1) = q_3, \tag{3}$$
$$P_\psi(Y_1 = +1, Y_2 = +1) = q_4 \tag{4}$$

A General Proof of Nonlocality without Inequalities for Bipartite States   97

where $q_1, q_2, q_3, q_4 \in [0,1]$. The issue whether and under which conditions a state ψ, a set of spin-observables $\{X_i, Y_i\}$ and a collection of probabilities $\{q_i\}$ exist and satisfy Eqs. (1)–(4) will be touched upon later on; for the moment we simply suppose that the previous equations are valid for an (unspecified) choice of the state, the observables and the ensuing quantum mechanical probabilities.

Consider now a mixed state σ and denote as ε its trace distance $D(\sigma, \psi)$ from the projection operator associated to the state $|\psi\rangle$ of the previous equations, that is, $D(\sigma, \psi) \equiv \frac{1}{2} Tr|\sigma - |\psi\rangle\langle\psi|| = \varepsilon$. A small trace distance implies that the states σ and ψ give rise to close probability distributions for every measurement outcome since the inequality

$$|Tr[Q\sigma] - Tr[Q|\psi\rangle\langle\psi|]| \le D(\sigma, \psi) \qquad (5)$$

holds for any projection operator Q. Using this property and considering such a mixed state σ, the probabilities (1)–(4) are modified as follows

$$P_\sigma(X_1 = +1, X_2 = +1) \in [q_1 - \varepsilon, q_1 + \varepsilon], \qquad (6)$$

$$P_\sigma(Y_1 = +1, X_2 = -1) \in [q_2 - \varepsilon, q_2 + \varepsilon], \qquad (7)$$

$$P_\sigma(X_1 = -1, Y_2 = +1) \in [q_3 - \varepsilon, q_3 + \varepsilon], \qquad (8)$$

$$P_\sigma(Y_1 = +1, Y_2 = +1) \in [q_4 - \varepsilon, q_4 + \varepsilon]. \qquad (9)$$

In order to face our basic question, namely if there is a way to locally account for the above mentioned set of joint probabilities, we need to make precise the idea of a hidden variable model for the state σ. Roughly speaking, it consists in any conceivable theory where (i) the measurement outcomes m, n of arbitrary single particle observables M_1, N_2 are predetermined given the variables (commonly referred to as hidden variables) $\lambda \in \Lambda$, Λ being a set, and where (ii) the quantum mechanical probabilities P_σ are obtained by averaging the predetermined values of the corresponding hidden variables probabilities P_λ over the (normalized to unity) positive distribution $\rho(\lambda)$ of such variables, according to

$$P_\sigma(M_1 = m, N_2 = n) = \int_\Lambda d\lambda \, \rho(\lambda) P_\lambda(M_1 = m, N_2 = n). \qquad (10)$$

The hidden variable model is called local if the joint probabilities P_λ of the model factorize whenever the measurements of the single-particle observables M_1 and N_2 refer to space-like separated events, that is

$$P_\lambda(M_1 = m, N_2 = n) = P_\lambda(M_1 = m) P_\lambda(N_2 = n) \quad \forall \lambda \in \Lambda. \qquad (11)$$

Here, $P_\lambda(M_1 = m)$ and $P_\lambda(N_2 = n)$ are the single particle probabilities of the indicated outcomes given by the model, and they are determined by the value of λ. In a deterministic hidden variable model, like the one we will be considering, such probabilities can assume only the values 0 or 1 for any given value of $\lambda \in \Lambda$ [1].

---

[1] Local and stochastic models, where P_λ belongs to the interval [0, 1], are completely equivalent to the local and deterministic models we are considering in this paper, as proven in Ref. [8].

In order to determine under which circumstances one gets a contradiction between the probabilities (6)–(9) and a local and deterministic hidden variable model accounting for them, we proceed as follows. First of all, we define A, B, C, and D, as the subsets of Λ where the (single-particle) probabilities $P_\lambda(X_1 = +1), P_\lambda(X_2 = +1), P_\lambda(Y_1 = +1)$ and $P_\lambda(Y_2 = +1)$ take the value $+1$, respectively—for example, $A = \{\lambda \in \Lambda \,|\, P_\lambda(X_1 = +1) = 1\}$.

It is then possible to rewrite Eqs. (6)–(9) in terms of the measures $\mu[Z] = \int_Z d\lambda\, \rho(\lambda)$ of appropriate subsets $Z \subseteq \Lambda$. In fact, let us rewrite, e.g., the relation of Eq. (7) taking into account Eqs. (10) and (11). We have

$$
\begin{aligned}
P_\sigma(Y_1 = +1, X_2 = -1) &= \int_\Lambda d\lambda\, \rho(\lambda) P_\lambda(Y_1 = +1) P_\lambda(X_2 = -1) \\
&= \int_\Lambda d\lambda\, \rho(\lambda) P_\lambda(Y_1 = +1)[1 - P_\lambda(X_2 = +1)] \\
&= \mu[C] - \mu[B \cap C] \in [q_2 - \varepsilon, q_2 + \varepsilon]
\end{aligned}
\tag{12}
$$

where use has been made of the relation $P_\lambda(X_2 = -1) + P_\lambda(X_2 = +1) = 1$ which must hold for any $\lambda \in \Lambda$. Proceeding in a similar way, Eqs. (6)–(9) are replaced by relations involving the measures of the indicated sets, i.e.,

$$
\mu[A \cap B] \in [q_1 - \varepsilon, q_1 + \varepsilon],
\tag{13}
$$

$$
\mu[C] - \mu[B \cap C] \in [q_2 - \varepsilon, q_2 + \varepsilon],
\tag{14}
$$

$$
\mu[D] - \mu[A \cap D] \in [q_3 - \varepsilon, q_3 + \varepsilon],
\tag{15}
$$

$$
\mu[C \cap D] \in [q_4 - \varepsilon, q_4 + \varepsilon].
\tag{16}
$$

If we follow the set-theoretic manipulations presented in Ref. [6] (which we omit here for the sake of brevity), starting from Eqs. (13)–(16) we end up with an inequality constraining the values of ε and $\{q_i\}$, whenever a local and deterministic hidden variable model exists yielding the quantum probabilities implied by σ, namely:

$$
-4\varepsilon \le q_1 + q_2 + q_3 - q_4.
\tag{17}
$$

Of course, the relevance of this result resides in the equivalent (and converse) statement: given a pure state ψ and a set of observables $\{X_i, Y_i\}$ satisfying Eqs. (1)–(4) and considering a mixed state σ, having trace distance $D(\sigma, \psi) = \varepsilon$ from ψ, if the relation

$$
q_1 + q_2 + q_3 - q_4 < -4\varepsilon
\tag{18}
$$

is satisfied, then there does not exist any local and deterministic hidden variable model reproducing the probability distributions implied by quantum mechanics for the state σ. Equivalently, without making any reference to a pure state ψ, we could have said: given a mixed state σ and a set of observables $\{X_i, Y_i\}$ satisfying Eqs. (6)–(9) for chosen values of $\{q_i\}$ and ε, if the relation of Eq. (18) is satisfied, then there cannot exist a local and deterministic hidden variable model for σ. Moreover, such a mixed state σ turns out to be nonseparable since, if it would be separable, it would admit a local model reproducing its predictions [9], contrary to our proof.

A General Proof of Nonlocality without Inequalities for Bipartite States 99

This procedure to identify states incompatible with the nonlocality assumption is general enough to encompass the usual nonlocality without inequalities proofs [1,4, 6,7] as its particular instances. For example, one can consider the following cases:

(i) $\varepsilon = 0, q_1 = q_2 = q_3 = 0$ and $q_4 > 0$: this is the situation considered in the original Hardy's proof [1]. In this case, given any entangled, but not maximally entangled, state of two-qubits ψ, one can explicitly exhibit four appropriate spin-observables $\{X_i, Y_i\}$, defined in terms of the coefficients appearing in the Schmidt decomposition of the state ψ, such that Eqs. (1)–(4) are satisfied. As a consequence, with our choice of the parameters ε and $\{q_i\}$, Eq. (18) is obviously satisfied and this automatically implies the nonlocal character of the state ψ. Nothing changes, as we have shown in Ref. [6], when one considers the more general case $\varepsilon > 0$: with the same choice of the state ψ and of the set of observables made by Hardy, one can still prove nonlocality for all mixed states σ for which $0 < 4\varepsilon < q_4$ is satisfied.

(ii) $\varepsilon = 0, q_2 = q_3 = 0$ and $0 < q_1 < q_4$: this is the situation considered in Ref. [7]. Also in this case, by following the calculations of the authors, one can identify some states and observables which satisfy both Eqs. (1)–(4) and the nonlocality condition of Eq. (18). Obviously, through our approach, one can also cover the case of mixed states σ by simply noticing that whenever $0 < 4\varepsilon < (q_4 - q_1)$ is satisfied nonlocality is established for the considered σ.

The procedure to discover quantum states exhibiting nonlocal effects can be further improved with a novel argument. It aims at enlarging the set of values of the parameters $\{q_i\}$ and ε for which no local and deterministic hidden variable model exists reproducing the quantum probabilities of the associated quantum states. To this end, let us consider four arbitrary sets A, B, C, and D,[2] and work out some useful relations between them. First of all, it is easy to prove that

$$D \subseteq \bar{A} \cup (A \cap D), \tag{19}$$
$$C \subseteq \bar{B} \cup (B \cap C), \tag{20}$$

where we have denoted with $\bar{Z}$ the complement of the set Z within Λ, i.e., $\bar{Z} = \Lambda - Z$. As a consequence of Eqs. (19)–(20) we obtain that $D \cup C \subseteq \bar{A} \cup \bar{B} \cup (A \cap D) \cup (B \cap C)$ and, being μ any measure defined on sets, we end up with the following set-theoretic inequality

$$\mu[D] + \mu[C] - \mu[D \cap C] = \mu[D \cup C]$$
$$\leq \mu[\bar{A} \cup \bar{B} \cup (A \cap D) \cup (B \cap C)]$$
$$\leq \mu[\bar{A} \cup \bar{B}] + \mu[A \cap D] + \mu[B \cap C], \tag{21}$$

where use have been made of the fact that $\mu[X] \leq \mu[Y]$ for any $X \subseteq Y$ (first inequality), and that $\mu[X \cup Y] \leq \mu[X] + \mu[Y]$, holding for any sets X, Y (second inequality).

---

[2] Obviously, later on we will identify A, B, C, D with the subsets of Λ we have considered previously but, at present, the relations we will exhibit are completely general, holding for any quadruple of sets.

Now, when the sets A,B,C,D, are precisely those appearing in Eqs. (13)–(16), one has

$$\mu[A\cap B]+\mu[C]-\mu[B\cap C]+\mu[D]-\mu[A\cap D]-\mu[C\cap D]\geq q_1+q_2+q_3-q_4-4\varepsilon \tag{22}$$

where the inequality derives from taking into account the bounds of Eqs. (13)–(16). Moreover, Eq. (21) and the relation $\mu[\bar{A}\cup\bar{B}]=\mu[\Lambda-(A\cap B)]=\mu[\Lambda]-\mu[A\cap B]$, lead to

$$\mu[A\cap B]+\mu[C]-\mu[B\cap C]+\mu[D]-\mu[A\cap D]-\mu[C\cap D]\leq\mu[\bar{A}\cup\bar{B}]+\mu[A\cap B]$$
$$=\mu[\Lambda]=1 \tag{23}$$

Concluding, by combining Eq. (22) with Eq. (23) and taking into account the relation of Eq. (17), we get the following theorem

Theorem I Consider a pure state $\psi\in\mathbb{C}^2\otimes\mathbb{C}^2$ and four spin-observables $\{X_i,Y_i\}_{i=1,2}$ such that Eqs. (1)–(4) are satisfied. Given a mixed state σ such that $D(\sigma,\psi)=\varepsilon$, if there exists a local and deterministic hidden variable model for σ then the inequality

$$-4\varepsilon\leq q_1+q_2+q_3-q_4\leq 1+4\varepsilon \tag{24}$$

holds true.

Once again, the usefulness of this result stems from the opposite statement: if the previous inequality is violated, that is, if either $q_1+q_2+q_3-q_4<-4\varepsilon$ or $q_1+q_2+q_3-q_4>1+4\varepsilon$, then there cannot exist any local and deterministic hidden variable model which can reproduce the joint probabilities of Eqs. (13)–(16).

The interval of values of the parameters $\{q_i\}$ and ε implying nonlocal effects for the considered mixed states has been considerably enlarged with respect to our previous work on the subject [6]. In fact, in Ref. [6], evidence of nonlocality has been related only to a violation of the lower bound of Eq. (24), while nothing has been proved concerning a violation of the upper bound. For example, as is well known, the original Hardy's criterion [1] does not allow to prove nonlocality for maximally entangled pure states, just because there does not exist any set of spin-observables giving rise to probabilities $\{q_i\}$ such that $q_{i\neq 4}=0$ and $q_4>0$. In particular, in the case of the singlet state and within the restricted scenario envisaged by Hardy, the joint probabilities of any conceivable choice of spin-observables cannot violate the lower bound of Eq. (24), assuming $\varepsilon=0$. A similar negative result was conjectured [7] to hold also for the case $q_{1,3}=0$ and $q_1-q_4<0$. On the contrary, according to the present argument, given the singlet state, we can choose $\{X_1,Y_2,Y_1,X_2\}$ as spin-observables lying in the same plane (e.g., in the $x-y$ plane) and forming the angles $0,\pi/4,\pi/2$ and $3\pi/4$ with respect to the positive direction of the x axis, respectively. In this case, one easily sees that the upper bound of Eq. (24) is violated just because the quantity $q_1+q_2+q_3-q_4$ equals $(1+\sqrt{2})/2$ and it exceeds 1, thus establishing nonlocality for a maximally entangled state. Thus, our generalized argument is powerful enough to account for the nonlocality of every (pure) entangled state of two spin-1/2 particles, contrary to the similar proofs existing in the literature.

A General Proof of Nonlocality without Inequalities for Bipartite States 101

Before concluding, let us briefly mention how the whole nonlocality argument can be generalized to the case of a Hilbert space $\mathbb{C}^{d_1} \otimes \mathbb{C}^{d_2}$, where d_1 and d_2 are integers strictly greater than 2. First of all, given any state ψ, let us define, in place of the previously considered observables, four new observables X_i and Y_i ($i = 1, 2$), such that the spectrum of X_i is the set $\{-1, 0 + 1\}$, while the one of Y_i is only required to contain the value $+1$. We then add to the set of Eqs. (1)–(4) two other relations

$$P_\psi(Y_1 = +1, X_2 = 0) = q_5, \tag{25}$$
$$P_\psi(X_1 = 0, Y_2 = +1) = q_6. \tag{26}$$

with $q_5, q_6 \in [0, 1]$. As a consequence, for a mixed state σ, having trace distance equal to ε from ψ, one has

$$P_\sigma(Y_1 = +1, X_2 = 0) \in [q_5 - \varepsilon, q_5 + \varepsilon], \tag{27}$$
$$P_\sigma(X_1 = 0, Y_2 = +1) \in [q_6 - \varepsilon, q_6 + \varepsilon]. \tag{28}$$

Now suppose that a local and deterministic model exists for σ and proceed as before. Using the relations Eqs. (27)–(28) and the fact that $P_\lambda(X_i = -1) + P_\lambda(X_i = 0) + P_\lambda(X_i = +1) = 1$ for any λ, Eqs. (14) and (15) get modified in the following way

$$\mu[C] - \mu[B \cap C] \in [q_2 + q_5 - 2\varepsilon, q_2 + q_5 + 2\varepsilon], \tag{29}$$
$$\mu[D] - \mu[A \cap D] \in [q_3 + q_6 - 2\varepsilon, q_3 + q_6 + 2\varepsilon] \tag{30}$$

respectively. Since all other relations constraining the measures of the considered sets remain valid by simply replacing q_2 and q_3 appearing in Eq. (24) with $q_2 + q_5 + \varepsilon$ and $q_3 + q_6 + \varepsilon$, respectively, we can conclude that, if a local deterministic hidden variable model exists for σ then the relation

$$-6\varepsilon \leq q_1 + q_2 + q_3 + q_5 + q_6 - q_4 \leq 1 + 6\varepsilon \tag{31}$$

holds true. Equivalently, given the mixed state σ, if (i) there exist a pure state ψ and observables $\{X_i, Y_i\}$ satisfying Eqs. (1)–(4) together with Eqs. (25) and (26), and (ii) if the trace distance $D(\sigma, \psi) = \varepsilon$ is such that either $q_1 + q_2 + q_3 + q_5 + q_6 - q_4 < -6\varepsilon$ or $q_1 + q_2 + q_3 + q_5 + q_6 - q_4 > 1 + 6\varepsilon$ then no local and deterministic model can exist for σ.

Up to now we have not faced the issue whether there exist (pure) states ψ and observables X_i and Y_i such that the associated probabilities take the values (1)–(4) and (25)–(26) and such that $q_1 + q_2 + q_3 + q_5 + q_6 - q_4 < 0$ and $q_i > 0$ hold. The answer is affirmative and the easiest way to prove this is to start by considering the particular case $q_{i \neq 4} = 0$ and $q_4 > 0$ (that is, the Hardy's case). As proven in [10], in such a situation, given any entangled state ψ possessing at least two different coefficients in its Schmidt decomposition, there exist appropriate observables $\{X_i, Y_i\}$ satisfying Eqs. (1)–(4) and (25)–(26) for the considered choice of $\{q_i\}$. Quantum mechanical probability distributions are the squared moduli of the scalar product between appropriate vectors with a fixed state vector ψ and, as such, they are continuous functionals. This implies that there is a neighborhood of ψ in the Hilbert space, containing infinitely many vectors ϕ for which the probabilities $\bar{q}_i$ of Eqs. (1)–(4) and of Eqs. (25)–(26) are such that (i) there exists at least one $\bar{q}_i$ strictly greater than zero, and (ii) the inequality $\bar{q}_1 + \bar{q}_2 + \bar{q}_3 + \bar{q}_5 + \bar{q}_6 - \bar{q}_4 < 0$ is still satisfied. Then, for any such state ϕ, we may determine a precise neighborhood of it, in the trace-distance topology, containing mixed states σ exhibiting nonlocal effects.

3 Clauser-Horne Inequality

In the previous section we have presented a very general approach which includes, and remarkably generalizes, all known nonlocality without inequalities proofs for bipartite quantum states. Evidence of nonlocality is obtained by a violation of (any instance) of Eq. (31)—or, in a restricted scenario, of Eq. (24). Such a relation is an inequality constraining joint probabilities and, as such, it must be related to the Clauser-Horne inequality. In fact, as proven by Fine,[3] satisfaction of the Clauser-Horne inequality is a necessary and sufficient condition for a quantum state to admit a local and deterministic hidden variable model accounting for the joint probability distributions involving four observables (which is exactly the situation considered by the nonlocality without inequalities arguments). In a sense, Clauser-Horne is the only relevant inequality within this scenario, all others inequalities being equivalent or weaker than it—that is, unable to identify all nonlocal states which the Clauser-Horne criterion detects. This is precisely what happens with the inequality (31) which we will prove to be straightforwardly implied by the Clauser-Horne inequality.

To start with we consider a mixed state σ acting on $\mathcal{B}(\mathbb{C}^{d_1} \otimes \mathbb{C}^{d_2})$, while the observables X_i and Y_i are the same as in the previous section. Then, a straightforward result follows.

Theorem II. If a local hidden variable model exists for σ then the relation

$$P_\sigma(X_1 = +1, X_2 = +1) + P_\sigma(Y_1 = +1, X_2 = -1) + P_\sigma(Y_1 = +1, X_2 = 0) +$$
$$P_\sigma(X_1 = -1, Y_2 = +1) + P_\sigma(X_1 = 0, Y_2 = +1) - P_\sigma(Y_1 = +1, Y_2 = +1) \in [0, 1] \tag{32}$$

is satisfied.

Proof: the existence of a local deterministic model for σ implies that Eq. (32) can be written as $\int d\lambda \rho(\lambda)[P_\lambda(X_1 = +1)P_\lambda(X_2 = +1) - P_\lambda(Y_1 = +1)P_\lambda(X_2 = +1) - P_\lambda(X_1 = +1)P_\lambda(Y_2 = +1) - P_\lambda(Y_1 = +1)P_\lambda(Y_2 = +1) + P_\lambda(Y_1 = +1) + P_\lambda(Y_2 = +1)]$. The possible values attained by the integrand of the previous expression belong to the interval $[0, 1]$ for any $\lambda \in \Lambda$, since

$$0 \le x\bar{y} - xy - \bar{x}y - \bar{x}\bar{y} + \bar{x} + y \le 1 \tag{33}$$

---

[3] Local and stochastic models, where P_λ belongs to the interval $[0, 1]$, are completely equivalent to the local and deterministic models we are considering in this paper, as proven in Ref. [8].

is an algebraic expression satisfied by any variables $x, y, \bar{x}, \bar{y} \in [0, 1]$, such as the single particle probabilities P_λ. As a consequence, the integral belongs to $[0, 1]$, thus concluding the proof. ∎

We stress that Eq. (32) is nothing more than an equivalent version of the Clauser-Horne inequality [3]. In fact, as one can deduce from the arguments yielding the proof of the previous theorem, Eq. (32) turns out to be equal to the expression

$$P_\sigma(X_1 = +1, X_2 = +1) - P_\sigma(Y_1 = +1, X_2 = +1) - P_\sigma(X_1 = +1, Y_2 = +1)$$
$$-P_\sigma(Y_1 = +1, Y_2 = +1) + P_\sigma(Y_1 = +1) + P_\sigma(Y_2 = +1) \in [0, 1], \tag{34}$$

which is the inequality derived by Clauser-Horne [3]. It is now apparent that if we take into account Eqs. (6)–(9) together with Eqs. (27)–(28), the inequality of Eq. (32) trivially implies the relation of Eq. (31) which we have obtained in the previous section. Therefore, in the framework of the generalized nonlocality without inequalities proofs, the relevant constraint linking $\{q_i\}$ and ε when a local hidden variable model exists for σ could have been simply deduced by the Clauser-Horne inequality.

It is worth stressing this point by analyzing once again the extremal case represented by Hardy's argument [1]. In its usual presentation, evidence of nonlocality is obtained by ascertaining that three appropriately chosen joint-probabilities are (strictly) null while a fourth one is not. The absence of inequalities is only seeming, since if we plug such joint-probabilities into Eq. (32) (by neglecting the probabilities involving the outcome $X_i = 0$, since, in this case, the whole Hilbert space is only four-dimensional) we obtain exactly the condition $q_4 > 0$ considered by Hardy. It is clearly an inequality and, most important of all, it is a particular instance of a violation of the Clauser-Horne inequality. Therefore, the expression *"nonlocality without inequalities proof"* used to refer to the Hardy's argument appears to be a bit misleading, because the kind of nonlocality condition it provides is nothing else than a particular instance of a violated Clauser-Horne inequality. A similar remark can be obviously applied to all existing Hardy-like nonlocality arguments [4, 6, 7].

4 Conclusions

In this paper we have exhibited a general framework for the so-called nonlocality without inequalities proofs for bipartite states, which includes all known cases existing in the literature [1, 4, 6, 7] as particular cases. Subsequently, we have provided evidence how these nonlocality arguments, including the original one devised by Hardy [1], are particular instances of the violation of the Clauser-Horne inequality, thus proving that such an inequality is the one which actually grasps the features of any state of a bipartite system which exhibits nonlocal effects.

Acknowledgement Work supported in part by Istituto Nazionale di Fisica Nucleare, Sezione di Trieste, Italy.

References

1. L. Hardy, *Phys. Rev. Lett.* **71**, 1665 (1993).
2. J.S. Bell, *Physics* (Long Island City, N.Y.) **1**, 195 (1964); J.F. Clauser, M.A. Horne, A. Shimony, and R.A. Holt, *Phys. Rev. Lett.* **23**, 880 (1969).
3. J.F. Clauser, M.A. Horne, *Phys. Rev. D* **10**, 526 (1974).
4. S. Goldstein, *Phys. Rev. Lett.* **72**, 1951 (1994); T.F. Jordan, *Phys. Rev. A* **50**, 62 (1994); D. Boschi, S. Branca, F. De Martini, and L. Hardy, *Phys. Rev. Lett.* **79**, 2755 (1997).
5. G. Kar, *Phys. Rev. A* **56**, 1023 (1997); S. Ghosh, G. Kar, and D. Sarkar, *Phys. Lett. A* **243**, 249 (1998); Jose L. Cerceda, *Phys. Lett. A* **327**, 433 (2004); S. Kunkri and S.K. Choudhary, *Phys. Rev. A* **72**, 022348 (2005).
6. G.C. Ghirardi and L. Marinatto, *Phys. Rev. A* **73**, 032102 (2006); G.C. Ghirardi and L. Marinatto, *Phys. Rev. A* **74** 062107 (2006).
7. S. Kunkri, S.K. Choudhary, A. Ahanj, and P. Joag, *Phys. Rev. A* **73**, 022346 (2006).
8. A. Fine, *Phys. Rev. Lett.* **48**, 291 (1982).
9. R.F. Werner, *Phys. Rev. A* **40**, 4277 (1989).
10. G.C. Ghirardi and L. Marinatto, *Phys. Rev. A* **72**, 014105 (2005).

On the Separability of Physical Systems

Jon P. Jarrett

Abstract In the context of Bell-type experiments, two notions of "separability" emerge from the application of some simple considerations from information theory. The first of these applies to physical states construed as probability measures, while the second applies to states construed in terms of "elements of physical reality". The former is found to be logically equivalent to the "completeness" constraint (a.k.a. "outcome independence" or "factorizabilility") in Bell-type arguments. Moreover, it is found that there are theories that are separable in the first sense but which are empirically equivalent to no theory that is separable in the second sense. I offer some speculations about the significance of these and a few related results.

Bell's Theorem and the associated empirical tests of the Bell Inequalities are widely considered to reveal some strikingly non-classical features of our world. In the interest of trying to come to a fuller understanding of these features, I wish to propose some suggestions for how we might characterize the "separability" (or lack thereof) of systems in entangled states. For this purpose, I offer the following brief summary of the Bell milieu.[1,2]

J.P. Jarrett
Department of Philosophy, University of Illinois at Chicago, IL, USA
e-mail: jarrett@uic.edu

[1] I find it congenial to address these issues in a framework consisting of an idealization of actual Bell-type experimental apparatuses, an idealization devised by N. David Mermin. I have dubbed Mermin's idealized experimental setup "the Mermin Contraption". It debuted (as simply "the device") in Mermin [1].

[2] This is intended as a handy reference for what is to follow. While most of this notation has become standard in the Bell literature, some of it (and some of my nomenclature) is rather idiosyncratic. See Jarrett [2] for further discussion.

W.C. Myrvold and J. Christian (eds.), *Quantum Reality, Relativistic Causality, and Closing the Epistemic Circle,* The Western Ontario Series in Philosophy of Science 73,
© Springer Science+Business Media B.V. 2009

The Mermin Contraption

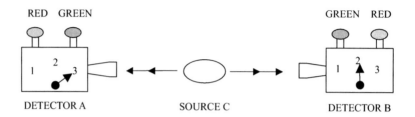

COMPOSITE SYSTEM STATES: $\lambda \in \Lambda$
DETECTOR SETTINGS: 1, 2, 3
MEASUREMENT OUTCOMES: GREEN (+1), RED (−1)

Joint Probability Functions

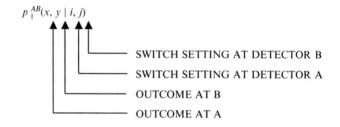

Determinism

$$\forall \lambda, x, y, i, j,$$
$$p_\lambda^{AB}(x,y|i,j) \in \{0,1\}$$

Theoretical Predictions

$$\forall \lambda, x, y, i, j,$$
$$P^{AB}(x,y|i,j) = \int_{\lambda \in \Lambda} \rho(\lambda) p_\lambda^{AB}(x,y|i,j) d\lambda,$$

where $\rho(\lambda)$ is the distribution of states.

The Mermin Contraption consists of three devices A, B and C. A and B are detectors of some sort, situated at diametrically opposed positions with respect to a

On the Separability of Physical Systems 107

source of particles, C. We make only some very limited assumptions regarding the nature of these devices and particles. Each detector can measure any one of three different things; the measurement outcomes consist of the illumination of either a red light or a green light at the detector. Distinct measurement-types correspond to distinct switch settings at a detector. Switch setting k at detector A corresponds to a measurement of the same type as switch setting k at detector B. A candidate theory for these experiments has to specify a set of states, Λ, for the two-particle system and must also provide an algorithm for computing joint probability functions of the form $p_\lambda(x, y \mid i, j)$. Here, λ is the state of the two particle system, x and y are the outcomes at site A and B, respectively; and i and j are the switch settings at site A and B, respectively. (A theory that satisfies DETERMINISM posits only states whose associated probability functions are restricted to $\{0,1\}$.) The testable predictions of the theory, then, are generated by summing these joint probability functions over all states, using a weighting function, $\rho(\lambda)$, the distribution of states, which the theory also must specify.

Marginal Probability Functions

$$(A) \ p_\lambda^A(x \mid i, j) \equiv p_\lambda^{AB}(x, +1 \mid i, j) + p_\lambda^{AB}(x, -1 \mid i, j)$$
$$(B) \ p_\lambda^B(y \mid i, j) \equiv p_\lambda^{AB}(+1, y \mid i, j) + p_\lambda^{AB}(-1, y \mid i, j)$$

Locality

$$\forall \lambda, x, i, i', j, j',$$
$$(A) \ p_\lambda^A(x \mid i, j) = p_\lambda^A(x \mid i, j')$$
$$(B) \ p_\lambda^B(y \mid i, j) = p_\lambda^B(y \mid i', j)$$

In terms of the joint probability functions, we can define marginal probabilities for A and B. I call *LOCALITY* the constraint that each of the marginals is independent of the switch setting at the distant site.[3] Relativity theory prohibits the superluminal transport of ma tter and energy. This relativistic "locality" constraint is by no means strictly equivalent to the foregoing LOCALITY. However, any theory that

---

[3] Abner Shimony prefers to call my LOCALITY condition *parameter independence*, and for good reason. This is a requirement that the probabilities at each detector site be independent of the switch setting (the "parameter") at the other detector site. "Locality" has been defined in various ways and taken to mean many different things; "parameter independence" however, is not only very descriptive, but also (in contrast to "locality") an expression whose meaning is univocal. For better or worse, I've chosen to retain this terminology, mainly out of deference to Einstein, Podolsky, and Rosen. I think that this constraint, while certainly not in any strict sense equivalent to the relativistic prohibition against superluminal signaling, does function in the Bell arguments in a manner that is very similar in spirit to the way that "locality," in the sense of EPR, was employed in their argument. Similar remarks apply to my use of the term "completeness."

108   J.P. Jarrett

violates LOCALITY faces at least a difficult reconciliation with relativity. (Note that quantum mechanics itself does satisfy LOCALITY.) Modulo a few related caveats, LOCALITY is a sufficient constraint from which to derive a Bell Inequality for theories that satisfy DETERMINISM. The results of actual experiments (in good agreement with the predictions of quantum mechanics) violate the Bell Inequality. Consequently, since relativity theory is so well confirmed independently, the experimental results are widely (though not universally) taken to provide rather direct evidence against DETERMINISM.[4]

Completeness

$$\forall \lambda, x, y, i, j,$$
$$p_\lambda^{AB}(x, y | i, j) = p_\lambda^A(x | i, j) \, p_\lambda^B(y | i, j)$$

or, equivalently,

$$(A) \; p_\lambda^A(x | i, j, y) \equiv \frac{p_\lambda^{AB}(x, y | i, j)}{p_\lambda^B(y | i, j)}, \, (p_\lambda^B(y | i, j) \neq 0)$$
$$= p_\lambda^A(x | i, j)$$
$$(B) \; p_\lambda^B(y | i, j, x) \equiv \frac{p_\lambda^{AB}(x, y | i, j)}{p_\lambda^A(x | i, j)}, \, (p_\lambda^A(x | i, j) \neq 0)$$
$$= p_\lambda^B(y | i, j)$$

Condition C

$$\forall \lambda, i, j, \; \det C(\lambda | i, j) = 0,$$
$$\text{where } C(\lambda | i, j) = \begin{bmatrix} p_\lambda^{AB}(+1, +1 | i, j) & p_\lambda^{AB}(+1, -1 | i, j) \\ p_\lambda^{AB}(-1, +1 | i, j) & p_\lambda^{AB}(-1, -1 | i, j) \end{bmatrix}$$

CONDITION C ⇔ COMPLETENESS

I call *COMPLETENESS* the requirement that for any specified state of the two-particle system and for any pair of measurement-types at the two detectors, the conditional probability of a specified outcome at either site given a specified outcome at the other site is just that (unconditioned) probability for getting the outcome at the first site.[5] To understand the sense in which this can be seen to be a kind of

---

[4] Most prominent among the dissenters are the advocates of a Bohmian mechanics. The Bohmians reject LOCALITY as a suitable constraint on physical theories.

[5] Shimony has called my COMPLETENESS constraint *outcome independence*, for it expresses the requirement that the probabilities at each detector site be independent of the outcome at the other detector site.

On the Separability of Physical Systems 109

completeness condition, consider the following: an independence constraint of this sort is precisely that which should be satisfied if the information associated with the outcome at A(B) is redundant with respect to the probability assignments at B(A). Because of this redundancy, conditionalization on the outcome at A(B) cannot alter the probabilities assigned at B(A). The conditional probability of such and such at B(A) given the outcome at A(B) is the same as the unconditioned probability at B(A) because the (complete) state description already includes that information; there is nothing more that one can learn regarding what one might find at B(A) based on what one finds at A(B).

One might think of it this way: Our two particles of a given pair—call them "A" and "B"—interact with each other at the source. As a result of this interaction, B might leave its "particle footprint" or some of its "particle DNA" or whatever on A. An observation of A, then, could conceivably reveal some trace of information about B's properties, so that conditionalization on the measurement outcome at A could permit us to revise the probability for getting a specified outcome at B. However, it is precisely all such information that we suppose to be included antecedently in a genuinely complete state description. Thus, if our state description *is* complete, there should be no correlation between such pairs of measurement outcomes, and we would expect the joint probability functions to reduce to the product of the two marginals. For this reason, in some of the literature on Bell's Theorem, COMPLETENESS also gets called "factorizability", for it constrains the joint probabilities to factor into this product of the two marginals.[6]

COMPLETENESS demands that states posited by the theory afford a representation of phenomena in terms of the interactions of entities possessing physical properties in a suitably independent manner. State descriptions that are "complete" in this sense encode information in a manner that insures the "screening off" of any correlations between sets of measurement outcomes at the two wings of the Mermin Contraption. Any theory (whether deterministic or not) that posits entities that possess properties (whatever those properties may be) in the ways familiar to us from our experience at the classical level might plausibly be expected to satisfy this constraint. It has the look of nothing more than a requirement that state descriptions include all causally relevant information. (Note that quantum mechanics violates COMPLETENESS.)

CONDITION C deserves mention here, if only as a curiosity. I will address its possible significance later. The elements of the matrix $C(\lambda|i,j)$ are the joint probabilities assigned by λ to the four possible pairs of measurement outcomes, for specified switch settings i and j. CONDITION C is the requirement that the determinant of C vanish for all λ, i, and j. As can easily be verified by substitution of the relevant expressions for the marginal probabilities (and by applying the standard normalization constraint, according to which the probabilities for all possible pairs of outcomes sum to unity), COMPLETENESS and CONDITION C are logically equivalent.[7]

---

[6] The term "factorizability" was introduced by Arthur Fine, long-time nemesis of its referent.

[7] It is useful to appeal to this equivalence between COMPLETENESS and CONDITION C in order to see that theories that satisfy DETERMINISM also satisfy COMPLETENESS, while the

Strong Locality

$$\forall \lambda, x, y, i, j, i', j',$$

$$p_\lambda^{AB}(x,y|i,j) = p_\lambda^A(x|i,j') \, p_\lambda^B(y|i',j)$$

$$\text{STRONG LOCALITY} \Longleftrightarrow \left\{ \begin{array}{c} \text{LOCALITY} \\ \& \\ \text{COMPLETENESS} \end{array} \right.$$

$$\Updownarrow$$

GENERALIZED BELL INEQUALITIES

Finally, STRONG LOCALITY is logically equivalent to the conjunction of LOCALITY and COMPLETENESS. Theories that satisfy STRONG LOCALITY represent the composite system as a pair of component systems that are mutually independent in both the sense of LOCALITY and that of COMPLETENESS. Bell's Theorem and the related experimental results provide good evidence that no theory satisfying STRONG LOCALITY is empirically adequate, for it is this constraint from which the generalized Bell Inequalities are derived. These generalized Bell Inequalities, too, are violated by empirical data that are in excellent accord with the predictions of quantum mechanics. By a broad consensus, COMPLETENESS is assumed to be the culprit in all of this. However, it is crucial to bear in mind that theories that violate COMPLETENESS need not *ipso facto* have omitted anything whatsoever from their representation of physical states of affairs. That COMPLETENESS is satisfied by no acceptable empirically adequate theory cannot be regarded as some eliminable artifact of our currently best theory; rather, it appears to be a reflection of some highly non-classical feature of the world itself. Quantum mechanics appears to hook on to the world in some fashion that accurately mirrors this feature. In this respect, one might argue that Einstein, Podolsky, and Rosen (EPR) were quite right in concluding that quantum mechanics is "incomplete" in some important sense. Nevertheless, Bohr also appears to have been right, at least to the extent that "incompleteness" in this sense is demonstrably not a defect in a theory.[8] If (as I am assuming here) our theory must satisfy LOCALITY, and if that theory is to be empirically adequate, then it ***must*** violate COMPLETENESS. Again, this is ***not*** to say that any empirically adequate theory that satisfies LOCALITY fails to incorporate into its state descriptions all causally relevant factors; but it ***is*** to say that even if the state descriptions of such a theory ***do*** include that information (and, so, might still be called "complete" by that measure), those state descriptions nevertheless fail to screen off Bell-type correlations. This situation constitutes a much more radical departure from the classical worldview than is required by the mere rejection of DETERMINISM.

---

converse does not hold in general. If a theory satisfies DETERMINISM, then normalization demands that three of the four probabilities in matrix C have the value 0, while the other has the value 1. Thus, the determinant of C vanishes. However, it is easy to construct a 2×2 matrix of numbers that are in the interval $[0,1)$ and which sum to 1 for which the determinant still vanishes.

[8] The classic papers here are Einstein, Podolsky, and Rosen [3] and Bohr [4].

On the Separability of Physical Systems   111

The results of Bell-type experiments may be seen to impose constraints on our conception of causality and the strictures of relativity. COMPLETENESS itself may be regarded as a locality condition of a special sort, one that is closely related to the notion of "separability" (of the two component systems in Bell-type experiments). Below, I will introduce a notion of "separability" that turns out to be logically equivalent to COMPLETENESS in the previously described sense. However, as a candidate for a criterion of the individuation of physical systems, this is a comparatively weak separability constraint, certainly one that is weaker than any that would have drawn the support of EPR. Consequently, the mere incorporation into one's worldview of the "causal holism" implicit in the rejection of what I shall call *EPR-SEPARABILITY* is not by itself sufficient to move us into the realm of viable theories. Put in terms of how far we are compelled to retreat from the worldview of classical physics, giving up DETERMINISM is not enough; giving up EPR-SEPARABILITY is not enough; COMPLETENESS (this even weaker form of separability) must be abandoned as well.

If we accept the popular view that *the* lesson (or at least *one* of the lessons) that we must learn from Bell's Theorem is that apart from the empirical adequacy of quantum mechanics, apart from its future successes or ultimate demise, any theory that supersedes quantum mechanics is also going to have to violate COMPLETENESS,[9] then we must ask just what precisely it *means* to say of a theory that it is "incomplete". In particular, is there anything *more* to say about such theories beyond that which has already been said?[10]

Subsequent to the EPR paper, Einstein offered his own argument for the incompleteness of quantum mechanics.[11] Guided by considerations raised therein, Don Howard has suggested that we understand "separability" as a requirement that each of the putative subsystems possess its own distinct physical state in such a way that they fix the state of the composite system in precisely the manner specified by COMPLETENESS. For specified values of i and j, (marginal at A) represents the state of the A component system, (marginal at B) represents the state of the B component system, and (joint AB probability) represents the state of the composite system. COMPLETENESS is the requirement that the joint probability associated with λ, i, and j be fixed by the product of the two marginals. Thus, Howard recommends that COMPLETENESS be construed simply *as* separability.[12]

---

[9] I trust that even the Bohmians would allow that it is at least of some interest to consider what follows from this supposition.

[10] Whether or not Bell-type correlations should be deemed in need of further "explanation", so as to render them less "mysterious" is a question that impinges on some of the most fundamental issues in the philosophy of science. I do have views about the issues that bear on this question, but I will refrain from unleashing them here. An extremely stimulating discussion of these matters may be found in Cushing and McMullin [5]. (See, especially, Arthur Fine [6], Bas van Fraassen [7], and R. I. G. Hughes [8].)

[11] Historical scholarship suggests the possibility that Einstein did not even *see* the EPR paper until it appeared in print (see Fine [9], pp. 35–36), and that in any case he was not altogether pleased with the finished product. Einstein's own "incompleteness" argument first appeared in [10] and was later more fully elaborated in [11].

[12] See Howard [12] and [13].

Before acceding to Howard's recommendation, there are two important questions that need to be considered:

(i) Are probability functions **alone** properly regarded as providing an exhaustive characterization of physical states?

(ii) Even if physical states are to be understood in this manner, why ought a constraint deserving of the name "separability" *require* that the state of the composite system be determined by the states of the two component systems in the particular manner specified by COMPLETENESS? More specifically, why couldn't it be determined by those states in some other way and still, then, be considered as expressing an acceptable "separability" constraint?[13]

I will begin with question (ii). One way to formalize the talk of the "information" contained in state descriptions is in terms the Shannon information. This suggests the following definition:

Separability

$$\forall \lambda, i, j,$$
$$I\left(p_{\lambda}^{AB} | i, j\right) = I\left(p_{\lambda}^{A} | i, j\right) + I\left(p_{\lambda}^{B} | i, j\right),$$

where $I(p) \equiv \sum p \ln p$ is the Shannon information associated with probability function p, and the sum is over all values of the outcome variable(s) in p.

SEPARABILITY is just the requirement that the Shannon information associated with the joint probability function (representing the state of the composite system) equal the sum of the Shannon information associated with the marginal probability function at A (representing the state of the A component system)[14] and the Shannon information associated with the marginal probability function at B (representing the state of the B component system). This constraint, then, has a natural interpretation as one way, of perhaps many ways, one might choose to make precise the *cliché* that the whole is equal to the sum of the parts, and thereby serves as *a* "separability" constraint.

Interpreted in this manner, SEPARABILITY presupposes that states be construed as probability functions. Quantum mechanical states are in one-to-one correspondence with probability measures over the set of closed linear subspaces of a Hilbert space. So quantum mechanics is one example of a theory that effectively identifies states with probability functions. Classical mechanics can be regarded as a theory of that type as well, but in classical mechanics, we can also represent states as points

---

[13] This latter question is one that has been raised explicitly by Erik Winsberg and Arthur Fine, about whom more anon. See Winsberg and Fine [14].

[14] Note that $I\left(p^{A} | i, j\right)$ is a function not only of λ, but also of i and j. So this quantity cannot be interpreted as the information associated merely with λ. This "contextual" character of SEPARABILITY attaches to COMPLETENESS and CONDITION C as well. I leave as a project for future research the investigation of relationships that hold among suitable counterparts to these three constraints, where the counterparts are characterized in terms of λ alone.

in a phase space, connecting states with a set of exact values of the position and momentum of each particle in the system. Suppose we were to try to generalize this: consider theories for which the states are given by a specification of exact values of some distinguished set of physical magnitudes. We might even choose to call these quantities "elements of physical reality", or "EPRs".[15] These EPRs need not be restricted to properties like the values of position and momentum, nor need the EPRs be restricted to properties in the ordinary sense at all. They might include, for example, irreducibly dispositional quantities, represented probabilistically as physical propensities. How then might we think about characterizing our "separability" constraint? This leads us back to question (i), and suggests something of the following sort:

EPR-Separability

A theory will be said to be *EPR-SEPARABLE* (or to satisfy *EPR-SEPARABILITY*) just in case it satisfies each of the following constraints:

(1) The theory distinguishes a collection of physical magnitudes, $\{\sigma_1, \sigma_2, \ldots, \sigma_K\}$, as the "elements of physical reality" or "EPRs", exact values of which uniquely specify the state. For all $\lambda \in \Lambda$, $\lambda = \lambda(<\sigma_1, \sigma_2, \ldots, \sigma_K>)$.

(2) For each state $\lambda \in \Lambda$, there exist unique subsystem states, λ_A and λ_B, such that $\lambda_A = \lambda_A(<\sigma_1^A, \sigma_2^A, \ldots, \sigma_K^A>)$, $\lambda_B = \lambda_B(<\sigma_1^B, \sigma_2^B, \ldots, \sigma_K^B>)$, and for each $k = 1, 2, \ldots, K$, $\sigma_k = \sigma_k(<\sigma_k^A, \sigma_k^B>)$.

(3) There exists a function $p_\omega(z|k_1, k_2)$ such that $p_\lambda^A(x|i, j) = p_{\lambda_A}(x|i, j)$ and $p_\lambda^B(y|i, j) = p_{\lambda_B}(y|j, i)$.

(4) $\overline{I}(\lambda|i, j) = \overline{I}(\lambda_A|i, j) + \overline{I}(\lambda_B|j, i)$, where

$$\overline{I}(\lambda|i, j) \equiv I(p_\lambda^{AB}|i, j) \equiv \sum_{x, y} p_\lambda^{AB}(x, y|i, j) \ln p_\lambda^{AB}(x, y|i, j)$$

and $\overline{I}(\lambda_\Gamma|k_1, k_2) \equiv \sum_z p_{\lambda_\Gamma}(z|k_m, k_{m'}) \ln p_{\lambda_\Gamma}(z|k_m, k_{m'})$, where

$$m = \begin{cases} 1, \Gamma = A \\ 2, \Gamma = B \end{cases} \quad \text{and} \quad m' = \begin{cases} 2, \Gamma = A \\ 1, \Gamma = B \end{cases}$$

(5) $p_\lambda^{AB}(x, y|i, j) = p_{\overline{\lambda}}^{AB}(y, x|j, i)$, where $\overline{\lambda} = \lambda(<\overline{\sigma}_1, \overline{\sigma}_2, \ldots, \overline{\sigma}_n>)$ and for each $k = 1, 2, \ldots, K$, $\overline{\sigma}_k = \sigma_k(<\sigma_k^B, \sigma_k^A>)$.

(6) $\rho(\lambda) = \phi_A(\lambda_A)\phi_B(\lambda_B)$, where $\phi_A = \phi_B \equiv \phi$.

(7) $\Lambda = \Lambda_A \times \Lambda_B$, where $\Lambda_A = \Lambda_B \equiv \Omega$.

(8) $P^{AB}(x, y|i, j) \equiv \int_\Lambda \rho(\lambda) p_\lambda^{AB}(x, y|i, j) d\lambda$ and $P^\Gamma(z|k_1, k_2) \equiv \int_{\Lambda_\Gamma} \phi_\Gamma(\lambda_\Gamma) p_{\lambda_\Gamma}(z|k_m, k_{m'}) d\lambda_\Gamma$, where m and m' are defined as in clause (4) above.

---

[15] Every ray of a Hilbert space, and so, every pure state of quantum mechanics, corresponds uniquely to some ("complete", in the quantum-mechanical sense) set of eigenvalues of a maximal set of commuting Hermitian operators. However, no one such set of Hermitian operators will serve for all pure states. So the states posited by quantum mechanics defy representation in terms of EPRs.

Clauses (1) and (2) of EPR-SEPARABILITY require that each state, for both the composite and the component systems, is fixed by appropriate sets of EPRs as indicated, so that each composite system state λ is decomposable into a pair of component system states, λ_A, to be associated with the A subsystem, and λ_B, to be associated with the B subsystem. Clause (3) requires that the probability functions determined by the subsystem states λ_A and λ_B, respectively, be equal to the corresponding marginal probability functions, as is needed for consistency with the definition of the joint probability functions associated with the composite system states. It also requires that each EPR associated with λ_A "play the same role" as the corresponding EPR associated with λ_B in generating their respective probability functions. Clause (4) is my proposal for imposing on this picture what I consider to be a central feature of any acceptable separability constraint. It is the requirement that the Shannon information associated with the probability function determined by the composite system state description be equal to the sum of the two corresponding quantities determined by the subsystem states. *The whole is equal to the sum of the parts.* SEPARABILITY *simpliciter* makes the analogous demand without reference to any EPR-conception of physical states.[16]

From clause (4), EPR-SEPARABILITY entails plain SEPARABILITY, while from clauses (1) and (2), the converse does not hold. The point here is simply that a theory may satisfy SEPARABILITY without positing anything whatsoever regarding "elements of physical reality" that provide a decomposition of a λ into a λ A and a λ B. (See the figure below.) If one prefers to think about this from the other direction, we might define HOLISM simply as the denial of SEPARABILITY, and EPR-HOLISM as the denial of EPR-SEPARABILITY.

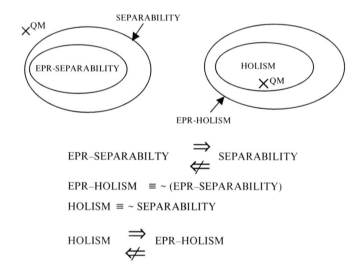

---

[16] Consequently, I presume that of those attracted to an information-theoretic characterization of "separability", those with a realist bent would favor something akin to EPR-SEPARABILITY, while those situated more toward the anti-realist end of the spectrum would prefer SEPARABILITY.

If we proceed, at least for the moment, under the assumption that we must reject SEPARABILITY, then this has consequences that are not as widely appreciated as I believe they ought to be. I will elaborate on this shortly. Returning to the characterization of EPR-SEPARABILITY, clauses (5), (6), and (7) impose some symmetry conditions on the two component systems based on the identity of the constituent entities, while clause (8) gives the standard prescription for generating testable predictions by summing over all states.[17]

If COMPLETENESS is to be abandoned, it should be of some interest to consider the relationship between this constraint and SEPARABILITY. As it happens, the two are logically equivalent.[18] This gives us the following chain of equivalences:[19]

SEPARABILITY

$\Updownarrow$

$(*) \forall \lambda, i, j,$

$$c_{11}{}^{c_{11}} c_{12}{}^{c_{12}} c_{21}{}^{c_{21}} c_{22}{}^{c_{22}} = (c_{11} - \Delta)^{c_{11}} (c_{12} + \Delta)^{c_{12}} (c_{21} + \Delta)^{c_{21}} (c_{22} - \Delta)^{c_{22}},$$

where $\Delta \equiv \det C(\lambda|i,j) = c_{11}c_{22} - c_{12}c_{21}$

$\Updownarrow$

$\forall \lambda, i, j, \Delta = 0$

$\Updownarrow$

CONDITION C

$\Updownarrow$

COMPLETENESS

Thus, in rejecting COMPLETENESS, we give up not just EPR-SEPARABILITY, but also that gap between EPR-SEPARABILITY and (the logically weaker) SEPARABILITY. Depending on one's views about the proper way to characterize states, one might say that the implications of Bell's Theorem and the experimental tests of the Bell Inequalities are even more drastic than previously thought. Our retreat

---

[17] For the sake of simplicity, I forego the readily available revisions needed to accommodate non-identical particles.

[18] This is an immediate consequence of the Gibbs Inequality (sometimes called "the Shannon-Gibbs Inequality". See Ash [15], pp. 16–19, for a proof. The Gibbs Inequality is a special case of a more general result that applies to all convex functions. For this more general result, see Billingsley [16], pp. 544–545. I am grateful to David Malament for calling my attention to the latter.

[19] I will have some things to say presently about the constraint denoted by the asterisk. For now, note that it may be seen to arise as follows: express SEPARABILITY in terms of the elements of the matrix $C(\lambda|i,j)$, and exponentiate both sides of the resulting equation. (*) emerges after a further bit of algebraic manipulation. That CONDITION C (and, so, COMPLETENESS) implies SEPARABILITY is now trivial. (The implication in the other direction is not.) For a generalization of each of these constraints, see Appendix 2.

from the classical worldview is even more extreme than perhaps is generally recognized. I wish to emphasize this point because the conventional wisdom has been that the deep meaning of Bell's Theorem and related experimental work is that not only deterministic theories, but even indeterministic theories that are not suitably "holistic", must be given up once and for all; and one *might* have thought that the holism in question, that associated with the "passion-at-a-distance" exhibited by the entangled states of systems in Bell-type experiments, in violation of COMPLETENESS, is really no more than that which follows from rejecting an EPR-style assumption of separability. In particular, one might have thought that every theory that satisfies COMPLETENESS (or, equivalently, SEPARABILITY), even those that do *not* posit states that are represented in terms of EPRs, is empirically equivalent to some such theory. If this were so, that logical gap between EPR-SEPARABILITY and SEPARABILITY *simpliciter* would be empirically vacuous, and most (certainly those who incline toward any form of anti-realism) would decline to acknowledge any meaningful distinction between the two classes of theories. However, it can be shown that there do exist COMPLETE (or, if one prefers, SEPARABLE) theories that are empirically equivalent to *no* EPR-SEPARABLE theory. (See Appendix 1.)

Consider that from a given composite-system state λ, COMPLETENESS alone determines neither λ_A nor λ_B. While it does afford a factorization of the joint probabilities into a product of the marginals, it remains utterly indifferent with regard to the manner of specification of the separate states themselves. Unless one simply rejects the EPR conception of state by *fiat* (in which case the marginal probabilities may be identified with the subsystem states), EPR-SEPARABILITY appears as a reasonable demand; and the constraint that states be represented in the form codified via that demand induces an *empirical* restriction that (merely) SEPARABLE theories need not meet.

I think it helpful to note that SEPARABILITY (and, so, COMPLETENESS) requires only that the *quantity* of information associated with the probability function determined by the composite system state description equal the sum of the corresponding *quantities* associated with the two marginal probability functions. But given this EPR conception of physical states, that requirement can well be met, even though the chunks of information appearing on the two sides of the equation that defines SEPARABILITY are *qualitatively* distinct.[20] Moreover, such qualitative distinctions can have empirical consequences. It is striking that empirical tests of the Bell Inequalities appear to demand that we reject even the weaker constraint (SEPARABILITY), the one having to do with information in this merely quantitative sense. This is what I meant when I suggested previously that Bell's Theorem and the associated experimental results force us to a more radical retreat from the classical world view than is perhaps generally recognized. I take EPR-SEPARABILITY to be a natural generalization of the classical worldview in the spirit of Einstein, Podolsky, and Rosen. It jettisons determinism as a requirement, but it retains the fundamental role of elements of physical reality. And yet, we see that the holism we

---

[20] For this insight I am indebted to Brandon Fogel, who suggested something similar to it in conversation.

On the Separability of Physical Systems   117

are compelled to embrace is even more extreme than that which follows from the mere rejection of this EPR-SEPARABILITY.

A few years ago, Winsberg and Fine proposed that we take as a suitable "separability" constraint, that the joint probability function (associated with the composite system) simply be some function of the marginals.[21] That it be the *product* of the marginals would then be a simple special case of that proposal; but to insist that the function in question take that *particular* form would be overly restrictive, on their view. However, it can now be seen that if one agrees to adopt the information theoretic conception of "separability", and if one chooses to characterize states as probability functions (**whether or not** those probability functions are taken to be the manifestations of some set of more fundamental elements of physical reality), then the "choice" of the product function is forced upon us.

One further point about the equivalence of SEPARABILITY and COMPLETENESS: We can define *mutual information* as indicated here:

Mutual Information

$\forall \lambda, i, j,$

$$I[X;Y]_{\lambda,i,j} \equiv \sum_{\substack{x=\pm 1 \\ y=\pm 1}} p_\lambda^{AB}(x,y|i,j) \ln \left[\frac{p_\lambda^{AB}(x,y|i,j)}{p_\lambda^A(x|i,j)\, p_\lambda^B(y|i,j)} \right] \equiv I[Y;X]_{\lambda,i,j}$$

$$= \sum_{\substack{x=\pm 1 \\ y=\pm 1}} p_\lambda^{AB}(x,y|i,j) \ln \left[\frac{p_\lambda^A(x|i,j,y)}{p_\lambda^A(x|i,j)} \right]$$

$$= \sum_{\substack{x=\pm 1 \\ y=\pm 1}} p_\lambda^{AB}(x,y|i,j) \ln \left[\frac{p_\lambda^B(y|i,j,x)}{p_\lambda^B(y|i,j)} \right],$$

$$\text{(for } p_\lambda^A(x|i,j)\, p_\lambda^B(y|i,j) \neq 0)$$

NOTE: COMPLETENESS $\Rightarrow \forall \lambda, i, j, I[X;Y]_{\lambda,i,j} = I[Y;X]_{\lambda,i,j} = 0$

That mutual information is a symmetric function is apparent from the definition. The conditional probabilities appearing in the definition reduce to the unconditioned probabilities if and only if COMPLETENESS is satisfied, so COMPLETENESS holds just in case the mutual information is zero. That means, in turn, that (for a specified pair of detector switch settings) the information in the composite system state description is such that conditionalization on the outcome at one measurement site has no bearing on the probabilities assigned at the other site, the former merely providing redundant information with respect to such a "complete" state description.

---

[21] See Winsberg and Fine [14]. These authors challenge the views developed by Howard. For a trenchant analysis of these issues, see Fogel [17].

118   J.P. Jarrett

I note this in an effort to add reinforcement to the earlier interpretation I suggested
for COMPLETENESS, where I also used the word "information", but there in an
informal, non-technical sense.

Consider now the following constraint:[22]

Condition S

$$\forall \lambda, i, j,$$
$$c_{11}^{c_{11}} c_{12}^{c_{12}} c_{21}^{c_{21}} c_{22}^{c_{22}} = (c_{11} - \Delta)^{c_{11}} (c_{12} + \Delta)^{c_{12}} (c_{21} + \Delta)^{c_{21}} (c_{22} - \Delta)^{c_{22}},$$

where $\Delta \equiv \det C(\lambda | i, j)$.

Suppose an imaginary someone who has a reasonably good familiarity with the
philosophical literature on Bell's Theorem, but who is blissfully unaware of any attempts toward an information-theoretic construal of "separability". What might be
this person's initial response upon being told that COMPLETENESS and SEPARA-
BILITY are logically equivalent? I daresay it might be something such as this, "So, I
see that someone has succeeded in expressing COMPLETENESS in a grossly more
complicated and obscure way." This might be followed by a sarcastic "(S)he has
my congratulations." However, if our imaginary friend were to reflect for a time and
play around with CONDITION S sufficiently, the response might continue in this
manner, "Wait a minute, that... that looks like entropy, I mean that's a constraint
on information; hey, that's a requirement that the information associated with the
probability function representing the composite system is equal to the sum... Could
we interpret that as a 'separability' condition of some sort?" Our friend just might
discern this link between the two conceptually distinct classes of theories: the class
of theories that satisfy COMPLETENESS on the one hand, and the class of theories
that satisfy SEPARABILITY on the other. That those two sets are coextensive might
be deemed to be of genuine significance.

This exercise of the imagination is intended to set the stage for a little speculation.
Might something similar to that which our friend achieved with CONDITION S be
done for CONDITION C as well? Of course, there is no reason why that would have
to be so; CONDITION C might be nothing more than a way of expressing COM-
PLETENESS (or, equivalently, SEPARABILITY) in a different mathematical form;
it might carry with it no suitably independent physical interpretation that would
help us to enrich our understanding of COMPLETENESS and SEPARABILITY.
However, the fact that CONDITION C is posed as the vanishing of a determinant
is very suggestive, even tantalizing; for this allows us to recognize CONDITION C
as the necessary and sufficient condition that there exist a non-trivial solution to a
particular set of equations.

---

[22] The alert reader will immediately recognize CONDITION S as the aforementioned asterisked
constraint.

On the Separability of Physical Systems 119

CONDITION C
⇕

$\forall \lambda, i, j$, the following pair of equations has a non-trivial solution:

$$z_1 p_\lambda^{AB}(+1,+1|i,j) + z_2 p_\lambda^{AB}(+1,-1|i,j) = 0$$
$$z_1 p_\lambda^{AB}(-1,+1|i,j) + z_2 p_\lambda^{AB}(-1,-1|i,j) = 0$$

[equivalently, $\forall \lambda, i, j$, the following pair of equations has a non-trivial solution:

$$z_1 p_\lambda^{AB}(+1,+1|i,j) + z_2 p_\lambda^{AB}(-1,+1|i,j) = 0$$
$$z_1 p_\lambda^{AB}(+1,-1|i,j) + z_2 p_\lambda^{AB}(-1,-1|i,j) = 0$$

Alternatively,

$$\text{CONDITION C} \iff \begin{cases} \forall \lambda, i, j, \text{ there exists a zero} - \text{eigenvalue} \\ \text{solution to the equation} \\ Cz = \gamma z \end{cases}$$

We can work out the solutions to these equations and put them in various forms and try to interpret them.[23] We can also note that CONDITION C asserts the existence of a zero-eigenvalue solution to a certain matrix equation. Does that matrix equation have an independent physical significance? What would it mean to demand of a theory that the corresponding matrix equation of that sort have *any* solution, let alone that it have a zero-eigenvalue solution?[24]

Success in the search for an independent physical interpretation of CONDITION C might yet deliver to us at least some small deepening of our understanding of COMPLETENESS and SEPARABILITY, and thereby help us to achieve a firmer grasp of precisely what it is that Bell's Theorem has to teach us about the way the world is not.

Acknowledgements I have had the great good fortune to discuss many of the ideas presented in this paper with generations of colleagues and students, my mentors and friends. It is no longer possible for me to give an accurate accounting of the many valuable contributions each has made to this work. However, I owe special thanks to Thomas Ferguson and Brandon Fogel, each of whom

---

[23] However these solutions are expressed, they suggest a symmetry condition of some sort (as does CONDITION C itself).

[24] For the quantum-mechanical case of two spin-1/2 particles in the singlet state, we have $c_{11} = c_{22} = (1/2) \sin^2(\theta/2)$, and $c_{12} = c_{21} = (1/2) \cos^2(\theta/2)$, where θ is the angle between the axes of alignment of the Stern-Gerlach magnets at A and B. This yields solutions to $Cz = \gamma z$ for eigenvalues $\gamma^\pm = (1/2) \sin^2(\theta/2)$, and $c_{12} = c_{21} = (1/2) [\sin^2(\theta/2) \pm \cos^2(\theta/2)]$. The eigenvectors associated with the γ^- eigenvalue have a curious similarity in form to those associated with the $\gamma^- = 0$ eigenvalue solutions to $Cz = \gamma z$ for the states posited by theories that satisfy CONDITION C (COMPLETENESS, SEPARABILITY). Perhaps this similarity in form will permit us to interpret CONDITION C as a symmetry constraint that quantum mechanics itself satisfies in some less restrictive, more general form. Perhaps.

120   J.P. Jarrett

has helped me to clarify my thinking in crucial ways. The value of the assistance rendered by Thomas Ferguson in working through all of the details and in preparing the manuscript has been inestimable.

Appendix 1

CLAIM: There exist SEPARABLE theories that are empirically equivalent to no EPR-SEPARABLE theory.

To establish the CLAIM, it will be helpful first to show the following:

$$(^{**})\ \text{EPR-SEPARABILITY} \implies \begin{cases} \forall x, y, i, j, \\ P^{AB}(x, y|i, j) = P^{AB}(y, x|j, i) \end{cases}$$

Note the following, straightforward consequence of EPR SEPARABILITY:

$$\forall \lambda, x, y, z, i, j,$$

(1) $p_\lambda^{AB}(x, y|i, j) = p_\lambda^A(x|i, j)p_\lambda^B(y|i, j)$

(2) $\overline{\Lambda} = \Lambda$ and $\rho(\lambda) = \rho(\overline{\lambda})$

(3) $P^{AB}(x, y|i, j) = P^A(x|i, j)P^B(y|i, j)$

(4) $P^A(z|i, j) = P^B(z|j, i)$

For arbitrary values of x, y, i, and j,[25]

$$P^{AB}(x, y|i, j) = P^A(x|i, j)P^B(y|i, j) \tag{3}$$

$$= \int_\Omega \phi(\omega)p_\omega(x|i, j)d\omega \int_\Omega \phi(\omega')p_{\omega'}(y|j, i)d\omega' \qquad <7>, <8>$$

$$= \int_{\Lambda_B} \phi_B(\lambda_B)p_{\lambda_B}(x|i, j)d\lambda_B \int_{\Lambda_A} \phi_A(\lambda_A)p_{\lambda_A}(y|j, i)d\lambda_A \quad <7>$$

$$= \int_{\Lambda_B \times \Lambda_A} \phi_B(\lambda_B)p_{\lambda_B}(x|i, j)\phi_A(\lambda_A)p_{\lambda_A}(y|j, i)d\lambda_B d\lambda_A$$

$$= \int_{\Lambda_B \times \Lambda_A} \phi_B(\lambda_B)\phi_A(\lambda_A)p_{\lambda_B}(x|i, j)p_{\lambda_A}(x|i, j)d\lambda_B d\lambda_A$$

$$= \int_{\overline{\Lambda}} \rho(\overline{\lambda})p_{\overline{\lambda}}^A(x|i, j)p_{\overline{\lambda}}^B(y|j, i)d\overline{\lambda} \qquad (2), <3>, <6>$$

---

[25] In what follows, numbers in angle brackets refer to the corresponding clauses in the definition of EPR-SEPARABILITY, while numbers in parentheses refer to the corresponding clauses of the aforementioned consequence of EPR-SEPARABILITY.

On the Separability of Physical Systems   121

$$= \int_{\bar{\Lambda}} \rho(\overline{\lambda}) p^{AB}_{\overline{\lambda}}(x, y | i, j) d\overline{\lambda} \quad (1)$$

$$= \int_{\Lambda} \rho(\lambda) p^{AB}_{\lambda}(x, y | i, j) d\lambda \quad (2)$$

$$= \int_{\Lambda} \rho(\lambda) p^{AB}_{\lambda}(y, x | j, i) d\lambda \quad <5>$$

$$= P^{AB}(y, x | j, i) \qquad\qquad <8>$$

This establishes (**).

Now let T be a theory that posits a set of states $\Lambda = \{\lambda_1, \lambda_2, \ldots, \lambda_m\}$ and uniform distribution function $\rho(\lambda_i) = \frac{1}{m}$ and suppose that for each $i = 1, 2, \ldots, m$, there are values α_i, β_i, γ_i such that for all k:

$$C(\lambda_i | k, k) = \frac{1}{(1 + \gamma_i)(\alpha_i + \beta_i)} \begin{bmatrix} \alpha_i & \beta_i \\ \gamma_i \alpha_i & \gamma_i \beta_i \end{bmatrix}, \text{ where } \beta_i > \gamma_i \alpha_i.$$

(Note: These states are normalized, and T satisfies CONDITION C.)

For T, then,

$$\forall k, P^{AB}(1, -1 | k, k) = \frac{1}{m} \sum_{i=1}^{m} \frac{\beta_i}{(1 + \gamma_i)(\alpha_i + \beta_i)}, \text{ and}$$

$$P^{AB}(-1, 1 | k, k) = \frac{1}{m} \sum_{i=1}^{m} \frac{\gamma_i \alpha_i}{(1 + \gamma_i)(\alpha_i + \beta_i)}$$

$$< \frac{1}{m} \sum_{i=1}^{m} \frac{\beta_i}{(1 + \gamma_i)(\alpha_i + \beta_i)} = P^{AB}(1, -1 | k, k)$$

Thus, by (**), T violates EPR-SEPARABILITY; but since T satisfies CONDI-TION C, it also satisfies SEPARABILITY. Moreover, since T requires that for all k, $P^{AB}(1, -1 | k, k) \neq P^{AB}(-1, 1 | k, k)$, this SEPARABLE theory makes predictions that are compatible with no EPR-SEPARABLE theory.

This establishes the CLAIM.

Appendix 2

COMPLETENESS, SEPARABILITY, CONDITION C, and equation (*) all admit of generalization to cases with arbitrary values of $s \geq 2$ and $n \geq 2$, where s is the number of component systems and n is the number of possible outcomes at each detector site. In the general case, we have the following:

OUTCOME VARIABLES: $x_1, x_2, \ldots, x_s$
POSSIBLE VALUES FOR VARIABLE x_k: $x_{k1}, x_{k2}, \ldots, x_{kn}$
DETECTOR SETTING VARIABLE FOR DETECTOR AT A_k: i_k

JOINT PROBABILITY FUNCTIONS:

$$p_\lambda^{A_1 A_2 \ldots A_s}(x_1, x_2, \ldots, x_s | i_1, i_2, \ldots, i_s)$$

MARGINALS:

$$p_\lambda^{A_K}(x_k | i_1, i_2, \ldots, i_s) \equiv \sum_{\substack{x_j \in X_j \\ j \neq k}} p_\lambda^{A_1 A_2 \ldots A_s}(x_1, x_2, \ldots, x_s | i_1, i_2, \ldots, i_s),$$

where $X_j \equiv \{x_{j1}, x_{j2}, \ldots, x_{js}\}$

and $\forall j, m, m', m < m' \Rightarrow x_{jm} \leq x_{jm'}$

COMPLETENESS

$\forall \lambda, x_j, i_j,$

$$p_\lambda^{A_1 A_2 \ldots A_s}(x_1, x_2, \ldots, x_s | i_1, i_2, \ldots, i_s) = \prod_{k=1}^{s} p_\lambda^{A_K}(x_k | i_1, i_2, \ldots, i_s)$$

SEPARABILITY

$\forall \lambda, i_j,$

$$I\left(p_\lambda^{A_1 A_2 \ldots A_s} | i_1, i_2, \ldots, i_s\right) = \sum_{k=1}^{s} I(p_\lambda^{A_k} | i_1, i_2, \ldots, i_s)$$

CONDITION C

For all $1 \leq j < m \leq s$, and for all $1 \leq j\prime < m\prime \leq s$,

$$\det \begin{bmatrix} c_{k_1 k_2 \ldots k_j \ldots k_m \ldots k_s} & c_{k_1 k_2 \ldots k_j \ldots k_{m'} \ldots k_s} \\ c_{k_1 k_2 \ldots k_{j'} \ldots k_m \ldots k_s} & c_{k_1 k_2 \ldots k_{j'} \ldots k_{m'} \ldots k_s} \end{bmatrix} = 0,$$

where $c_{k_1 k_2 \ldots k_s}(\lambda | i_1, i_2, \ldots, i_s) = p_\lambda^{A_1 A_2 \ldots A_s}(x_{1(k1)}, x_{2(k2)}, \ldots, x_{s(ks)} | i_1, i_2, \ldots, i_s).$

CONDITION Q

$$\forall \lambda, i_j, k_j,$$

$$q_{k_1 k_2 \ldots k_s}(\lambda | i_1, i_2, \ldots, i_s) = 0,$$

On the Separability of Physical Systems

where $q_{k_1 k_2 \ldots k_s}(\lambda | i_1, i_2, \ldots, i_s) =$

$$-\left\{ c_{k_1 k_2 \ldots k_s} \sum_{j=1}^{s-1} \left[\frac{(s-1)!(-1)^j}{j!(s-j-1)!} \left(\sum_{<k_1', k_2', \ldots, k_s'> \in M(k_1, k_2, \ldots, k_s)} c_{k_1' k_2' \ldots k_s'} \right)^j \right] \right.$$

$$\left. + \sum_{m=1}^{s} \left[(c_{k_1 k_2 \ldots k_s})^{s-m} \sum_{j_m > j_{m-1} > \ldots > j_1} \prod_{i=1}^{m} \left(\sum_{<k_1', k_2', \ldots, k_s'> \in K_{j_i}} c_{k_1' k_2' \ldots k_s'} \right) \right] \right\},$$

$$M(k_1, k_2, \ldots, k_s) \equiv \{ <k_1', k_2', \ldots, k_s'> \, | \, <k_1', k_2', \ldots, k_s'> \neq <k_1, k_2, \ldots, k_s> \},$$

$$K_j(k_1, k_2, \ldots, k_s) \equiv \{ <k_1', k_2', \ldots, k_s'> \, | \, k_j' = k_j \, \& \, <k_1', k_2', \ldots, k_s'>$$
$$\neq <k_1, k_2, \ldots, k_s> \},$$

and for $m = 1$,

$$\sum_{j_m > j_{m-1} > \ldots > j_1} \prod_{i=1}^{m} \left(\sum_{<k_1', k_2', \ldots, k_s'> \in K_{j_i}} c_{k_1' k_2' \ldots k_s'} \right) \equiv \sum_{j=1}^{s} \left(\sum_{<k_1', k_2', \ldots, k_s'> \in K_j} c_{k_1' k_2' \ldots k_s'} \right).$$

NOTE: The generalized form of $(*)$ is as follows:

$$\prod_{k_1, k_2, \ldots, k_s} c_{k_1 k_2 \ldots k_s}^{c_{k_1 k_2 \ldots k_s}} = \prod_{k_1, k_2, \ldots, k_s} \left(c_{k_1 k_2 \ldots k_s} - q_{k_1 k_2 \ldots k_s} \right)^{c_{k_1 k_2 \ldots k_s}}$$

For $n = s = 2$, we have $q_{11} = q_{22} = \det C = -q_{12} = -q_{21}$, and CONDITION Q reduces to $(*)$ in this case.

I offer the following assertion here without proof:

$\forall n \geq 2, \forall s \geq 2$, the following are logically equivalent:

COMPLETENESS
CONDITION C
SEPARABILITY

References

1. Mermin, N. David (1981) "Quantum mysteries for anyone." *Journal of Philosophy* **78**: 397–408.
2. Jarrett, Jon P. (1989) "Bell's Theorem: a guide to the implications." In [5] 60–79.
3. Einstein, Albert, Boris Podolsky, and Nathan Rosen (1935) "Can quantum-mechanical description of physical reality be considered complete?" *Physical Review* **47**: 777–780.
4. Bohr, Niels (1935) "Can quantum-mechanical description of physical reality be considered complete?" *Physical Review* **48**: 696–702.
5. Cushing, James T. and Ernan McMullin, editors (1989) *Philosophical Consequences of Quantum Theory: Reflections on Bell's Theorem*. Notre Dame, IN: University of Notre Dame Press.
6. Fine, Arthur (1989) "Do correlations need to be explained?" In [5], 175–194.

7. van Fraassen, Bas (1989) "The Charybdis of realism: epistemological implications of Bell's Inequality." In [5], 97–113.
8. Hughes, R. I. G. (1989) "Bell's Theorem, ideology, and structural explanation." In [5], 195–207.
9. Fine, Arthur (1986) *The Shaky Game: Einstein, Realism, and the Quantum Theory*. Chicago: University of Chicago Press.
10. Einstein, Albert (1936) "Physik und Realität." *Journal of the Franklin Institute* **221**: 313–347.
11. Einstein, Albert (1948) "Quantenmechanik und Wirklichkeit." *Dialectica* **2**: 320–324.
12. Howard, Don (1985) "Einstein on locality and separability." *Studies in History and Philosophy of Science* **16**: 171–201.
13. Howard, Don (1989) "Holism, separability, and the metaphysical implications of the Bell experiments." In [5], 224–253.
14. Winsberg, Eric, and Arthur Fine (2003) "Quantum life: interaction, entanglement, and separation." *Journal of Philosophy*, **C(2)**: 80–97.
15. Ash, Robert B. (1990) *Information Theory*. New York: Dover Publications. (Originally published in 1965 by Interscience Publishers, New York.)
16. Billingsley, Patrick (1995) *Probability and Measure*, 3rd edition. New York: Wiley. (First edition published in 1979.)
17. Fogel, Brandon (2007) "Formalizing the separability condition in Bell's Theorem". *Studies in History and Philosophy of Modern Physics*, **38**, 920–937.

Bell Inequalities: Many Questions, a Few Answers

Nicolas Gisin

Abstract What can be more fascinating than *experimental metaphysics*, to quote one of Abner Shimony's enlightening expressions? Bell inequalities are at the heart of the study of nonlocality. I present a list of open questions, organised in three categories: fundamental; linked to experiments; and exploring nonlocality as a resource. New families of inequalities for binary outcomes are presented.

1 Introduction

This Festschrift in honor of Abner Shimony is the ideal occasion to review some of the many questions about Bell inequalities that remain open, despite more than four decades of active research and a vast number of publications on this fascinating subject. Indeed, Abner was—in modern terminology—an early adaptor of the product *Bell inequality*. At that time, in the 1960s and 1970s, it required quite some courage and independence of thought, two qualities characterizing Abner, to recognize the value of Bell's work on the foundations of Quantum Physics. Even in the 1980s, after Aspect's experiments, Bell inequality was still considered a dirty work. "Bohr sorted out all that years ago", was the standard answer. In those days, if you wanted your work published in PRL or similar high-standard journals you had better avoid terms like Bell inequality and (even worse) quantum nonlocality.

Starting with Artur Ekert's PRL relating Bell inequalities with quantum key distribution things have drastically changed [1]. Today it would be hard to find an issue of PRL without a mention of Bell inequality, nonlocality and—on top of it all—"the potential relevance of the presented work for quantum information processing". It is nice to see how human physicists are! And who is more human, in the most noble

---

N. Gisin
Group of Applied Physics, University of Geneva, 1211 Geneva 4, Switzerland
e-mail: nicolas.gisin@physics.unige.ch

W.C. Myrvold and J. Christian (eds.), *Quantum Reality, Relativistic Causality, and Closing the Epistemic Circle*, The Western Ontario Series in Philosophy of Science 73,
© Springer Science+Business Media B.V. 2009

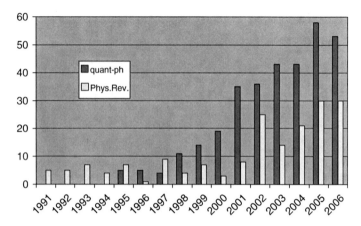

Fig. 1 (Color online) Number of occurrences of the words *Bell inequality* or *Bell inequalities* in the title or abstract of papers published during the last 16 years on the quant-ph preprint server and in Physical Review (PRL+PRA+PRB+PRC+PRD+PRE)

sense of the word, than Abner? Abner, you helped me tremendously; moreover, you did so at a time when I really needed it. Thank you Abner!

Let's return to the product *Bell inequalities*. Today it is fashionable, see Fig. 1, although I suspect that a large majority of physicists would still be unable to properly derive any Bell inequality. I bet that in a few decades Bell inequalities will be taught at high school, because of their mathematical simplicity, their force as an example of the scientific methodology and their huge impact on our world view. Yet, there remains a surprisingly large number of open questions, several of which are listed in Section 3. Section 4 presents a new family of Bell inequalities for an arbitrary even number of settings and binary outcomes. In appendix B an elegant Bell inequality for qubits is presented; its optimal quantum violation requires measurements of all three Pauli matrices σ_x, σ_y and σ_z. However, let's start in Section 2 by defining the notation.

2 Bell Inequalities

Bell inequalities are relations between conditional probabilities valid under the locality assumption. Hence, *a priori* they have nothing to do with quantum physics (and thus should not be written using quantum operators). However, it is the fact that quantum physics predicts a violation of these relations that makes them interesting. The purpose here is not to present yet another derivation of Bell inequalities, but merely to fix notation. Let $p(a,b,c,...|x,y,z,...)$ denote the conditional probability that players $A,B,C,...$ produce the outcome $a,b,c,...$ when they receive the input $x,y,z,....$ Typically the players are physicists that perform measurements $x,y,z,...$ with results $a,b,c,....$ Note that $a,b,c,...$ need not be numbers. We call the conditional probabilities $p(a,b,c,...|x,y,z,...)$ correlations.

Bell Inequalities: Many Questions, a Few Answers

We assume the numbers of players, inputs and outcomes are all finite. Under the assumption of locality (i.e., there is a probability distribution $p(\lambda)$ such that $p(a,b,c,...|x,y,z,...) = \sum_{\lambda} p(\lambda) \cdot p(a|x,\lambda) \cdot p(b|y,\lambda) \cdot p(c|z,\lambda) \cdot ...)$ the set of all correlations is convex with finitely many vertices. Such sets are called polytopes [2]. Thus, for any given finite number of players, inputs and outcomes, the set of local correlation $p(a,b,c,...|x,y,z,...)$ is called the *local polytope* [2]. These polytopes are bounded by facets (hyperplanes). Each facet can be described by a linear equation: $\sum_{a,b,c,...x,y,z,...} C^{xyz...}_{abc...} p(a,b,c,...|x,y,z,...) = S_{lhv}$ with real coefficients $C^{xyz...}_{abc...}$ and S_{lhv}. All local correlations lie on one side of the facet, hence they necessarily satisfy the inequality:

$$\sum_{a,b,c,...,x,y,z,...} C^{xyz...}_{abc...} p(a,b,c,...|x,y,z,...) \leq S_{lhv} \tag{1}$$

Such inequalities are called tight Bell inequalities (for an elegant, but not tight Bell inequality, see appendix B). We say that a quantum state ρ is nonlocal iff there are measurements on ρ that produce a correlation that violates a Bell inequality.

The famous CHSH inequality [3] reads

$$E(x=0,y=0) + E(x=0,y=1) + \\ E(x=1,y=0) - E(x=1,y=1) \leq 2 \tag{2}$$

where in our notations $E(x,y) = p(a=b|x,y) - p(a \neq b|x,y)$. It is convenient to use the following self-explanatory matrix notation:

$$CHSH \doteq \begin{pmatrix} +1 & +1 \\ +1 & -1 \end{pmatrix} \leq 2 \tag{3}$$

This CHSH inequality is the only tight Bell inequality for the bipartite case (i.e., two players) with binary inputs and outcomes (up to local symmetries).

Let us emphasize that the entire game consists for each player in producing, for a given situation, a classical outcome with some probability for any possible input. In the quantum case this implies performing measurements with classical outcomes on a given quantum state ρ. Accordingly, the players can't combine several instances, i.e., several quantum states ρ, and perform quantum information processing on them, i.e., exploit coherent measurement on $\rho^{\otimes n}$ for $n \geq 2$. Note that this does not exclude the situation where the players receive a fixed number of states, like e.g., $\rho^{\otimes 3}$, but this is a different game from the one based on ρ. Clearly, *a priori* a state ρ can be local, while $\rho^{\otimes n}$ is nonlocal for all $n \geq n_{threshold} > 1$.

3 Open Questions

The open questions can be organized in three groups. First, the fundamental questions, most in the spirit of Bell. Next, questions more related to experiments, in the spirit of Abner's works (e.g., the famous CHSH-Bell inequality and the detection loophole). Finally, Bell-like inequalities for nonlocal resources, the most timely research on nonlocality.

Note that many open questions in quantum information theory are listen on the web page [4].

3.1 Fundamental Questions

There are infinitely many Bell inequalities. Even if one is restricted to tight Bell inequalities corresponding to facets of the polytope of local correlations, the number of Bell inequality is infinite. Restricting the given number of inputs and outcomes limits the number of Bell inequalities, but it is a computationally hard problem to list them [2].

1. Why is the CHSH inequality almost always the most efficient one to prove a quantum state to be nonlocal? Until 2004 there was no example of a quantum state not violating the CHSH inequality, but violating some other Bell inequality [5]. Still today, no natural example, i.e., a state with some natural symmetry, has been found. This leads to the concept of *relevant Bell inequalities*: an inequality is relevant with respect to a given set of inequalities if there is a quantum state violating it, but not violating any of the inequalities in the set.
2. Is there a finite set of inequalities such that no other inequality is relevant with respect to that set? What if one limits the dimension of the Hilbert space?
3. Find an inequality that is more efficient than the CHSH one for the Werner states [6] or prove it is impossible. In dimension two, Werner states are simply mixtures of a maximally entangled pure state ψ with noise (i.e., the identity operator): $\rho_W = W|\psi\rangle\langle\psi| + (1-W)\mathbb{1}/4$, where W is the visibility. A local model exists for $W \lesssim 0.66$ [7], the CHSH inequality proves Werner states to be nonlocal for $W > 1/\sqrt{2}$. The region in between is unknown. The same question for the isotropic state (mixture of maximally pure state and noise) has been answered in part in [8, 9] where a generalization of the CHSH inequality to arbitrary numbers of outcomes has been shown to be more efficient. But for the isotropic states there remains also a gap in between the best known local model [10] and the proven nonlocality visibility threshold.
4. Is hidden nonlocality generic for all entangled quantum states, including mixed states? In dimension ≥ 5, Popescu proved that the Werner states, although admitting local models, have hidden nonlocality, i.e., there are local filters such that if the Werner state passes the filters, then the resulting state violates the CHSH inequality [11], see also [12] for a simple example of hidden nonlocality. In the same vein, one should ask whether for all quantum states with hidden nonlocality there is a Bell inequality, possibly with more inputs and outcomes, that can be violated by this state? Finally, is there an example of hidden nonlocality that requires a sequence of local filters rather than a single one (the local model should reproduce all intermediate results)?

Bell Inequalities: Many Questions, a Few Answers

5. Prove some entangled quantum states to be local. This requires one to prove the existence of a local model. This has been done for Werner states (see [6] for projective measurements and [13] for general POVMs) and very recently for isotropic states [10]. A weaker form of this question asks for a proof that a state can't violate any Bell inequality with less than a given number of inputs and/or outcomes. There is only one general result to this question, see the elegant construction in [14].

6. Why are almost all known Bell inequalities for more than 2 outcomes maximally violated by states that are not maximally entangled [15]? There is quite a lot of evidence that entanglement and nonlocality are different resources [16].

7. Can all Bell inequalities with d outcomes be maximally violated by a quantum state of dimension d? Or is there an example requiring states of dimension larger than the number of outcomes? In reference [17], a Bell inequality with m outcomes on Alice's side and binary outcomes on Bob's side is presented. It is maximally violated by the maximally entangled state in dimension m.

8. Is there a local quantum state ρ such that ρ^n violates some Bell inequality? Note that if the state ρ is distillable, then ρ^n, for large enough n, contains hidden nonlocality.

9. Find genuine n-party inequalities violated by all n-party pure entangled states. In the case of two parties, the CHSH inequality is such an example, i.e., it can be violated by any pure entangled state of whatever dimension [18, 19]. In the case of three parties there are entangled states that do not violate the MBK inequality [20, 21]. In [22, 23], a Bell inequality is presented that shows numerical evidence that all 3-party pure entangled state violate it. But the case of arbitrarily many parties is still open. Note that all n-party pure entangled states can always be projected onto a 2-party pure entangled state by projecting $n-2$ parties onto appropriate local pure states [24]. This can be formulated as a tight Bell inequality where $n-2$ parties have only a single input. Hence, there is a set of $\binom{n}{n-2}$ inequalities that does the job. But is there a single inequality?

10. There is no known Bell inequality that requires POVMs for optimal violation on some quantum states. For binary outcomes, one can prove that POVMs are never relevant [25], but for larger a number of outcomes the question is open.

11. Almost all Bell inequalities are maximally violated by quantum states and measurements that can all be written, in an appropriate basis, using only real numbers. This is surprising since interference, a basic quantum property, "requires" complex numbers. It would be nice to find Bell inequalities suitable for distinguishing real Hilbert spaces from complex ones (i.e., an inequality that can only be violated by states and settings that require complex numbers). An example is [17].

12. Is there a bound entangled state that violates some Bell inequality? In [26] Masanes proves that no bound entangled state violates the CHSH inequality. But what about other Bell inequalities? Note that in the case of three players or more, it is important to distinguish different meanings of bound entanglement: bound means that the players can't distill a maximally entangled states between

all of them; while totally bound means that even if some parties join into groups, they still can't distil entanglement between the groups. Dürr found a bound entangled state of 8 qubits that violates the MKB inequality [27]. However the violation is small, indicating that there is no 8-party entanglement [28]. Actually it was then demonstrated for qubits that any violation of a Bell inequality, with two inputs per player implies that the players, can join into groups such that the groups can distill a maximally entangled state [29, 30].

13. In the case of more than 2 parties, find inequalities testing models that assume bi-partite nonlocality but no arbitrary multi-partite nonlocality. A first example was presented already in 1987 by Svetlichny [31] and generalized in [32–34].

14. Find families of Bell inequalities valid for any number of inputs and outcomes. An example of such a family is presented in [5]. Another example is presented in this paper, see Section 4, though valid only for binary outcomes and even numbers of settings. The MKB inequality [20, 21] is an example of a family of Bell inequalities with fixed numbers of inputs and outcomes, but for arbitrarily many parties. See also the recent [35].

15. Given a multi-party quantum state ρ, how can one know whether ρ is nonlocal, i.e., whether there is a Bell inequality and measurements such that quantum physics predicts a violation of the inequality? For pairs of qubits and the CHSH inequality this problem has been solved in 1995 by the Horodecki family [36], but the general problem seems exceedingly hard.

3.2 Questions Relevant for Experiments

The original Bell inequality [37] is, strictly speaking, not a Bell inequality according to the modern terminology that we use here. Indeed, the original inequality required, besides locality, another assumption about perfect correlations. Abner immediately recognized that this auxiliary assumption made the entire enterprize non testable and searched for an inequality involving only measurable quantities. This led him and his co-workers to find the CHSH inequality. Interestingly, the CHSH paper [3] already mentions the detection loophole, again underlying the importance the authors gave to the experimental issues. Concerning the detection loophole, see also [38].

1. Find Bell inequalities easier to test experimentally with today's technology, while avoiding all known loopholes. Quantum nonlocality is so fundamental for our world view that it deserves to be tested in the most convincing way. It is thus surprising and annoying that no experiment to date has managed to close simultaneously the locality loophole (space-like separation from the choice of settings until the classical data are secured) and the detection loophole. The latter consists in assuming that the detection efficiency is independent of the hypothetical local variables (for example, if polarization would be unknown, one would assume that all detectors are polarization insensitive, a clearly wrong assumption). Reference [39] presents a simple model reproducing all quantum correlations on maximally entangled qubits assuming detection efficiencies of 2/3

Bell Inequalities: Many Questions, a Few Answers 131

(and projective measurements). Violation of the CHSH inequality requires detection efficiencies of at least 82.84%, for maximally entangled states. There is only a single known inequality with few settings that does better, though only marginally better, $\eta_{threshold} = \sqrt{2/3} \approx 81.65\%$ [40]. This inequality has three settings on each side and is *not* a facet of the polytope of local correlations. For numbers of settings larger than 100 a better inequality has been derived from communication complexity arguments [41]. Interestingly, Philippe Eberhard noticed that partially entangled states are less sensitive to the detection loophole [42].

2. A timely variation of the previous question addresses situations where the detection efficiency differs from one side of the experiment to the other. This is natural for experiments on entanglement between quantum systems of different kinds, like, e.g., an atom and a photon [43–45].

3. Find inequalities suitable for a Bell test with simple quantum-optics states and homodyne detectors. Indeed, the homodyne detection technique is well developed and always produces an outcomes. But simple cases likes, e.g., a delocalized photon in state $|0,1\rangle + |1,0\rangle$, although clearly entangled, does not violate the CHSH inequality with homodyne detection and a simple binarisation of the measurement results. More complicated states could violate the CHSH, but only by a tiny amount, see [46] and references therein.

4. Find inequalities for many settings. Experimentally one rarely measures precisely the four probabilities that appear in the CHSH inequality. Most of the time a series of points is measured and fitted with a sinus. Hence, an inequality for such series of points could be more appropriate. Examples are given in [47, 48] and in Section 4.

3.3 Bell-Like Inequalities for Nonlocal Resources

This subsection presents recently opened questions and moves away from the traditional work on Bell inequalities. It starts by admitting quantum nonlocality and aims at better quantifying it and at understanding it as a new kind of resource. These questions investigate nonlocal but non-signaling correlations [49]. Recall that a correlation $p(a,b,c,...|x,y,z,...)$ is non-signaling iff all the marginals are independent of the other players's inputs: $\sum_{b,c,...} p(a,b,c,...|x,y,z,...) = p(a|x)$, $\sum_{a,c,d,...} p(a,b,c,...|x,y,z,...) = p(b,|y)$, etc.

Bell inequalities are tests for correlations that can be simulated using only local resources and shared randomness (a modern terminology for the obsolete *local hidden variables*). This view raises the question of correlations that can be simulated using, in addition to shared randomness, some finite amount of some given nonlocal resource. For example, it is known that any pair of projective (Von Neumann) measurements on any maximally entangled state of two 2-level quantum systems can be simulated using only shared randomness and a single PR-box (a sort of unit of nonlocality) [50–53]. Hence, it is interesting to characterize all correlations that can't

132   N. Gisin

be simulated using shared randomness and one PR-box. Surprisingly, some correlations resulting from quantum measurements on partially entangled 2-level systems are of that kind.

1. Is there a Bell-like inequality valid for all correlations simulable with a single bit of communication and violated by some partially entangled 2-qubit states? Actually, the entire field of research considered in this subsection started with a paper presenting Bell-like inequalities valid for 1 bit of communication [54]. However, the presented inequalities can't be violated by any 2-qubit states. We know that maximally entangled 2-qubit states can be simulated with a single bit of communication; thus such states don't violate any of the considered Bell-like inequalities. However, the question remains open for partially entangled states.

2. Are all partially entangled qubit pairs not simulable by a single PR-box? A few Bell-like inequalities satisfied by all correlations simulable by a single PR-box and shared randomness are known [55, 56]. From these one knows that very poorly entangled states can't be simulated with one PR-box, but the case of high-but-not-maximally entangled states is open.

3. Find inequalities satisfied by all correlation that can be simulated by two PR-boxes. Two bits of communication suffice to simulate any two qubit state. Is the same true for two PR-boxes?

4. Find any non-signaling box [49] with finitely many inputs and outcomes with which one can simulate partially entangled states.

5. Find the *Quantum-Bell inequalities* that bound the correlations achievable with quantum measurements and states? An example is the Tsirelson bound [57] stating that quantum correlations can't violate the CHSH inequality by more than the well known factor $2\sqrt{2}$, see also [58].

6. Can a secret key be distilled out of any nonlocal correlation, (secret against any non-signaling adversary performing arbitrary individual attacks) [59–61]? This question may appear to move away from *Bell questions*, but it concerns the power of the nonlocal resources as witnessed by Bell inequalities. It also addresses the question of the existence of bound information [62, 63], a classical analog to bound entanglement.

4 The AS-Bell Inequality Family

I know of only a single family of bipartite Bell inequalities valid for any number of inputs and outcomes [5]. In this section I briefly present a new family of bipartite Bell inequalities for any even number of inputs and binary outcomes. I found this family by looking for correlation Bell inequalities with a few inputs and binary outcomes. Recall that a correlation inequality involves only expectation values: $E(x,y) = p(a = b|x,y) - p(a \neq b|x,y)$. For binary inputs, the CHSH is the only inequality. For ternary inputs, there is no new correlation inequality [5,64]. For four inputs on each side, I searched numerically all possibilities assuming small integer coefficient. I found only two new inequalities (the coefficient in the matrix indicate the coefficients of the corresponding expectation values):

$$AS_4 \doteq \begin{pmatrix} +1 & +1 & +1 & +1 \\ +1 & +1 & +1 & -1 \\ +1 & +1 & -2 & 0 \\ +1 & -1 & 0 & 0 \end{pmatrix} \leq 6 \tag{4}$$

$$D_4 \doteq \begin{pmatrix} +2 & +1 & +1 & +2 \\ +1 & +1 & +2 & -2 \\ +1 & +2 & -2 & -1 \\ +2 & -2 & -1 & -1 \end{pmatrix} \leq 10 \tag{5}$$

Avis and co-workers demonstrated that these are indeed the only correlation inequalities for four inputs [65]. Inspired by inequality AS_4, it is not difficult to guess the form of the next inequalities:

$$AS_6 \doteq \begin{pmatrix} +1 & +1 & +1 & +1 & +1 & +1 \\ +1 & +1 & +1 & +1 & +1 & -1 \\ +1 & +1 & +1 & +1 & -2 & 0 \\ +1 & +1 & +1 & -3 & 0 & 0 \\ +1 & +1 & -2 & 0 & 0 & 0 \\ +1 & -1 & 0 & 0 & 0 & 0 \end{pmatrix} \leq 12 \tag{6}$$

$$AS_8 \doteq \begin{pmatrix} +1 & +1 & +1 & +1 & +1 & +1 & +1 & +1 \\ +1 & +1 & +1 & +1 & +1 & +1 & +1 & -1 \\ +1 & +1 & +1 & +1 & +1 & +1 & -2 & 0 \\ +1 & +1 & +1 & +1 & +1 & -3 & 0 & 0 \\ +1 & +1 & +1 & +1 & -4 & 0 & 0 & 0 \\ +1 & +1 & +1 & -3 & 0 & 0 & 0 & 0 \\ +1 & +1 & -2 & 0 & 0 & 0 & 0 & 0 \\ +1 & -1 & 0 & 0 & 0 & 0 & 0 & 0 \end{pmatrix} \leq 20 \tag{7}$$

The generalization to arbitrary even number of inputs is straightforward. Note that AS_2 is nothing but the CHSH inequality. Numerically, these AS_n inequalities are tight and maximally violated by maximally entangled qubit states for visibilities larger than V_n, with $V_2 = 1/\sqrt{2} \approx 0.7071$, $V_4 \approx 0.7348$, $V_{10} \approx 0.7469$, $V_{32} \approx 0.7497$, $V_{50} \approx 0.7499$. Apparently $V_\infty \approx 0.75$; this contrasts with the I_{nn22} family presented in [5] where for binary outcomes and large numbers of inputs the threshold visibility appears to tend to 1. All settings can be chosen to lie on a grand circle of the Poincaré sphere.

5 Conclusion

We are lucky to live at the time where physics discovers and explores the nonlocal characteristics of Nature. Contrary to the nonlocality of Newtonian gravitation, quantum nonlocality is with us for ever [66, 67]. Future historians of Science will describe our epoch as that of the great discovery of nonlocality. The name of Abner Shimony will forever be associated with this fascinating epoch.

The choice of questions listed in this contribution to Abner's Festschrift is necessarily somewhat subjective. Others may like to add their favorite ones or to formulate the questions differently. Important is the fact that there are many interesting open questions of very different kinds. The basic maths is simple, but a deeper understanding requires concepts ranging from combinatorial and complexity theories to algebra and geometry in high dimensions. Hence, it is likely that most of the listed problems are hard. But their solutions, even partial solutions, will be valuable contributions to one of the most fascinating research fields of the 21st century.

Note added in proof. Since the writing of this paper, posted on the arXiv as quant-ph/0702021, several of the problems listed in this contribution have been (partially) solved. In particular, for problem 3.1.3 see arXiv:0806.0096, for 3.1.4 see *Phys. Rev. Lett.* **100**, 090403 (2008), for 3.1.7 see arXiv:0712.4320, for 3.1.11 (and the question raised in appendix B) see arXiv:0810.1923, for 3.3.4 see *Phys. Rev. A* **78**, 052111 (2008).

Acknowledgment This work has been supported by the EC under project QAP (contract n. IST-015848) and by the Swiss NCCR *Quantum Photonics*. Thanks are due to Rob Thew, Toni Acin, André Méthot, Sandu Popescu and Valerio Sacarani for their comments on previous versions of this paper.

Appendix A: Some Diagonal Bell Inequalities

Correlation Bell inequalities of a form similar to D4 (eq. 5) can easily be found numerically. For five inputs on each side there seems to exist only two such inequalities (at least I found only two). They are entirely defined by their first line and the permutation rule as in (5) (from one line to the next: shift each entry to the left, the entry that falls out is re-introduced on the right hand side with the opposite sign):

$$D5_1 \doteq (11011) \leq 8 \tag{8}$$
$$D5_2 \doteq (32113) \leq 20 \tag{9}$$

For six inputs I found:

$$D6_1 \doteq (101011) \leq 10 \tag{10}$$
$$D6_2 \doteq (311124) \leq 28 \tag{11}$$
$$D6_3 \doteq (422125) \leq 36 \tag{12}$$
$$D6_4 \doteq (422136) \leq 42 \tag{13}$$

For more inputs, the numbers of such D-inequalities seems to grow rapidly.

Bell Inequalities: Many Questions, a Few Answers   135

Appendix B: An Elegant Bell Inequalities

In [17] Helle Bechmann-Pasquinucci and myself presented a Bell inequality tailored
for quantum cryptography in high dimension Hilbert spaces. Since this inequality
seems to have a few original features, like being optimally violated by states and
quantum measurements requiring complex numbers and Hilbert spaces of dimension larger than the number of outcomes on Bob's side (but equal to the number
of outcomes on Alice's side), I recall it in this appendix with the notations used
throughout this contribution. Moreover, this new way of looking at this inequality
underlines its similarity with communication complexity [68].

In this game, Alice receives as input a number $x \in \{0, 1, ..., n-1\}$, while Bob's
input consists of n numbers $y_0, y_1, ..., y_{n-1}$ with each $y_j \in \{0, 1, ..., m-1\}$. Basically,
the goal is that Alice outputs $a = y_x$. As such this would be merely an example
of a communication complexity game. But in our game, Bob can use a joker and
refuse that this instance of the game counts. Accordingly, Bob's outcome is binary.
Whenever $b = 0$, the score is null, whatever Alice's outcome. Whenever $b = 1$ the
score is $+1$ if $a = y_x$ and -1 if $a \neq y_x$. Explicitly, the Bell inequality reads:

$$SHB =$$

$$\sum_{\substack{x = 0...n-1 \\ y_0...y_{n-1} = 0...m-1}} \begin{pmatrix} p(a = y_x, b = 1 | x, y_0, ..., y_{n-1}) \\ -p(a \neq y_x, b = 1 | x, y_0, ..., y_{n-1}) \end{pmatrix}$$

$$\leq S_{local} \qquad (14)$$

The optimal local strategy consists of Alice and Bob agreeing in advance on a sequence $y_0^g, ..., y_{n-1}^g$ and Alice producing $a = y_x^g$ while Bob accepts the game only for
the inputs $y_0, ..., y_{n-1}$ for which the averaged score is positive:

$$S_{local} = \sum_{r=0}^{\lceil \frac{n-1}{2} \rceil} (n - 2r) \binom{n}{r} \qquad (15)$$

Let us concentrate on the case $n = 2$ for which $S_{local} = 2$. The optimal quantum
strategy requires Alice and Bob to share a maximally entangled state of dimension
m. Alice measures her quantum system in one out of two mutually conjugated bases,
depending on her input $x = 0$ or $x = 1$. Bob receives two symbols as input, y_0 and
y_1, corresponding to two quantum states, one in each of the two bases. He applies to
his quantum system a measurement described by the projector onto the state which
lies precisely in between the two states that correspond to y_0 and y_1 (Since the two
states belong to two mutually conjugated bases, such an intermediate state is always
uniquely define. Take for instance the eigenstate with maximal eigenvalue of the
density matrix obtained by a 50–50% mixture of the two states.) If Bob's projection
is successful, this projects Alice's state onto the state that maximizes her chance of

finding the correct outcome. In such a case Bob outputs $b = 1$. In the alternative case, i.e., failure of his projection measurement, he outputs $b = 0$; that is, Bob's outcome is his measurement result. With this quantum strategy, Alice and Bob beat the optimal local strategy by a factor $\sqrt{m}$:

$$S_{quantum} = 2\sqrt{m} > S_{local} = 2 \qquad (16)$$

Note that for $m = 2$, this reduces to the well studied CHSH inequality. Indeed, although in this case Bob has formally four possible inputs, the corresponding four projectors form two bases. Explicitly, Alice measures one of the two operators σ_x or σ_z, depending on her input, and Bob measures in the intermediate bases σ_{+45^0} (for his inputs 0,0 and 1,1) or σ_{-45^0} (for inputs 0,1 and 1,0).

For $n = 2$ and $m = 3$ the quantum optimum of $2\sqrt{3} \approx 3.464$ is reached by the strategy summarised above and presented in [17]. Numerical evidence suggests that if one restricts oneself to settings that can be expressed using only real numbers, the maximum is slightly lower: $10/3 \approx 3.333$ [17]. Moreover, this maximum is reached for a non-maximally entangled state. But it is unknown whether a higher score can be achieved using only real numbers in larger Hilbert spaces.

The case $n = 3$, $m = 2$ appears also to be interesting. Indeed, the quantum maximum is $4\sqrt{3} \approx 6.928$, while the maximum using only real numbers is reached by the singlet state at $2 + 2\sqrt{5} \approx 6.472$. This might open the possibility to test correlations requiring complex Hilbert spaces (however, here again it remains to test the inequality in higher dimensions).

Note that this inequality $n = 3, m = 2$ can also be written as a correlation inequality. Indeed, Bob's $m^n = 8$ inputs can be grouped into four projective measurements. In this form, this inequality reads:

$$S_{3\times 4} \doteq \begin{pmatrix} +1 & +1 & +1 \\ +1 & -1 & -1 \\ -1 & +1 & -1 \\ -1 & -1 & +1 \end{pmatrix} \leq 6 \qquad (17)$$

Another elegant feature of this case $n = 3$, $m = 2$ is seen when the optimal settings are represented on the Poincaré sphere: for Alice the three vectors are mutually orthogonal, while Bob's four vectors are on the vertices of the tetrahedron, see Fig. 2.

To conclude, let us note that most inequalities presented in this appendix, in particular the elegant $S_{3\times 4}$, are not facets of the local polytope. This indicates that the geometry of the local polytope doesn't match the symmetries of elegant quantum states and measurements. In the case of three and four inputs on Alice and Bob's side, respectively, all facets are known [5], hence one shouldn't be surprised that the new inequality $S_{3\times 4}$ is not a facet.

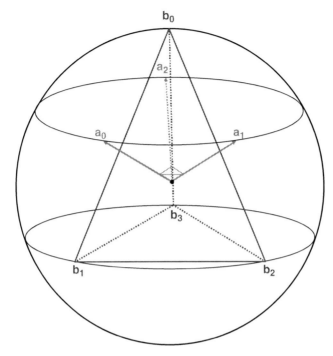

Fig. 2 (Color online) Measurement settings represented on the Poincaré sphere for the elegant inequality S_{3x4} defined by eq. (17). Alice's three settings are represented by three mutually orthogonal vectors, and Bob's four settings by the vertices of the tetrahedron

References

1. A.K. Ekert, *Phys. Rev. Lett.* **67**, 661–663 (1991).
2. I. Pitowski, *Quantum Probability — Quantum Logic*, (Lecture Notes in Physics, Vol. 321). Berlin, Springer-Verlag 1989.
3. J.F. Clauser, M.A. Horne, A. Shimony and R.A. Holt, *Phys. Rev. Lett.* **23**, 880 (1969).
4. http://www.imaph.tu–bs.de/qi/problems.
5. D. Collins and N. Gisin, *J. Phys. A: Math. Gen.* **37**, 1775 (2004).
6. R.F. Werner, *Phys. Rev. A* **40**, 4277 (1989).
7. A. Acin, N. Gisin and B. Toner, *Phys. Rev. A* **73**, 062105 (2006).
8. D. Kaszlikowski et al., *Phys. Rev. Lett.* **85**, 4418 (2000).
9. D. Collins, N. Gisin, N. Linden, S. Popescu and S. Massar, *Phys. Rev. Lett.* **88**, 040404 (2002).
10. M.L. Almeida et al, quant-ph/0703018.
11. S. Popescu, *Phys. Rev. Lett.* **74**, 2619–2622 (1995).
12. N. Gisin, *Phys. Lett. A* **210**, 151–156 (1996); ibid. **224**, 317–318 (1997).
13. J. Barrett, *Phys. Rev. A* bf65, 042302 (2002).
14. B.M. Therhal, A. Doherty and D. Schwab, *Phys. Rev. Lett.* **90**, 157903 (2003).
15. A. Acin, T. Durt, N. Gisin and J.I. Latorre, *Phys. Rev. A* **65**, 052325 (2002).
16. A.A. Méthot and V. Scarani, *Quant. Inf. Comput.* **7**, 157 (2007).
17. H. Bechmann-Pasquinucci and N. Gisin, *Phys. Rev. A* **67**, 062310, 2003. See also appendix B.
18. N. Gisin, *Phys. Lett. A* **154**, 201–202 (1991).
19. N. Gisin and A. Peres, *Phys. Lett. A* **162**, 15–17 (1992).

20. N.D. Mermin, *Phys. Rev. Lett.* **65**, 1838 (1990).
21. A.V. Belinskii and D.N. Klyshko, *Phys. Usp.* **36**, 653 (1993).
22. A. Acin et al., *Phys. Rev. Lett.* **92**, 250404 (2004).
23. J.-L. Chen, *Phys. Rev. Lett.* **93**, 140407 (2004).
24. S. Popescu and D. Rohrlich, *Phys. Lett. A* **166**, 293–297 (1992).
25. R. Cleve, P. Hoyer, B. Toner and J. Watrous, quant-ph/0404076.
26. L. Masanes, *Phys. Rev. Lett.* **97**, 050503 (2006).
27. W. Dür, *Phys. Rev. Lett.* **87**, 230402 (2001).
28. N. Gisin and H. Bechmann-Pasquinucci, *Phys. Lett. A* **246**, 1–6 (1998).
29. A. Acin, *Phys. Rev. Lett.* **88**, 027901 (2001).
30. A. Acin, V. Scarani and M. Wolf, *Phys. Rev. A* **66**, 04323 (2002).
31. G. Svetlichny, *Phys. Rev. D* **35**, 3066 (1987).
32. D. Collins et al., *Phys. Rev. Lett.* **88**, 170405 (2002).
33. M. Seevinck and G. Svetlichny, *Phys. Rev. Lett.* **89**, 060401 (2002).
34. J.L. Cereceda, *Phys. Rev. A* **66**, 024102 (2002).
35. K. Nagata, W. Laskowski and T. Paterek, *Phys. Rev. A* **74**, 062109 (2006).
36. M. Horodecki, P. Horodecki and R. Horodecki, *Phys. Lett. A* **200**, 340 (1995).
37. J.S Bell, *Physics* **1**, 195–200 (1964); reprinted in: J.S. Bell, *Speakable and Unspeakable in Quantum Mechanics: Collected Papers on Quantum Philosophy*, (Cambridge University Press, Cambridge, 1987, revised edition 2004).
38. Ph. Pearle, *Phys.Rev. D* **2**, 1418 (1970).
39. B. Gisin and N. Gisin, *Phys. Lett. A* **260**, 323–327 (1999).
40. S. Massar, S. Pironio, J. Roland and B. Gisin, *Phys. Rev. A* **66**, 052112 (2002).
41. H. Buhrman, P. Hoyer, S. Massar and H. Roehrig, *Phys. Rev. Lett.* **91**, 047903 (2003).
42. Ph. Eberhard, *Phys. Rev. A* **47**, R747 (1993).
43. J. Volz et al., *Phys. Rev. Lett.* **96**, 030404 (2006).
44. A. Cabello and J.-A. Larsson, quant-ph/0701191, *Phys. Rev. Lett.* **98**, 220402 (2007).
45. N. Brunner, N. Gisin, V. Scarani and Ch. Simon, quant-ph/0702130, *Phys. Rev. Lett.* **98**, 220403 (2007).
46. R. Garca-Patrón, J. Fiurácek, N.J. Cerf, J. Wenger, R. Tualle-Brouri and Ph. Grangier, *Phys. Rev. Lett.* **93**, 130409 (2004).
47. M. Zukowski, *Phys. Lett. A* **177**, 290 (1993). See also quant-ph/9908009.
48. N. Gisin, *Phys. Lett. A* **260**, 1–3 (1999).
49. J. Barrett et al., *Phys. Rev. A* **71**, 022101 (2005).
50. S. Popescu and D. Rohrlich, *Found. Phys.* **24**, 379–385, (1994).
51. N. Cerf, N. Gisin and S. Massar, *Phys. Rev. Lett.* **94**, 220403 (2005).
52. J. Barrett and S. Pironio, *Phys. Rev. Lett.* **95**, 140401 (2005).
53. F. Dupuis et al., quant-ph/0701142.
54. D. Bacon and B. Toner, *Phys. Rev. Lett.* **90**, 157904 (2003).
55. N. Brunner, N. Gisin and V. Scarani, *New J. Phys.* **7**, 1–14 (2005).
56. N. Brunner, V. Scarani and N. Gisin, *J. Math. Phys.* **47**, 112101 (2006).
57. B.S. Tsirelson, *Lett. Math. Phys.* **4**, 83, 1980.
58. M. Navascues, S. Pironio and A. Acin, quant-ph/0607119.
59. J. Barrett, L. Hardy and A. Kent, *Phys. Rev. Lett.* **95**, 010503 (2005).
60. V. Scarani, N. Gisin, N. Brunner, L. Masanes, S. Pironio and A. Acin, *Phys. Rev. A* **74**, 042339 (2006).
61. A. Acin, N. Gisin and Lluis Masanes, *Phys. Rev. Lett.* **97**, 120405 (2006).
62. N. Gisin and S. Wolf, *Phys. Rev. Lett.* **83**, 4200–4203 (1999).
63. N. Gisin, R. Renner and S. Wolf, *Algorithmica* **34**, 389–412 (2002).
64. C. Sliwa, *Phys. Lett. A* **317**, 165–168 (2003).
65. D. Avis, H. Imai and T. Ito, *J. Phys. A* **39**, 11283–11299 (2006).
66. Isaac Newton, Papers & Letters on Natural Philosophy and related documents, page 302, Edited, with a general introduction, by Bernard Cohen, assisted by Robert E. Schofield Harvard University Press, Cambridge, Massachusetts, 1958.
67. N. Gisin, *Can relativity be considered complete? From Newtonian nonlocality to quantum nonlocality and beyond*, quant-ph/0512168.
68. Brassard, G., *Q communication complexity*, quant-ph/0101005.

B. Experiment

Do Experimental Violations of Bell Inequalities Require a Nonlocal Interpretation of Quantum Mechanics? II: Analysis à la Bell

Edward S. Fry, Xinmei Qu, and Marlan O. Scully

Abstract Bell inequalities are derived assuming (i) hidden variables, (ii) positive probabilities for seemingly physical correlations, and (iii) locality. The over-riding role of assumption (ii) has generally not been emphasized. Since results of Bell inequality experiments show a violation of the inequality and agreement with quantum mechanical predictions, one or more of these assumptions is wrong. Thus, in the physical world, we cannot have hidden variables, and/or we must accept negative probabilities, and/or we must accept non-locality. Equivalently, the experiments tell us that any hidden variable theory (with associated non-negative probabilities) must be non-local; on the other hand, if a theory encompasses no hidden variables (e.g. quantum mechanics), the experiments do not make a statement about locality. Of course, the definition of "locality" plays a critical role, and that will be reviewed. In a previous paper (Phys. Lett. A *347*, 56–61, 2005), it was shown that the assumption of hidden variables (e.g. seemingly physical correlations) leads directly to negative (non-physical) probabilities in the Wigner–Bell model. In this paper, we provide analyses based both on Bell's derivation of the inequality and on the Clauser–Horne version for inherently stochastic theories. We examine probabilities that must be non-negative in these derivations and show how to evaluate them within the framework of quantum mechanics. We repeatedly show that the assumption of hidden variables in the derivation of a Bell inequality leads to supposedly non-negative probabilities whose quantum mechanical counterparts are, in fact, negative under some conditions.

E.S. Fry and X. Qu
Department of Physics, Texas A&M University, College Station, TX 77843-4242, USA
e-mail: fry@physics.tamu.edu, xinmei-qu@neo.tamu.edu

M.O. Scully
PRISM and the Departments of Mechanical Engineering, Aerospace Engineering, and Chemistry, Princeton University, Princeton, NJ 08544, USA

Institute for Quantum Studies and the Departments of Physics, Chemical Engineering, and Electrical Engineering, Texas A&M University, College Station, TX 77843-4242, USA
e-mail: scully@physics.tamu.edu

W.C. Myrvold and J. Christian (eds.), *Quantum Reality, Relativistic Causality, and Closing the Epistemic Circle*, The Western Ontario Series in Philosophy of Science 73,
© Springer Science+Business Media B.V. 2009

1 Introduction

The interpretation of quantum physics represents one of the more intriguing problems of modern science. On the one hand, quantum mechanics is a superbly successful computational tool with unprecedented predictive powers. On the other hand, the foundations of the subject and its relationship to relativity theory are still a source of vigorous debate and counter-intuitive notions.

A major focus has been the fascinating arguments of Einstein–Podolsky–Rosen (EPR) [1] and the associated work of John Bell that resulted in the inequalities bearing his name [2]. A modification of the Bell inequality by Clauser, Horne, Shimony, and Holt led to some of the first experimental tests [3–6], followed by other landmark experiments [7–11], as well as proposed "loophole free" experiments [12–14]. The history of Bell inequalities has been well-documented; of special note is the book "*Search for a Naturalistic World View: Volume 2*" by Shimony [15] and "*Speakable and Unspeakable in Quantum Mechanics*" by Bell [16].

In this paper, we provide an analysis based on Bell's derivation of the inequality for hidden variable theories and another based on the derivation of the Clauser–Horne inequality for stochastic hidden variables. We analyze probabilities that must be non-negative in both of these derivations. We show how to evaluate these probabilities within the framework of quantum mechanics, and show that they can, in fact, be negative. Thus, in these two examples as well as one in the previous paper [17], the assumption of hidden variables inevitably leads to classical (supposedly non-negative) probabilities whose counterparts in quantum mechanics are negative, i.e. hidden variables and positive probabilities for the corresponding seemingly physical correlations are jointly untenable within quantum mechanics.

2 Locality and Nonlocality

It had long been recognized that Bell's concept of locality was essentially a conjunction of two concepts. Jarrett was first to clearly enunciate them [18]; he defined the two components and named them "locality" and "completeness". He coined the name "strong locality" for the conjunction of these two components; "strong locality" is equivalent to the locality concept used by Bell. Shimony provides a superb, in depth discussion that will only be briefly summarized [19]. In order to provide a more descriptive nomenclature, Shimony renamed the two components "*Parameter Independence*" and "*Outcome Independence*". These can be defined as follows:

Parameter Independence: The outcome of a measurement on particle 1 is independent of the analyzer parameters of a spatially separate analyzer apparatus for particle 2. Specifically, this requires

$$p_\lambda^1(m|a,b) = p_\lambda^1(m|a),$$
$$p_\lambda^2(n|a,b) = p_\lambda^2(n|b),$$

(1)

where the right hand sides are independent of b and a, respectively.

Analysis à la Bell   143

Outcome Independence: The outcome of a measurement on particle 2 is independent of the outcome of a measurement on particle 1. This requires

$$p_\lambda^1(m|a,b,n) = p_\lambda^1(m|a,b),$$
$$p_\lambda^2(n|a,b,m) = p_\lambda^2(n|a,b),$$

(2)

where the right hand sides are independent of n and m, respectively. The notation in Eqs. (1) and (2) can be explained via the example, $p_\lambda^1(m|a,b,n)$. This is the conditional probability of the outcome m in a measurement on particle 1 given the apparatus parameters (e.g. Stern–Gerlach orientations) are a for the apparatus used to measure particle 1 and b for the apparatus used to measure particle 2, and also given that the outcome of a measurement on particle 2 is n.

Jarrett's theorem is that the conjunction of *Parameter Independence* and *Outcome Independence* is equivalent to Bell's locality assumption,

$$p_\lambda(m,n|a,b) = p_\lambda^1(m|a)\, p_\lambda^2(n|b),$$

(3)

which explicitly displays the assumption that $p_\lambda^1(m|a)$ is independent of both n and b, and that $p_\lambda^2(n|b)$ is independent of both m and a.

As discussed by Shimony [19], a violation of *Parameter Independence* implies an in-principle possibility of superluminal communication. Shimony points to three independent proofs [20–22] that quantum mechanics does not violate *Parameter Independence*, but notes that these depend on the linearity of quantum dynamics. Gisin [23] has shown that introducing non-linear elements (e.g. Weinberg [24]) does enable a violation of *Parameter Independence*.

On the other hand, *Outcome Independence* does not permit superluminal communication [19]. But, it is violated by quantum mechanical entanglement as can be seen by direct examination of the quantum state for two spin one-half particles in a singlet state,

$$|\Psi\rangle = \frac{1}{\sqrt{2}}\{|\uparrow\rangle_1\,|\downarrow\rangle_2 - |\downarrow\rangle_1\,|\uparrow\rangle_2\}.$$

(4)

The probability of a measurement of spin down along some axis for particle 2 is zero or unity depending on whether the outcome of a measurement along the same axis for particle 1 is down or up, respectively.

Quantum mechanics inherently violates the *Outcome Independence* aspect of Bell's locality assumption. Unfortunately, the word "non-local" generally conjures up thoughts such as superluminal communication; that is not justified in the case of a violation of *Outcome Independence*. When Tittle et al. [11] begin their paper "Violation of Bell Inequalities by Photons More Than 10 km Apart" with the statement that quantum theory is non-local, they are specifically referring to its violation of *Outcome Independence*, not superluminal communication.

It is appropriate to restate the assumptions used in the derivation of Bell inequalities, explicitly listing the two components of locality:

  (i) Hidden variables
 (ii) Positive probabilities for some seemingly physical correlations

(iii) *Parameter Independence*
(iv) *Outcome Independence*

Quantum mechanics inherently violates (iv). We are emphasizing that, in addition, (i) and (ii), are also jointly incompatible in quantum mechanics, although the importance of (ii) has not been generally recognized. There is no justification to argue a violation of *Parameter Independence* (i.e. superluminal communication) based on the results of Bell inequality experiments.

3 Classical (Hidden Variable) Preliminaries

Consider two spin 1/2 particles, in the entangled state given by Eq. (4). The particles are spatially well separated and their spins are measured by Stern–Gerlach apparata (SGA) whose z-directions are defined by unit vectors $\hat{\mathbf{a}}$, $\hat{\mathbf{b}}$, or $\hat{\mathbf{c}}$ e.g. Fig. 1. We introduce the following notation for the classical probability P_{ab} of measuring the component of spin of particle 1 to be $+1/2$ in the direction $\hat{\mathbf{a}}$ and the spin of particle 2 to be $+1/2$ in the direction $\hat{\mathbf{b}}$,

$$P_{ab} = P(+_a \quad d_b \quad d_c | d_a \quad +_b \quad d_c). \tag{5}$$

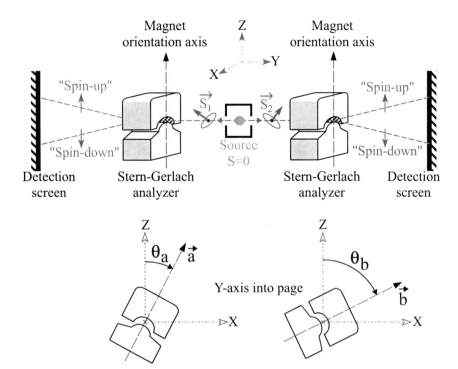

Fig. 1 Bohm's version of the Einstein–Podolsky–Rosen gedankenexperiment

Analysis à la Bell   145

In this six-element symbol, the left side of the partition refers to particle 1 and the right side to particle 2. Each element of the symbol shows whether a measurement of the component of the spin of the particle gives $+$ or $-$ in the corresponding direction $\hat{a}$, $\hat{b}$, or $\hat{c}$; b_j indicates no determination in the direction $\hat{j}$. This notation inherently assumes locality, e.g. we make the assignment $+_b$ for particle 2 independent of any direction $\hat{a}$ or outcome, $+$ or $-$, for particle 1.

Now, in a classical (hidden variable) theory, if the spin component of particle 1 is measured in the direction $\hat{a}$, and found to be $+1/2$, then a measurement of the spin of particle 2 in the same direction $\hat{a}$ can be predicted to be $-1/2$ with certainty. Hence, we include this information and write,

$$P_{ab} = P(+ - d_c | - + d_c). \tag{6}$$

Similarly, we have

$$P_{bc} = P(d_a + - | d_a - +), \tag{7}$$

$$P_{ac} = P(+ d_b - | - d_b +). \tag{8}$$

In a classical probability theory, the remaining undetermined components, b_j, will have specific values. That is, in any fixed direction, the spin component of every particle must have a definite value, either $+1/2$ or $-1/2$. Consequently, we can write Eq. (6) as a sum of probabilities for these two possibilities for b_c

$$\begin{aligned} P_{ab} &= P(+ - + | - + -) + P(+ - - | - + +) \\ &\equiv P(+ - +) \qquad + P(+ - -), \end{aligned} \tag{9}$$

where on the second line we have dropped the redundant information involving the second particle by writing $P(+ - +)$ for $P(+ - + | - + -)$; that is, for every direction, the signs for the second particle must be opposite to those for the first particle. Thus, we can write Eqs. (6)–(8) as

$$P_{ab} = P(+ - +) + P(+ - -), \tag{10}$$

$$P_{bc} = P(+ + -) + P(- + -), \tag{11}$$

$$P_{ac} = P(+ + -) + P(+ - -). \tag{12}$$

The $P(+ - +)$, etc. describe measurements that, contrary to the actual experimental situation, were never carried out. In fact, no prescription is known for measuring the separate classical probabilities $P(+ - +)$, $P(+ - -)$, etc. on the right hand sides of Eqs. (10)–(12). But, their sums, the P_{ab}, etc. on the left hand sides are indeed physical measurable probabilities.

4 Quantum Mechanics Preliminaries

For calculational purposes, the state vector, Eq. (4), for two spin 1/2 particles in an entangled singlet state can be written as

$$|\Psi\rangle = \frac{1}{\sqrt{2}} \left\{ \begin{pmatrix} 1 \\ 0 \end{pmatrix}_1 \begin{pmatrix} 0 \\ 1 \end{pmatrix}_2 - \begin{pmatrix} 0 \\ 1 \end{pmatrix}_1 \begin{pmatrix} 1 \\ 0 \end{pmatrix}_2 \right\}. \tag{13}$$

We now set the stage for our quantum mechanical considerations by recalling that for a spin 1/2 particle in a state $|\Psi\rangle$, the probability of observing spin up in a direction at angle θ to the z-axis is

$$|\langle\theta|\Psi\rangle|^2 = \langle\Psi|\theta\rangle\langle\theta|\Psi\rangle = \langle\Psi|\hat{\pi}_\theta|\Psi\rangle, \tag{14}$$

where $\hat{\pi}_\theta = |\theta\rangle\langle\theta|$ is the projection operator, and $|\theta\rangle = \exp[-(i/2)\sigma_y\theta]|\uparrow\rangle$. Thus, the projection operator for spin up in the direction of a unit vector $\hat{r}$ at angle θ to the z-axis is

$$\hat{\pi}_\theta = |\theta\rangle\langle\theta| = \frac{1}{2}(I + \sigma_z\cos\theta + \sigma_x\sin\theta) \equiv \frac{1}{2}(I + \vec{\sigma}\cdot\hat{r}), \tag{15}$$

where $\vec{\sigma} = (\sigma_x, \sigma_y, \sigma_z)$ are Pauli matrices and I is the identity matrix. We can also explicitly rewrite $\hat{\pi}_\theta$ as a matrix,

$$\hat{\pi}_\theta = \frac{1}{2} \begin{pmatrix} 1+\cos\theta & \sin\theta \\ \sin\theta & 1-\cos\theta \end{pmatrix}. \tag{16}$$

We define the physical observable P_{ab} to be the joint probability that if particle 1 passes through an SGA oriented in direction $\hat{a}$, it will be deflected in the $+\hat{a}$ ("up") direction and if particle 2 passes through an SGA oriented in direction $\hat{b}$, it will be deflected in the $+\hat{b}$ direction

$$P_{ab} = \left\langle \Psi \left| \hat{\pi}_{\hat{a}}^{(1)} \hat{\pi}_{\hat{b}}^{(2)} \right| \Psi \right\rangle, \tag{17}$$

where each projection operator has a superscript identifying the particle and a vector subscript defining the measurement direction in the x–z plane. (A calligraphy font P is used for quantum mechanical probabilities to contrast with the plain font P for classical probabilities.) Similarly,

$$P_{bc} = \left\langle \Psi \left| \hat{\pi}_{\hat{b}}^{(1)} \hat{\pi}_{\hat{c}}^{(2)} \right| \Psi \right\rangle, \tag{18}$$

$$P_{ac} = \left\langle \Psi \left| \hat{\pi}_{\hat{a}}^{(1)} \hat{\pi}_{\hat{c}}^{(2)} \right| \Psi \right\rangle, \tag{19}$$

Evaluation of Eq. (17) using Eq. (16) and the state vector Eq. (13) gives,

$$P_{ab} = \frac{1}{4}\{1 - \cos(\theta_a - \theta_b)\} = \frac{1}{4}\{1 - \hat{a}\cdot\hat{b}\}, \tag{20}$$

Analysis à la Bell   147

where $\theta_a(\theta_b)$ is the angle of vector $\hat{\mathbf{a}}$ ($\hat{\mathbf{b}}$) with respect to the z-axis. This is the well-known result for the quantum mechanical prediction. Other forms can be readily obtained from this expression; for example, P_{-ab} is the joint probability when the SGA for particle 1 is in the direction $\hat{\mathbf{a}}$ and that for particle 2 is in the direction $\hat{\mathbf{b}}$,

$$P_{-ab} = \frac{1}{4}\left\{1 + \hat{\mathbf{a}} \cdot \hat{\mathbf{b}}\right\}. \tag{21}$$

Similarly, for example, we immediately see from Eq. (20) that $P_{-a-b} = P_{ab}$. For other directions such as P_{ac}, one would simply change b to c in Eq. (20).

Let us now present a simple, quantum mechanically sound argument that will also provide guidance later. In particular, it will enable us to cast the quantum mechanical joint passage probability P_{ab} into forms that are in direct correspondence with the classical probability expressions, Eqs. (10)–(12). First, recall the completeness expression,

$$I = \hat{\pi}_{\hat{\mathbf{a}}}^{(1)} + \hat{\pi}_{-\hat{\mathbf{a}}}^{(1)}. \tag{22}$$

In this expression, note that if the angle between $\hat{\mathbf{a}}$ and the +z-axis is θ_a, then for $-\hat{\mathbf{a}}$ this angle is $\theta_a + \pi$. By inserting this identity operator I into Eqs. (17)–(19), we can obtain the quantum mechanical analogs of Eqs. (10)–(12),

$$P_{ab} = \left\langle \Psi \left| \hat{\pi}_{+\hat{\mathbf{a}}}^{(1)} \hat{\pi}_{+\hat{\mathbf{b}}}^{(2)} I^{(1)} \right| \Psi \right\rangle = \left\langle \Psi \left| \hat{\pi}_{+\hat{\mathbf{a}}}^{(1)} \hat{\pi}_{+\hat{\mathbf{b}}}^{(2)} \hat{\pi}_{+\hat{\mathbf{c}}}^{(1)} \right| \Psi \right\rangle + \left\langle \Psi \left| \hat{\pi}_{+\hat{\mathbf{a}}}^{(1)} \hat{\pi}_{+\hat{\mathbf{b}}}^{(2)} \hat{\pi}_{-\hat{\mathbf{c}}}^{(1)} \right| \Psi \right\rangle \tag{23}$$
$$\equiv \qquad P(+-+) \qquad + \qquad P(+--),$$

$$P_{bc} = \left\langle \Psi \left| I^{(1)} \hat{\pi}_{+\hat{\mathbf{b}}}^{(1)} \hat{\pi}_{+\hat{\mathbf{c}}}^{(2)} \right| \Psi \right\rangle = \left\langle \Psi \left| \hat{\pi}_{+\hat{\mathbf{a}}}^{(1)} \hat{\pi}_{+\hat{\mathbf{b}}}^{(2)} \hat{\pi}_{+\hat{\mathbf{c}}}^{(1)} \right| \Psi \right\rangle + \left\langle \Psi \left| \hat{\pi}_{-\hat{\mathbf{a}}}^{(1)} \hat{\pi}_{+\hat{\mathbf{b}}}^{(2)} \hat{\pi}_{+\hat{\mathbf{c}}}^{(1)} \right| \Psi \right\rangle \tag{24}$$
$$\equiv \qquad P(++-) \qquad + \qquad P(-+-),$$

$$P_{ac} = \left\langle \Psi \left| \hat{\pi}_{+\hat{\mathbf{b}}}^{(1)} I^{(1)} \hat{\pi}_{+\hat{\mathbf{c}}}^{(2)} \right| \Psi \right\rangle = \left\langle \Psi \left| \hat{\pi}_{+\hat{\mathbf{a}}}^{(1)} \hat{\pi}_{+\hat{\mathbf{b}}}^{(2)} \hat{\pi}_{+\hat{\mathbf{c}}}^{(1)} \right| \Psi \right\rangle + \left\langle \Psi \left| \hat{\pi}_{+\hat{\mathbf{a}}}^{(1)} \hat{\pi}_{-\hat{\mathbf{b}}}^{(2)} \hat{\pi}_{+\hat{\mathbf{c}}}^{(1)} \right| \Psi \right\rangle \tag{25}$$
$$\equiv \qquad P(++-) \qquad + \qquad P(+--),$$

where $P(+-+)$, $P(+--)$, etc. are the quantum mechanical analogs of the $P(+-+)$, $P(+--)$, etc. in Eqs. (10)–(12). Note that the concept of locality plays no role in the derivation of Eqs. (23)–(25).

Using the projection operator, Eq. (16), and the state vector, Eq. (13), it is a straight-forward calculation to evaluate the terms on the right hand sides of Eqs. (23)–(25). We explicitly present the first one,

$$\left\langle \Psi \left| \hat{\pi}_{+\hat{\mathbf{a}}}^{(1)} \hat{\pi}_{+\hat{\mathbf{b}}}^{(2)} \hat{\pi}_{+\hat{\mathbf{c}}}^{(1)} \right| \Psi \right\rangle = P(+-+) = \frac{1}{8}[1 - \hat{\mathbf{a}} \cdot \hat{\mathbf{b}} + \hat{\mathbf{a}} \cdot \hat{\mathbf{c}} - \hat{\mathbf{b}} \cdot \hat{\mathbf{c}}], \tag{26}$$

and note that knowledge of $P(+-+)$, is sufficient to obtain the others. For example, $P(+--) = \frac{1}{8}[1 - \hat{\mathbf{a}} \cdot \hat{\mathbf{b}} - \hat{\mathbf{a}} \cdot \hat{\mathbf{c}} + \hat{\mathbf{b}} \cdot \hat{\mathbf{c}}]$ is obtained from Eq. (26) by simply changing $\hat{\mathbf{c}}$ to $-\hat{\mathbf{c}}$. Likewise, an exchange of particle labels (e.g. $1 \leftrightarrow 2$) corresponds to a change of sign of the vector direction ($\hat{\mathbf{a}}$, $\hat{\mathbf{b}}$, or $\hat{\mathbf{c}}$), since spin "up" for one particle corresponds to spin "down" for the other. Such considerations are "useful" because they make clear the derived results that,

$$P(+--) = \left\langle \Psi \left| \hat{\pi}^{(1)}_{+\hat{a}} \hat{\pi}^{(1)}_{-\hat{b}} \hat{\pi}^{(2)}_{+\hat{c}} \right| \Psi \right\rangle = \left\langle \Psi \left| \hat{\pi}^{(1)}_{+\hat{a}} \hat{\pi}^{(2)}_{+\hat{b}} \hat{\pi}^{(1)}_{-\hat{c}} \right| \Psi \right\rangle$$

$$= \left\langle \Psi \left| \hat{\pi}^{(1)}_{+\hat{a}} \hat{\pi}^{(1)}_{-\hat{b}} \hat{\pi}^{(1)}_{-\hat{c}} \right| \Psi \right\rangle = \dots \quad , \tag{27}$$

where, for example, the third term involves only particle 1. Furthermore, in spite of the fact that projection operators for the same particle do not generally commute, the P (.. .. ..), are independent of the ordering of the operators as may be shown by explicit evaluation, e.g.

$$P(+--) = \left\langle \Psi \left| \hat{\pi}^{(1)}_{+\hat{a}} \hat{\pi}^{(1)}_{-\hat{b}} \hat{\pi}^{(2)}_{+\hat{c}} \right| \Psi \right\rangle = \left\langle \Psi \left| \hat{\pi}^{(1)}_{-\hat{b}} \hat{\pi}^{(1)}_{+\hat{a}} \hat{\pi}^{(2)}_{+\hat{c}} \right| \Psi \right\rangle$$

$$= \left\langle \Psi \left| \hat{\pi}^{(1)}_{+\hat{a}} \hat{\pi}^{(1)}_{-\hat{c}} \hat{\pi}^{(1)}_{-\hat{b}} \right| \Psi \right\rangle = \dots, \tag{28}$$

In a previous paper [17], we have shown that in a classical hidden variable theory, the probabilities given in Eqs. (10)–(12) lead to the expression

$$P_{ab} + P_{bc} = P_{ac} + P(+-+) + P(-+-). \tag{29}$$

Since it is assumed every P(.. .. ..) ≥ 0, the Bell–Wigner inequality is immediately obtained,

$$P_{ab} + P_{bc} \geq P_{ac}. \tag{30}$$

The expressions shown in the second lines of Eqs. (23)–(25) can be combined to obtain

$$P_{ab} + P_{bc} = P_{ac} + P(+-+) + P(-+-). \tag{31}$$

By comparing this to Eq. (29), one can easily identify the quantum analogs of the classical probabilities, P(+-+). But, the quantities P (+-+) and P(-+-), which are the quantum counterparts of P(+-+) , and P(-+-), are not always positive. For example, at the angles $\theta_a = 0$, $\theta_b = \pi/3$, and $\theta_c = 2\pi/3$, we find P(+-+) = P (-+-) = -0.0625. Thus, the introduction of hidden variables leads to probabilities that must be positive to obtain a Bell inequality, but whose quantum counterparts can be negative; i.e. they cannot be physical probabilities. This is the first example that shows the assumptions (i) hidden variables and (ii) corresponding positive probabilities are jointly untenable in quantum mechanics.

We will make use of the foregoing to examine the assumption of positive probabilities in Bell's original derivation of an inequality [2]. We will then further extend this examination to the strong Bell inequality of Clauser–Horne [25]. The latter does not require auxiliary assumptions for a doable experiment and also tests inherently stochastic theories.

Analysis à la Bell 149

5 Bell's Original Inequality

In his pioneering paper [2] Bell considers an EPR state given by Eq. (13), and assumes a "more complete specification effected by means of parameters λ" with probability distribution $\rho(\lambda)$. Then the expectation value of the product of the two components $\vec{\sigma}_1 \cdot \hat{\mathbf{a}}$ and $\vec{\sigma}_2 \cdot \hat{\mathbf{b}}$ of particles 1 and 2 in the directions $\hat{\mathbf{a}}$ and $\hat{\mathbf{b}}$ respectively, is

$$E_{ab} \equiv E(\hat{\mathbf{a}}, \hat{\mathbf{b}}) = \int d\lambda \rho(\lambda) A(\hat{\mathbf{a}}, \lambda) B(\hat{\mathbf{b}}, \lambda), \tag{32}$$

where $A = \pm 1$ and $B = \pm 1$ are the results of measuring $\vec{\sigma}_1 \cdot \hat{\mathbf{a}}$ and $\vec{\sigma}_2 \cdot \hat{\mathbf{b}}$, respectively. The quantum mechanical expectation value of $(\vec{\sigma}_1 \cdot \hat{\mathbf{a}})(\vec{\sigma}_2 \cdot \hat{\mathbf{b}})$ in the singlet state, Eq. (13), is

$$E_{ab} = \langle \Psi | (\vec{\sigma}_1 \cdot \hat{\mathbf{a}}) (\vec{\sigma}_2 \cdot \hat{\mathbf{b}}) | \Psi \rangle = -\hat{\mathbf{a}} \cdot \hat{\mathbf{b}}. \tag{33}$$

(Note, Bell used P for expectation value; we use E and E to avoid confusion with the probabilities P and P.) The relation between the quantum mechanical quantities P_{ab} and E_{ab} is easily determined. First, from Eq. (15) we find,

$$\hat{\pi}^{(1)}_{+\hat{\mathbf{a}}} - \hat{\pi}^{(1)}_{-\hat{\mathbf{a}}} = \frac{1}{2}\left(I + \vec{\sigma}_1 \cdot \hat{\mathbf{a}}\right) - \frac{1}{2}\left(I - \vec{\sigma}_1 \cdot \hat{a}\right) = \vec{\sigma}_1 \cdot \hat{\mathbf{a}}. \tag{34}$$

Hence, from Eq. (33) we have

$$
\begin{aligned}
E_{ab} &= \left\langle \Psi \left| \left(\hat{\pi}^{(1)}_{+\hat{\mathbf{a}}} - \hat{\pi}^{(1)}_{-\hat{\mathbf{a}}}\right) \left(\hat{\pi}^{(2)}_{+\hat{\mathbf{b}}} - \hat{\pi}^{(1)}_{-\hat{\mathbf{b}}}\right) \right| \Psi \right\rangle \\
&= \left\langle \Psi \left| \hat{\pi}^{(1)}_{+\hat{\mathbf{a}}}\hat{\pi}^{(2)}_{+\hat{\mathbf{b}}} \right| \Psi \right\rangle - \left\langle \Psi \left| \hat{\pi}^{(1)}_{-\hat{\mathbf{a}}}\hat{\pi}^{(2)}_{+\hat{\mathbf{b}}} \right| \right\rangle - \left\langle \Psi \left| \hat{\pi}^{(1)}_{+\hat{\mathbf{a}}}\hat{\pi}^{(2)}_{-\hat{\mathbf{b}}} \right| \Psi \right\rangle + \left\langle \Psi \left| \hat{\pi}^{(1)}_{-\hat{\mathbf{a}}}\hat{\pi}^{(2)}_{-\hat{\mathbf{b}}} \right| \Psi \right\rangle.
\end{aligned} \tag{35}
$$

From the definition, Eq. (17), this gives

$$E_{ab} = P_{ab} - P_{-ab} - P_{a-b} + P_{-a-b}. \tag{36}$$

Now, since the space $\rho(\lambda)$ in Eq. (32) is spanned by 4 regions with classical probabilities $P_{ab}, P_{-ab}, P_{a-b}, P_{-a-b}$, in which A and B have values ± 1, the integral in Eq. (32) can be explicitly evaluated as

$$E_{ab} = (+1)(+1)P_{ab} + (-1)(+1)P_{-ab} + (+1)(-1)P_{a-b} + (-1)(-1)P_{-a-b}, \tag{37}$$

$$E_{ab} = P_{ab} - P_{-ab} - P_{a-b} + P_{-a-b}, \tag{38}$$

in exact correspondence with the quantum mechanical result, Eq. (36).

To derive the Bell inequality, relationships between expectation values for combinations among three directions, e.g. $\hat{\mathbf{a}}, \hat{\mathbf{b}}$ and $\hat{\mathbf{c}}$ are required. Consequently, we use Eqs. (10)–(12) to rewrite measurable expectation values E_{jk} in terms of the hidden variable probabilities $P(\ldots\ldots)$. Using Eq. (10) in Eq. (38) gives

$$E_{ab} = P(+-+) + P(+--) - P(--+) - P(---) - P(+++)$$
$$- P(++-) + P(-++) + P(-+-). \tag{39}$$

Together with the normalization condition $\sum P(\pm\pm\pm) = 1$, this becomes

$$E_{ab} = 2\{P(+-+) + P(+--) + P(-++) + P(-+-)\} - 1. \tag{40}$$

Similarly, we find

$$E_{ac} = 2\{P(++-) + P(+--) + P(-++) + P(--+)\} - 1. \tag{41}$$
$$E_{bc} = 2\{P(++-) + P(-+-) + P(+-+) + P(--+)\} - 1. \tag{42}$$

Comparison of Eqs. (40)–(42) shows that

$$E_{ab} - E_{ac} = 1 + E_{bc} - 4P(++-) - 4P(--+) \tag{43}$$
$$E_{ac} - E_{ab} = 1 + E_{bc} - 4P(-+-) - 4P(+-+) \tag{44}$$

If, and only if, the $P(\ldots)$'s are ≥ 0, then Eqs. (43) and (44) can be written

$$E_{ab} - E_{ac} \leq 1 + E_{bc}, \tag{45}$$
$$E_{ac} - E_{ab} \leq 1 + E_{bc}, \tag{46}$$

respectively. The right hand sides are identical and the left-hand sides have opposite sign; consequently we immediately have Bell's original inequality,

$$|E_{ac} - E_{ab}| \leq 1 + E_{bc}. \tag{47}$$

Thus, the Bell hidden variable proof requires the assumption that classical probabilities like $P(+-+)$ are ≥ 0; but, their quantum mechanical counterparts can, in fact, be negative. One can only expect that the experimental tests of Bell inequalities should violate the inequalities when the experimental conditions are such that the quantum mechanical counterparts of the classical probabilities are negative. Again, since the results of experimental tests agree with the quantum mechanical predictions, hidden variables and their corresponding positive probabilities cannot co-exist in quantum mechanics.

We demonstrate the necessity of this positive probability assumption one more time by explicitly examining the derivation of Eq. (15) in Bell's original paper [2]. Specifically, in the second equation preceding his Eq. (15), Bell writes,

$$E(\hat{\mathbf{a}}, \hat{\mathbf{b}}) - E(\hat{\mathbf{a}}, \hat{\mathbf{c}}) = \int d\lambda \rho(\lambda) A(\hat{\mathbf{a}}, \lambda) B(\hat{\mathbf{b}}, \lambda) \left[A(\hat{\mathbf{b}}, \lambda) A(\hat{\mathbf{c}}, \lambda) - 1 \right]. \tag{48}$$

The space $\rho(\lambda)$ in Eq. (48) is spanned by 8 regions with probabilities $P(\pm\pm\pm)$ in which $A(\hat{\mathbf{a}}, \lambda)$, $A(\hat{\mathbf{b}}, \lambda)$, and $A(\hat{\mathbf{c}}, \lambda)$ have values ± 1. Evaluation of the integral in Eq. (48) over these eight regions gives

$$E(\hat{\mathbf{a}}, \hat{\mathbf{b}}) - E(\hat{\mathbf{a}}, \hat{\mathbf{c}}) = -2P(++-) + 2P(+-+) + 2P(-+-) - 2P(--+). \tag{49}$$

Analysis à la Bell   151

Bell takes the absolute value of both sides to obtain his eq. 15,

$$\left| E(\hat{a}, \hat{b}) - E(\hat{a}, \hat{c}) \right| \leq +2P(++-) +2P(+-+) +2P(-+-) +2P(--+). \quad (50)$$

Using Eq. (42) gives

$$\left| E(\hat{a}, \hat{b}) - E(\hat{a}, \hat{c}) \right| \leq 1 + E(\hat{b}, \hat{c}). \quad (51)$$

To obtain Eq. (50) from Eq. (49), one must assume all the required classical $P(\pm \pm \pm)$ are real, non-negative probabilities. In fact, we know their quantum counterparts can be negative.

6 The Bell Inequality of Clauser–Horne

Clauser and Horne [25] proved a Bell inequality for general local realistic theories, including inherently stochastic ones. It had the distinct advantage of defining a doable experiment that did not require auxiliary assumptions. Clauser and Shimony [26] derived the Clauser–Horne (CH) inequality using a method invented by Wigner [27] and then independently by Belinfante [28]. We will first very briefly sketch their derivation, but with two changes in notation to match the rest of our paper. We replace their $\rho(ij; kl)$ with $P(ij; kl)$; their $p_{12}(a, b)$ with P_{ab}; their $p_1(a)$ with P_{1a}; and their $p_2(b)$ with P_{2b}. We also make the discussion in the context of Bohm's version of the Einstein–Podolsky–Rosen *gedankenexperiment*, see Fig.1.

Consider two orientations, a and a', of the analyzer for particle 1, and two orientations, b and b', of the analyzer for particle 2. At each orientation, there are two possible results of the measurement, $+1$ and -1. (We will denote these as simply $+$ and $-$). Thus the space of possible measurement results consists of 16 disjoint subspaces, i.e. two possible results for each orientation: a, a', b and b' gives $2^4 = 16$ subspaces. The probability of each subspace is defined to be $P(ij; kl)$, where i is $+ (-)$ if the spin of particle 1 is up (down) in the direction a; j is $+ (-)$ if the spin of particle 1 is up (down) in the direction a'; k is $+ (-)$ if the spin of particle 2 is up (down) in the direction b; and l is $+ (-)$ if the spin of particle 2 is up (down) in the direction b'. The authors state "Clearly all $P(ij; kl)$ are non-negative", where we have written P instead of their ρ; this statement is crucial to the derivation of the Bell inequality and it is these $P(ij; kl)$ whose quantum mechanical counterparts will be shown to have negative values. Since the 16 subspaces are exhaustive and disjoint, we have

$$\sum_{ijkl} P(ij; kl) = 1. \quad (52)$$

The probability P_{ab} for measurement results of spin up in the direction a for particle 1 and direction b for particle 2 is given by the sum of all the probabilities $P(ij; kl)$ for which $i = +$ and $k = +$, similarly for $P_{a'b}$, etc. The probability $P_{1a'}$ for measurement results of spin up in the direction a for particle 1 is given by the sum of all the probabilities $P(ij; kl)$ for which $j = +$, the result for P_{2b} is analogous. Thus we have:

$$P_{ab} = P(++;++) + P(++;+-) + P(+-;++) + P(+-;+-) \tag{53}$$

$$P_{ab'} = P(++;++) + P(++;-+) + P(+-;++) + P(+-;-+) \tag{54}$$

$$P_{a'b} = P(++;++) + P(++;+-) + P(-+;++) + P(-+;+-) \tag{55}$$

$$P_{a'b'} = P(++;++) + P(++;-+) + P(-;++) + P(-+;-+) \tag{56}$$

$$
\begin{aligned}
P_{1a'} = {}&P(++;++) + P(++;+-) + P(++;-+) + P(++;--) \\
&+ P(-+;++) + P(-+;+-) + P(-+;-+) + P(-+;--)
\end{aligned}
\tag{57}
$$

$$
\begin{aligned}
P_{2b} = {}&P(++;++) + P(++;+-) + P(+-;++) + P(+-;+-) \\
&+ P(-+;++) + P(-+;+-) + P(--;++) + P(--;+-)
\end{aligned}
\tag{58}
$$

It follows that

$$
\begin{aligned}
&P_{ab} - P_{ab'} + P_{a'b} + P_{a'b'} - P_{1a'} - P_{2b} \\
&= -P(++;-+) - P(++;--) - P(+-;++) - P(+-;-+) \\
&\quad - P(-+;+-) - P(-+;--) - P(--;++) + P(--;+-)
\end{aligned}
\tag{59}
$$

The right hand side (RHS) of Eq. (59) is a sum over only 8 of the 16 subspaces, therefore we have

$$\text{RHS} \geq -\sum_{ijkl} P(ij; kl) = -1. \tag{60}$$

From Eqs. (59) and (60) the Bell inequality of CH for the more general stochastic case is obtained,

$$-1 \leq P_{ab} - P_{ab'} + P_{a'b} + P_{a'b'} - P_{1a'} - P_{2b} \leq 0. \tag{61}$$

But this result is critically dependent all the $P(\dots)$S in the RHS of Eq. (59) being non-negative; we now evaluate their quantum mechanical counterparts.

Again, in concert with the definition of P_{ab}, Eq. (53), we define the quantum mechanical observable P_{ab} to be the joint probability that if particle 1 passes through an SGA oriented in the direction $\hat{a}$, it will be deflected in the $+\hat{a}$ ("up") direction and if particle 2 passes through an SGA oriented in the direction $\hat{b}$, it will be deflected in the $+\hat{b}$ direction, Eq. (17),

$$P_{ab} = \left\langle \Psi \left| \hat{\pi}_{\hat{a}}^{(1)} \hat{\pi}_{\hat{b}}^{(2)} \right| \Psi \right\rangle. \tag{62}$$

We now write two identity operators analogous to Eq. (22)

$$I^{(1)} = \hat{\pi}_{\hat{a}'}^{(1)} + \hat{\pi}_{-\hat{a}'}^{(1)} \quad \text{and} \quad I^{(2)} = \hat{\pi}_{\hat{b}'}^{(2)} + \hat{\pi}_{-\hat{b}'}^{(2)}, \tag{63}$$

Analysis à la Bell   153

and insert them in Eq. (62),

$$P_{ab} = \left\langle \Psi \left| \hat{\pi}_{\hat{a}}^{(1)} I^{(1)} \hat{\pi}_{\hat{b}}^{(2)} I^{(2)} \right| \Psi \right\rangle = \left\langle \Psi \left| \hat{\pi}_{\hat{a}}^{(1)} \left(\hat{\pi}_{\hat{a}'}^{(1)} + \hat{\pi}_{-\hat{a}'}^{(1)} \right) \hat{\pi}_{\hat{b}}^{(2)} \left(\hat{\pi}_{\hat{b}'}^{(2)} + \hat{\pi}_{-\hat{b}}^{(2)} \right) \right| \Psi \right\rangle.$$
(64)

Expanding the projection operator products gives

$$P_{ab} = \left\langle \Psi \left| \hat{\pi}_{\hat{a}}^{(1)} \hat{\pi}_{\hat{a}'}^{(1)} \hat{\pi}_{\hat{b}}^{(2)} \hat{\pi}_{\hat{b}'}^{(2)} \right| \Psi \right\rangle + \left\langle \Psi \left| \hat{\pi}_{\hat{a}}^{(1)} \hat{\pi}_{\hat{a}'}^{(1)} \hat{\pi}_{\hat{b}}^{(2)} \hat{\pi}_{-\hat{b}'}^{(2)} \right| \Psi \right\rangle$$
$$+ \left\langle \Psi \left| \hat{\pi}_{\hat{a}}^{(1)} \hat{\pi}_{-\hat{a}'}^{(1)} \hat{\pi}_{\hat{b}}^{(2)} \hat{\pi}_{-\hat{b}'}^{(2)} \right| \Psi \right\rangle + \left\langle \Psi \left| \hat{\pi}_{\hat{a}}^{(1)} \hat{\pi}_{-\hat{a}'}^{(1)} \hat{\pi}_{\hat{b}}^{(2)} \hat{\pi}_{-\hat{b}'}^{(2)} \right| \Psi \right\rangle$$
(65)

The four terms on the right hand side of Eq. (65) are the quantum mechanical counterparts of the four corresponding classical probabilities on the right hand side of Eq. (53). We therefore write

$$P_{ab} = P(++;++) + P(++;+-) + P(+-;++) + P(+-;+-) \qquad (66)$$

We evaluate the first term on the right hand side of Eqs. (65) or (66) using the entangled singlet state of Eq. (13) and the projection operators given by Eq. (16),

$$P(++;++) = \left\langle \Psi \left| \hat{\pi}_{\hat{a}}^{(1)} \hat{\pi}_{\hat{a}'}^{(1)} \hat{\pi}_{\hat{b}}^{(2)} \hat{\pi}_{\hat{b}'}^{(2)} \right| \Psi \right\rangle$$
$$= \frac{1}{16} \{ 1 + \cos(\theta_a - \theta_{a'}) - \cos(\theta_a - \theta_b) - \cos(\theta_{a'} - \theta_b) - \cos(\theta_a - \theta_{b'})$$
$$- \cos(\theta_{a'} - \theta_{b'}) + \cos(\theta_b - \theta_{b'}) + \cos(\theta_a - \theta_{a'}) \cos(\theta_b - \theta_{b'})$$
$$+ \sin(\theta_a - \theta_{a'}) \sin(\theta_b - \theta_{b'}) \}$$
$$\equiv P(\theta_a, \theta_{a'}, \theta_b, \theta_{b'}),$$
(67)

where θ_a, $\theta_{a'}$, θ_b, and $\theta_{b'}$ are the angles between the z-axis and the unit vectors $\hat{a}, \hat{a}', \hat{b}$ and $\hat{b}'$, respectively, and we have defined the function $p(\theta_a, \theta_{a'}, \theta_b, \theta_{b'})$. The function $p(\theta_a, \theta_{a'}, \theta_b, \theta_{b'})$ can be used to evaluate the probabilities $P(ij; kl)$ on all 16 subspaces by noting that if $\hat{a}$ is at angle θ_a to the z = axis, then $-\hat{a}$ is at angle $\pi + \theta_a$ to the z = axis. Thus, for example

$$P(-+;++) = p(\pi + \theta_a, \theta_{a'}, \theta_b, \theta_{b'}). \qquad (68)$$

Equation (67) can also be written directly in terms of the unit vectors $\hat{a}$, $\hat{a}'$, $\hat{b}$ and $\hat{b}'$

$$P(++;++) = \frac{1}{16} \{ 1 + \hat{a} \cdot \hat{a}' - \hat{a} \cdot \hat{b} - \hat{a}' \cdot \hat{b} - \hat{a}' \cdot \hat{b}' - \hat{a}' \cdot \hat{b}' + \hat{b} \cdot \hat{b}' + (\hat{a} \cdot \hat{b})(\hat{a}' \cdot \hat{b}')$$
$$- (\hat{a}' \cdot \hat{b})(\hat{a} \cdot \hat{b}') - (\hat{a} \cdot \hat{a}')(\hat{b} \cdot \hat{b}') \}. \qquad (69)$$

All 16 of the quantum counterparts, $P(ij; kl)$, can be immediately obtained from Eq. (69) by changing the sign of the appropriate vector(s) in the result; this is equivalent to adding π to the angle as in Eq. (68). For example, we have

$$P(+-;++) = \frac{1}{16}\left\{1 - \hat{a}\cdot\hat{a}' - \hat{a}\cdot\hat{b} + \hat{a}'\cdot\hat{b} - \hat{a}'\cdot\hat{b}' + \hat{a}'\cdot\hat{b}\prime + \hat{b}\cdot\hat{b}' - (\hat{a}\cdot\hat{b})(\hat{a}'\cdot\hat{b}')\right.$$
$$\left. + (\hat{a}'\cdot\hat{b})(\hat{a}\cdot\hat{b}') + (\hat{a}\cdot\hat{a}\prime)(\hat{b}\cdot\hat{b}')\right\}. \tag{70}$$

Using Eq. (69) together with the appropriate modifications for the other 15 expressions, e.g. Eq. (70), one can show by direct evaluation that

$$\sum_{ijkl} P(ij;kl) = 1, \tag{71}$$

independent of the unit vectors $\hat{a}, \hat{a}', \hat{b}$ and $\hat{b}'$. Equation (71) is the quantum counterpart of Eq. (52).

We can evaluate all $P(ij;kl)$ for arbitrary $\hat{a}, \hat{a}', \hat{b}$ and $\hat{b}'$ using $p(\theta_a, \theta_{a'}, \theta_b, \theta_{b'})$ as defined in Eq. (67) and with the caveat of9adding π to the angle when the corresponding i, j, k, or l is negative. The crucial discovery is that, although the classical $P(ij;kl)$ must be non-negative in order to derive a Bell inequality, their quantum counterparts $P(ij;kl)$ can be negative. As an example,

$$p\left(\theta_a, \frac{\pi}{2}, \frac{\pi}{2}, 0\right) = -\frac{\cos\theta_a}{8}. \tag{72}$$

This is clearly negative for a wide range of angles.

7 Summary

We have repeatedly found that the introduction of hidden variables leads to classical probabilities that must be non-negative in order to derive a Bell inequality, but whose quantum counterparts can, in fact, be negative. (It should be noted that formulation of those non-negative classical probabilities assumes Bell's locality.) It is not surprising that quantum mechanical predictions violate Bell inequalities; quantum mechanics predicts some quantities are negative that the Bell inequality derivations assume are non-negative! The assumption of non-negative probabilities for seemingly physical correlations is in direct conflict with the predictions of quantum mechanics. The fact that experimental tests of Bell inequalities violate the inequalities and agree with quantum mechanical predictions is strong evidence for quantum mechanics without hidden variables. The Bell inequality derivations make a locality assumption that is inherently violated by quantum mechanics from the aspect of *Outcome Independence*; but, the Bell inequality experiments offer no justification for non-locality in the sense of superluminal communication.

To conclude, it is worthwhile to recall some previous work that bears directly on the present discussion of negative probabilities in classical models that reproduce quantum mechanical predictions. These include: (i) the Belinfante–Scully [28, 29] hidden variable theory that requires negative probabilities in order to reproduce quantum predictions; (ii) a response by Aspect [30] showing that a classical model

Analysis à la Bell

of Barut invokes negative probabilities; (iii) the observation by Meystre [31] that the introduction of negative probabilities permits construction of local hidden variable theories that reproduce quantum predictions.

Acknowledgements We gratefully acknowledge support from the Robert A. Welch Foundation (grants A-1261 and A-1218), and the Office of Naval Research. We also thank the following for many valuable discussions and suggestions: A. Aspect, P. Berman, R. Feinberg, R. Garisto, R. Glauber, D. Greenberger, R. Griffiths, P. Meystre, M. Rubin, Y. Shih, J. Sipes, and Th. Walther; the early critical and very extensive comments by A. Shimony and by J. Clauser, were especially valuable.

References

1. A. Einstein, B. Podolsky, and N. Rosen, "Can Quantum-Mechanical Description of Physical Reality Be Considered Complete?" *Phys. Rev.* **47**, 777–780 (1935).
2. J. S. Bell, "On the Einstein-Podolsky-Rosen Paradox," *Physics* **1**, 195–200 (1964).
3. S. J. Freedman and J. F. Clauser, "Experimental Test of Local Hidden-Variable Theories," *Phys. Rev. Lett.* **28**, 938–941 (1972).
4. E. S. Fry and R. C. Thompson, "Experimental Test of Local Hidden-Variable Theories," *Phys. Rev. Lett.* **37**, 465–468 (1976).
5. A. Aspect, P. Grangier, and G. Roger, "Experimental Realization of Einstein-Podolsky-Rosen-Bohm Gedankenexperiment: A New Violation of Bell's Inequalities," *Phys. Rev. Lett.* **49**, 91–94 (1982).
6. A. Aspect, J. Dalibard, and G. Roger, "Experimental Test of Bell's Inequalities Using Time-Varying Analyzers," *Phys. Rev. Lett.* **49**, 1804–1807 (1982).
7. Z. Y. Ou and L. Mandel, "Violation of Bell's Inequality and Classical Probability in a Two-Photon Correlation Experiment," *Phys. Rev. Lett.* **61**, 50–53 (1988).
8. Z. Y. Ou, S. F. Pereira, H. J. Kimble, and K. C. Peng, "Realization of the Einstein-Podolsky-Rosen Paradox for Continuous Variables," *Phys. Rev. Lett.* **68**, 3663–3666 (1992).
9. Y. H. Shih and C. O. Alley, "New Type of Einstein-Podolsky-Rosen-Bohm Experiment Using Pairs of Light Quanta Produced by Optical Parametric Down Conversion," *Phys. Rev. Lett.* **61**, 2921–2924 (1988).
10. P. R. Tapster, J. G. Rarity, and P. C. M. Owens, "Violation of Bell's Inequality over 4 km of Optical Fiber," *Phys. Rev. Lett.* **73**, 1923–1926 (1994).
11. W. Tittel, J. Brendel, H. Zbinden, and N. Gisin, "Violation of Bell Inequalities by Photons More Than 10 km Apart," *Phys. Rev. Lett.* **81**, 3563–3566 (1998).
12. E. S. Fry and T. Walther, "A Bell Inequality Experiment Based on Molecular Dissociation – Extension of the Lo-Shimony Proposal to ^{199}Hg (Nuclear Spin 1/2) Dimers", in R.S. Cohen, M.A. Horne, and J. Stachel, eds., *Experimental Metaphysics—Quantum Mechanical Studies for Abner Shimony, Vol. I.* Kluwer Academic: Dordrecht, The Netherlands. p. 61–71 (1997).
13. E. S. Fry, T. Walther, and S. Li, "Proposal for a Loophole Free Test of the Bell Inequalities," *Phys. Rev. A* **52**, 4381–4395 (1995).
14. P. G. Kwiat, P. H. Eberhard, A. M. Steinberg, and R. Y. Chiao, "Proposal for a Loophole-Free Bell Inequality Experiment," *Phys. Rev. A* **49**, 3209–3220 (1994).
15. A. Shimony, *Search for a Naturalistic World View: Volume 2.* Cambridge University Press: New York (1993).
16. J. S. Bell, *Speakable and Unspeakable in Quantum Mechanics.* Cambridge University Press: Cambridge (1987).
17. M. O. Scully, N. Erez, and E. S. Fry, "Do EPR-Bell Correlations Require a Non-Local Interpretation of Quantum Mechanics? I: Wigner Approach," *Phys. Lett. A* **347**, 56–61 (2005).

18. J. P. Jarrett, "On the Physical Significance of the Locality Conditions in the Bell Arguments," *Noûs* **18**, 569–589 (1984).
19. A. Shimony, "New Aspects of Bell's Theorem", in J. Ellis and D. Amati, eds., *Quantum Reflections*. Cambridge University Press: Cambridge. p. 136–164 (2000).
20. P. H. Eberhard, "Bell's Theorem without Hidden Variables," *Nuovo Cimento* **38**B, 75–80 (1977).
21. G. C. Ghirardi, A. Rimini, and T. Weber, "A General Argument against Superluminal Transmission through the Quantum Mechanical Measurement Process," *Nuovo Cimento Letters* **27**, 293–298 (1980).
22. D. N. Page, "The Einstein-Podolsky-Rosen Physical Reality Is Completely Described by Quantum Mechanics," *Phys. Lett.* **91**A, 57–60 (1982).
23. N. Gisin, "Weinberg's Non-Linear Quantum Mechanics and Supraluminal Communications," *Phys. Lett. A* **143**, 1–2 (1990).
24. S. Weinberg, "Testing Quantum Mechanics," *Ann. Phys.* **194**, 336–386 (1989).
25. J. F. Clauser and M. A. Horne, "Experimental Consequences of Objective Local Theories," *Phys. Rev. D* **10**, 526–535 (1974).
26. J. F. Clauser and A. Shimony, "Bell's Theorem: Experimental Tests and Implications," *Rep. Prog. Phys.* **41**, 1881–1927 (1978).
27. E. P. Wigner, "On Hidden Variables and Quantum Mechanical Probabilities," *Am. J. Phys.* **38**, 1005–1009 (1970).
28. F. J. Belinfante, A *Survey of Hidden-Variable Theories*. Pergamon: New York (1973).
29. M. O. Scully, "How to Make Quantum Mechanics Look Like a Hidden-Variable Theory and Vice Versa," *Phys. Rev. D* **28**, 2477–2484 (1983).
30. A. Aspect, "Comment on a Classical Model of EPR Experiment with Quantum Mechanical Correlations and Bell Inequalities", in G.T. Moore and M.O. Scully, eds., *Frontiers of Nonequilibrium Statistical Physics*. Plenum: New York. p. 185–189 (1986).
31. P. Meystre, "Is reality really real? An Introduction to Bell's Inequalities", in A.O. Barut, ed., *Quantum Electrodynamics and Quantum Optics*. Plenum: New York. p. 443–458 (1984).

The Physics of $2 \neq 1 + 1$

Yanhua Shih

Abstract One of the most surprising consequences of quantum mechanics is the entanglement of two or more distant particles. In an entangled EPR two-particle system, the value of the momentum (position) for neither single subsystem is determined. However, if one of the subsystems is measured to have a certain momentum (position), the other subsystem is determined to have a unique corresponding value, despite the distance between them. This peculiar behavior of an entangled quantum system has been observed experimentally, such as in two-photon temporal correlation measurements and in two-photon imaging experiments. This article addresses the fundamental concerns behind these experimental observations and explores the nonclassical nature of two-photon superposition by emphasizing the physics of $2 \neq 1 + 1$.

1 Introduction

In quantum theory, *a particle* is allowed to exist in a set of orthogonal states simultaneously. A vivid picture of this concept might be Schrödinger's cat, where his cat is in a state of both alive and dead simultaneously. In mathematics, the concepts of "alive" and "dead" are expressed through the idea of orthogonality. In quantum mechanics, the superpositions of these orthogonal states are used to describe the physical reality of a quantum object. In this respect the superposition principle is indeed a mystery when compared with our everyday experience.

In this article, we discuss another surprising consequence of quantum mechanics, namely that of quantum entanglement. Quantum entanglement involves a multi-particle system in a coherent superposition of orthogonal states. Here again Schrödinger's cat is a nice way of emphasizing the strangeness of quantum entanglement. Now imagine two Schrödinger's cats propagating to separate distant

Y. Shih
Department of Physics, University of Maryland, Baltimore County, Baltimore, MD 21250, USA

W.C. Myrvold and J. Christian (eds.), *Quantum Reality, Relativistic Causality, and Closing the Epistemic Circle*, The Western Ontario Series in Philosophy of Science 73,
© Springer Science+Business Media B.V. 2009

locations. The two cats are nonclassical by means of the following two criteria: (1) each of the cats is in a state of alive and dead simultaneously; (2) the two must be observed to be both alive or both dead whenever we observe them, despite their separation. There would probably be no concern if our observations were based on a large number of alive–alive or dead–dead twin cats, pair by pair, with say a 50% chance to observe a dead–dead or alive–alive pair. However, we are talking about a single pair of cats with this single pair being in the state of alive–alive and dead–dead simultaneously, and, in addition each of the cats in the pair must be alive and dead simultaneously. The superposition of multi-particle states with these entangled properties represents a troubling concept to classical theory. These concerns derive not only from the fact that the superposition of multi-particle states has no classical counterpart, but also because it represents a nonlocal behavior which may never be understood classically.

The concept of quantum entanglement in 1935 was emphasized by Einstein, Podolsky and Rosen [1], who suggested a *gedankenexperiment* and introduced an entangled two-particle system based on the superposition of two-particle wavefunctions. The EPR system is composed of two distant interaction-free particles which are characterized by the following wavefunction:

$$\Psi(x_1, x_2) = \frac{1}{2\pi\hbar} \int dp_1 dp_2 \, \delta(p_1 + p_2) \, e^{ip_1(x_1 - x_0)/\hbar} e^{ip_2 x_2/\hbar} = \delta(x_1 - x_2 - x_0) \quad (1)$$

where $e^{ip_1(x_1-x_0)/\hbar}$ and $e^{ip_2 x_2/\hbar}$ are the eigenfunctions with eigenvalues $p_1 = p$ and $p_2 = -p$ of the momentum operators $\hat{p}_1$ and $\hat{p}_2$ associated with particles 1 and 2, respectively. x_1 and x_2 are the coordinate variables to describe the positions of particles 1 and 2, respectively; and x_0 is a constant. The EPR state is very peculiar. Although there is no interaction between the two distant particles, the two-particle superposition cannot be factorized into a product of two individual superpositions of two particles. Remarkably, quantum theory permits such states.

What can we learn from the EPR state of Eq. (1)?

(1) In coordinate representation, the wavefunction is a delta function $\delta(x_1 - x_2 - x_0)$. The two particles are separated in space with a constant value of $x_1 - x_2 = x_0$, although the coordinates x_1 and x_2 of the two particles are both unspecified.
(2) The delta wavefunction $\delta(x_1 - x_2 - x_0)$ is the result of the superposition of plane wavefunctions for free particle one, $e^{ip_1(x_1-x_0)/\hbar}$, and free particle two, $e^{ip_2 x_2/\hbar}$, with a particular distribution $\delta(p_1 + p_2)$. It is $\delta(p_1 + p_2)$ that made the superposition special. Although the momentum of particle one and particle two may take on any values, the delta function restricts the superposition to only those terms in which the total momentum of the system takes a constant value of zero.

Now, we transfer the wavefunction from coordinate representation to momentum representation:

$$\Psi(p_1, p_2) = \frac{1}{2\pi\hbar} \int dx_1 dx_2 \, \delta(x_1 - x_2 - x_0) \, e^{-ip_1(x_1-x_0)/\hbar} e^{-ip_2 x_2/\hbar} = \delta(p_1 + p_2). \quad (2)$$

The Physics of $2 \neq 1 + 1$ 159

What can we learn from the EPR state of Eq. (2)?

(1) In momentum representation, the wavefunction is a delta function $\delta(p_1 + p_2)$. The total momentum of the two-particle system takes a constant value of $p_1 + p_2 = 0$, although the momenta p_1 and p_2 are both unspecified.

(2) The delta wavefunction $\delta(p_1 + p_2)$ is the result of the superposition of plane wavefunctions for free particle one, $e^{-ip_1(x_1 - x_0)/\hbar}$, and free particle two, $e^{-ip_2 x_2/\hbar}$, with a particular distribution $\delta(x_1 - x_2 - x_0)$. It is $\delta(x_1 - x_2 - x_0)$ that made the superposition special. Although the coordinates of particle one and particle two may take on any values, the delta function restricts the superposition to only those terms in which $x_1 - x_2$ is a constant value of x_0.

In an EPR system, the value of the momentum (position) for neither single subsystem is determined. However, if one of the subsystems is measured to be at a certain momentum (position), the other one is determined with a unique corresponding value, despite the distance between them. An idealized EPR state of a two-particle system is therefore characterized by $\Delta(p_1 + p_2) = 0$ and $\Delta(x_1 - x_2) = 0$ simultaneously, even if the momentum and position of each individual free particle are completely undefined, i.e., $\Delta p_j \sim \infty$ and $\Delta x_j \sim \infty$, $j = 1, 2$. In other words, each of the subsystems may have completely random values or all possible values of momentum and position in the course of their motion, but the correlations of the two subsystems are determined with certainty whenever a joint measurement is performed.

The EPR states of Eqs. (1) and (2) are simply the results of the quantum mechanical *superposition of two-particle states*. The physics behind EPR states is far beyond the acceptable limit of Einstein.

Does a free particle have a defined momentum and position in the state of Eqs. (1) and (2), regardless of whether we measure it or not? On one hand, the momentum and position of neither independent particle is specified and the superposition is taken over all possible values of the momentum and position. We may have to believe that the particles do not have any defined momentum and position, or have all possible values of momentum and position within the superposition, during the course of their motion. On the other hand, if the measured momentum (position) of one particle uniquely determines the momentum (position) of the other distant particle, it would be hard for anyone who believes no action-at-a-distance to imagine that the momenta (position) of the two particles are not predetermined with defined values before the measurement. EPR thus put us into a paradoxical situation. It seems reasonable for us to ask the same question that EPR had asked in 1935: "Can quantum-mechanical description of physical reality be considered complete?" [1].

In their 1935 article, Einstein, Podolsky and Rosen argued that the existence of the entangled two-particle state of Eqs. (1) and (2), a straightforward quantum mechanical superposition of two-particle states, led to the violation of the uncertainty principle of quantum theory. To draw their conclusion, EPR started from the following criteria.

Locality: there is no action-at-a-distance;

Reality: "if, without in any way disturbing a system, we can predict with certainty the value of a physical quantity, then there exist an element of physical reality corresponding to this quantity." According to the delta wavefunctions, we can predict with certainty the result of measuring the momentum (position) of particle 1 by measuring the momentum (position) of particle 2, and the measurement of particle 2 cannot cause any disturbance to particle 1, if the measurements are space-like separated events. Thus, both the momentum and position of particle 1 must be elements of physical reality regardless of whether we measure it or not. This, however, is not allowed by quantum theory. Now consider:

Completeness: "every element of the physical reality must have a counterpart in the complete theory." This led to the question as the title of their 1935 article: "Can Quantum-Mechanical Description of Physical Reality Be Considered Complete?"

The EPR argument was never appreciated by Copenhagen. Bohr criticized EPR's criterion of physical reality [2]: "it is too narrow". However, it is perhaps not easy to find a wider criterion. A memorable quote from Wheeler, "No elementary quantum phenomenon is a phenomenon until it is a recorded phenomenon", summarizes what Copenhagen has been trying to teach us [3]. By 1927, most physicists accepted the Copenhagen interpretation as the standard view of quantum formalism. Einstein, however, refused to compromise. As Pais recalled in his book, during a walk around 1950, Einstein suddenly stopped and "asked me if I really believed that the moon (pion) exists only if I look at it." [4]

There has been arguments considering $\Delta(p_1 + p_2)\Delta(x_1 - x_2) = 0$ a violation of the uncertainty principle. This argument is false. It is easy to find that $p_1 + p_2$ and $x_1 - x_2$ are not conjugate variables. As we know, non-conjugate variables correspond to commuting operators in quantum mechanics, if the corresponding operators exist.[1] To have $\Delta(p_1 + p_2) = 0$ and $\Delta(x_1 - x_2) = 0$ simultaneously, or to have $\Delta(p_1 + p_2)\Delta(x_1 - x_2) = 0$, is not a violation of the uncertainty principle. This point can easily be seen from the following two dimensional Fourier transforms:

$$\Psi(x_1, x_2) = \frac{1}{2\pi\hbar} \int dp_1\, dp_2\, \delta(p_1 + p_2)\, e^{ip_1(x_1-x_0)/\hbar}\, e^{ip_2 x_2/\hbar}$$

$$= \frac{1}{2\pi\hbar} \int d(p_1 + p_2)\, \delta(p_1 + p_2)\, e^{i(p_1+p_2)(x_1'+x_2)/2\hbar} \int d(p_1 - p_2)/2\, e^{i(p_1-p_2)(x_1'-x_2)/2\hbar}$$

$$= 1 \times \delta(x_1 - x_2 - x_0)$$

where $x' = x_1 - x_0$;

$$\Psi(p_1, p_2) = \frac{1}{2\pi\hbar} \int dx_1\, dx_2\, \delta(x_1 - x_2 - x_0)\, e^{-ip_1(x_1-x_0)/\hbar}\, e^{-ip_2 x_2/\hbar}$$

$$= \frac{1}{2\pi\hbar} \int d(x_1' + x_2)\, e^{-i(p_1+p_2)(x_1'+x_2)/2\hbar} \int d(x_1' - x_2)/2\, \delta(x_1' - x_2)\, e^{-i(p_1-p_2)(x_1'-x_2)/2\hbar}$$

$$= \delta(p_1 + p_2) \times 1.$$

The Fourier conjugate variables are $(x_1 + x_2) \Leftrightarrow (p_1 + p_2)$ and $(x_1 - x_2) \Leftrightarrow (p_1 - p_2)$. Although it is possible to have $\Delta(x_1 - x_2) \sim 0$ and $\Delta(p_1 + p_2) \sim 0$ simultaneously,

---

[1] It is possible that no quantum mechanical operator is associated with a measurable variable, such as time t. From this perspective, an uncertainty relation based on variables rather than operators is more general.

The Physics of $2 \neq 1 + 1$ 161

the uncertainty relations must hold for the Fourier conjugates $\Delta(x_1 + x_2)\Delta(p_1 + p_2) \geq \hbar$, and $\Delta(x_1 - x_2)\Delta(p_1 - p_2) \geq \hbar$; with $\Delta(p_1 - p_2) \sim \infty$ and $\Delta(x_1 + x_2) \sim \infty$.

In fact, in their 1935 paper, Einstein–Podolsky–Rosen never questioned $\Delta(x_1 - x_2)\Delta(p_1 + p_2) = 0$ as a violation of the uncertainty principle. The violation of the uncertainty principle was probably not Einstein's concern at all, although their 1935 paradox was based on the argument of the uncertainty principle. What really bothered Einstein so much? For all of his life, Einstein, a true believer of realism, never accepted that a particle does not have a defined momentum and position during its motion, but rather is specified by a probability amplitude of certain a momentum and position. "God does not play dice" was the most vivid criticism from Einstein to refuse the Schrödinger's cat. The entangled two-particle system was used as an example to clarify and to reinforce Einstein's realistic opinion. To Einstein, the acceptance of Schrödinger's cat perhaps means action-at-a-distance or an inconsistency between quantum mechanics and the theory of relativity, when dealing with the entangled EPR two-particle system. Let us follow Copenhagen to consider that *each particle* in an EPR pair has no defined momentum and position, or has all possible momentum and position within the superposition state, i.e., imagine $\Delta p_j \neq 0$, $\Delta x_j \neq 0$, $j = 1, 2$, for *each single-particle* until the measurement. Assume the measurement devices are particle counting devices able to identify the position of each particle among an ensemble of particles. For each registration of a particle the measurement device records a value of its position. No one can predict what value is registered for each measurement; the best knowledge we may have is the probability to register that value. If we further assume no physical interaction between the two distant particles and believe no action-at-a-distance exist in nature, we would also believe that no matter how the two particles are created, the two registered values must be independent of each other. Thus, the value of $x_1 - x_2$ is unpredictable within the uncertainties of Δx_1 and Δx_2. The above statement is also valid for the momentum measurement. Therefore, after a set of measurements on a large number of particle pairs, the statistical uncertainty of the measurement on $p_1 + p_2$ and $x_1 - x_2$ must obey the following inequalities:

$$\Delta(p_1 + p_2) = \sqrt{(\Delta p_1)^2 + (\Delta p_2)^2} > Max(\Delta p_1, \Delta p_2) \tag{3}$$

$$\Delta(x_1 - x_2) = \sqrt{(\Delta x_1)^2 + (\Delta x_2)^2} > Max(\Delta x_1, \Delta x_2).$$

Equation (3) is obviously true in statistics, especially when we are sure that no disturbance is possible between the two independent-local measurements. This condition can be easily realized by making the two measurement events space-like separated events. The classical inequality of Eq. (3) would not allow $\Delta(p_1 + p_2) = 0$ and $\Delta(x_1 - x_2) = 0$ as required in the EPR state, unless $\Delta p_1 = 0$, $\Delta p_2 = 0$, $\Delta x_1 = 0$ and $\Delta x_2 = 0$, simultaneously. Unfortunately, the assumption of $\Delta p_1 = 0$, $\Delta p_2 = 0$, $\Delta x_1 = 0$, $\Delta x_2 = 0$ cannot be true because it violates the uncertainty relations $\Delta p_1 \Delta x_1 \geq \hbar$ and $\Delta p_2 \Delta x_2 \geq \hbar$.

162   Y. Shih

In a non-perfect entangled system, the uncertainties of $p_1 + p_2$ and $x_1 - x_2$ may differ from zero. Nevertheless, the measurements may still satisfy the EPR inequalities [5]:

$$\Delta(p_1 + p_2) < min(\Delta p_1, \Delta p_2) \tag{4}$$
$$\Delta(x_1 - x_2) < min(\Delta x_1, \Delta x_2).$$

The apparent contradiction between the classical inequality Eq. (3) and the EPR inequality Eq. (4) deeply troubled Einstein. While one sees the measurements of $p_1 + p_2$ and $x_1 - x_2$ of the two distant individual free particles satisfying Eq. (4), but believing Eq. (3), one might easily be trapped into concluding either there is a violation of the uncertainty principle or there exists action-at-a-distance.

Is it possible to have a realistic theory which provides correct predictions of the behavior of a particle similar to quantum theory and, at the same time, respects the description of physical reality by EPR as "complete"? Bohm and his followers have attempted a "hidden variable theory", which seemed to satisfy these requirements [6]. The hidden variable theory was successfully applied to many different quantum phenomena until 1964, when Bell proved a theorem to show that an inequality, which is violated by certain quantum mechanical statistical predictions, can be used to distinguish local hidden variable theory from quantum mechanics [7]. Since then, the testing of Bell's inequalities became a standard instrument for the study of fundamental problems of quantum theory [8]. The experimental testing of Bell's inequality started from the early 1970's. Most of the historical experiments concluded the violation of the Bell's inequalities and thus disproved the local hidden variable theory [8–10].

In the following, we examine a simple yet popular realistic model to simulate the behavior of the entangled EPR system. This model concerns an ensemble of classically correlated particles instead of the quantum mechanical superposition of a particle. In terms of "cats", this model is based on the measurement of a large number of twin cats in which 50% are alive–alive twins and 50% are dead–dead twins. This model refuses the concept of Schrödinger's cat which requires *a cat* to be alive and dead simultaneously, and *each pair* of cats involved in a joint detection event is in the state of alive–alive and dead–dead simultaneously.

In this model, we may have three different states:

(1) State one, each single pair of particles holds defined momenta p_1 = constant and p_2 = constant with $p_1 + p_2 = 0$. From pair to pair, the values of p_1 and p_2 may vary significantly. The sum of p_1 and p_2, however, keeps a constant of zero. Thus, each joint detection of the two distant particles measures precisely the constant values of p_1 and p_2 and measures $p_1 + p_2 = 0$. The uncertainties of Δp_1 and Δp_2 only have statistical meaning in terms of the measurements of an ensemble. This model successfully simulated $\Delta(p_1 + p_2) = 0$ based on the measurement of a large number of classically correlated particle pairs. This is, however, only half of the EPR story. Can we have $\Delta(x_1 - x_2) = 0$ simultaneously in this model? We do have $\Delta x_1 \sim \infty$ and $\Delta x_2 \sim \infty$, otherwise the uncertainty

The Physics of $2 \neq 1+1$ 163

principle will be violated. The position correlation, however, can never achieve $\Delta(x_1 - x_2) = 0$ by any means.

(2) State two, each single pair of particles holds a well defined position $x_1 = $ constant and $x_2 = $ constant with $x_1 - x_2 = x_0$. From pair to pair, the values of x_1 and x_2 may vary significantly. The difference of x_1 and x_2, however, maintains a constant of x_0. Thus, each joint detection of the two distant particles measures precisely the constant values of x_1 and x_2 and measures $x_1 - x_2 = x_0$. The uncertainties of Δx_1 and Δx_2 only have statistical meaning in terms of the measurements of an ensemble. This model successfully simulated $\Delta(x_1 - x_2) = 0$ based on the measurement of a large number of classically correlated particle pairs. This is, however, only half of the EPR story. Can we have $\Delta(p_1 + p_2) = 0$ simultaneously in this model? We do have $\Delta p_1 \sim \infty$ and $\Delta p_2 \sim \infty$, otherwise the uncertainty principle will be violated. The momentum correlation, however, can never achieve $\Delta(p_1 + p_2) = 0$ by any means.

The above two models of classically correlated particle pairs can never achieve both $\Delta(p_1 + p_2) = 0$ and $\Delta(x_1 - x_2) = 0$. What would happen if we combine the two parts together? This leads to the third model of classical simulation.

(3) State three, among a large number of classically correlated particle pairs, we assume 50% to be in state one and the other 50% state two. The $p_1 + p_2$ measurements would have 50% chance with $p_1 + p_2 = 0$ and 50% chance with $p_1 + p_2 = $ random value. On the other hand, the $x_1 - x_2$ measurements would have 50% chance with $x_1 - x_2 = x_0$ and 50% chance with $x_1 - x_2 = $ random value. What are the statistical uncertainties on the measurements of $(p_1 + p_2)$ and $(x_1 - x_2)$ in this case? If we focus on only these events of state one, the statistical uncertainty on the measurement of $(p_1 + p_2)$ is $\Delta(p_1 + p_2) = 0$, and if we focus on these events of state two, the statistical uncertainty on the measurement of $(x_1 - x_2)$ is $\Delta(x_1 - x_2) = 0$; however, if we consider all the measurements together, the statistical uncertainties on the measurements of $(p_1 + p_2)$ and $(x_1 - x_2)$, are both infinity: $\Delta(p_1 + p_2) = \infty$ and $\Delta(x_1 - x_2) = \infty$.

In conclusion, classically correlated particle pairs may partially simulate EPR correlation with three types of optimized observations:

(1) $\Delta(p_1 + p_2) = 0$ (100%) & $\Delta(x_1 - x_2) = \infty$ (100%);

(2) $\Delta(x_1 - x_2) = 0$ (100%) & $\Delta(p_1 + p_2) = \infty$ (100%);

(3) $\Delta(p_1 + p_2) = 0$ (50%) & $\Delta(x_1 - x_2) = 0$ (50%);

Within one setup of experimental measurements, only the entangled EPR states result in the simultaneous observation of

$$\Delta(p_1 + p_2) = 0 \ (100\%) \ \& \ \Delta(x_1 - x_2) = 0 \ (100\%)$$
$$\Delta p_1 \sim \infty, \quad \Delta p_2 \sim \infty, \quad \Delta x_1 \sim \infty, \quad \Delta x_2 \sim \infty.$$

We thus have a tool, besides the testing of Bell's inequality, to distinguish quantum entangled states from classically correlated particle pairs.

164   Y. Shih

2 Entangled State

The entangled state of a two-particle system was mathematically formulated by
Schrödinger [11]. Consider a pure state for a system composed of two distinguishable subsystems

$$|\Psi\rangle = \sum_{a,b} c(a,b) |a\rangle |b\rangle \tag{5}$$

where $\{|a\rangle\}$ and $\{|b\rangle\}$ are two sets of orthogonal vectors for subsystems 1 and 2,
respectively. If $c(a,b)$ does not factor into a product of the form $f(a) \times g(b)$, then it
follows that the state does not factor into a product state for subsystems 1 and 2:

$$\hat{\rho} = |\Psi\rangle\langle\Psi| = \sum_{a,b} c(a,b)|a\rangle|b\rangle \sum_{a',b'} c^*(a',b')\langle b'|\langle a'| \neq \hat{\rho}_1 \times \hat{\rho}_2, \tag{6}$$

where $\hat{\rho}$ is the density operator, the state was defined by Schrödinger as an entangled
state.

Following this notation, the first classic entangled state of a two-particle system,
the EPR state of Eqs. (1) and (2), is thus written as:

$$|\Psi\rangle_{EPR} = \sum_{x_1,x_2} \delta(x_1 - x_2 + x_0) |x_1\rangle|x_2\rangle = \sum_{p_1,p_2} \delta(p_1 + p_2) |p_1\rangle|p_2\rangle, \tag{7}$$

where we have described the entangled two-particle system as the coherent superposition of the momentum eigenstates as well as the coherent superposition of
the position eigenstates. The two δ-functions in Eq. (7) represent, respectively and
simultaneously, the perfect position–position and momentum–momentum correlation. Although the two distant particles are interaction-free, the superposition selects
only the eigenstates which are specified by the δ-function. We may use the following statement to summarize the surprising feature of the EPR state: *the values of the
momentum and the position for neither interaction-free single subsystem is deter-
minated. However, if one of the subsystems is measured to be at a certain value of
momentum and/or position, the momentum and/or position of the other one is 100%
determined, despite the distance between them.*

It should be emphasized again that Eq. (7) is true, simultaneously, in the conjugate space of momentum and position. This is different from classically correlated
states

$$\hat{\rho} = \sum_{p_1,p_2} \delta(p_1 + p_2) |p_1\rangle|p_2\rangle\langle p_2|\langle p_1|, \tag{8}$$

or

$$\hat{\rho} = \sum_{x_1,x_2} \delta(x_1 - x_2 + x_0) |x_1\rangle|x_2\rangle\langle x_2|\langle x_1|. \tag{9}$$

Equations (8) and (9) represent mixed states. Equations (8) and (9) cannot be true simultaneously as we have discussed earlier. Thus, we can distinguish entangled states
from classically correlated states through the measurements of the EPR inequalities
of Eq. (4).

The Physics of $2 \neq 1 + 1$ 165

2.1 Two-Photon State of Spontaneous Parametric Down-Conversion

The state of a signal-idler photon pair created in spontaneous parametric down-conversion (SPDC) is a typical EPR state [12, 13]. Roughly speaking, the process of SPDC involves sending a pump laser beam into a nonlinear material, such as a non-centrosymmetric crystal. Occasionally, the nonlinear interaction leads to the annihilation of a high frequency pump photon and the simultaneous creation of a pair of lower frequency signal-idler photons forming an entangled two-photon state:

$$|\Psi\rangle = \Psi_0 \sum_{s,i} \delta\left(\omega_s + \omega_i - \omega_p\right) \delta\left(\mathbf{k}_s + \mathbf{k}_i - \mathbf{k}_p\right) a_s^\dagger(\mathbf{k}_s)\, a_i^\dagger(\mathbf{k}_i) \,|\, 0\rangle \qquad (10)$$

where ω_j, $\mathbf{k}_j$ (j = s, i, p) are the frequency and wavevector of the signal (s), idler (i), and pump (p), $a_s^\dagger$ and $a_i^\dagger$ are creation operators for the signal and the idler photon, respectively, and Ψ_0 is the normalization constant. We have assumed a CW monochromatic laser pump, i.e., ω_p and $\mathbf{k}_p$ are considered as constants. The two delta functions in Eq. (10) are technically named as the phase matching condition [12, 14]:

$$\omega_p = \omega_s + \omega_i, \qquad \mathbf{k}_p = \mathbf{k}_s + \mathbf{k}_i. \qquad (11)$$

The names *signal* and *idler* are historical leftovers. The names perhaps came about due to the fact that in the early days of SPDC, most of the experiments were done with non-degenerate processes. One radiation was in the visible range (and thus easily observable, the signal), while the other was in the IR range (usually not measured, the idler). We will see in the following discussions that the role of the idler is no any less important than that of the signal. The SPDC process is referred to as type-I if the signal and idler photons have identical polarizations, and type-II if they have orthogonal polarizations. The process is said to be *degenerate* if the SPDC photon pair has the same free space wavelength (e.g. $\lambda_i = \lambda_s = 2\lambda_p$), and *nondegenerate* otherwise. In general, the pair exit the crystal *non-collinearly*, that is, propagate to different directions defined by the second equation in Eq. (11) and Snell's law. In addition, the pair may also exit *collinearly*, in the same direction, together with the pump.

The state of the signal–idler pair can be derived, quantum mechanically, by the first order perturbation theory with the help of the nonlinear interaction Hamiltonian. The SPDC interaction arises in a nonlinear crystal driven by a pump laser beam. The polarization, i.e., the dipole moment per unit volume, is given by

$$P_i = \chi_{i,j}^{(1)} E_j + \chi_{i,j,k}^{(2)} E_j E_k + \chi_{i,j,k,l}^{(3)} E_j E_k E_l + \dots \qquad (12)$$

where $\chi^{(m)}$ is the *mth* order electrical susceptibility tensor. In SPDC, it is the second order nonlinear susceptibility $\chi^{(2)}$ that plays the role. The second order nonlinear interaction Hamiltonian can be written as

$$H = \varepsilon_0 \int_V d\mathbf{r} \, \chi_{ijk}^{(2)} \, E_i E_j E_k \tag{13}$$

where the integral is taken over the interaction volume V.

It is convenient to use the Fourier representation for the electrical fields in Eq. (13):

$$\mathbf{E}(\mathbf{r}, t) = \int d\mathbf{k} \, [\mathbf{E}^{(-)}(\mathbf{k})e^{-i(\omega(\mathbf{k})t - \mathbf{k} \cdot \mathbf{r})} + \mathbf{E}^{(+)}(\mathbf{k})e^{i(\omega(\mathbf{k})t - \mathbf{k} \cdot \mathbf{r})}]. \tag{14}$$

Substituting Eq. (14) into Eq. (13) and keeping only the terms of interest, we obtain the SPDC Hamiltonian in the interaction representation:

$$
\begin{aligned}
&H_{int}(t) \\
&= \varepsilon_0 \int_V d\mathbf{r} \int d\mathbf{k}_s \, d\mathbf{k}_i \, \chi_{lmn}^{(2)} E_{pl}^{(+)} e^{i(\omega_p t - \mathbf{k}_p \cdot \mathbf{r})} E_{sm}^{(-)} e^{-i(\omega_s(\mathbf{k}_s)t - \mathbf{k}_s \cdot \mathbf{r})} E_{in}^{(-)} e^{-i(\omega_i(\mathbf{k}_i)t - \mathbf{k}_i \cdot \mathbf{r})} + h.c.,
\end{aligned}
\tag{15}
$$

where $h.c.$ stands for Hermitian conjugate. To simplify the calculation, we have also assumed the pump field to be a monochromatic plane wave with wave vector $\mathbf{k}_p$ and frequency ω_p.

It is easily noticeable that in Eq. (15), the volume integration can be done for some simplified cases. At this point, we assume that V is infinitely large. Later, we will see that the finite size of V in longitudinal and/or transversal directions may have to be taken into account. For an infinite volume V, the interaction Hamiltonian Eq. (15) is written as

$$H_{int}(t) = \varepsilon_0 \int d\mathbf{k}_s \, d\mathbf{k}_i \, \chi_{lmn}^{(2)} E_{pl}^{(+)} E_{sm}^{(-)} E_{in}^{(-)} \, \delta(\mathbf{k}_p - \mathbf{k}_s - \mathbf{k}_i) e^{i(\omega_p - \omega_s(\mathbf{k}_s) - \omega_i(\mathbf{k}_i))t} + h.c. \tag{16}$$

It is reasonable to consider the pump field to be classical, which is usually a laser beam, and quantize the signal and idler fields, which are both at the single-photon level:

$$E^{(-)}(\mathbf{k}) = i\sqrt{\frac{2\pi\hbar\omega}{V}} a^\dagger(\mathbf{k}), \quad E^{(+)}(\mathbf{k}) = i\sqrt{\frac{2\pi\hbar\omega}{V}} a(\mathbf{k}), \tag{17}$$

where $a^\dagger(\mathbf{k})$ and $a(\mathbf{k})$ are photon creation and annihilation operators, respectively. The state of the emitted photon pair can be calculated by applying the first order perturbation

$$|\Psi\rangle = -\frac{i}{\hbar} \int dt \, H_{int}(t) \, |0\rangle. \tag{18}$$

By using vacuum $|0\rangle$ for the initial state in Eq. (18), we assume that there is no input radiation in any signal and idler modes, that is, we have a spontaneous parametric down conversion (SPDC) process.

Further assuming an infinite interaction time, evaluating the time integral in Eq. (18) and omitting altogether the constants and slow (square root) functions of ω, we obtain the *entangled* two-photon state of Eq. (10) in the form of an integral [13]:

$$|\Psi\rangle = \Psi_0 \int d\mathbf{k}_s d\mathbf{k}_i \, \delta[\omega_p - \omega_s(\mathbf{k}_s) - \omega_i(\mathbf{k}_i)] \delta(\mathbf{k}_p - \mathbf{k}_s - \mathbf{k}_i) a_s^\dagger(\mathbf{k}_s) a_i^\dagger(\mathbf{k}_i)|0\rangle \quad (19)$$

where Ψ_0 is a normalization constant which has absorbed all omitted constants.

The way of achieving phase matching, i.e., the delta functions, in Eq. (19) basically determines how the signal–idler pair "looks". For example, in a negative uniaxial crystal, one can use a linearly polarized pump laser beam as an extraordinary ray of the crystal to generate a signal–idler pair both polarized as the ordinary rays of the crystal, which is defined as type-I phase matching. One can alternatively generate a signal–idler pair with one ordinary polarized and another extraordinary polarized, which is defined as type II phase matching. Figure 1 shows three examples of an SPDC two-photon source. All three schemes have been widely used for different experimental purposes. Technical details can be found in text books and research references in nonlinear optics.

The two-photon state in the forms of Eq. (10) or Eq. (19) is a pure state, which mathematically describes the behavior of a signal–idler photon pair. The surprise comes from the coherent superposition of the two-photon modes:

> Does the signal or the idler photon in the EPR state of Eq. (10) or Eq. (19) have a defined energy and momentum regardless of whether we measure it or not? Quantum mechanics answers: No! However, if one of the subsystems is measured with a certain energy and momentum, the other one is determined with certainty, despite the distance between them.

It is indeed a mystery from a classical point of view. There has been, nevertheless, classical models to avoid the surprises. One of the classical realistic models insists that the state of Eq. (10) or Eq. (19) only describes the behavior of an ensemble of photon pairs. In this model, the energy and momentum of the signal photon and the idler photon in each individual pair are defined with certain values and the

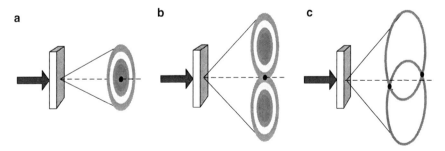

Fig. 1 Three widely used SPDC setups. (**a**) Type-I SPDC. (**b**) Collinear degenerate type-II SPDC. Two rings overlap at one region. (**c**) Non-collinear degenerate type-II SPDC. For clarity, only two degenerate rings, one for e-polarization and the other for o-polarization, are shown. Notice, the color rainbows represent the distribution function of a signal–idler pair. One signal–idler pair yields the entire rainbow

168   Y. Shih

resulting state is a statistical mixture. Mathematically, it is incorrect to use a pure state to characterize a statistical mixture. The concerned statistical ensemble should be characterized by the following density operator

$$\hat{\rho} = \int d\mathbf{k}_s \, d\mathbf{k}_i \, \delta(\omega_p - \omega_s - \omega_i) \, \delta(\mathbf{k}_p - \mathbf{k}_s - \mathbf{k}_i) \, a_s^\dagger(\mathbf{k}_s) \, a_i^\dagger(\mathbf{k}_i) |0\rangle\langle 0| a_s(\mathbf{k}_s) \, a_i(\mathbf{k}_i)$$
(20)

which is very different from the pure state of SPDC. We will show later that a statistical mixture of Eq. (20) can never have delta-function-like two-photon temporal and/or spatial correlation that is shown by the measurement of SPDC.

For finite dimensions of the nonlinear interaction region, the entangled two-photon state of SPDC may have to be estimated in a more general format. Following the earlier discussions, we write the state of the signal–idler photon pair as

$$|\Psi\rangle = \int d\mathbf{k}_s \, d\mathbf{k_i} \, F(\mathbf{k}_s, \mathbf{k}_i) \, a_i^\dagger(\mathbf{k}_s) \, a_s^\dagger(\mathbf{k}_i) |0\rangle$$
(21)

where

$$F(\mathbf{k}_s, \mathbf{k}_i) = \varepsilon \, \delta(\omega_p - \omega_s - \omega_i) \, f(\Delta_z L) \, h_{tr}(\vec{\kappa}_1 + \vec{\kappa}_2)$$
$$f(\Delta_z L) = \int_L dz \, e^{-i(k_p - k_{sz} - k_{iz})z}$$
$$h_{tr}(\vec{\kappa}_1 + \vec{\kappa}_2) = \int_A d\vec{\rho} \, \tilde{h}_{tr}(\vec{\rho}) \, e^{-i(\vec{\kappa}_s + \vec{\kappa}_i)\cdot\vec{\rho}}$$
$$\Delta_z = k_p - k_{sz} - k_{iz}$$
(22)

where ε is named as the parametric gain index. ε is proportional to the second order electric susceptibility $\chi^{(2)}$ and is usually treated as a constant, L is the length of the nonlinear interaction, the integral in $\vec{\kappa}$ is evaluated over the cross section A of the nonlinear material illuminated by the pump, $\vec{\rho}$ is the transverse coordinate vector, $\vec{\kappa}_j$ (with $j = s, i$) is the transverse wavevector of the signal and idler, and $f(|\vec{\rho}|)$ is the transverse profile of the pump, which can be treated as a Gaussion in most of the experimental conditions. The functions $f(\Delta_z L)$ and $h_{tr}(\vec{\kappa}_1 + \vec{\kappa}_2)$ turn to δ-functions for an infinitely long ($L \sim \infty$) and wide ($A \sim \infty$) nonlinear interaction region. The reason we have chosen the form of Eq. (22) is to separate the "longitudinal" and the "transverse" correlations. We will show that $\delta(\omega_p - \omega_s - \omega_i)$ and $f(\Delta_z L)$ together can be rewritten as a function of $\omega_s - \omega_i$. To simplify the mathematics, we assume near co-linearly SPDC. In this situation, $|\vec{\kappa}_{s,i}| \ll |\mathbf{k}_{s,i}|$.

Basically, the function $f(\Delta_z L)$ determines the "longitudinal" space-time correlation. Finding the solution of the integral is straightforward:

$$f(\Delta_z L) = \int_0^L dz \, e^{-i(k_p - k_{sz} - k_{iz})z} = e^{-i\Delta_z L/2} \, sinc(\Delta_z L/2).$$
(23)

Now, we consider $f(\Delta_z L)$ with $\delta(\omega_p - \omega_s - \omega_i)$ together, and taking advantage of the δ-function in frequencies by introducing a detuning frequency Ω to evaluate function $f(\Delta_z L)$:

The Physics of $2 \neq 1 + 1$ 169

$$\omega_s = \omega_s^0 + \Omega$$
$$\omega_i = \omega_i^0 - \Omega \tag{24}$$
$$\omega_p = \omega_s + \omega_i = \omega_s^0 + \omega_i^0.$$
$$\Omega = (\omega_s - \omega_i)/2.$$

The dispersion relation $k(\omega)$ allows us to express the wave numbers through the frequency detuning Ω:

$$k_s \approx k(\omega_s^0) + \Omega \frac{dk}{d\omega}\Big|_{\omega_s^0} = k(\omega_s^0) + \frac{\Omega}{u_s},$$

$$k_i \approx k(\omega_i^0) - \Omega \frac{dk}{d\omega}\Big|_{\omega_i^0} = k(\omega_i^0) - \frac{\Omega}{u_i} \tag{25}$$

where u_s and u_i are group velocities for the signal and the idler, respectively. Now, we connect Δ_z with the detuning frequency Ω:

$$\begin{aligned}
\Delta_z &= k_p - k_{sz} - k_{iz} \\
&= k_p - \sqrt{(k_s)^2 - (\vec{\kappa}_s)^2} - \sqrt{(k_i)^2 - (\vec{\kappa}_i)^2} \\
&\cong k_p - k_s - k_i + \frac{(\vec{\kappa}_s)^2}{2k_s} + \frac{(\vec{\kappa}_i)^2}{2k_i} \\
&\cong k_p - k(\omega_s^0) - k(\omega_i^0) + \frac{\Omega}{u_s} - \frac{\Omega}{u_i} + \frac{(\vec{\kappa}_s)^2}{2k_s} + \frac{(\vec{\kappa}_i)^2}{2k_i} \\
&\cong D\Omega
\end{aligned} \tag{26}$$

where $D \equiv 1/u_s - 1/u_i$. We have also applied $k_p - k(\omega_s^0) - k(\omega_i^0) = 0$ and $|\vec{\kappa}_{s,i}| \ll |\mathbf{k}_{s,i}|$. The "longitudinal" wavevector correlation function is rewritten as a function of the detuning frequency $\Omega = (\omega_s - \omega_i)/2$: $f(\Delta_z L) \cong f(\Omega DL)$. In addition to the above approximations, we have inexplicitly assumed the angular independence of the wavevector $k = n(\theta)\omega/c$. For type II SPDC, the refraction index of the extraordinary-ray depends on the angle between the wavevector and the optical axis and an additional term appears in the expansion. Making the approximation valid, we have restricted our calculation to a near-collinear process. Thus, for a good approximation, in the near-collinear experimental setup

$$\Delta_z L \cong \Omega DL = (\omega_s - \omega_i)DL/2. \tag{27}$$

Type-I degenerate SPDC is a special case. Due to the fact that $u_s = u_i$, and hence, $D = 0$, the expansion of $k(\omega)$ should be carried out up to the second order. Instead of (27), we have

$$\Delta_z L \cong -\Omega^2 D'L = -(\omega_s - \omega_i)^2 D'L/4 \tag{28}$$

where

$$D' \equiv \frac{d}{d\omega}\left(\frac{1}{u}\right)\Big|_{\omega^0}.$$

The two-photon state of the signal–idler pair is then approximated as

$$|\Psi\rangle = \int d\Omega \, d\vec{\kappa}_s \, d\vec{\kappa}_i \, f(\Omega) \, h_{tr}(\vec{\kappa}_s + \vec{\kappa}_i) \, a_s^\dagger(\omega_s^0 + \Omega, \vec{\kappa}_s) \, a_i^\dagger(\omega_i^0 - \Omega, \vec{\kappa}_s)|0\rangle \quad (29)$$

where the normalization constant has been absorbed into $f(\Omega)$.

3 Correlation Measurement of Entangled State

EPR state is a pure state which characterizes the behavior of a pair of entangled particles. In principle, one EPR pair contains all information of the correlation. A question naturally arises: Can we then observe the EPR correlation from the measurement of one EPR pair? The answer is no. Generally speaking, we may never learn any meaningful physics from the measurement of one particle or one pair of particles. To learn the correlation, an ensemble of a large number of *identical* pairs is necessary, where "identical" means that all pairs which are involved in the ensemble measurement must be prepared in the same state, except for an overall phase factor. This is a basic requirement of quantum measurement theory.

Correlation measurements are typically statistical and involve a large number of measurements of individual quanta. Quantum mechanics does not predict a precise outcome for a measurement. Rather, quantum mechanics predicts the probabilities for certain outcomes. In photon counting measurements, the outcome of a measurement is either a *"yes"* (a count or a "click") or a *"no"* (no count). In a joint measurement of two photon counting detectors, the outcome of *"yes"* means a *"yes–yes"* or a "click–click" joint registration. If the outcome of a joint measurement shows 100% *"yes"* for a certain set of values of a physical observable or a certain relationship between physical variables, the measured quantum system is correlated in that observable. As a good example, EPR's *gedankenexperiment* suggested to us a system of quanta with perfect correlation $\delta(x_1 - x_2 + x_0)$ in position. To examine the EPR correlation, we need to have a 100% *"yes"* when the positions of the two distant detectors satisfy $x_1 - x_2 = x_0$, and 100% *"no"* otherwise, when $x_1 - x_2 \neq x_0$. To show this experimentally, a realistic approach is to measure the correlation function of $|f(x_1 - x_2)|^2$ by observing the joint detection counting rates of $R_{1,2} \propto |f(x_1 - x_2)|^2$ while scanning all possible values of $x_1 - x_2$. In quantum optics, this means the measurement of the second-order correlation function, or $G^{(2)}(\mathbf{r}_1, t_1; \mathbf{r}_2, t_2)$, in the form of longitudinal correlation $G^{(2)}(\tau_1 - \tau_2)$ and/or transverse correlation $G^{(2)}(\vec{\rho}_1 - \vec{\rho}_2)$, where $\tau_j = t_j - z_j/c$, $j = 1, 2$, and $\vec{\rho}_j$ is the transverse coordinate of the jth point-like photon counting detector.

Now, we study the two-photon correlation of the entangled photon pair of SPDC. The probability of jointly detecting the signal and idler at space-time points $(\mathbf{r}_1, t_1)$ and $(\mathbf{r}_2, t_2)$ is given by the Glauber theory [15]:

$$G^{(2)}(\mathbf{r}_1, t_1; \mathbf{r}_2, t_2) = \langle E^{(-)}(\mathbf{r}_1, t_1) E^{(-)}(\mathbf{r}_2, t_2) E^{(+)}(\mathbf{r}_2, t_2) E^{(+)}(\mathbf{r}_1, t_1) \rangle \quad (30)$$

where $E^{(-)}$ and $E^{(+)}$ are the negative-frequency and the positive-frequency field operators of the detection events at space-time points $(\mathbf{r}_1, t_1)$ and $(\mathbf{r}_2, t_2)$. The expectation value of the joint detection operator is calculated by averaging over the quantum states of the signal–idler photon pair. For the two-photon state of SPDC,

$$G^{(2)}(\mathbf{r}_1,t_1;\mathbf{r}_2,t_2) = |\langle 0|E^{(+)}(\mathbf{r}_2,t_2)E^{(+)}(\mathbf{r}_1,t_1)|\Psi\rangle|^2 = |\psi(\mathbf{r}_1,t_1;\mathbf{r}_2,t_2)|^2 \quad (31)$$

where $|\Psi\rangle$ is the two-photon state, and $\Psi(\mathbf{r}_1,t_1;\mathbf{r}_2,t_2)$ is named the effective two-photon wavefunction. To evaluate $G^{(2)}(\mathbf{r}_1,t_1;\mathbf{r}_2,t_2)$ and $\psi(\mathbf{r}_1,t_1;\mathbf{r}_2,t_2)$, we need to propagate the field operators from the two-photon source to space-time points $(\mathbf{r}_1,t_1)$ and $(\mathbf{r}_2,t_2)$.

In general, the field operator $E^{(+)}(\mathbf{r},t)$ at space-time point $(\mathbf{r},t)$ can be written in terms of the Green's function, which propagates a quantized mode from space-time point $(\mathbf{r}_0,t_0)$ to $(\mathbf{r},t)$ [16, 17]:

$$E^{(+)}(\mathbf{r},t) = \sum_{\mathbf{k}} g(\mathbf{k},\mathbf{r}-\mathbf{r}_0,t-t_0) E^{(+)}(\mathbf{k},\mathbf{r}_0,t_0). \quad (32)$$

where $g(\mathbf{k},\mathbf{r}-\mathbf{r}_0,t-t_0)$ is the Green's function, which is also named the optical transfer function. For a different experimental setup, $g(\mathbf{k},\mathbf{r}-\mathbf{r}_0,t-t_0)$ can be quite different. To simplify the notation, we have assumed one polarization.

Considering an idealized simple experimental setup, shown in Fig. 2, in which collinear propagated signal and idler pairs are received by two point photon counting detectors D_1 and D_2, respectively, for longitudinal $G^{(2)}(\tau_1 - \tau_2)$ and transverse $G^{(2)}(\vec{\rho}_1 - \vec{\rho}_2)$ correlation measurements. To simplify the mathematics, we further assume paraxial experimental condition. It is convenient, in the discussion of longitudinal and transverse correlation measurements, to write the field $E^{(+)}(\mathbf{r}_j,t_j)$

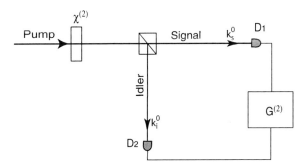

Fig. 2 Collinear propagated signal–idler photon pair, either degenerate or non-degenerate, are received by two distant point photo-detectors D_1 and D_2, respectively, for longitudinal $G^{(2)}(\tau_1 - \tau_2)$ and transverse $G^{(2)}(\vec{\rho}_1 - \vec{\rho}_2)$ correlation measurements. To simplify the mathematics, we assume paraxial approximation is applicable to the signal–idler fields. The z_1 and z_2 are chosen along the central wavevector $\mathbf{k}_s^0$ and $\mathbf{k}_i^0$

172   Y. Shih

in terms of its longitudinal and transversal space-time variables under the Fresnel paraxial approximation:

$$E^{(+)}(\vec{\rho}_j, z_j, t_j) \cong \int d\omega \, d\vec{\kappa} \, g(\vec{\kappa}, \omega; \vec{\rho}_j, z_j) e^{-i\omega t_j} a(\omega, \vec{\kappa})$$

$$\cong \int d\omega \, d\vec{\kappa} \, \gamma(\vec{\kappa}, \omega; \vec{\rho}_j, z_j) e^{-i\omega \tau_j} a(\omega, \vec{\kappa}) \qquad (33)$$

where $g(\vec{\kappa}, \omega; \vec{\rho}_j, z_j) = \gamma(\vec{\kappa}, \omega; \vec{\rho}_j, z_j) e^{i\omega z_j/c}$ is the spatial part of the Green's function, $\vec{\rho}_j$ and z_j are the transverse and longitudinal coordinates of the jth photo-detector and $\vec{\kappa}$ is the transverse wavevector. We have chosen $z_0 = 0$ and $t_0 = 0$ at the output plane of the SPDC. For convenience, all constants associated with the field are absorbed into $g(\vec{\kappa}, \omega; \vec{\rho}_j, z_j)$.

The two-photon effective wavefunction $\Psi(\vec{\rho}_1, z_1, t_1; \vec{\rho}_2, z_2, t_2)$ is thus calculated as follows

$$\Psi(\vec{\rho}_1, z_1, t_1; \vec{\rho}_2, z_2, t_2)$$
$$= \langle 0| \int d\omega' \, d\vec{\kappa}' \, g(\vec{\kappa}', \omega'; \vec{\rho}_2, z_2) e^{-i\omega' t_2} a(\omega', \vec{\kappa}')$$
$$\times \int d\omega'' \, d\vec{\kappa}'' \, g(\vec{\kappa}'', \omega''; \vec{\rho}_1, z_1) e^{-i\omega'' t_1} a(\omega'', \vec{\kappa}'')$$
$$\times \int d\Omega \, d\vec{\kappa}_s \, d\vec{\kappa}_i \, f(\Omega) h_{tr}(\vec{\kappa}_s + \vec{\kappa}_i) a_s^\dagger(\omega_s^0 + \Omega, \vec{\kappa}_s) a_i^\dagger(\omega_i^0 - \Omega, \vec{\kappa}_i) |0\rangle$$
$$= \Psi_0 e^{-i(\omega_s^0 \tau_1 + \omega_i^0 \tau_2)}$$
$$\times \int d\Omega \, d\vec{\kappa}_s \, d\vec{\kappa}_i \, f(\Omega) h_{tr}(\vec{\kappa}_s + \vec{\kappa}_i) e^{-i\Omega(\tau_1 - \tau_2)} \gamma(\vec{\kappa}_s, \Omega; \vec{\rho}_1, z_1) \gamma(\vec{\kappa}_i, -\Omega; \vec{\rho}_2, z_2).$$
$$(34)$$

Although Eq. (34) cannot be factorized into a trivial product of longitudinal and transverse integrals, it is not difficult to measure the temporal correlation and the transverse correlation separately by choosing suitable experimental conditions.

Experiments may be designed for measuring either temporal (longitudinal) or spatial (transverse) correlation only. Thus, based on different experimental setups, we may simplify the calculation to either the temporal (longitudinal) part:

$$\Psi(\tau_1; \tau_2) = \Psi_0 e^{-i(\omega_s^0 \tau_1 + \omega_i^0 \tau_2)} \int d\Omega \, f(\Omega) e^{-i\Omega(\tau_1 - \tau_2)} = \Psi_0 e^{-i(\omega_s^0 \tau_1 + \omega_i^0 \tau_2)} \mathcal{F}_{\tau_1 - \tau_2}\{f(\Omega)\}$$
$$(35)$$

or the spatial part:

$$\Psi(\vec{\rho}_1, z_1; \vec{\rho}_2, z_2) = \Psi_0 \int d\vec{\kappa}_s \, d\vec{\kappa}_i \, h_{tr}(\vec{\kappa}_s + \vec{\kappa}_i) g(\vec{\kappa}_s, \omega_s; \vec{\rho}_1, z_1) g(\vec{\kappa}_i, \omega_i; \vec{\rho}_2, z_2). \quad (36)$$

In Eq. (35), $\mathcal{F}_{\tau_1 - \tau_2}\{f(\Omega)\}$ is the Fourier transform of the spectrum amplitude function $f(\Omega)$. In Eq. (36), we may treat $h_{tr}(\vec{\kappa}_s + \vec{\kappa}_i) \sim \delta(\vec{\kappa}_s + \vec{\kappa}_i)$ by assuming certain experimental conditions.

3.1 Two-Photon Temporal Correlation

To measure the two-photon temporal correlation of SPDC, we select a pair of transverse wavevectors $\vec{\kappa}_s = -\vec{\kappa}_i$ in Eq. (34) by using appropriate optical apertures. The effective two-photon wavefunction is thus simplified to that of Eq. (35)

$$\Psi(\tau_1;\tau_2) \cong \Psi_0 e^{-i(\omega_s^0 \tau_1 + \omega_i^0 \tau_2)} \int d\Omega\, f(\Omega) e^{-i\Omega(\tau_1 - \tau_2)} \quad (37)$$

$$= \left[\Psi_0 e^{-\frac{i}{2}(\omega_s^0 + \omega_i^0)(\tau_1 + \tau_2)} \right] \left[\mathcal{F}_{\tau_1 - \tau_2}\{f(\Omega)\} e^{-\frac{i}{2}(\omega_s^0 - \omega_i^0)(\tau_1 - \tau_2)} \right]$$

where, again, $\mathcal{F}_{\tau_1 - \tau_2}\{f(\Omega)\}$ is the Fourier transform of the spectrum amplitude function $f(\Omega)$. Equation (37) indicates a 2-D wavepacket: a narrow envelope along the $\tau_1 - \tau_2$ axis with constant amplitude along the $\tau_1 + \tau_2$ axis. In certain experimental conditions, the function $f(\Omega)$ of SPDC can be treated as constant from $-\infty$ to ∞ and thus $\mathcal{F}_{\tau_1 - \tau_2} \sim \delta(\tau_1 - \tau_2)$. In this case, for fixed positions of D_1 and D_2, the 2-D wavepacket means the following: the signal–idler pair may be jointly detected at any time; however, if the signal is registered at a certain time t_1, the idler must be registered at a unique time of $t_2 \sim t_1 - (z_1 - z_2)/c$. In other words, although the joint detection of the pair may happen at any times of t_1 and t_2 with equal probability ($\Delta(t_1 + t_2) \sim \infty$), the registration time difference of the pair must be a constant $\Delta(t_1 - t_2) \sim 0$. A schematic of the two-photon wavepacket is shown in Fig. 3. It is a non-factorizeable 2-D wavefunction indicating the entangled nature of the two-photon state. The longitudinal correlation function $G^{(2)}(\tau_1 - \tau_2)$ is thus

$$G^{(2)}(\tau_1 - \tau_2) \propto |\mathcal{F}_{\tau_1 - \tau_2}\{f(\Omega)\}|^2,$$

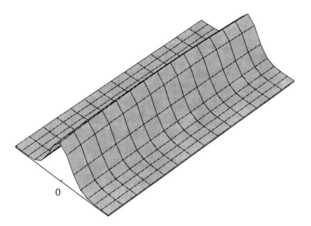

Fig. 3 A schematic envelope of a two-photon wavepacket with a Gaussian shape along $\tau_1 - \tau_2$ corresponding to a Gaussian function of $f(\Omega)$. In the case of SPDC, the envelope is close to a δ-function in $\tau_1 - \tau_2$ corresponding to a broad-band $f(\Omega) =$ constant. The wavepacket is uniformly distributed along $\tau_1 + \tau_2$ due to the assumption of $\omega_p =$ constant

which is a δ-function-like function in the case of SPDC. Thus, we have shown the entangled signal–idler photon pair of SPDC hold a typical EPR correlation in energy and time:

$$\Delta(\omega_s + \omega_i) \sim 0 \ \& \ \Delta(t_1 - t_2) \sim 0$$

$$\text{with } \Delta\omega_s \sim \infty, \quad \Delta\omega_i \sim \infty, \quad \Delta t_1 \sim \infty, \quad \Delta t_2 \sim \infty.$$

Now we examine a statistical model of SPDC for temporal correlation. As we have discussed earlier, realistic statistical models have been proposed to simulate the EPR two-particle state. Recall that for a mixed state in the form of

$$\hat{\rho} = \sum_j P_j |\Psi_j\rangle\langle\Psi_j|$$

where P_j is the probability for specifying a given set of state vectors $|\Psi_j\rangle$, the second-order correlation function of fields $E(\mathbf{r}_1, t_1)$ and $E(\mathbf{r}_2, t_2)$ is given by

$$
\begin{aligned}
G^{(2)}(\mathbf{r}_1, t_1; \mathbf{r}_2, t_2) &= Tr[\hat{\rho}\, E^{(-)}(\mathbf{r}_1, t_1)\, E^{(-)}(\mathbf{r}_2, t_2)\, E^{(+)}(\mathbf{r}_2, t_2)\, E^{(+)}(\mathbf{r}_1, t_1)] \\
&= \sum_j P_j \langle\Psi_j| E^{(-)}(\mathbf{r}_1, t_1)\, E^{(-)}(\mathbf{r}_2, t_2)\, E^{(+)}(\mathbf{r}_2, t_2)\, E^{(+)}(\mathbf{r}_1, t_1)|\Psi_j\rangle \\
&= \sum_j P_j\, G_j^{(2)}(\mathbf{r}_1, t_1; \mathbf{r}_2, t_2),
\end{aligned}
$$

which is a weighted sum over all individual contributions of $G_j^{(2)}$. Considering the following simplified version of Eq. (20) to simulate the state of SPDC as a mixed state:

$$\hat{\rho} = \int d\Omega |f(\Omega)|^2\, a^\dagger(\omega_s^0 + \Omega)\, a^\dagger(\omega_i^0 - \Omega)|0\rangle\langle 0|a(\omega_i^0 - \Omega)\, a(\omega_s^0 + \Omega), \quad (38)$$

with

$$|\Psi_\Omega\rangle = a^\dagger(\omega_s^0 + \Omega)\, a^\dagger(\omega_i^0 - \Omega)|0\rangle, \quad P_j = d\Omega |f(\Omega)|^2. \quad (39)$$

It is easy to find $G_\Omega^{(2)}(\tau_1 - \tau_2) = \text{constant}$, and thus $G^{(2)}(\tau_1 - \tau_2) = \text{constant}$. This means that the uncertainty of the measurement on $t_1 - t_2$ for the mixed state of Eq. (38) is infinite: $\Delta(t_1 - t_2) \sim \infty$. Although the energy (frequency) or momentum (wavevector) for each photon may be defined with constant values pair by pair, the corresponding temporal correlation measurement of the ensemble can never achieve a δ-function-like relationship. In fact, the correlation is undefined, i.e., taking an infinite uncertainty. Thus, the statistical model of SPDC cannot satisfy the EPR inequalities of Eq. (4).

3.2 Two-Photon Spatial Correlation

Similar to that of the two-photon temporal correlation, as an example, we analyze the effective two-photon wavefunction of the signal–idler pair of SPDC. To

The Physics of $2 \neq 1 + 1$ 175

emphasize the spatial part of the two-photon correlation, we choose a pair of fre-
quencies ω_s and ω_i with $\omega_s + \omega_i = \omega_p$. In this case, the effective two-photon wavefunction of Eq. (34) is simplified to that of Eq. (36)

$$\Psi(\vec{\rho}_1, z_1; \vec{\rho}_2, z_2) = \Psi_0 \int d\vec{\kappa}_s \, d\vec{\kappa}_i \, \delta(\vec{\kappa}_s + \vec{\kappa}_i) g(\vec{\kappa}_s, \omega_s, \vec{\rho}_1, z_1) g(\vec{\kappa}_i, \omega_i, \vec{\rho}_2, z_2)$$

where we have assumed $h_{tr}(\vec{\kappa}_s + \vec{\kappa}_i) \sim \delta(\vec{\kappa}_s + \vec{\kappa}_i)$, which is reasonable by assuming
a large enough transverse cross-session laser beam of pump.

We now design a simple joint detection measurement between two point photon
counting detectors D_1 and D_2 located at $(\vec{\rho}_1, z_1)$ and $(\vec{\rho}_2, z_2)$, respectively, for the
detection of the signal and idler photons. We have assumed that the two-photon
source has a finite but large transverse dimension. Under this simple experimental
setup, the Green's function, or the optical transfer function describing arm-j, $j =
1, 2$, in which the signal and the idler freely propagate to photodetector D_1 and D_2,
respectively, is given by Eq. (A.5) of the Appendix. Substitute the $g_j(\omega, \vec{\kappa}; z_j, \vec{\rho}_j)$,
$j = 1, 2$, into Eq. (36), the effective wavefunction is then given by

$$\Psi(\vec{\rho}_1, z_1; \vec{\rho}_2, z_2) = \Psi_0 \int d\vec{\kappa}_s \, d\vec{\kappa}_i \, \delta(\vec{\kappa}_s + \vec{\kappa}_i) \left(\frac{-i\omega_s}{2\pi c z_1} e^{i\frac{\omega_s}{c} z_1} \right) \left(\frac{-i\omega_i}{2\pi c z_2} e^{i\frac{\omega_i}{c} z_2} \right)$$
$$\times \int_A d\vec{\rho}_s \, d\vec{\rho}_i \, G\left(|\vec{\rho}_1 - \vec{\rho}_s|, \frac{\omega_s}{c z_1} \right) e^{i\vec{\kappa}_s \cdot \vec{\rho}_s} G\left(|\vec{\rho}_2 - \vec{\rho}_i|, \frac{\omega_i}{c z_2} \right) e^{i\vec{\kappa}_i \cdot \vec{\rho}_i}$$
$$(40)$$

where $\vec{\rho}_s$ ($\vec{\kappa}_s$) and $\vec{\rho}_i$ ($\vec{\kappa}_i$) are the transverse coordinates (wavevectors) for the
signal and the idler fields, respectively, defined on the output plane of the two-
photon source. The integral of $d\vec{\rho}_s$ and $d\vec{\rho}_i$ is over area A, which is determined
by the transverse dimension of the nonlinear interaction. The Gaussian function
$G(|\vec{\alpha}|, \beta) = e^{i(\beta/2)|\vec{\alpha}|^2}$ represents the Fresnel phase factor that is defined in the Appendix. The integral of $d\vec{\kappa}_s$ and $d\vec{\kappa}_i$ can be evaluated easily with the help of the EPR
type two-phonon transverse wavevector distribution function $\delta(\vec{\kappa}_s + \vec{\kappa}_i)$:

$$\int d\vec{\kappa}_s \, d\vec{\kappa}_i \, \delta(\vec{\kappa}_s + \vec{\kappa}_i) e^{i\vec{\kappa}_s \cdot \vec{\rho}_s} e^{i\vec{\kappa}_i \cdot \vec{\rho}_i} \sim \delta(\vec{\rho}_s - \vec{\rho}_i). \tag{41}$$

Thus, we have shown that the entangled signal–idler photon pair of SPDC holds a
typical EPR correlation in transverse momentum and position while the correlation
measurement is on the output plane of the two-photon source, which is very close
to the original proposal of EPR:

$$\Delta(\vec{\kappa}_s + \vec{\kappa}_i) \sim 0 \ \& \ \Delta(\vec{\rho}_s - \vec{\rho}_i) \sim 0$$
$$\text{with } \Delta\vec{\kappa}_s \sim \infty, \ \Delta\vec{\kappa}_i \sim \infty, \ \Delta\vec{\rho}_s \sim \infty, \ \Delta\vec{\rho}_i \sim \infty.$$

In EPR's language, we may never know where the signal photon and the idler photon
are emitted from the output plane of the source. However, if the signal (idler) is
found at a certain position, the idler (signal) must be observed at a corresponding

176   Y. Shih

unique position. The signal and the idler may have also any transverse momentum. However, if the transverse momentum of the signal (idler) is measured at a certain value in a certain direction, the idler (signal) must be of equal value but pointed to a certain opposite direction. In *collinear* SPDC, the signal–idler pair is always emitted from the same point in the output plane of the two-photon source, $\vec{\rho}_s = \vec{\rho}_i$, and if one of them propagates slightly off from the collinear axes, the other one must propagate to the opposite direction with $\vec{\kappa}_s = -\vec{\kappa}_i$.

The interaction of spontaneous parametric down-conversion is nevertheless a local phenomenon. The nonlinear interaction coherently creates mode-pairs that satisfy the phase matching conditions of Eq. (11) which are also named as energy and momentum conservation. The signal–idler photon pair can be excited to any of these coupled modes or in all of these coupled modes simultaneously, resulting in a particular two-photon superposition. It is this superposition among those particular "selected" two-photon states which allows the signal–idler pair to come out from the same point of the source and propagate to opposite directions with $\vec{\kappa}_s = -\vec{\kappa}_i$.

The two-photon superposition becomes more interesting when the signal–idler is separated and propagated to a large distance, either by free propagation or guided by optical components such as a lens. A classical picture would consider the signal photon and the idler photon independent whenever the pair is released from the two-photon source because there is no interaction between the distant photons in free space. Therefore, the signal photon and the idler photon should have independent and random distributions in terms of their transverse position $\vec{\rho}_1$ and $\vec{\rho}_2$. This classical picture, however, is incorrect. It is found that the signal–idler two-photon system would not lose its entangled nature in the transverse position. This interesting behavior has been experimentally observed in quantum imaging by means of an EPR type correlation in transverse position. The sub-diffraction limit spatial resolution observed in the "quantum lithography" experiment and the nonlocal correlation observed in the "ghost imaging" experiment are both the results of this peculiar superposition among those "selected" two-photon amplitudes, namely that of two-photon superposition, corresponding to different yet indistinguishable alternative ways of triggering a joint photo-electron event at a distance. Two-photon superposition does occur in a distant joint detection event of a signal–idler photon pair. There is no surprise that one has difficulties facing this phenomenon. The two-photon superposition is a nonlocal concept in this case. There is no counterpart for such a concept in classical theory and it may never be understood classically.

Now we consider propagating the signal–idler pair away from the source to $(\vec{\rho}_1, z_1)$ and $(\vec{\rho}_2, z_2)$, respectively, and taking the result of Eq. (41), i.e., $\vec{\rho}_s = \vec{\rho}_i = \vec{\rho}_0$ on the output plane of the SPDC source, the effective two-photon wavefunction becomes

$$
\Psi(\vec{\rho}_1, z_1; \vec{\rho}_2, z_2)
$$
$$
= -\frac{\omega_s \, \omega_i}{(2\pi c)^2 z_1 z_2} \, e^{i(\frac{\omega_s}{c} z_1 + \frac{\omega_i}{c} z_2)} \int_A d\vec{\rho}_0 \, G\left(|\vec{\rho}_1 - \vec{\rho}_0|, \frac{\omega_s}{c z_1}\right) G\left(|\vec{\rho}_2 - \vec{\rho}_0|, \frac{\omega_i}{c z_2}\right)
$$
$$
(42)
$$

The Physics of $2 \neq 1 + 1$ 177

where $\vec{\rho}_0$ is defined on the output plane of the two-photon source. Equation (42) indicates that the propagation-diffraction of the signal and the idler cannot be considered as independent. The signal–idler photon pair are created and diffracted together in a peculiar entangled manner. This point turns out to be both interesting and useful when the two photodetectors coincided, or are replaced by a two-photon sensitive material. Taking $z_1 = z_2$ and $\vec{\rho}_1 = \vec{\rho}_2$, Eq. (42) becomes

$$\Psi(\vec{\rho}, z; \vec{\rho}, z) = -\frac{\omega_s \, \omega_i}{(2\pi c z)^2} \, e^{i(\frac{\omega_p}{c} z)} \int_A d\vec{\rho}_0 \, G(|\vec{\rho} - \vec{\rho}_0|, \frac{\omega_p}{cz}) \qquad (43)$$

where ω_p is the pump frequency, which means that the signal–idler pair is diffracted as if they have twice the frequency or half the wavelength. This effect is named as "two-photon diffraction". This effect is useful for enhancing the spatial resolution of imaging.

4 Quantum Imaging

Although questions regarding fundamental issues of quantum theory still exist, quantum entanglement has started to play important roles in practical engineering applications. Quantum imaging is one of these exciting areas [18]. Taking advantage of entangled states, Quantum imaging has so far demonstrated two peculiar features: (1) enhancing the spatial resolution of imaging beyond the diffraction limit, and (2) reproducing ghost images in a "nonlocal" manner. Both the apparent "violation" of the uncertainty principle and the "nonlocal" behavior of the momentum–momentum position–position correlation are due to the two-photon coherent effect of entangled states, which involves the superposition of two-photon amplitudes, a nonclassical entity corresponding to different yet indistinguishable alternative ways of triggering a joint-detection event in the quantum theory of photodetection. In this section, we will focus our discussion on the physics of imaging resolution enhancement. The nonlocal phenomenon of ghost imaging will be discussed in the following section.

The concept of imaging is well defined in classical optics. Figure 4 schematically illustrates a standard imaging setup. A lens of finite size is used to image the object onto an image plane which is defined by the "Gaussian thin lens equation"

$$\frac{1}{s_i} + \frac{1}{s_o} = \frac{1}{f} \qquad (44)$$

where s_o is the distance between object and lens, f is the focal length of the lens, and s_i is the distance between the lens and image plane. If light always follows the laws of geometrical optics, the image plane and the object plane would have a perfect point-to-point correspondence, which means a perfect image of the object, either magnified or demagnified. Mathematically, a perfect image is the result of a convolution of the object distribution function $f(\vec{\rho}_o)$ and a δ-function. The δ-function characterizes the perfect point-to-point relationship between the object plane and

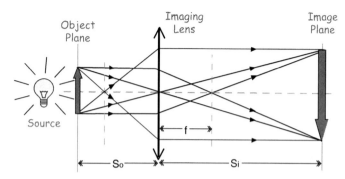

Fig. 4 A lens produces an *image* of an object in the plane defined by the Gaussian thin lens equation $1/s_i + 1/s_o = 1/f$. The concept of an image is based on the existence of a point-to-point relationship between the object plane and the image plane

the image plane:

$$F(\vec{\rho}_i) = \int_{obj} d\vec{\rho}_o f(\vec{\rho}_o)\, \delta(\vec{\rho}_o + \frac{\vec{\rho}_i}{m}) = f(\vec{\rho}_o) \otimes \delta(\vec{\rho}_o + \frac{\vec{\rho}_i}{m}) \quad (45)$$

where $\vec{\rho}_o$ and $\vec{\rho}_i$ are 2-D vectors of the transverse coordinate in the object plane and the image plane, respectively, and m is the magnification factor. The symbol $\otimes$ means convolution.

Unfortunately, light behaves like a wave. The diffraction effect turns the point-to-point correspondence into a point-to-"spot" relationship. The δ-function in the convolution of Eq. (45) will be replaced by a point-spread function.

$$F(\vec{\rho}_i) = \int_{obj} d\vec{\rho}_o f(\vec{\rho}_o)\, somb\left[\frac{R}{s_o}\frac{\omega}{c}\left|\vec{\rho}_o + \frac{\vec{\rho}_i}{m}\right|\right] = f(\vec{\rho}_o) \otimes somb\left[\frac{R}{s_o}\frac{\omega}{c}\left|\vec{\rho}_o + \frac{\vec{\rho}_i}{m}\right|\right]$$
(46)

where

$$somb(x) = \frac{2J_1(x)}{x},$$

and $J_1(x)$ is the first-order Bessel function, R is the radius of the imaging lens. R/s_o is named as the numerical aperture of the imaging system. The finite size of the spot, which is defined by the point-spread function, determines the spatial resolution of the imaging setup, and thus, limits the ability of making demagnified images. It is clear from Eq. (46), the use of a larger imaging lens and shorter wavelength light of source will result in a narrower point-spread function. To improve the spatial resolution, one of the efforts in the lithography industry is the use of shorter wavelengths. This effort is, however, limited to a certain level because of the inability of lenses to effectively work beyond a certain "cutoff" wavelength.

Equation (46) imposes a diffraction limited spatial resolution on an imaging system while the aperture size of the imaging system and the wavelength of the light source are both fixed. This limit is fundamental in both classical optics and in quantum mechanics. Any violation would be considered as a violation of the uncertainty principle.

Surprisingly, the use of quantum entangled states gives a different result: by replacing classical light sources in Fig. 5 with entangled N-photon states, the spatial resolution of the image can be improved by a factor of N, despite the Rayleigh diffraction limit. Is this a violation of the uncertainty principle? The answer is no! The uncertainty relation for an entangled N-particle system is radically different from that of N independent particles. In terms of the terminology of imaging, what we have found is that the $somb(x)$ in the convolution of Eq. (46) has a different form in the case of an entangled state. For example, an entangled two-photon system has

$$x = \frac{R}{s_o} \frac{2\omega}{c} |\vec{\rho}_o + \frac{\vec{\rho}_i}{m}|.$$

Comparing with Eq. (46), the factor of 2ω yields a point-spread function half the width of that from Eq. (46) and results in a doubling spatial resolution for imaging.

It should be further emphasized that one must not confuse a "projection" with an image. A projection is the shadow of an object, which is obviously different from the image of an object. Figure 6 distinguishes a projection shadow from an image. In a projection, the object-shadow correspondence is essentially a "momentum" correspondence, which is defined only by the propagation direction of the light rays.

We now analyze classical imaging. The analysis starts with the propagation of the field from the object plane to the image plane. In classical optics, such propagation

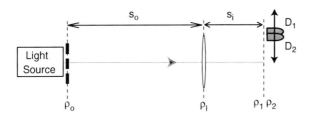

Fig. 5 Typical imaging setup. A lens of finite size is used to produce a demagnified image of a object with limited spatial resolution. Replacing classical light with an entangled N-photon system, the spatial resolution can be improved by a factor of N, despite the Rayleigh diffraction limit

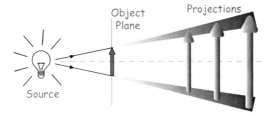

Fig. 6 Projection: a light source illuminates an object and no image forming system is present, no image plane is defined, and only projections, or shadows, of the object can be observed

180   Y. Shih

is described by an optical transfer function $h(\mathbf{r} - \mathbf{r}_0, t - t_0)$, which accounts for the propagation of all modes of the field. To be consistent with quantum optics calculations, we prefer to work with the single-mode propagator $g(\mathbf{k}, \mathbf{r} - \mathbf{r}_0, t - t_0)$, and to write the field $E(\mathbf{r}, t)$ in terms of its longitudinal (z) and transverse ($\vec{\rho}$) coordinates under the Fresnel paraxial approximation:

$$E(\vec{\rho}, z, t) = \int d\omega \, d\vec{\kappa} \, \tilde{E}(\vec{\kappa}, \omega) \, g(\vec{\kappa}, \omega; \vec{\rho}, z) \, e^{-i\omega t} \qquad (47)$$

where $\tilde{E}(\omega, \vec{\kappa})$ is the complex amplitude of frequency ω and transverse wavevector $\vec{\kappa}$. In Eq. (47) we have taken $z_0 = 0$ and $t_0 = 0$ at the object plane as usual. To simplify the notation, we have assumed one polarization.

Based on the experimental setup of Fig. 5, $g(\vec{\kappa}, \omega; \vec{\rho}, z)$ is found to be

$$g(\vec{\kappa}, \omega; \vec{\rho}_i, s_o + s_i)$$

$$= \int_{obj} d\vec{\rho}_o \int_{lens} d\vec{\rho}_l \left\{ A(\vec{\rho}_o) e^{i\vec{\kappa}\cdot\vec{\rho}_o} \right\} \left\{ \frac{-i\omega}{2\pi c} \frac{e^{i\frac{\omega}{c} s_o}}{s_o} G(|\vec{\rho}_l - \vec{\rho}_o|, \frac{\omega}{cs_o}) \right\}$$

$$\times \left\{ G(|\vec{\rho}_l|, -\frac{\omega}{cf}) \right\} \left\{ \frac{-i\omega}{2\pi c} \frac{e^{i\frac{\omega}{c} s_i}}{s_i} G(|\vec{\rho}_i - \vec{\rho}_l|, \frac{\omega}{2cs_i}) \right\} \qquad (48)$$

where $\vec{\rho}_o$, $\vec{\rho}_l$, and $\vec{\rho}_i$ are two-dimensional vectors defined, respectively, on the object, the lens, and the image planes. The first curly bracket includes the object-aperture function $A(\vec{\rho}_o)$ and the phase factor $e^{i\vec{\kappa}\cdot\vec{\rho}_o}$ contributed to the object plane by each transverse mode $\vec{\kappa}$. Here we have assumed a far-field finite size source. Thus, a phase factor $e^{i\vec{\kappa}\cdot\vec{\rho}_o}$ appears on the object plane of $z = 0$. If a collimated laser beam is used, this phase factor turns out to be a constant. The terms in the second and the fourth curly brackets describe free-space Fresnel propagation-diffraction from the source/object plane to the imaging lens, and from the imaging lens to the detection plane, respectively. The Fresnel propagator includes a spherical wave function $e^{i\frac{\omega}{c}(z_j - z_k)}/(z_j - z_k)$ and a Fresnel phase factor $G(|\vec{\alpha}|, \beta) = e^{i(\beta/2)|\vec{\alpha}|^2} = e^{i\omega|\vec{\rho}_j - \vec{\rho}_k|^2/2c(z_j - z_k)}$. The third curly bracket adds the phase factor, $G(|\vec{\rho}_l|, -\frac{\omega}{cf}) = e^{-i\frac{\omega}{2cf}}$, which is introduced by the imaging lens.

Applying the properties of the Gaussian function, Eq. (48) can be simplified into the following form

$$g(\vec{\kappa}, \omega; \vec{\rho}_i, z = s_o + s_i)$$

$$= \frac{-\omega^2}{(2\pi c)^2 s_o s_i} e^{i\frac{\omega}{c}(s_o + s_i)} G(|\vec{\rho}_i|, \frac{\omega}{cs_i}) \int_{obj} d\vec{\rho}_o A(\vec{\rho}_o) G(|\vec{\rho}_o|, \frac{\omega}{cs_o}) e^{i\vec{\kappa}\cdot\vec{\rho}_o}$$

$$\times \int_{lens} d\vec{\rho}_l G(|\vec{\rho}_l|, \frac{\omega}{c}[\frac{1}{s_o} + \frac{1}{s_i} - \frac{1}{f}]) e^{-i\frac{\omega}{c}(\frac{\vec{\rho}_o}{s_o} + \frac{\vec{\rho}_i}{s_i})\cdot\vec{\rho}_l}. \qquad (49)$$

The image plane is defined by the Gaussian thin-lens equation of Eq. (44). Hence, the second integral in Eq. (49) simplifies and gives, for a finite sized lens of radius R, the so called point-spread function of the imaging system: $somb(x) = 2J_1(x)/x$,

The Physics of $2 \neq 1 + 1$
181

where $x = [\frac{R}{s_o} \frac{\omega}{c} |\vec{\rho}_o + \rho_i/m|]$, $J_1(x)$ is the first-order Bessel function and $m = s_i/s_o$ is the magnification of the imaging system.

Substituting the result of Eq. (49) into Eq. (47) enables one to obtain the classical self-correlation of the field, or, equivalently, the intensity on the image plane

$$I(\vec{\rho}_i, z_i, t_i) = \langle E^*(\vec{\rho}_i, z_i, t_i) E(\vec{\rho}_i, z_i, t_i) \rangle \tag{50}$$

where $\langle ... \rangle$ denotes an ensemble average. We assume monochromatic light for classical imaging as usual.[2]

Case (I): *incoherent imaging.* The ensemble average of $\langle \tilde{E}^*(\vec{\kappa}, \omega) \tilde{E}(\vec{\kappa}', \omega) \rangle$ yields zeros except when $\vec{\kappa} = \vec{\kappa}'$. The image is thus

$$I(\vec{\rho}_i) \propto \int d\vec{\rho}_o |A(\vec{\rho}_o)|^2 |somb[\frac{R}{s_o} \frac{\omega}{c} |\vec{\rho}_o + \frac{\vec{\rho}_i}{m}|]|^2. \tag{51}$$

An incoherent image, magnified by a factor of m, is thus given by the convolution between the squared moduli of the object aperture function and the point-spread function. The spatial resolution of the image is thus determined by the finite width of the $|somb|^2$-function.

Case (II): *coherent imaging.* The coherent superposition of the $\vec{\kappa}$ modes in both $E^*(\vec{\rho}_i, \tau)$ and $E(\vec{\rho}_i, \tau)$ results in a wavepacket. The image, or the intensity distribution on the image plane, is thus

$$I(\vec{\rho}_i) \propto \left| \int_{obj} d\vec{\rho}_o A(\vec{\rho}_o) e^{i\frac{\omega}{2cs_o} |\vec{\rho}_o|^2} somb[\frac{R}{s_o} \frac{\omega}{c} |\vec{\rho}_o + \frac{\vec{\rho}_i}{m}|] \right|^2. \tag{52}$$

A coherent image, magnified by a factor of m, is thus given by the squared modulus of the convolution between the object aperture function (multiplied by a Fresnel phase factor) and the point-spread function.

For $s_i < s_o$ and $s_o > f$, both Eqs. (51) and (52) describe a real demagnified inverted image. In both cases, a narrower *somb*-function yields a higher spatial resolution. Thus, the use of shorter wavelengths allows for improvement of the spatial resolution of an imaging system.

To demonstrate the working principle of quantum imaging, we replace classical light with an entangled two-photon source such as spontaneous parametric down-conversion (SPDC) and replace the ordinary film with a two-photon absorber, which is sensitive to two-photon transition only, on the image plane. We will show that, in the same experimental setup of Fig. 5, an entangled two-photon system gives rise, on a two-photon absorber, to a point-spread function half the width of the one obtained in classical imaging at the same wavelength. Then, without employing shorter wavelengths, entangled two-photon states improve the spatial resolution of a *two-photon image* by a factor of 2 [19, 20]. We will also show that the entangled two-photon system yields a peculiar Fourier transform function as if it is produced by a light source with $\lambda/2$.

---

[2] Even if assuming a perfect lens without chromatic aberration, Fresnel diffraction is wavelength dependent. Hence, large broadband ($\Delta\omega \sim \infty$) would result in blurred images in classical imaging. Surprisingly, the situation is different in quantum imaging: no aberration blurring.

182   Y. Shih

In order to cover two different measurements, one on the image plane and one on the Fourier transform plane, we generalize the Green's function of Eq. (48) from the image plane of $z = s_o + s_i$ to an arbitrary plane of $z = s_o + d$, where d may take any values for different experimental setups:

$$
\begin{aligned}
g(\vec{\kappa}_j, &\omega_j; \vec{\rho}_k, z = s_o + d) \\
&= \int_{obj} d\vec{\rho}_o \int_{lens} d\vec{\rho}_l A(\vec{\rho}_o) \left\{ \frac{-i\omega_j}{2\pi c s_o} e^{i\vec{\kappa}_j \cdot \vec{\rho}_o} e^{i\frac{\omega_j}{c} s_o} G(|\vec{\rho}_o - \vec{\rho}_l|, \frac{\omega_j}{c s_o}) \right\} \\
&\times G(|\vec{\rho}_l|, -\frac{\omega_j}{cf}) \left\{ \frac{-i\omega_j}{2\pi c d} e^{i\frac{\omega_j}{c} d} G(|\vec{\rho}_l - \vec{\rho}_k|, \frac{\omega_j}{cd}) \right\},
\end{aligned}
\tag{53}
$$

where $\vec{\rho}_o$, $\vec{\rho}_l$, and $\vec{\rho}_j$ are two-dimensional vectors defined, respectively, on the (transverse) output plane of the source (which coincide with the object plane), on the transverse plane of the imaging lens and on the detection plane; and $j = s, i$, labels the signal and the idler; $k = 1, 2$, labels the photodetector D_1 and D_2. The function $A(\vec{\rho}_o)$ is the object-aperture function, while the terms in the first and second curly brackets of Eq. (53) describe, respectively, free propagation from the output plane of the source/object to the imaging lens, and from the imaging lens to the detection plane.

Similar to the earlier calculation, by employing the second and third expressions given in Eq. (A.3), Eq. (53) simplifies to

$$
\begin{aligned}
g(\vec{\kappa}_j, &\omega_j; \vec{\rho}_k, z = s_o + d) \\
&= \frac{-\omega_j^2}{(2\pi c)^2 s_o d} e^{i\frac{\omega_j}{c}(s_o + d)} G(|\vec{\rho}_k|, \frac{\omega_j}{cd}) \int_{obj} d\vec{\rho}_o A(\vec{\rho}_o) G(|\vec{\rho}_o|, \frac{\omega_j}{c s_o}) e^{i\vec{\kappa}_j \cdot \vec{\rho}_o} \\
&\times \int_{lens} d\vec{\rho}_l G(|\vec{\rho}_l|, \frac{\omega_j}{c}[\frac{1}{s_o} + \frac{1}{d} - \frac{1}{f}]) e^{-i\frac{\omega_j}{c}(\frac{\vec{\rho}_o}{s_o} + \frac{\vec{\rho}_k}{d}) \cdot \vec{\rho}_l}.
\end{aligned}
\tag{54}
$$

Substituting the Green's functions into Eq. (34), the effective two-photon wavefunction $\Psi(\vec{\rho}_1, z; \vec{\rho}_2, z)$ is thus

$$
\begin{aligned}
\Psi(\vec{\rho}_1, z; \vec{\rho}_2, z) = \Psi_0 & \int d\Omega f(\Omega) G(|\vec{\rho}_1|, \frac{\omega_s}{cd}) G(|\vec{\rho}_2|, \frac{\omega_i}{cd}) \\
&\times \int_{obj} d\vec{\rho}_o A(\vec{\rho}_o) G(|\vec{\rho}_o|, \frac{\omega_s}{c s_o}) \int_{obj} d\vec{\rho}'_o A(\vec{\rho}'_o) G(|\vec{\rho}'_o|, \frac{\omega_i}{c s_o}) \\
&\times \int_{lens} d\vec{\rho}_l G(|\vec{\rho}_l|, \frac{\omega_s}{c}[\frac{1}{s_o} + \frac{1}{d} - \frac{1}{f}]) e^{-i\frac{\omega_s}{c}(\frac{\vec{\rho}_o}{s_o} + \frac{\vec{\rho}_1}{d}) \cdot \vec{\rho}_l} \\
&\times \int_{lens} d\vec{\rho}'_l G(|\vec{\rho}'_l|, [\frac{\omega_i}{c}[\frac{1}{s_o} + \frac{1}{d} - \frac{1}{f}]) e^{-i\frac{\omega_i}{c}(\frac{\vec{\rho}'_o}{s_o} + \frac{\vec{\rho}_2}{d}) \cdot \vec{\rho}'_l} \\
&\times \int d\vec{\kappa}_s d\vec{\kappa}_i \, \delta(\vec{\kappa}_s + \vec{\kappa}_i) e^{i(\vec{\kappa}_s \cdot \vec{\rho}_o + \vec{\kappa}_i \cdot \vec{\rho}'_o)}
\end{aligned}
\tag{55}
$$

where we have absorbed all constants into Ψ_0, including the phase

$$
e^{i\frac{\omega_s}{c}(s_o + d)} e^{i\frac{\omega_i}{c}(s_o + d)} = e^{i\frac{\omega_p}{c}(s_o + d)}.
$$

The Physics of $2 \neq 1 + 1$ 183

The double integral of $d\vec{\kappa}_s$ and $d\vec{\kappa}_i$ yields a δ-function of $\delta(\vec{\rho}_o - \vec{\rho}'_o)$, and Eq. (55) is simplified as:

$$
\begin{aligned}
\Psi(\vec{\rho}_1, z; \vec{\rho}_2, z) \\
= \Psi_0 \int d\Omega\, f(\Omega)\, G(|\vec{\rho}_1|, \frac{\omega_s}{cd})\, G(|\vec{\rho}_2|, \frac{\omega_i}{cd}) \int_{obj} d\vec{\rho}_o\, A^2(\vec{\rho}_o)\, G(|\vec{\rho}_o|, \frac{\omega_p}{cs_o}) \\
\times \int_{lens} d\vec{\rho}_l\, G(|\vec{\rho}_l|, \frac{\omega_s}{c}[\frac{1}{s_o} + \frac{1}{d} - \frac{1}{f}])\, e^{-i\frac{\omega_s}{c}(\frac{\vec{\rho}_o}{s_o} + \frac{\vec{\rho}_1}{d})\cdot\vec{\rho}_l} \\
\times \int_{lens} d\vec{\rho}'_l\, G(|\vec{\rho}'_l|, [\frac{\omega_i}{c}[\frac{1}{s_o} + \frac{1}{d} - \frac{1}{f}])\, e^{-i\frac{\omega_i}{c}(\frac{\vec{\rho}_o}{s_o} + \frac{\vec{\rho}_2}{d})\cdot\vec{\rho}'_l}.
\end{aligned}
\tag{56}
$$

We consider the following two cases:

Case (I) on the imaging plane and $\vec{\rho}_1 = \vec{\rho}_2 = \vec{\rho}$.
  In this case, Eq. (56) is simplified as

$$
\begin{aligned}
\Psi(\vec{\rho}, z; \vec{\rho}, z) \propto \int_{obj} d\vec{\rho}_o\, A^2(\vec{\rho}_o) G(|\vec{\rho}_o|, \frac{\omega_p}{cs_o}) \int d\vec{\rho}_l\, e^{-i\frac{\omega_p}{2c}(\frac{\vec{\rho}_o}{s_o} + \frac{\vec{\rho}}{s_i})\cdot\vec{\rho}_l} \int d\vec{\rho}'_l\, e^{-i\frac{\omega_p}{2c}(\frac{\vec{\rho}_o}{s_o} + \frac{\vec{\rho}}{s_i})\cdot\vec{\rho}'_l} \\
\times \left\{ \int d\Omega\, f(\Omega)\, e^{-i\Omega[(\frac{\vec{\rho}_o}{cs_o} + \frac{\vec{\rho}}{cs_i})\cdot(\vec{\rho}_l - \vec{\rho}'_l)]} \right\}
\end{aligned}
\tag{57}
$$

where we have used $\omega_s = \omega_p/2 + \Omega$ and $\omega_s = \omega_p/2 - \Omega$ following $\omega_s + \omega_i = \omega_p$. The integral of $d\Omega$ gives a δ-function of $\delta[(\frac{\vec{\rho}_o}{cs_o} + \frac{\vec{\rho}}{cs_i})(\vec{\rho}_l - \vec{\rho}'_l)]$ while taking the integral to infinity with a constant $f(\Omega)$. This result indicates again that the propagation-diffraction of the signal and the idler are not independent. The "two-photon diffraction" couples the two integrals in $\vec{\rho}_o$ and $\vec{\rho}'_o$ as well as the two integrals in $\vec{\rho}_l$ and $\vec{\rho}'_l$ and thus gives the $G^{(2)}$ function

$$
G^{(2)}(\vec{\rho}, \vec{\rho}) \propto \left| \int_{obj} d\vec{\rho}_o\, A^2(\vec{\rho}_o)\, e^{i\frac{\omega_p}{2cs_o}|\vec{\rho}_o|^2} \frac{2J_1\left(\frac{R}{s_o}\frac{\omega_p}{c}|\vec{\rho}_o + \frac{\vec{\rho}}{m}|\right)}{\left(\frac{R}{s_o}\frac{\omega_p}{c}|\vec{\rho}_o + \frac{\vec{\rho}}{m}|\right)^2} \right|^2
\tag{58}
$$

which indicates that a coherent image (see Eq. (52)) magnified by a factor of $m = s_i/s_o$ is reproduced on the image plane by joint-detection or by two-photon absorption.

In Eq. (58), the point-spread function is characterized by the pump wavelength $\lambda_p = \lambda_{s,i}/2$; hence, the point-spread function is half the width of the (first order) classical case (Eqs. (52) and (51)). An entangled two-photon state thus gives an image in joint-detection with double spatial resolution when compared to the image obtained in classical imaging. Moreover, the spatial resolution of the two-photon image obtained by perfect SPDC radiation is further improved because it is determined by the function $2J_1(x)/x^2$, which is much narrower than the $somb(x)$.

It is interesting to see that, different from the classical case, the frequency integral over $\Delta\omega_s \sim \infty$ does not give any blurring, but rather enhances the spatial resolution of the two-photon image.

184   Y. Shih

Case (II): on the Fourier transform plane and $\vec{\rho}_1 = \vec{\rho}_2 = \vec{\rho}$.

The detectors are now placed in the focal plane, i.e., $d = f$. In this case, the spatial effective two-photon wavefunction $\Psi(\vec{\rho}, z; \vec{\rho}, z)$ becomes:

$$\Psi(\vec{\rho},z;\vec{\rho},z) \propto \int d\Omega \, f(\Omega) \int_{obj} d\vec{\rho}_o \, A^2(\vec{\rho}_o) \, G(|\vec{\rho}_o|, \frac{\omega_p}{cs_o}) \int_{lens} d\vec{\rho}_l \, G(|\vec{\rho}_l|, \frac{\omega_s}{cs_o}) \, e^{-i\frac{\omega_s}{c}(\frac{\vec{\rho}_o}{s_o} + \frac{\vec{\rho}}{f})\cdot\vec{\rho}_l}$$

$$\times \int_{lens} d\vec{\rho'}_l \, G(|\vec{\rho'}_l|, \frac{\omega_i}{cs_o}) \, e^{-i\frac{\omega_i}{c}(\frac{\vec{\rho}_o}{s_o} + \frac{\vec{\rho}}{f})\cdot\vec{\rho'}_l}. \tag{59}$$

We will first evaluate the two integrals over the lens. To simplify the mathematics we approximate the integral to infinity. Differing from the calculation for imaging resolution, the purpose of this evaluation is to determine the Fourier transform. Thus, the approximation of an infinite lens is appropriate. By applying Eq. (A.3), the two integrals over the lens contribute the following function of $\vec{\rho}_o$ to the integral of $d\vec{\rho}_o$ in Eq. (59):

$$C \, G(|\vec{\rho}_o|, -\frac{\omega_p}{cs_o}) \, e^{-i\frac{\omega_p}{cf}\vec{\rho}_o\cdot\vec{\rho}}$$

where C absorbs all constants including a phase factor $G(|\vec{\rho}|, -\frac{\omega_p}{cf^2/s_o})$. Replacing the two integrals of $d\vec{\rho}_l$ and $d\vec{\rho'}_l$ in Eq. (59) with this result, we obtain:

$$\Psi(\vec{\rho},z;\vec{\rho},z) \propto \int d\Omega \, f(\Omega) \int_{obj} d\vec{\rho}_o \, A^2(\vec{\rho}_o) \, e^{-i\frac{\omega_p}{cf}\vec{\rho}\cdot\vec{\rho}_o} \propto \mathcal{F}_{[\frac{\omega_p}{cf}\vec{\rho}]}\{A^2(\vec{\rho}_o)\}, \tag{60}$$

which is the Fourier transform of the object-aperture function. When the two photodetectors scan together (i.e., $\vec{\rho}_1 = \vec{\rho}_2 = \vec{\rho}$), the second-order transverse correlation $G^{(2)}(\vec{\rho}, z; \vec{\rho}, z)$, where $z = s_o + f$, is reduced to:

$$G^{(2)}(\vec{\rho},z;\vec{\rho},z) \propto \left| \mathcal{F}_{[\frac{\omega_p}{cf}\vec{\rho}]}\{A^2(\vec{\rho}_o)\} \right|^2. \tag{61}$$

Thus, by replacing classical light with entangled two-photon sources, in the double-slit setup of Fig. 5, a Young's double-slit interference/diffraction pattern with twice the interference modulation and half the pattern width, compared to that of classical light at wavelength $\lambda_{s,i} = 2\lambda_p$, is observed in the joint detection. This effect has also been examined in a recent "quantum lithography" experiment [20].

Due to the lack of two-photon sensitive material, the first experimental demonstration of quantum lithography was measured on the Fourier transform plane, instead of the image plane. Two point-like photon counting detectors were scanned jointly, similar to the setup illustrated in Fig. 5, for the observation of the interference/diffraction pattern of Eq. (61). The published experimental result is shown in Fig. 7 [20]. It is clear that the two-photon Young's double-slit interference-diffraction pattern has half the width with twice the interference modulation compared to that of the classical case although the wavelengths are both 916 nm.

Following linear Fourier optics, it is not difficult to see that, with the help of another lens (equivalently building a microscope), one can transform the Fourier transform function of the double-slit back onto its image plane to observe its image with twice the spatial resolution.

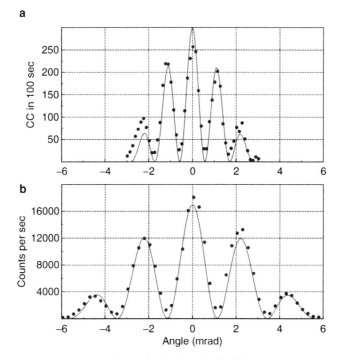

Fig. 7 (a) Two-photon Fourier transform of a double-slit. The light source was a collinear degenerate SPDC of $\lambda_{s,i} = 916$ nm. (b) Classical Fourier transform of the same double-slit. A classical light source of $\lambda = 916$ nm was used

The key to understanding the physics of this experiment is again through entangled nature of the signal–idler two-photon system. As we have discussed earlier, the pair is always emitted from the same point on the output plane of the source, thus always passing the same slit together if the double-slit is placed close to the surface of the nonlinear crystal. There is no chance for the signal–idler pair to pass different slits in this setup. In other words, each point of the object is "illuminated" by the pair "together" and the pair "stops" on the image plane "together". The point-"spot" correspondence between the object and image planes are based on the physics of two-photon diffraction, resulting in a twice narrower Fourier transform function in the Fourier transform plane and twice the image resolution in the image plane. The unfolded schematic setup, which is shown in Fig. 8, may be helpful for understanding the physics. It is not difficult to calculate the interference-diffraction function under the experimental condition indicated in Fig. 8. The non-classical observation is due to the superposition of the two-photon amplitudes, which are indicated by the straight lines connecting D_1 and D_2. The two-photon diffraction, which restricts the spatial resolution of a two-photon image, is very different from that of classical light. Thus, there should be no surprise in having an improved spatial resolution even beyond the classical limit.

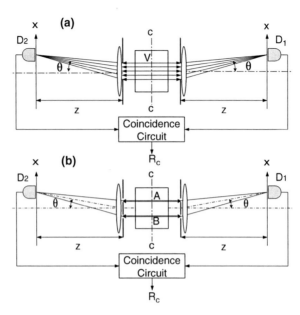

Fig. 8 Unfolded experimental setup. The joint measurement is on the Fourier transform plane. Each point of the object is "illuminated" by the signal–idler pair "together", resulting in twice narrower interference-diffraction pattern width in the Fourier transform plane through the joint detection of the signal–idler pair, equivalent to the use of classical light of $\lambda/2$

It is worthwhile to emphasize the following important aspects of physics in this simplified illustration:

(1) The goal of lithography is the reproduction of demagnified images of complicated patterns. The sub-wavelength interference feature does not necessarily translate into an improvement of the lithographic performance. In fact, the Fourier transform argument works for *imaging setups* only; sub-wavelength interference in a Mach-Zehnder type interferometer, for instance, does not necessarily lead to an image.

(2) In the imaging setup, it is the peculiar nature of the entangled N-photon system that allows one to generate an image with N-times the spatial resolution: the entangled photons come out from one point of the object plane, undergo N-photon diffraction, and stop in the image plane within a N-times narrower spot than that of classical imaging. The historical experiment by D'Angelo et al., in which the working principle of quantum lithography was first demonstrated, has taken advantage of the entangled two-photon state of SPDC: the signal–idler photon pair comes out from either the upper slit or the lower slit that is in the object plane, undergoes two-photon diffraction, and stops in the image plane within a twice narrower image than that of the classical one. It is easy to show that a second Fourier transform, by means of the use of a second lens to set up a simple microscope, will produce an image on the image plane with double spatial resolution.

(3) Certain "clever" tricks allow the production of doubly modulated interference patterns by using classical light in joint photo-detection. These tricks, however, may never be helpful for imaging. Thus, they may never be useful for lithography.

5 Ghost Imaging

The *nonlocal* position–position and momentum–momentum EPR correlation of the entangled two-photon state of SPDC was successfully demonstrated in 1995 [21] inspired by the theory of Klyshko [22]. The experiment was immediately named as "ghost imaging" in the physics community due to its surprising nonlocal nature. The important physics demonstrated in the experiment, however, may not be the so called "ghost". Indeed, the original purpose of the experiment was to study the EPR correlation in position and in momentum and to test the EPR inequality of Eq. (4) for the entangled signal-idler photon pair of SPDC [18, 23]. The experiments of "ghost imaging" [21] and "ghost interference" [24] together stimulated the foundation of quantum imaging in terms of geometrical and physical optics.

The schematic setup of the "ghost" imaging experiment is shown in Fig. 9. A CW laser is used to pump a nonlinear crystal, which is cut for degenerate type-II phase matching to produce a pair of orthogonally polarized signal (e-ray of the crystal) and idler (o-ray of the crystal) photons. The pair emerges from the crystal as collinear, with $\omega_s \cong \omega_i \cong \omega_p/2$. The pump is then separated from the signal–idler pair by a dispersion prism, and the remaining signal and idler beams are sent in different directions by a polarization beam splitting Thompson prism. The signal beam passes

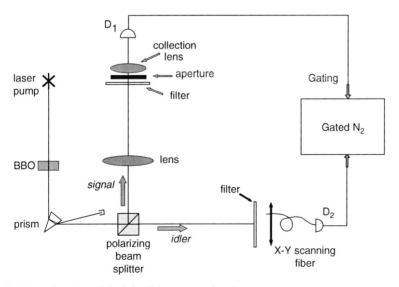

Fig. 9 Schematic set-up of the "ghost" image experiment

through a convex lens with a 400 mm focal length and illuminates a chosen aperture (mask). As an example, one of the demonstrations used the letters "UMBC" for the object mask. Behind the aperture is the "bucket" detector package D_1, which consists of a short focal length collection lens in whose focal spot is an avalanche photodiode. D_1 is mounted in a fixed position during the experiment. The idler beam is met by detector package D_2, which consists of an optical fiber whose output is mated with another avalanche photodiode. The input tip of the fiber is scanned in the transverse plane by two step motors. The output pulses of each detector, which are operating in photon counting mode, are sent to a coincidence counting circuit for the signal–idler joint detection.

By recording the coincidence counts as a function of the fiber tip's transverse plane coordinates, the image of the chosen aperture (for example, "UMBC") is observed, as reported in Fig. 10. It is interesting to note that while the size of the "UMBC" aperture inserted in the signal beam is only about 3.5 mm × 7 mm, the observed image measures 7 mm × 14 mm. The image is therefore magnified by a factor of 2. The observation also confirms that the focal length of the imaging lens, f, the aperture's optical distance from the lens, S_o, and the image's optical distance from the lens, S_i (which is from the imaging lens going backward along the signal photon path to the two-photon source of the SPDC crystal then going forward along the path of idler photon to the image), satisfy the Gaussian thin lens equation. In this experiment, S_o was chosen to be $S_o = 600$ mm, and the twice magnified clear image was found when the fiber tip was on the plane of $S_i = 1200$ mm. While D_2 was scanned on other transverse planes not defined by the Gaussian thin lens equation, the images blurred out.

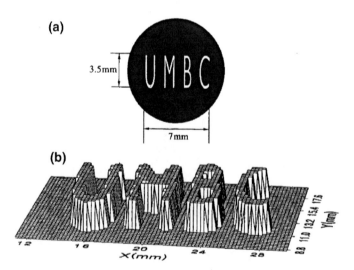

Fig. 10 (a) A reproduction of the actual aperature "UMBC" placed in the signal beam. (b) The image of "UMBC": coincidence counts as a function of the fiber tip's transverse plane coordinates. The step size is 0.25 mm. The image shown is a "slice" at the half maximum value

The measurement of the signal and the idler subsystem themselves are very different. The single photon counting rate of D_2 was recorded during the scanning of the image and was found fairly constant in the entire region of the image. This means that the transverse coordinate uncertainty of either signal or idler is considerably large compared to that of the transverse correlation of the entangled signal–idler photon pair: Δx_1 (Δy_1) and Δx_2 (Δy_2) are much greater than $\Delta(x_1 - x_2)$ ($\Delta(y_1 - y_2)$).

The EPR δ-functions, $\delta(\vec{\rho}_s - \vec{\rho}_i)$ and $\delta(\vec{\kappa}_s + \vec{\kappa}_i)$ in transverse dimension, are the key to understanding this interesting phenomenon. In degenerate SPDC, although the signal–idler photon pair has equal probability to be emitted from any point on the output surface of the nonlinear crystal, the transverse position δ-function indicates that if one of them is observed at one position, the other one must be found at the same position. In other words, the pair is always emitted from the same point on the output plane of the two-photon source. The transverse momentum δ-function, defines the angular correlation of the signal–idler pair: the transverse momenta of a signal–idler amplitude are equal but pointed in opposite directions: $\vec{\kappa}_s = -\vec{\kappa}_i$. In other words, the two-photon amplitudes are always existing at roughly equal yet opposite angles relative to the pump. This then allows for a simple explanation of the experiment in terms of "usual" geometrical optics in the following manner: we envision the nonlinear crystal as a "hinge point" and "unfold" the schematic of Fig. 9 into that shown in Fig. 11. The signal–idler two-photon amplitudes can then be represented by straight lines (but keep in mind the different propagation directions) and therefore, the image is well produced in coincidences when the aperture, lens, and fiber tip are located according to the Gaussian thin lens equation of Eq. (5). The image is exactly the same as one would observe on a screen placed at the fiber tip if

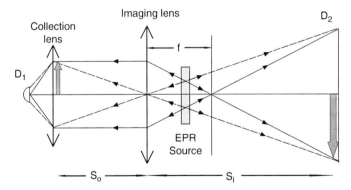

Fig. 11 An unfolded setup of the "ghost" imaging experiment, which is helpful for understanding the physics. Since the two-photon "light" propagates along "straight-lines", it is not difficult to find that any geometrical light point on the subject plane corresponds to an unique geometrical light point on the image plane. Thus, a "ghost" image of the subject is made nonlocally in the image plane. Although the placement of the lens, the object, and detector D_2 obeys the Gaussian thin lens equation, it is important to remember that the geometric rays in the figure actually represent the two-photon amplitudes of an entangled photon pair. The point to point correspondence is the result of the superposition of these two-photon amplitudes

detector D_1 were replaced by a point-like light source and the nonlinear crystal by a reflecting mirror.

Following a similar analysis in geometric optics, it is not difficult to find that any geometrical "light spot" on the subject plane, which is the intersection point of all possible two-photon amplitudes coming from the two-photon light source, corresponds to a unique geometrical "light spot" on the image plane, which is another intersection point of all the possible two-photon amplitudes. This point to point correspondence made the "ghost" image of the subject-aperture possible. Despite the completely different physics from classical geometrical optics, the remarkable feature is that the relationship between the focal length of the lens, f, the aperture's optical distance from the lens, S_o, and the image's optical distance from the lens, S_i, satisfy the Gaussian thin lens equation:

$$\frac{1}{s_o} + \frac{1}{s_i} = \frac{1}{f}.$$

Although the placement of the lens, the object, and the detector D_2 obeys the Gaussian thin lens equation, it is important to remember that the geometric rays in the figure actually represent the two-photon amplitudes of a signal–idler photon pair and the point to point correspondence is the result of the superposition of these two-photon amplitudes. The "ghost" image is a realization of the 1935 EPR *gedankenexperiment*.

Now we calculate $G^{(2)}(\vec{\rho}_o, \vec{\rho}_i)$ for the "ghost" imaging experiment, where $\vec{\rho}_o$ and $\vec{\rho}_i$ are the transverse coordinates on the object plane and the image plane. We will show that there exists a δ-function like point-to-point relationship between the object plane and the image plane, i.e., if one measures the signal photon at a position of $\vec{\rho}_o$ on the object plane the idler photon can be found only at a certain unique position of $\vec{\rho}_i$ on the image plane satisfying $\delta(m\vec{\rho}_o - \vec{\rho}_i)$, where $m = -(s_i/s_o)$ is the image-object magnification factor. After demonstrating the δ-function, we show how the object-aperture function of $A(\vec{\rho}_o)$ is transfered to the image plane as a magnified image $A(\vec{\rho}_i/m)$. Before showing the calculation, it is worthwhile to emphasize again that the "straight lines" in Fig. 11 schematically represent the two-photon amplitudes belonging to a pair of signal–idler photon. A "click–click" joint measurement at $(\mathbf{r}_1, t_1)$, which is on the object plane, and $(\mathbf{r}_2, t_2)$, which is on the image plane, in the form of an EPR δ-function, is the result of the coherent superposition of all these two-photon amplitudes.

We follow the unfolded experimental setup shown in Fig. 12 to establish the Green's functions $g(\vec{\kappa}_s, \omega_s, \vec{\rho}_o, z_o)$ and $g(\vec{\kappa}_i, \omega_i, \vec{\rho}_2, z_2)$. In arm-1, the signal propagates freely over a distance d_1 from the output plane of the source to the imaging lens, then passes an object aperture at distance s_o, and then is focused onto photon counting detector D_1 by a collection lens. We will evaluate $g(\vec{\kappa}_s, \omega_s, \vec{\rho}_o, z_o)$ by propagating the field from the output plane of the two-photon source to the object plane. In arm-2, the idler propagates freely over a distance d_2 from the output plane of the two-photon source to a point-like detector D_2. $g(\vec{\kappa}_i, \omega_i, \vec{\rho}_2, z_2)$ is thus a free propagator.

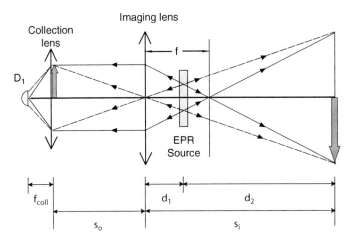

Fig. 12 In arm-1, the signal propagates freely over a distance d_1 from the output plane of the source to the imaging lens, then passes an object aperture at distance s_o, and then is focused onto photon counting detector D_1 by a collection lens. In arm-2, the idler propagates freely over a distance d_2 from the output plane of the source to a point-like photon counting detector D_2

(I) Arm-1 (source to object):

The optical transfer function or Green's function in arm-1, which propagates the field from the source plane to the object plane, is given by:

$$g(\vec{\kappa}_s, \omega_s; \vec{\rho}_o, z_o = d_1 + s_o)$$
$$= e^{i\frac{\omega_s}{c}z_o} \int_{lens} d\vec{\rho}_l \int_A d\vec{\rho}_s \left\{ \frac{-i\omega_s}{2\pi c d_1} e^{i\vec{\kappa}_s \cdot \vec{\rho}_s} G(|\vec{\rho}_s - \vec{\rho}_l|, \frac{\omega_s}{cd_1}) \right\}$$
$$\times \left\{ G(|\vec{\rho}_l|, \frac{\omega_s}{cf}) \right\} \left\{ \frac{-i\omega_s}{2\pi c s_o} G(|\vec{\rho}_l - \vec{\rho}_o|, \frac{\omega_s}{cs_o}) \right\}, \quad (62)$$

where $\vec{\rho}_s$ and $\vec{\rho}_l$ are the transverse vectors defined, respectively, on the output plane of the source and on the plane of the imaging lens. The terms in the first and third curly brackets in Eq. (62) describe free space propagation from the output plane of the source to the imaging lens and from the imaging lens to the object plane, respectively. The function $G(|\vec{\rho}_l|, \frac{\omega}{cf})$ in the second curly brackets is the transformation function of the imaging lens. Here, we treat it as a thin-lens: $G(|\vec{\rho}_l|, \frac{\omega}{cf}) \cong e^{-i\frac{\omega}{2cf}|\vec{\rho}_l|^2}$.

(II) Arm-2 (from source to image):

In arm-2, the idler propagates freely from the source to the plane of D_2, which is also the plane of the image. The Green's function is thus:

$$g(\vec{\kappa}_i, \omega_i; \vec{\rho}_2, z_2 = d_2) = \frac{-i\omega_i}{2\pi c d_2} e^{i\frac{\omega_i}{c}d_2} \int_A d\vec{\rho}'_s G(|\vec{\rho}'_s - \vec{\rho}_2|, \frac{\omega_i}{cd_2}) e^{i\vec{\kappa}_i \cdot \vec{\rho}'_s} \quad (63)$$

192   Y. Shih

where $\vec{\rho}'_s$ and $\vec{\rho}_2$ are the transverse vectors defined, respectively, on the output plane of the source, and on the plane of the photo-detector D_2.

(III) $\Psi(\vec{\rho}_o, \vec{\rho}_i)$ (object plane – image plane):

To simplify the calculation and to focus on the transverse correlation, in the following calculation we assume degenerate ($\omega_s = \omega_i = \omega$) and collinear SPDC. The transverse two-photon effective wavefunction $\Psi(\vec{\rho}_o, \vec{\rho}_2)$ is then evaluated by substituting the Green's functions $g(\vec{\kappa}_s, \omega; \vec{\rho}_o, z_o)$ and $g(\vec{\kappa}_i, \omega; \vec{\rho}_2, z_2)$ into the expression given in Eq. (36):

$$\Psi(\vec{\rho}_o, \vec{\rho}_2)$$

$$\propto \int d\vec{\kappa}_s \, d\vec{\kappa}_i \, \delta(\vec{\kappa}_s + \vec{\kappa}_i) \, g(\vec{\kappa}_s, \omega; \vec{\rho}_o, z_o) \, g(\vec{\kappa}_i, \omega; \vec{\rho}_2, z_2)$$

$$\propto e^{i\frac{\omega}{c}(s_0 + s_i)} \int d\vec{\kappa}_s \, d\vec{\kappa}_i \, \delta(\vec{\kappa}_s + \vec{\kappa}_i) \int_{lens} d\vec{\rho}_l \int_A d\vec{\rho}_S \, e^{i\vec{\kappa}_s \cdot \vec{\rho}_S} G(|\vec{\rho}_S - \vec{\rho}_l|, \frac{\omega}{cd_1})$$

$$\times \, G(|\vec{\rho}_l|, \frac{\omega}{cf}) \, G(|\vec{\rho}_l - \vec{\rho}_o|, \frac{\omega}{cs_o}) \int_A d\vec{\rho}'_S \, e^{i\vec{\kappa}_i \cdot \vec{\rho}'_S} G(|\vec{\rho}'_S - \vec{\rho}_2|, \frac{\omega}{cd_2}) \qquad (64)$$

where we have ignored all the proportional constants. Completing the double integral of $d\vec{\kappa}_s$ and $d\vec{\kappa}_s$

$$\int d\vec{\kappa}_s \, d\vec{\kappa}_i \, \delta(\vec{\kappa}_s + \vec{\kappa}_i) \, e^{i\vec{\kappa}_s \cdot \vec{\rho}_S} \, e^{i\vec{\kappa}_i \cdot \vec{\rho}'_S} \sim \delta(\vec{\rho}_S - \vec{\rho}'_S), \qquad (65)$$

Eq. (64) becomes:

$$\Psi(\vec{\rho}_o, \vec{\rho}_2)$$

$$\propto \int_{lens} d\vec{\rho}_l \int_A d\vec{\rho}_S \, G(|\vec{\rho}_2 - \vec{\rho}_S|, \frac{\omega}{cd_2}) G(|\vec{\rho}_S - \vec{\rho}_l|, \frac{\omega}{cd_1}) G(|\vec{\rho}_l|, \frac{\omega}{cf}) G(|\vec{\rho}_l - \vec{\rho}_o|, \frac{\omega}{cs_o}).$$

We then apply the properties of the Gaussian functions of Eq. (A.3) and complete the integral on $d\vec{\rho}_S$ by assuming the transverse size of the source is large enough to be treated as infinity.

$$\Psi(\vec{\rho}_o, \vec{\rho}_2) \propto \int_{lens} d\vec{\rho}_l \, G(|\vec{\rho}_2 - \vec{\rho}_l|, \frac{\omega}{cs_i}) G(|\vec{\rho}_l|, \frac{\omega}{cf}) G(|\vec{\rho}_l - \vec{\rho}_o|, \frac{\omega}{cs_o}). \qquad (66)$$

Although the signal and idler propagate to different directions along two optical arms, Interestingly, the Green function in Eq. (66) is equivalent to that of a classical imaging setup, if we imagine the fields start propagating from a point $\vec{\rho}_o$ on the object plane to the lens and then stop at point $\vec{\rho}_2$ on the imaging plane. The mathematics is consistent with our previous qualitative analysis of the experiment.

The integral on $d\vec{\rho}_l$ yields a point-to-point relationship between the object plane and the image plane that is defined by the Gaussian thin-lens equation:

$$\int_{lens} d\vec{\rho}_l \, G(|\vec{\rho}_l|, \frac{\omega}{c}[\frac{1}{s_o} + \frac{1}{s_i} - \frac{1}{f}]) e^{-i\frac{\omega}{c}(\frac{\vec{\rho}_o}{s_o} + \frac{\vec{\rho}_i}{s_i}) \cdot \vec{\rho}_l} \propto \delta(\vec{\rho}_o + \frac{\vec{\rho}_i}{m}) \qquad (67)$$

The Physics of $2 \neq 1 + 1$ 193

where the integral is approximated to infinity and the Gaussian thin-lens equation of $1/s_o + 1/s_i = 1/f$ is applied. We have also defined $m = s_i/s_o$ as the magnification factor of the imaging system. The function $\delta(\vec{\rho}_o + \vec{\rho}_i/m)$ indicates that a point $\vec{\rho}_o$ on the object plane corresponds to a unique point $\vec{\rho}_i$ on the image plane. The two vectors point in opposite directions and the magnitudes of the two vectors hold a ratio of $m = |\vec{\rho}_i|/|\vec{\rho}_o|$.

If the finite size of the imaging lens has to be taken into account (finite diameter D), the integral yields a point-spread function of $somb(x)$:

$$\int_{lens} d\vec{\rho}_l \, e^{-i\frac{\omega}{c}(\frac{\vec{\rho}_o}{s_o} + \frac{\vec{\rho}_i}{s_i}) \cdot \vec{\rho}_l} \propto somb\left(\frac{R}{s_o} \frac{\omega}{c}[\vec{\rho}_o + \frac{\vec{\rho}_i}{m}]\right) \tag{68}$$

where $somb(x) = 2J_1(x)/x$, $J_1(x)$ is the first-order Bessel function and R/s_o is named as the numerical aperture. The point-spread function turns the point-to-point correspondence between the object plane and the image plane into a point-to-"spot" relationship and thus limits the spatial resolution. This point has been discussed in detail in the last section.

Therefore, by imposing the condition of the Gaussian thin-lens equation, the transverse two-photon effective wavefunction is approximated as a δ function

$$\Psi(\vec{\rho}_o, \vec{\rho}_i) \propto \delta(\vec{\rho}_o + \frac{\vec{\rho}_i}{m}) \tag{69}$$

where $\vec{\rho}_o$ and $\vec{\rho}_i$, again, are the transverse coordinates on the object plane and the image plane, respectively, defined by the Gaussian thin-lens equation. Thus, the second-order spatial correlation function $G^{(2)}(\vec{\rho}_o, \vec{\rho}_i)$ turns out to be:

$$G^{(2)}(\vec{\rho}_o, \vec{\rho}_i) = |\Psi(\vec{\rho}_o, \vec{\rho}_i)|^2 \propto |\delta(\vec{\rho}_o + \frac{\vec{\rho}_i}{m})|^2. \tag{70}$$

Equation (70) indicates a point to point EPR correlation between the object plane and the image plane, i.e., if one observes the signal photon at a position $\vec{\rho}_o$ on the object plane, the idler photon can only be found at a certain unique position $\vec{\rho}_i$ on the image plane satisfying $\delta(\vec{\rho}_o + \vec{\rho}_i/m)$ with $m = s_i/s_o$.

We now include an object-aperture function, a collection lens and a photon counting detector D_1 into the optical transfer function of arm-1 as shown in Fig. 9.

We will first treat the collection-lens-D_1 package as a "bucket" detector. The "bucket" detector integrates all $\Psi(\vec{\rho}_o, \vec{\rho}_2)$ which passes the object aperture $A(\vec{\rho}_o)$ as a joint photo-detection event. This process is equivalent to the following convolution:

$$R_{1,2} \propto \int_{obj} d\vec{\rho}_o \, |A(\vec{\rho}_o)|^2 \, |\Psi(\vec{\rho}_o, \vec{\rho}_i)|^2 \simeq |A(\frac{-\vec{\rho}_i}{m})|^2 \tag{71}$$

where, again, D_2 is scanning in the image plane, $\vec{\rho}_2 = \vec{\rho}_i$. Equation (71) indicates a magnified (or demagnified) image of the object-aperture function by means of the joint-detection events between distant photodetectors D_1 and D_2. The "−" sign in $A(-\vec{\rho}_i/m)$ indicates opposite orientation of the image. The model of the "bucket" detector is a good and realistic approximation.

194   Y. Shih

Now we consider a detailed evaluation by including the object-aperture function, the collection lens and the photon counting detector D_1 into arm-1. The Green's function of Eq. (62) becomes:

$$
\begin{aligned}
g(\vec{\kappa}_s, \omega_s; \vec{\rho}_1, z_1 &= d_1 + s_o + f_{coll}) \\
&= e^{i\frac{\omega_s}{c}z_1} \int_{obj} d\vec{\rho}_o \int_{lens} d\vec{\rho}_l \int_A d\vec{\rho}_S \left\{ \frac{-i\omega_s}{2\pi c d_1} e^{i\vec{\kappa}_s \cdot \vec{\rho}_S} G(|\vec{\rho}_S - \vec{\rho}_l|, \frac{\omega_s}{c d_1}) \right\} \\
&\quad \times G(|\vec{\rho}_l|, \frac{\omega_s}{cf}) \left\{ \frac{-i\omega_s}{2\pi c s_o} G(|\vec{\rho}_l - \vec{\rho}_o|, \frac{\omega_s}{c s_o}) \right\} A(\vec{\rho}_o) \\
&\quad \times G(|\vec{\rho}_o|, \frac{\omega_s}{cf_{coll}}) \left\{ \frac{-i\omega_s}{2\pi c f_{coll}} G(|\vec{\rho}_o - \vec{\rho}_1|, \frac{\omega_s}{c f_{coll}}) \right\}
\end{aligned}
\tag{72}
$$

where f_{coll} is the focal-length of the collection lens and D_1 is placed on the focal point of the collection lens. Repeating the previous calculation, we obtain the transverse two-photon effective wavefunction:

$$
\Psi(\vec{\rho}_1, \vec{\rho}_2) \propto \int_{obj} d\vec{\rho}_o A(\vec{\rho}_o) \delta(\vec{\rho}_o + \frac{\vec{\rho}_2}{m}) = A(\vec{\rho}_o) \otimes \delta(\vec{\rho}_o + \frac{\vec{\rho}_2}{m})
\tag{73}
$$

where $\otimes$ means convolution. Notice, in Eq. (73) we have ignored the phase factors which have no contribution to the formation of the image. The joint detection counting rate, $R_{1,2}$, between photon counting detectors D_1 and D_2 is thus:

$$
R_{1,2} \propto G^{(2)}(\vec{\rho}_1, \vec{\rho}_2) \propto \left| A(\vec{\rho}_o) \otimes \delta(\vec{\rho}_o + \frac{\vec{\rho}_2}{m}) \right|^2 = \left| A(\frac{-\vec{\rho}_2}{m}) \right|^2
\tag{74}
$$

where, again, $\vec{\rho}_2 = \vec{\rho}_i$.

As we have discussed earlier, the point-to-point EPR correlation is the result of the coherent superposition of a special selected set of two-photon states. In principle, one signal–idler pair contains all the necessary two-photon amplitudes that generate the ghost image—a nonclassical characteristic which we name as a *two-photon coherent* image.

6 Popper's Experiment

In quantum mechanics, one can never expect to measure both the precise position and momentum of a particle simultaneously. It is prohibited. We say that the quantum observable "position" and "momentum" are "complementary" because the precise knowledge of the position (momentum) implies that all possible outcomes of measuring the momentum (position) are equally probable.

Karl Popper, being a "metaphysical realist", however, took a different point of view. In his opinion, the quantum formalism *could* and *should* be interpreted realistically: a particle must have a precise position and momentum [25]. This view was shared by Einstein. In this regard, he invented a thought experiment in the

The Physics of $2 \neq 1+1$

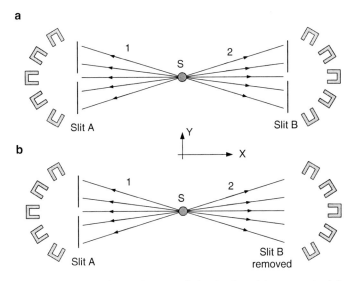

Fig. 13 Popper's thought experiment. An entangled pair of particles are emitted from a point source with momentum conservation. A narrow slit on screen A is placed in the path of particle 1 to provide the precise knowledge of its position on the y-axis and this also determines the precise y-position of its twin, particle 2, on screen B. (**a**) Slits A and B are both adjusted very narrowly. (**b**) Slit A is kept very narrow and slit B is left wide open

early 1930's aimed to support his realistic interpretation of quantum mechanics [26]. What Popper intended to show in his thought experiment is that a particle can have both precise position and momentum simultaneously through the correlation measurement of an entangled two-particle system.

Similar to EPR's *gedankenexperiment*, Popper's thought experiment is also based on the feature of *two-particle entanglement*: if the position or momentum of particle 1 is known, the corresponding observable of its twin, particle 2, is then 100% determined. Popper's original thought experiment is schematically shown in Fig. 13. A point source S, positronium as Popper suggested, is placed at the center of the experimental arrangement from which entangled pairs of particles 1 and 2 are emitted in opposite directions along the respective positive and negative x-axes towards two screens A and B. There are slits on both screens parallel to the y-axis and the slits may be adjusted by varying their widths Δy. Beyond the slits on each side stand an array of Geiger counters for the joint measurement of the particle pair as shown in the figure. The entangled pair could be emitted to any direction in 4π solid angles from the point source. However, if particle 1 is detected in a certain direction, particle 2 is then known to be in the opposite direction due to the momentum conservation of the pair.

First, let us imagine the case in which slits A and B are both adjusted very narrowly. In this circumstance, particle 1 and particle 2 experience diffraction at slit A and slit B, respectively, and exhibit greater Δp_y for smaller Δy of the slits. There seems to be no disagreement in this situation between Copenhagen and Popper.

Next, suppose we keep slit A very narrow and leave slit B wide open. The main purpose of the narrow slit A is to provide the precise knowledge of the position y of particle 1 and this subsequently determines the precise position of its twin (particle 2) on side B through quantum entanglement. Now, Popper asks, in the absence of the physical interaction with an actual slit, does particle 2 experience a greater uncertainty in Δp_y due to the precise knowledge of its position? Based on his beliefs, Popper provides a straightforward prediction: *particle 2 must not experience a greater Δp_y unless a real physical narrow slit B is applied*. However, if Popper's conjecture is correct, this would imply the product of Δy and Δp_y of particle 2 could be smaller than h ($\Delta y \Delta p_y < h$). This may pose a serious difficulty for Copenhagen and perhaps for many of us. On the other hand, if particle 2 going to the right does scatter like its twin, which has passed though slit A, while slit B is wide open, we are then confronted with an apparent *action-at-a-distance*!

The use of a "point source" in Popper's proposal has been criticized historically as the fundamental mistake Popper made [27]. It is true that a point source can never produce a pair of entangled particles which preserves the EPR correlation in momentum as Popper expected. However, notice that a "point source" is *not* a necessary requirement for Popper's experiment. What is required is a precise position–position EPR correlation: if the position of particle 1 is precisely known, the position of particle 2 is 100% determined. As we have shown in the last section, "ghost" imaging is a perfect tool to achieve this.

In 1998, Popper's experiment was realized with the help of two-photon "ghost" imaging [28]. Figure 14 is a schematic diagram that is useful for comparison with the original Popper's thought experiment. It is easy to see that this is a typical "ghost" imaging experimental setup. An entangled photon pair is used to image slit A onto the distant image plane of "screen" B. In the setup, s_o is chosen to be twice the focal length of the imaging lens LS, $s_o = 2f$. According to the Gaussian thin lens equation, an equal size "ghost" image of slit A appears on the two-photon image plane at $s_i = 2f$. The use of slit A provides a precise knowledge of the position of photon 1 on the y-axis and also determines the precise y-position of its twin, photon 2, on screen B by means of the two-photon "ghost" imaging. The experimental condition specified in Popper's experiment is then achieved. When slit A is adjusted to a certain narrow width and slit B is wide open, slit A provides precise knowledge about the position of photon 1 on the y-axis up to an accuracy Δy which equals the width of slit A, and the corresponding "ghost image" of pinhole A at screen B determines the precise position y of photon 2 to within the same accuracy Δy. Δp_y of photon 2 can be independently studied by measuring the width of its "diffraction pattern" at a certain distance from "screen" B. This is obtained by recording coincidences between detectors D_1 and D_2 while scanning detector D_2 along its y-axis, which is behind screen B at a certain distance.

Figure 15 is a conceptual diagram to connect the modified Popper's experiment with two-photon "ghost" imaging. In this unfolded "ghost" imaging setup, we assume the entangled signal–idler photon pair holds a perfect transverse momentum correlation with $\vec{k}_s + \vec{k}_i \sim 0$, which can be easily realized in SPDC. In this experiment, we have chosen $s_o = s_i = 2f$. Thus, an equal size "ghost" image of slit A is expected to appear on the image plane of screen B.

The Physics of $2 \neq 1+1$

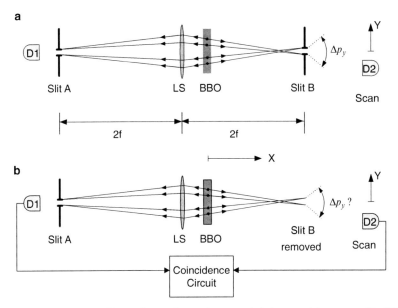

Fig. 14 Modified version of Popper's experiment. An entangled photon pair is generated by SPDC. A lens and a narrow slit A are placed in the path of photon 1 to provide the precise knowledge of its position on the y-axis and also to determine the precise y-position of its twin, photon 2, on screen B by means of two-photon "ghost" imaging. Photon counting detectors D_1 and D_2 are used to scan in y-directions for joint detections. (**a**) Slits A and B are both adjusted very narrowly. (**b**) Slit A is kept very narrow and slit B is left wide open

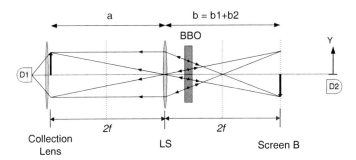

Fig. 15 An unfolded schematic of ghost imaging. We assume the entangled signal–idler photon pair holds a perfect momentum correlation $\delta(\mathbf{k}_s + \mathbf{k}_i) \sim 0$. The locations of the slit A, the imaging lens LS, and the "ghost" image must be governed by the Gaussian thin lens equation. In this experiment, we have chosen $s_o = s_i = 2f$. Thus, the "ghost" image of slit A is expected to be the same size as that of slit A

The detailed experimental setup is shown in Fig. 16 with indications of the various distances. A CW Argon ion laser line of $\lambda_p = 351.1$ nm is used to pump a 3 mm long beta barium borate (BBO) crystal for type-II SPDC to generate an orthogonally polarized signal–idler photon pair. The laser beam is about 3 mm in diameter

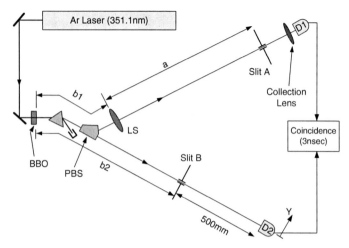

Fig. 16 Schematic of the experimental setup. The laser beam is about 3mm in diameter. The "phase-matching condition" is well reinforced. Slit A (0.16mm) is placed 1,000mm = $2f$ behind the converging lens, LS ($f = 500$mm). The one-to-one "ghost image" (0.16mm) of slit A is located at B. The optical distance from LS in the signal beam taken as back through PBS to the SPDC crystal ($b_1 = 255$mm) and then along the idler beam to "screen B" ($b_2 = 745$mm) is 1,000mm = $2f$ ($b = b_1 + b_2$)

with a diffraction limited divergence. It is important to keep the pump beam a large size so that the transverse phase-matching condition, $\vec{k}_s + \vec{k}_i \sim 0$ ($\vec{k}_p = 0$), is well reinforced in the SPDC process, where $\vec{k}_j$ ($j = s, i$) is the transverse wavevector of the signal (s) and idler (i), respectively. The collinear signal-idler beams, with $\lambda_s = \lambda_i = 702.2$ nm $= 2\lambda_p$ are separated from the pump beam by a fused quartz dispersion prism, and then split by a polarization beam splitter PBS. The signal beam (photon 1) passes through the converging lens LS with a 500mm focal length and a 25mm diameter. A 0.16mm slit is placed at location A which is 1000mm ($= 2f$) behind the lens LS. A short focal length lens is used with D_1 for focusing the signal beam that passes through slit A. The point-like photon counting detector D_2 is located 500mm behind "screen B". "Screen B" is the image plane defined by the Gaussian thin lens equation. Slit B, either adjusted as the same size as that of slit A or opened completely, is placed to coincide with the "ghost" image. The output pulses from the detectors are sent to a coincidence circuit. During the measurements, detector D_1 is fixed behind slit A while detector D_2 is scanned on the y-axis by a step motor.

Measurement 1: Measurement 1 studied the case in which both slits A and B were adjusted to be 0.16mm. The y-coordinate of D_1 was chosen to be 0 (center) while D_2 was allowed to scan along its y-axis. The circled dot data points in Fig. 17 show the *coincidence* counting rates against the y-coordinates of D_2. It is a typical single-slit diffraction pattern with $\Delta y \Delta p_y = h$. Nothing is special in this measurement except that we have learned the width of the diffraction pattern for the 0.16mm slit and this

The Physics of $2 \neq 1+1$

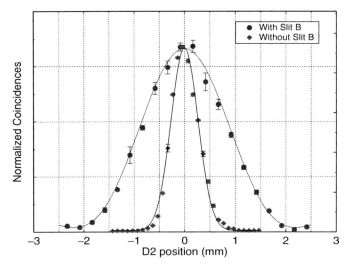

Fig. 17 The observed coincidence patterns. The y-coordinate of D_1 was chosen to be 0 (center) while D_2 was allowed to scan along its y-axis. Circled dot points: *Slit A = Slit B = 0.16mm*. Diamond dot points: *Slit A = 0.16mm, Slit B wide open*. The width of the *sinc* function curve fitted by the circled dot points is a measure of the minimum Δp_y diffracted by a 0.16mm slit

represents the minimum uncertainty of Δp_y. We should emphasize at this point that the *single* detector counting rate of D_2 as a function of its position y is basically the same as that of the coincidence counts except for a higher counting rate.

Measurement 2: The same experimental conditions were maintained except that slit B was left wide open. This measurement is a test of Popper's prediction. The y-coordinate of D_1 was chosen to be 0 (center) while D_2 was allowed to scan along its y-axis. Due to the entangled nature of the signal–idler photon pair and the use of a coincidence measurement circuit, only those twins which have passed through slit A and the "ghost image" of slit A at screen B with an uncertainty of $\Delta y = 0.16\,\text{mm}$ (which is the same width as the real slit B we have used in measurement 1) would contribute to the coincidence counts through the joint detection of D_1 and D_2. The diamond dot data points in Fig. 17 report the measured coincidence counting rates against the y coordinates of D_2. The measured width of the pattern is narrower than that of the diffraction pattern shown in measurement 1. It is also interesting to notice that the single detector counting rate of D_2 keeps constant in the entire scanning range, which is very different from that in measurement 1. The experimental data has provided a clear indication of $\Delta y \Delta p_y < h$ in the joint measurements of the entangled photon pairs.

Given that $\Delta y \Delta p_y < h$, is this a violation of the uncertainty principle? Does quantum mechanics agree with this peculiar experimental result? If quantum mechanics does provide a solution with $\Delta y \Delta p_y < h$ for photon 2. We would indeed be forced to face a paradox as EPR had pointed out in 1935.

200   Y. Shih

Quantum mechanics does provide a solution that agrees with the experimental result. However, the solution is for a joint measurement of an entangled photon pair that involves both photon 1 and photon 2, but not just for photon 2 itself .

We now examine the experimental results with the quantum mechanical calculation by adopting the formalisms from the ghost image experiment with two modifications:

Case (I): slits A $= 0.16$ mm, slit B $= 0.16$ mm.

This is the experimental condition for measurement one: slit B is adjusted to be the same as slit A. There is nothing surprising about this measurement. The measurement simply provides us with the knowledge for Δp_y of photon 2 caused by the diffraction of slit B ($\Delta y = 0.16$ mm). The experimental data shown in Fig. 17 agrees with the calculation. Notice that slit B is about 745 mm away from the 3 mm two-photon source, the angular size of the light source is roughly the same as $\lambda/\Delta y$, $\Delta\theta \sim \lambda/\Delta y$, where $\lambda = 702$ nm is the wavelength and $\Delta y = 0.16$ mm is the width of the slit. The calculated diffraction pattern is very close to that of the "far-field" Fraunhofer diffraction of a 0.16 mm single-slit.

Case (II): slit A $= 0.16$ mm, slits B $\sim \infty$ (wide open).

Now we remove slit B from the ghost image plane. The calculation of the transverse effective two-photon wavefunction and the second-order correlation is the same as that of the ghost image except the observation plane of D_2 is moved behind the image plane to a distance of 500 mm. The two-photon image of slit A is located at a distance $s_i = 2f = 1,000$ mm ($b_1 + b_2$) from the imaging lens, in this measurement D_2 is placed at $d = 1,500$ mm from the imaging lens. The measured pattern is simply a "blurred" two-photon image of slit A. The "blurred" two-photon image can be calculated from Eq. (75) which is a slightly modified version of Eq. (66)

$$
\begin{aligned}
\Psi(\vec{\rho}_o, \vec{\rho}_2) &\propto \int_{lens} d\vec{\rho}_l \, G(|\vec{\rho}_2 - \vec{\rho}_l|, \frac{\omega}{cd}) \, G(|\vec{\rho}_l|, \frac{\omega}{cf}) \, G(|\vec{\rho}_l - \vec{\rho}_o|, \frac{\omega}{cs_o}) \\
&\propto \int_{lens} d\vec{\rho}_l \, G(|\vec{\rho}_l|, \frac{\omega}{c}[\frac{1}{s_o} + \frac{1}{d} - \frac{1}{f}]) \, e^{-i\frac{\omega}{c}(\frac{\vec{\rho}_o}{s_o} + \frac{\vec{\rho}_2}{d})\cdot\vec{\rho}_l}
\end{aligned}
\qquad (75)
$$

where d is the distance between the imaging lens and D_2. In this measurement, D_2 was placed 500 mm behind the image plane, i.e., $d = s_i + 500$ mm. The numerically calculated "blurred" image, which is narrower than that of the diffraction pattern of the 0.16 mm slit B, agrees with the measured result of Fig. 17 within experimental error.

The measurement does show a result of $\Delta y \Delta p_y < h$. The measurement, however, has nothing to do with the uncertainty relation, which governs the behavior of photon 2 (the idler). Popper and EPR were correct in the prediction of the outcomes of their experiments. Popper and EPR, on the other hand, made the same error by applying the results of two-particle physics to the explanation of the behavior of an individual subsystem.

In both the Popper and EPR experiments, the measurements are "joint detection" between two detectors applied to entangled states. Quantum mechanically, an entangled two-particle state only provides *the precise knowledge of the correlations*

The Physics of $2 \neq 1 + 1$ 201

of the pair. The behavior of "photon 2" observed in the joint measurement is conditioned upon the measurement of its twin. A quantum must obey the uncertainty principle but the "conditional behavior" of a quantum in an entangled two-particle system is different in principle. We believe paradoxes are unavoidable if one insists the *conditional behavior* of a particle is the *behavior* of the particle. This is the central problem in the rationale behind both Popper and EPR. $\Delta y \Delta p_y \geq h$ is not applicable to the *conditional behavior* of either "photon 1" or "photon 2" in the cases of Popper and EPR.

The behavior of photon 2 being conditioned upon the measurement of photon 1 is well represented by the two-photon amplitudes. Each of the *straight lines* in the above discussion corresponds to a two-photon amplitude. Quantum mechanically, the superposition of these two-photon amplitudes are responsible for a "click–click" measurement of the entangled pair. A "click–click" joint measurement of the two-particle entangled state projects out certain two-particle amplitudes, and only these two-particle amplitudes are featured in the quantum formalism. In the above analysis we never consider "photon 1" or "photon 2" *individually*. Popper's question about the momentum uncertainty of photon 2 is then inappropriate.

Once again, the demonstration of Popper's experiment calls our attention to the important message: the physics of an entangled two-particle system must be inherently very different from that of individual particles.

7 Subsystem in an Entangled Two-Photon State

The entangled EPR two-particle state is a pure state with zero entropy. The precise correlation of the subsystems is completely described by the state. The measurement, however, is not necessarily always on the two-photon system. It is an experimental choice to study a single subsystem and to ignore the other. What can be learn about a subsystem from these kinds of measurements? Mathematically, it is easy to show that by taking a partial trace of a two-particle pure state, the state of each subsystem is in a mixed state with entropy greater than zero. One can only learn statistical properties of the subsystems in this kind of measurement.

In the following, again, we use the signal–idler pair of SPDC as an example to study the physics of a subsystem. The two-photon state of SPDC is a pure state that satisfies

$$\hat{\rho}^2 = \hat{\rho}, \quad \hat{\rho} \equiv |\Psi\rangle \langle \Psi|$$

where $\hat{\rho}$ is the density operator corresponding to the two-photon state of SPDC. The single photon states of the signal and idler

$$\hat{\rho}_s = tr_i |\Psi\rangle \langle \Psi|, \quad \hat{\rho}_i = tr_s |\Psi\rangle \langle \Psi|$$

are not pure states. To calculate the signal (idler) state from the two-photon state, we take a partial trace, as usual, summing over the idler (signal) modes.

202   Y. Shih

We assume a type II SPDC. The orthogonally polarized signal and idler are degenerate in frequency around $\omega_s^0 = \omega_i^0 = \omega_p/2$. To simplify the discussion, by assuming appropriate experimental conditions, we trivialize the transverse part of the state and write the two-photon state in the following simplified form:

$$|\Psi\rangle = \Psi_0 \int d\Omega \, \Phi(\mathrm{DL}\Omega) a_s^\dagger(\omega_s^0 + \Omega) a_i^\dagger(\omega_i^0 - \Omega)|0\rangle$$

where $\Phi(\mathrm{DL}\Omega)$ is a *sinc*-like function:

$$\Phi(\mathrm{DL}\Omega) = \frac{1 - e^{-i\mathrm{DL}\Omega}}{i\mathrm{DL}\Omega}$$

which is a function of the crystal length L, and the difference of inverse group velocities of the signal (ordinary) and the idler (extraordinary), $D \equiv 1/u_o - 1/u_e$. The constant Ψ_0 is calculated from the normalization $tr\hat{\rho} = \langle \Psi \mid \Psi \rangle = 1$. It is easy to calculate and to find $\hat{\rho}^2 = \hat{\rho}$ for the two-photon state of the signal–idler pair.

Summing over the idler modes, the density matrix of signal is given by

$$\hat{\rho}_s = \Psi_0^2 \int d\Omega \, |\Phi(\Omega)|^2 \, a_s^\dagger(\omega_s^0 + \Omega)|0\rangle \langle 0| \, a_s(\omega_s^0 + \Omega) \qquad (76)$$

with

$$|\Phi(\Omega)|^2 = \mathrm{sinc}^2 \frac{\mathrm{DL}\Omega}{2}$$

where all constants coming from the integral have been absorbed into Ψ_0. First, we find immediately that $\hat{\rho}_s^2 \neq \hat{\rho}_s$. It means the state of the signal is a mixed state (as is the idler). Second, it is very interesting to find that the spectrum of the signal depends on the group velocity of the idler. This, however, should not come as a surprise, because the state of the signal photon is calculated from the two-photon state by summing over the idler modes.

The spectrum of the signal and idler has been experimentally verified by Strekalov et al using a Michelson interferometer in a standard Fourier spectroscopy type measurement [29]. The measured interference pattern is shown in Fig. 18. The envelope of the sinusoidal modulations (in segments) is fitted very well by two "notch" functions (upper and lower part of the envelope). The experimental data agrees with the theoretical analysis of the experiment.

The following is a simple calculation to explain the observed "notch" function. We first define the field operators:

$$E^{(+)}(t, z_d) = E^{(+)}\left(t - \frac{z_1}{c}, z_0\right) + E^{(+)}\left(t - \frac{z_2}{c}, z_0\right)$$

where z_d is the position of the photo-detector, z_0 is the input point of the interferometer, $t_1 = t - \frac{z_1}{c}$ and $t_2 = t - \frac{z_2}{c}$, respectively, are the early times before propagating to the photodetector at time t with time delays of z_1/c and z_2/c, where z_1 and z_2 are the optical paths in arm 1 and arm 2 of the interferometer. We have defined a very general field operator which is a superposition of two early fields propagated

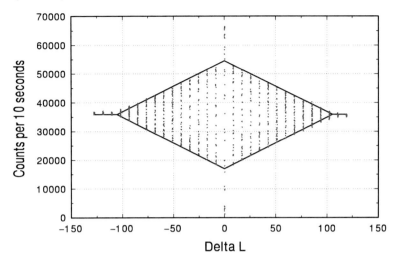

Fig. 18 Experimental data indicated a "double notch" envelope. Each of the dotted single vertical line contains many cycles of sinusoidal modulation

individually through arm 1 and arm 2 of any type of interferometer. The counting rate of the photon counting detector is thus

$$R_d = tr\left[\hat{\rho}_s E^{(-)}(t,z_d) E^{(+)}(t,z_d)\right]$$
$$= \Psi_0^2 \int d\Omega \, |\Phi(\Omega)|^2 \, |\langle 0|E^{(+)}(t,z_d) a_s^\dagger(\omega_s^0+\Omega)|0\rangle|^2$$
$$= \Psi_0^2 \int d\Omega \, |\Phi(\Omega)|^2 \, |\langle 0|\left[E^{(+)}(t-\frac{z_1}{c},z_0) + E^{(+)}(t-\frac{z_2}{c},z_0)\right] a_s^\dagger(\omega_s^0+\Omega)|0\rangle|^2$$
$$\propto 1 + Re\left[e^{-i\omega^0 \tau} \int d\Omega \, \text{sinc}^2\frac{DL\Omega}{2} e^{-i\Omega\tau}\right] \quad (77)$$

where $\tau = (z_1 - z_2)/c$. The Fourier transform of $\text{sinc}^2(DL\Omega/2)$ has a "notch" shape. It is noticed that the base of the "notch" function is determined by parameter DL of the SPDC, which is easily confirmed from the experiment.

Now we turn to another interesting aspect of physics, namely the physics of entropy. In classical information theory, the concept of entropy, named as Von Neuman entropy, is defined by [30]

$$S = -tr(\hat{\rho} \log \hat{\rho}) \quad (78)$$

where $\hat{\rho}$ is the density operator. It is easy to find that the entropy of the entangled two-photon pure state is zero. The entropy of its subsystems, however, are both greater than zero. The value of the Von Neuman entropy can be numerically evaluated from the measured spectrum. Note that the density operator of the subsystem is diagonal. Taking its trace is simply performing an integral over the frequency spectrum with the measured spectrum function. It is straightforward to find the entropy of the subsystems $S_s > 0$. This is an expected result due to the statistical mixture nature of the subsystem. Considering that the entropy of the two-photon system is zero and the entropy of the subsystems are both greater than zero, does this mean that negative entropy is present somewhere in the entangled two-photon system? According to classical "information theory", for the entangled two-photon system, $S_s + S_{s|i} = 0$, where $S_{s|i}$ is the conditional entropy. It is this conditional entropy that must be negative, which means that *given the result of a measurement over one particle, the result of a measurement over the other must yield negative information* [31]. This paradoxical statement is similar and, in fact, closely related to the EPR "paradox". It comes from the same philosophy as that of the EPR.

8 Summary

The physics of an entangled system is very different from that of either classically independent or correlated systems. We use $2 \neq 1 + 1$ to emphasize the nonclassical behavior of an entangled two-particle system. The entangled system is characterized by the properties of an entangled state which does not specify the state of an individual system, but rather describes the correlation between the subsystems. An entangled two particle state is a pure state which involves the superposition of a set of "selected" two-particle states, or two-particle quantum mechanical amplitudes. Here, the term "selection" stems from the physical laws which govern the creation of the subsystems in the source, such as energy or momentum conservation. Interestingly, quantum mechanics allows for the superposition of these local two-particle states which have been observed in nature. However, the most surprising physics arises from the joint measurement of the two particles when they are released form the source and propagated a large distance apart. The two well separated interaction-free particles do not lose their entangled properties, i.e., they maintain their "selected" set of two-particle superposition. In this sense quantum mechanics allows for the two-particle superposition of well separated particles which has, remarkably, also been observed to exist in nature.

The two-photon state of SPDC is a good example. The nonlinear interaction of spontaneous parametric down-conversion coherently creates a set of mode in pairs that satisfy the phase matching conditions of Eq. (11) which is also characteristic of energy and momentum conservation. The signal–idler photon pair can be excited to any or all of these coupled modes simultaneously, resulting in a superposition of these coupled modes inside of the nonlinear crystal. The physics behind the two-photon superposition becomes even more interesting when the signal–idler pair is separated and propagated a large distance apart outside the nonlinear crystal, either through free propagation or guided by optical components. Remarkably the entangled pair does not lose its entangled properties once the subsystems are interaction free. As a result the properties of the entangled two-photon system, such as the EPR correlation or the EPR inequalities, are still observable in the joint detection counting rate of the pair, regardless of the distance between the two photons as well as the two individual photo-detection events. In this situation the superposition of the two-photon amplitudes, corresponding to different yet indistinguishable alternative ways of triggering a joint photo-electron event at any distance can be regarded as nonlocal. There is no counterpart to such a concept in classical theory and this behavior may never be understood in any classical sense. It is with this intent that we use $2 \neq 1+1$ to emphasize that the physics of a two-photon is not the same as that of two photons.

Acknowledgment The author would like to thank A. Shimony for many years of encouragement on the study of the fundamental problems of quantum theory and thank C.O. Alley, D.N. Klyshko, and M.H. Rubin for many years of research collaboration. Thanks also goes to T.B. Pittman, D.V. Strekalov, Y.H. Kim, M. D'Angelo, A. Valencia, and G. Scarcelli for their beautiful Ph.D. thesis work. This research was supported in part by AFOSR, ARO-MURI and by the NASA-CASPR program.

Appendix: Fresnel Propagation-Diffraction

In Fig. A.1, the field is freely propagated from the source plane σ_0 to an arbitrary plane σ. It is convenient to describe such a propagation in the form of Eq. (33). We now evaluate $g(\vec{\kappa}, \omega; \vec{\rho}, z)$, namely the Green's function for free-space Fresnel propagation-diffraction.

According to the Huygens-Fresnel principle, the field at a space-time point $(\vec{\rho}, z, t)$ is the result of a superposition of the spherical secondary wavelets originated from each point on the σ_0 plane, see Fig. A.1,

$$E^{(+)}(\vec{\rho}, z, t) = \int d\omega d\vec{\kappa}\, a(\omega, \vec{\kappa}) \int_{\sigma_0} d\vec{\rho}_0 \frac{\tilde{A}(\vec{\rho}_0)}{r'} e^{-i(\omega t - kr')} \qquad (A.1)$$

where $\tilde{A}(\vec{\rho}_0)$ is the complex amplitude, or distribution function, in terms of the transverse coordinate $\vec{\rho}_0$, which may be a constant, a simple aperture function, or a

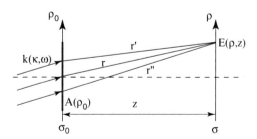

Fig. A.1 Schematic of free-space Fresnel propagation. The complex amplitude $\tilde{A}(\vec{\rho}_0)$ is composed by a real function $A(\vec{\rho}_0)$ and a phase $e^{-i\vec{\kappa}\cdot\vec{\rho}_0}$ associated with each of the transverse wavevector $\vec{\kappa}$ on the plane of σ_0. Notice: only one mode of wavevector $\mathbf{k}(\vec{\kappa}, \omega)$ is shown in the figure

combination of the two. In Eq. (A.1), we have taken $z_0 = 0$ and $t_0 = 0$ on the source plane of σ_0 as usual.

In a paraxial approximation, we take the first-order expansion of r' in terms of z and $\vec{\rho}$

$$r' = \sqrt{z^2 + |\vec{\rho} - \vec{\rho}_0|^2} \simeq z(1 + \frac{|\vec{\rho} - \vec{\rho}_0|^2}{2z^2}).$$

$E^{(+)}(\vec{\rho}, z, t)$ is thus approximated as

$$E^{(+)}(\vec{\rho}, z, t) \simeq \int d\omega \, d\vec{\kappa} \, a(\omega, \vec{\kappa}) \int d\vec{\rho}_0 \frac{\tilde{A}(\vec{\rho}_0)}{z} e^{i\frac{\omega}{c}z} e^{i\frac{\omega}{2cz}|\vec{\rho} - \vec{\rho}_0|^2} e^{-i\omega t}$$

where $e^{i\frac{\omega}{2cz}|\vec{\rho} - \vec{\rho}_0|^2}$ is named as the Fresnel phase factor.

Assuming the complex amplitude $\tilde{A}(\vec{\rho}_0)$ is composed of a real function $A(\vec{\rho}_0)$ and a phase $e^{-i\vec{\kappa} \cdot \vec{\rho}_0}$, associated with the transverse wavevector and the transverse coordinate on the plane of σ_0, which is reasonable for the setup of Fig. A.1, $E(\vec{\rho}, z, t)$ can be written in the following form

$$E^{(+)}(\vec{\rho}, z, t) = \int d\omega \, d\vec{\kappa} \, a(\omega, \vec{\kappa}) e^{-i\omega t} \frac{e^{i\frac{\omega}{c}z}}{z} \int d\vec{\rho}_0 A(\vec{\rho}_0) e^{i\vec{\kappa} \cdot \vec{\rho}_0} e^{i\frac{\omega}{2cz}|\vec{\rho} - \vec{\rho}_0|^2}.$$

The Green's function $g(\vec{\kappa}, \omega; \vec{\rho}, z)$ for free-space Fresnel propagation is thus

$$g(\vec{\kappa}, \omega; \vec{\rho}, z) = \frac{e^{i\frac{\omega}{c}z}}{z} \int_{\sigma_0} d\vec{\rho}_0 A(\vec{\rho}_0) e^{i\vec{\kappa} \cdot \vec{\rho}_0} G(|\vec{\rho} - \vec{\rho}_0|, \frac{\omega}{cz}). \qquad (A.2)$$

In Eq. (A.2) we have defined a Gaussian function $G(|\vec{\alpha}|, \beta) = e^{i(\beta/2)|\alpha|^2}$, namely the Fresnel phase factor. It is straightforward to find that the Gaussian function $G(|\vec{\alpha}|, \beta)$ has the following properties:

$$G^*(|\vec{\alpha}|, \beta) = G(|\vec{\alpha}|, -\beta),$$
$$G(|\vec{\alpha}|, \beta_1 + \beta_2) = G(|\vec{\alpha}|, \beta_1) G(|\vec{\alpha}|, \beta_2),$$
$$G(|\vec{\alpha}_1 + \vec{\alpha}_2|, \beta) = G(|\vec{\alpha}_1|, \beta) G(|\vec{\alpha}_2|, \beta) e^{i\beta \vec{\alpha}_1 \cdot \vec{\alpha}_2},$$
$$\int d\vec{\alpha} \, G(|\vec{\alpha}|, \beta) e^{i\vec{\gamma} \cdot \vec{\alpha}} = i\frac{2\pi}{\beta} G(|\vec{\gamma}|, -\frac{1}{\beta}). \qquad (A.3)$$

Notice that the last equation in Eq. (A.3) is the Fourier transform of the $G(|\vec{\alpha}|, \beta)$ function. As we shall see in the following, these properties are very useful in simplifying the calculations of the Green's functions $g(\vec{\kappa}, \omega; \vec{\rho}, z)$.

Now, we consider inserting an imaginary plane σ' between σ_0 and σ. This is equivalent having two consecutive Fresnel propagations with a diffraction-free σ' plane of infinity. Thus, the calculation of these consecutive Fresnel propagations should yield the same Green's function as that of the above direct Fresnel propagation shown in Eq. (A.2):

$$g(\vec{\kappa}, \omega; \vec{\rho}, z)$$

$$= C^2 \frac{e^{i\frac{\omega}{c}(d_1+d_2)}}{d_1 d_2} \int_{\sigma'} d\vec{\rho'} \int_{\sigma_0} d\vec{\rho}_0 \tilde{A}(\vec{\rho}_0) G(|\vec{\rho'} - \vec{\rho}_0|, \frac{\omega}{cd_1}) G(|\vec{\rho} - \vec{\rho'}|, \frac{\omega}{cd_2})$$

$$= C \frac{e^{i\frac{\omega}{c}z}}{z} \int_{\sigma_0} d\vec{\rho}_0 \tilde{A}(\vec{\rho}_0) G(|\vec{\rho} - \vec{\rho}_0|, \frac{\omega}{cz}) \tag{A.4}$$

where C is a necessary normalization constant for a valid Eq. (A.4), and $z = d_1 + d_2$. The double integral of $d\vec{\rho}_0$ and $d\vec{\rho'}$ in Eq. (A.4) can be evaluated as

$$\int_{\sigma'} d\vec{\rho'} \int_{\sigma_0} d\vec{\rho}_0 \tilde{A}(\vec{\rho}_0) G(|\vec{\rho'} - \vec{\rho}_0|, \frac{\omega}{cd_1}) G(|\vec{\rho} - \vec{\rho'}|, \frac{\omega}{cd_2})$$

$$= \int_{\sigma_0} d\vec{\rho}_0 \tilde{A}(\vec{\rho}_0) G(\vec{\rho}_0, \frac{\omega}{cd_1}) G(\vec{\rho}, \frac{\omega}{cd_2}) \int_{\sigma'} d\vec{\rho'} G(\vec{\rho'}, \frac{\omega}{c}(\frac{1}{d_1} + \frac{1}{d_2})) e^{-i\frac{\omega}{c}(\frac{\vec{\rho}_0}{d_1} + \frac{\vec{\rho}}{d_2}) \cdot \vec{\rho'}}$$

$$= \frac{i2\pi c}{\omega} \frac{d_1 d_2}{d_1 + d_2} \int_{\sigma_0} d\vec{\rho}_0 \tilde{A}(\vec{\rho}_0) G(\vec{\rho}_0, \frac{\omega}{cd_1}) G(\vec{\rho}, \frac{\omega}{cd_2}) G(|\frac{\vec{\rho}_0}{d_1} + \frac{\vec{\rho}}{d_2}|, \frac{\omega}{c}(\frac{d_1 d_2}{d_1 + d_2}))$$

$$= \frac{i2\pi c}{\omega} \frac{d_1 d_2}{d_1 + d_2} \int_{\sigma_0} d\vec{\rho}_0 \tilde{A}(\vec{\rho}_0) G(|\vec{\rho} - \vec{\rho}_0|, \frac{\omega}{c(d_1 + d_2)})$$

where we have applied Eq. (A.3), and the integral of $d\vec{\rho'}$ has been taken to infinity. Substituting this result into Eq. (A.4), we thus have

$$g(\vec{\kappa}, \omega; \vec{\rho}, z) = C^2 \frac{i2\pi c}{\omega} \frac{e^{i\frac{\omega}{c}(d_1+d_2)}}{d_1 + d_2} \int_{\sigma_0} d\vec{\rho}_0 \tilde{A}(\vec{\rho}_0) G(|\vec{\rho} - \vec{\rho}_0|, \frac{\omega}{c(d_1 + d_2)})$$

$$= C \frac{e^{i\frac{\omega}{c}z}}{z} \int_{\sigma_0} d\vec{\rho}_0 \tilde{A}(\vec{\rho}_0) G(|\vec{\rho} - \vec{\rho}_0|, \frac{\omega}{cz}).$$

Therefore, the normalization constant C must take the value of $C = -i\omega/2\pi c$. The normalized Green's function for free-space Fresnel propagation is thus

$$g(\vec{\kappa}, \omega; \vec{\rho}, z) = \frac{-i\omega}{2\pi c} \frac{e^{i\frac{\omega}{c}z}}{z} \int_{\sigma_0} d\vec{\rho}_0 \tilde{A}(\vec{\rho}_0) G(|\vec{\rho} - \vec{\rho}_0|, \frac{\omega}{cz}). \tag{A.5}$$

References

1. A. Einstein, B. Podolsky, and N. Rosen, *Phys. Rev.* **35**, 777 (1935).
2. N. Bohr, *Phys. Rev.* **48**, 696 (1935).
3. J. Wheeler, "The 'Past' and the 'Delayed-Choice Double-Slit Experiment'," in A.R. Maslow, ed., *Mathematical Foundations of Quantum Theory*, Academic Press, 1978, pp. 9–48.
4. A. Pais, *'Subtle is the lord...' The Science and the Life of Albert Einstein*, Oxford University Press, Oxford and New York, 1982.
5. M. D'Angelo, Y.H. Kim, S.P. Kulik, and Y.H. Shih, *Phys. Rev. Lett.* **92**, 233601 (2004).
6. D. Bohm, *Phys. Rev.* **85**, 166 180 (1952); D. Bohm, *Causality and Chance in Modern Physics*, D. Van Nostrand Co., Princeton, 1957; D. Bohm and Y. Aharonov, *Phys. Rev.* **108**, 1070 (1957).

7. J.S. Bell, Physics **1**, 195 (1964); *Speakable and Unspeakable in Quantum Mechanics*, Cambridge University Press, New York, 1987.
8. J.F. Clauser and A. Shimony, *Rep. Prog. Phys.* **41**, 1883 (1978).
9. A. Aspect, J. Dalibard, and G. Roger, *Phys. Rev. Lett.*, **47**, 460 (1981); A. Aspect, J. Dalibard, and G. Roger, *Phys. Rev. Lett.*, **49**, 91 (1981); A. Aspect, J. Dalibard, and G. Roger, *Phys. Rev. Lett.*, **49**, 1804 (1981).
10. Y.H. Shih and C.O. Alley, *Phys. Rev. Lett.* **61**, 2921 (1988); Z.Y. Ou and L. Mandel, *Phys. Rev. Lett.* **62**, 50 (1988); T.E. Kiess, Y.H. Shih, A.V. Sergienko, and C.O. Alley, *Phys. Rev. Lett.* **71**, 3893 (1993); P.G. Kwiat, et al., *Phys. Rev. Lett.* **75**, 4337 (1995).
11. E. Schrödinger, *Naturwissenschaften* **23**, 807, 823, 844 (1935); English translations appear in ref. [3].
12. D.N. Klyshko, *Photon and Nonlinear Optics*, Gordon and Breach Science, New York, 1988.
13. Y.H. Shih, *IEEE J. Selected Topics in Quantum Electronics* **9**, 1455 (2003).
14. A. Yariv, *Quantum Electronics*, Wiley, New York (1989).
15. R.J. Glauber, *Phys. Rev.* **130**, 2529 (1963); *Phys. Rev.* **131**, 2766 (1963).
16. M.H. Rubin, *Phys. Rev. A* **54**, 5349 (1996).
17. J.W. Goodman, *Introduction to Fourier Optics*, McGraw-Hill, New York, NY, 1968.
18. Y.H. Shih, *IEEE Journal of Selected Topics in Quantum Electronics*, **13**, 1016 (2007).
19. This effect was first proposed for lithography application, namely quantum lithography, by A.N. Boto, et al., *Phys. Rev. Lett.* **85**, 2733 (2000).
20. M. D'Angelo, M.V. Chekhova, and Y.H. Shih, *Phys. Rev. Lett.* **87**, 013603 (2001). Note: Due to the lack of a two-photon absorber, the joint-detection measurement in this experiment was on the Fourier transform plane rather than on the image plane. It might be helpful to point out that the observation of sub-wavelength interference in a Mach Zehnder type interferometer cannot lead to sub-diffraction-limited images, except a set of double modulated interference pattern. The Fourier transform argument works only for imaging setups as is in this experiment.
21. T.B. Pittman, Y.H. Shih, D.V. Strekalov, and A.V. Sergienko, *Phys. Rev. A* **52**, R3429 (1995).
22. D.N. Klyshko, *Usp. Fiz. Nauk*, **154**, 133 (1988); *Sov. Phys. Usp* **31**, 74 (1988); *Phys. Lett. A* **132**, 299 (1988).
23. M. D'Angelo, A. Valencia, M.H. Rubin, and Y.H. Shih, *Phys. Rev. A* **72**, 013810 (2005).
24. D.V. Strekalov, A.V. Sergienko, D.N. Klyshko and Y.H. Shih, *Phys. Rev. Lett.* **74**, 3600 (1995).
25. K.R. Popper, in *Open Questions in Quantum Physics*, G. Tarozzi and A. van der Merwe, eds., D. Reidel Publishing Co., Dordrecht, 1985; K.R. Popper, in *Determinism in Physics*, E.I. Bitsakis and N. Tambakis, eds., Gutenberg Publishing, Athens, 1985.
26. K.R. Popper, *Naturwissenschaften* **22**, 807 (1934); K.R. Popper, *Quantum Theory and the Schism in Physics*, Hutchinson, London, 1982.
27. For criticisms of Popper's experiment, see for example, D. Bedford and F. Selleri, *Lett. Nuovo Cimento*, **42**, 325 (1985); M.J. Collett and R. Loudon, *Nature* **326**, 671 (1987); A. Sudberg, *Philosophy of Science*, **52**, 470 (1985); A. Sudberg, in A. van der Merwe, et al., eds., *Microphysical Reality*, Kluwer Academic, Dordrecht, 1988; M. Horne, *Experimental Metaphysics*, R.S. Cohen, M. Horne and J. Stachel, eds., Kluwer Academic, Dordrecht, 1997.
28. Y.H. Kim and Y.H. Shih, *Found. Phys.*, **29**, 1849 (1999).
29. D.V. Strekalov, Y.H. Kim, and Y.H. Shih, *Phys. Rev. A* **60**, 2685 (1999).
30. C.E. Shannon and W. Weaver, *The Mathematical Theory of Communication*, University of Illinois Press, Urbana, 1949.
31. N.J. Cerf and C. Adami, *Phys. Rev. Lett.* **79**, 5194 (1997).

Part IV
Probability, Uncertainty, and Stochastic Modifications of Quantum Mechanics

Interpretations of Probability in Quantum Mechanics: A Case of "Experimental Metaphysics"

Geoffrey Hellman

Abstract After reviewing paradigmatic cases of "experimental metaphysics" basing inferences against local realism and determinism on experimental tests of Bells theorem (and successors), we concentrate on clarifying the meaning and status of "objective probability" in quantum mechanics. The terms "objective" and "subjective" are found ambiguous and inadequate, masking crucial differences turning on the question of what the numerical values of probability functions measure vs. the question of the nature of the "events" on which such functions are defined. This leads naturally to a 2×2 matrix of types of interpretations, which are then illustrated with salient examples. (Of independent interest are the splitting of "Copenhagen interpretation" into "objective" and "subjective" varieties in one of the dimensions and the splitting of Bohmian hidden variables from (other) modal interpretations along that same dimension.) It is then explained why Everett interpretations are difficult to categorize in these terms. Finally, we argue that Bohmian mechanics does not seriously threaten the experimental-metaphysical case for ultimate randomness and purely physical probabilities.

1 Introduction

One of the most philosophically important and fruitful ideas emerging from the work of Abner Shimony et al. relating to the Bell theorems, named and highlighted by Shimony, is that of "experimental metaphysics". (See e.g. Shimony [1].) Although the general view that there is no sharp boundary between metaphysics and natural science and that questions in the former domain are affected by empirical evidence bearing directly on the latter is not new, and indeed forms a central tenet of mid-twentieth-century Quinean philosophy of science, the links between experimental tests of Bell inequalities, for example, bear far more directly

---

G. Hellman
Department of Philosophy, University of Minnesota

W.C. Myrvold and J. Christian (eds.), *Quantum Reality, Relativistic Causality, and Closing the Epistemic Circle*, The Western Ontario Series in Philosophy of Science 73,
© Springer Science+Business Media B.V. 2009

on matters standardly called "metaphysical" than even that sophisticated philosophy could ever have anticipated. Those links—for example, between experiments of Aspect, et al. [2] and theses of "local realism" and "local determinism"—stand independently of any appeal to Duhemian-Quinean "holism" of testing (according to which it is really whole theories, including various metaphysical background principles, that are tested by experimental evidence, rather than individual statements). What is truly extraordinary about the tests of Bell-type inequalities is the directness of the role of the metaphysical theses, e.g. Einstein's principle of separability of physical states according to space-time location, leading to mathematically precise conditions constraining assignments of values of relevant quantities of local hidden variables theories. The accumulated wealth of evidence confirming quantum correlations between separated subsystems (e.g. paired photons in atomic cascades), thus violating relevant Bell-type inequalities, tells quite directly that aspects of physical nature as we understand it violate separability. The same holds, *mutatis mutandis*, with respect to the conclusion that certain systems exhibit (temporally or dynamically) indeterministic behavior, that certain "actualization" phenomena (e.g. of a value of polarization or spin in a specified orientation) occur "randomly", violating the entrenched rationalist principle of causality (or "sufficient reason").

But what about "loopholes"? Testing a Bell-type inequality always involves special assumptions pertaining to the experimental setup, and the tenacious devil's advocate is bound to find some narrow crack somewhere through which a hidden variable or two might slip. In some cases, improvements in the experimental apparatus have sealed a crack shut or have promised to do so (were one to try hard enough, along a course that would merely continue a pattern of improvements, say of the efficiency of photon detectors); in other cases, a new style of proof of inequalities has bypassed a putative gap in earlier derivations between a "metaphysical" motivating premise (e.g. separability) and a mathematical condition (e.g. factorizability of certain joint probabilities relative to hypothetical, hidden, "physically complete" states) taken to "precisify" the premise.[1] But surely it is in the nature of the beast that there will always be "loopholes", i.e. some wiggle-room "in principle" for the die-hard hidden-variables advocate, some uncertainty in the case based on experimental tests of Bell-type inequalities, however carefully and sturdily it may have been erected. Experimental metaphysics is, after all, *experimental*, and—as Abner Shimony has often emphasized—we must be fallibilists, recognising the possibility of error but without that at all muting the voice of reason.

To motivate the main focus of this essay, let us recall the case against (dynamical) *"determinism-in-nature"* based on experimental tests of Bell-type inequalities (including the Clauser–Horne–Shimony–Holt [3] and related inequalities). That case rests on an argument beginning, *per reductio,* with the assumption of *"local determinism"*, that the actual outcomes of (say, for simplicity) spin experiments on each of a pair of spin-$\frac{1}{2}$ particles prepared in the singlet state, carried out under circumstances such that the analyzer-setting (orientation of magnetic field of Stern-Gerlach devices) and outcome events at opposite wings of the setup are space-like separated,

---

[1] See e.g. [4].

Interpretations of Probability in Quantum Mechanics 213

are determined by physical conditions on a space-like slice restricted to the past light cones, respectively, of the individual analyzer-setting and outcome events. Those conditions may, of course, pertain to the local measuring apparatus. But it is required that physical conditions of the apparatus at the opposite wing *not* be included. Outcomes at each wing are assumed to be physically and statistically independent of parameter settings at the other wing ("*parameter independence*"). This is the "*locality*" part of "local determinism". (We will assume that this is well-motivated by various arguments based on limitations on the speed of energy-momentum transfer inherent in the special and general theories of relativity.) Further, in the ideal case, *conservation* (of angular momentum) requires that opposite outcomes will be found in parallel experiments (i.e. same orientation of magnetic fields of Stern-Gerlach apparati set up along a common axis at the two respective wings of the experiment).[2] Still these assumptions are not sufficient for a test of "determinism-in-nature". For each (sub-)system can be tested only once for a particular orientation of spin and setting of parameters at the opposite wing. What then is to block a deterministic account (theory) of any ensemble of such systems-*cum*-experiments you like which just manages to deliver the right actual outcomes for each experiment, one-by-one, as it were? By "one-by-one" we don't mean that such a theory simply provides a *list* of outcomes, and therefore could be excluded for not being "well-systematized". Rather we mean that the theory somehow manages to take into account—for all we know, perhaps in a unified way—only the actual conditions obtaining for the individual systems involved. The short answer is that such a theory is not "*robust*" unless it also supports counterfactuals telling us what outcomes *would have emerged* at a given wing *had the parameter settings been different* at the *opposite* wing. This, of course, is the same requirement that the Einstein–Podolsky–Rosen [5] argument invoked in their famous case that quantum mechanics must be "incomplete". In the present setting, it is used to infer that a *robust* or *respectable* deterministic theory must deliver (counterfactual) predictions of outcomes at a given wing for *various* parameter settings at the opposite wing. (Three directions altogether suffice, set e.g. $120°$ apart, for deriving a Bell inequality discriminating local, deterministic hidden variables' predictions from those of quantum mechanics.)[3] It is assumed that, as in experiments of Aspect, et al., the preparation of apparatus at an opposite wing being considered is space-like related to the opposite subsystem's measurement at the other wing so that it is reasonable to assume that no physical interaction occurs between these "events" (or space-time regions). Then we can proceed as in the standard proofs of Bell's theorem, that the theory must deliver a set of (actually and counterfactually predicted) "values" to the given subsystems' spins that respect parameter-independence and thus necessarily satisfy Bell-type inequalities in cases in which quantum mechanically predicted statistics violate them.[4]

---

[2] In the case of polarization in photon cascade experiments, (exact) conservation requires that passage or non-passage through analyzers set at the same orientation at opposite wings be directly strictly correlated.

[3] For an elegant presentation and further simplification of an argument due to D. Mermin (itself simplifying one given by E. Wigner), see [6], 143–48.

[4] For more explicit and rigorous derivations, see e.g. [7, 8].

This reasoning makes it clear that the "metaphysics" in experimental metaphysics is mediated by requirements that we are led to impose on putative *theories* that would transcend quantum mechanics but account for the observed statistics. That should not be surprising, however, since the metaphysical words (such as "local determinism") must be spelled out carefully if we are to carry out a mathematical argument constraining possible explanations of the observations, and of course explanations in physics, at any rate, typically involve some theory.[5]

If the physical world, at least at the quantum level, is really indeterministic in the ways described by the Bell results just outlined, it is natural to speak of individual outcomes in tests for spin or polarization as "objectively random", in that literally nothing in nature causes any of those particular outcomes (as opposed to the opposite value of the relevant two-valued observable). If we think of trying to connect this with various mathematical definitions of "random sequence" (of digits), we can imagine generating sequences of outcomes (coded by, say, 0's and 1's, respectively, for the two possible outcomes at each wing, L and R, taken separately) by repeating "identically prepared" experiments many times and checking the relevant formal properties exhibited by the outcomes at a given wing. (Clearly, this will be rather easier if the mathematical definition of "random sequence", such as that of Kolmogorov-Chaitin, applies to finite sequences!) But we will not expect such sequences to be entirely random or chaotic in an intuitive sense. Rather we expect that, in almost every case, they will exhibit convergence of ratios presented in the initial segments to probability values given by quantum theory. (For example, in the case of spin-$\frac{1}{2}$ particles of singlet-state pairs, we expect convergence to $\frac{1}{2}$ of the ratio of (occurrences of) one of the possible outcomes to the total number of all outcomes in initial segments of longer and longer sequences of outcomes at each wing, L or R, taken separately.) But in making such connections we think that we only re-enforce the view that we are here dealing with "objective probabilities", fundamentally different from the probabilities found useful in classical statistical mechanics or in applications to everyday life in which we are prepared to grant that causal determinism reigns—at least with practical certainty—, with respect to the macroscopic systems involved, in spite of the quantum mechanical nature of their micro-constituents.

This brings us to the main questions we would like to address: How, more precisely, are we to understand quantum probabilities as "objective" or not? Furthermore, as different interpretations of the quantum formalism interpret probabilities differently, it should be useful to classify them according to their treatment of the central concept of *probability*. How shall this be done? We propose a scheme in the next section, and then illustrate how it helps in assessing the reasoning of experimental metaphysics in central cases such as that of indeterminism-in-nature as just reviewed.

---

[5] For certain purposes, it may be useful to think of "theories" as classes of models, according to some version of the so-called "semantic view" of theories. But when it comes to *explanations*, especially in physics, the earlier sentence- or statement-based notion of "theory" has its point.

Interpretations of Probability in Quantum Mechanics 215

2 Interpretations of QM Probabilities

We take for granted that the mathematical apparatus for treating probabilities in quantum mechanics is well understood, due to the work of von Neumann, Lüders, Mackey, and culminating in the famous theorem of Gleason [9] characterizing measures on the closed linear subspaces of Hilbert space (of dimension ≥ 3) as given by the quantum algorithm via trace-class statistical operators. However, this machinery is open to a wide variety of interpretations bearing on physics and experiment which it is our purpose here to classify and survey briefly with the aim of clarifying the meaning and place of so-called "objective" interpretations of quantum probability.

It is unfortunate, we maintain, that interpretations of quantum probability have been labelled simply "objective" and "subjective", for this encourages conflation of issues that must be kept distinct if serious confusions are to be avoided. These issues pertain to *two* dimensions integral to the very concept of probability. The first issue concerns the *values* of probability functions, the real numbers assigned lying in the interval $[0, 1]$. The question here is *what*, according to a given interpretation, *quantum probabilities* **measure**. For example, do they measure actual relative frequencies of experimental outcomes in *ensembles* of systems, or limiting values of such frequencies taken over idealized (infinite) sequences of such outcomes? Or do they measure strengths of physical dispositions of *individual* systems to behave in various ways *if* they undergo, or were to undergo, various interactions with other systems ("propensities" is a term for such dispositions)? Or do they measure degrees of belief that a rational betting agent given certain specified information would have in this or that prediction about the system? That is *dimension 1*. And one could reasonably classify these possible answers as "objective" or "subjective": for example, strengths of physical dispositions of individual quantum systems are an objective matter, whereas degrees of belief or certainty of agents are not unreasonably termed "subjective". Relative frequencies in ensembles are more complex: while the frequencies themselves are an objective matter, if the ensembles are selected according to states of knowledge, we tend to speak of the associated probabilities as "subjective", whereas if ensembles are selected according to, say, a (putatively) complete set of physical properties or physical state, we would classify the associated probability as "objective".

Perhaps this is the primary meaning of the "objective/subjective" distinction that discussants of the subject have had in mind. But there is a second dimension which, from a foundational point of view, is equally important and which intrudes itself upon all the examples given so far. That concerns not the values of probability functions but rather their *domain of definition*, the "events" on which they are defined, the *bearers* of probability as it were, or *what the probabilities are probabilities of*.[6] Thus, in connection with the options mentioned above for interpreting what

---

[6] As indicated, the "event space" of quantum probabilities in a purely mathematical sense is perfectly definite (the lattice of closed subspaces of the Hilbert space representing the system). Subspaces typically correspond to "properties" of the form "the value of observable O lies in Borel set I". However, interpretations differ as to just how these properties are related to physical systems and the experiments performed on them, e.g. whether they are "objectively possessed", "found in

quantum probabilities *measure,* we may ask, strength of dispositions to do *what,* described *how?* Or degree of belief in *what,* described *how?* Even in the case of relative frequencies in ensembles selected in a manner already classified as above (e.g. as relative to epistemic states of agents or not), we may further ask, "ensembles of systems" *doing what, described how?* In classical mechanics, these issues are not problematic: probabilities are assigned to measurable regions of phase space and these are understood as collections of physical states in which certain physical magnitudes are possessed by the systems in question. Probabilities, even though they are based on our ignorance of precise details of the systems involved, are still probabilities of *possession of properties.* Indeed states can be considered essentially as "lists" of key physical properties. That of course is notoriously not the case in quantum mechanics, except under certain non-standard interpretations, and it is decidedly not the case in *textbook* quantum mechanics. Indeed, in order to avoid contradictions that naturally arise in the peculiarly quantum mechanical context of *incompatible observables,* one has resort to talk of properties, not simply *possessed* by quantum systems, but *found to be possessed if suitably measured,* which is the essential move in the Bohrian doctrine of "complementarity". Pure quantum states can be taken as extremal probability measures on closed subspaces of Hilbert space (equivalently projection operators) which specify how systems would behave, what properties they would exhibit, if observed in this or that specified way. Even this is controversial, resting on inference from the observed *apparatus* system in a measurement to properties of the *quantum* system itself, and interpretations appealing to complementarity (of the "Copenhagen" variety, broadly speaking) range from "more objective" in licensing such inferences to "completely operational" in banning them entirely. So now the simple classical language of property possession, a purely objective matter, has been replaced with a complicated reference to a variety of possible outcomes of interactions, and these themselves are described with language frequently bringing in "big, bad words" like "measurement", "observation", etc., which are not yet explained physically and which refer obliquely to cognitive agents. That is, the "events" assigned probabilities have "subjective" elements in their common descriptions.[7] Thus, this *dimension 2* can also be divided broadly into "objective" and "subjective" sides, where "objective" applies to probability bearers described in physical language without reference to "measurement", "observation", or "appearance", etc., and "subjective" applies to bearers whose description does make such reference.

This leads then to a two-by-two matrix of interpretative possibilities, with, say, dimension 1 labelling the rows and dimension 2 labelling the columns:

	Obj 2	Subj 2
Obj 1	Modal Interps	Textbook (e.g. Bohm '51)
Subj 1	Bohmian Mech	Instrumentalist CI, Bayesian

---

appropriate measurements", etc. It is "bearers of probablities" in this extra-mathematical sense that we are concerned with here.

[7] "Anthropocentric" would be a more accurate term than "subjective". But it has too many syllables, so we acquiesce in the more common terminology.

Interpretations of Probability in Quantum Mechanics 217

Let us comment briefly on the cell occupants and why they are where they are.

Modal interpretations give up the eigenvalue-eigenstate link, assigning some values to systems beyond what that rule allows (i.e. to observables pertaining to systems *not* in an eigenstate of those observables). Conflicts with "no go"results ruling out sufficiently many such value assignments (at a time), based e.g. on Gleason's theorem or the Kochen and Specker theorem, are avoided by severely restricting value assignments to special situations, e.g. to operators ("observables") appearing in the polar decomposition of the pure state of a whole, typically complex system[8]; or to the form of Hamiltonian operators appearing in the dynamical description of interactions typified in "good measurements" of a given observable. Relative to such value assignments, however, *probabilities* are *of possession of properties*, just as in classical physics (Objective 2), although the properties themselves are characteristically quantum mechanical (based on the closed linear subspaces of Hilbert space). But note that, while good measurements are taken to reveal such properties, officially modal interpretations avoid terms like "measurement" as primitive, speaking instead of interactions described with Hamiltonian operators meeting certain formal conditions.[9] However, *ultimate physical randomness* is also recognized: just which properties will be revealed or actualized in an interaction involving an individual quantum system is not determined by anything in nature; rather quantum (pure) states give measures of the strength of dispositions to actualize various properties depending on the interaction. Thus, these interpretations seek objectivity in both senses. Although an attractive solution to the notorious measurement problem is provided, challenges remain especially in connection with relativity, where value assignments of modal interpretations can readily violate Lorentz invariance,[10] and with extension to quantum field theory, where modal rules for assigning properties yield only trivial results in fairly common situations.[11] It remains to be seen whether a minimalist modal interpretation can isolate a class of genuine "measurement type" interactions, described in QFT, which admit non-trivial property ascriptions.

In contrast with modal interpretations, textbook treatments, such as Bohm's [10] *Quantum Theory*, respect the eigenvalue–eigenstate link. Moreover, probabilities are *of* measurement outcomes, classically described in a classical background framework. The notion of "measurement" or "recording apparatus with many degrees of freedom", in effect leading to decoherence, is taken as given, hence the placement in column 2. This is thus a version of "Copenhagen interpretation", although of a decidedly "objective" variety, because of the treatment of the first dimension of probability, what the numbers measure. Again, like modal interpretations, it is the *strengths of complex physical dispositions* of the quantum systems themselves, dispositions to reveal this or that value of given observable *if* a suitable measurement is or were carried out. These are conceptually new, quantum mechanical properties, requiring the mastery of new scientific ideas and language (sharing

---

[8] See [11, 12].

[9] See [13], Ch. 9.

[10] See e.g. [14]; but also [15].

[11] See [16].

also this feature of modal interpretations). Again, "ultimate physical randomness" makes sense on this view and is taken as a remarkable, non-classical feature of the physical world. Pure quantum states are physically complete, and the probabilities they provide (when lying strictly between 0 and 1), while they indeed reflect our ignorance of actualizations, also describe these complex, non-classical physical properties, thought of as "tendencies" or "propensities".[12] This aspect is thoroughly objective (row 1), even though "subjective" elements may enter in saying "what these tendencies are toward".

A remark on sources: Bohm's 1951 [10] text is the most reflective, sustained, and consistent effort to work out these ideas in detail that I am familiar with. Perhaps Bohr scholars can judge the extent to which it represents Bohr's own considered views. In any case, it strikes me as still the most defensible presentation of Copenhagen around, one whose main themes are echoed in many other texts and contexts. Its principal drawbacks are two: in requiring a classical background—even with the cut varying with context, as it must, since "recording apparati" can also be treated as quantum systems—it does not readily lend itself to the notion of a "wave-function of the whole universe" as needed in quantum cosmology. Secondly, in its appeal to randomized phase factors entering into the wave function of a system interacting with a measuring device with many degrees of freedom, it provides at best an approximate solution—good "for all practical purposes" (FAPP solutions, as John Bell called a whole class of attempts along these lines)—of the measurement problem.

Moving to the second row, Bohmian mechanics based on Bohm's hidden variables theory of 1952 is the exact reversal of the (partially) "objective" Copenhagen interpretation just considered.[13] Here probabilities are *of* objective position properties of systems, to which all quantum observables are ultimately reduced. But

---

[12] Bohm frequently uses the term 'tendency' although not 'propensity', which was used prominently by Popper. Popper's own understanding, however, was quite at odds with the interpretation we are describing, as he thought quantum-mechanical probabilities and randomness were not essentially different from what is encountered in classical statistical physics. Popper's "propensity interpretation" of QM was trenchantly criticized by Feyerabend [17] and by Bub [18], essentially for ignoring the peculiarly quantum phenomenon of incompatible observables giving rise to non-classical methods of evaluating conditional probabilities. In effect, Popper did not attend to the crucial distinction we are labelling dimension 2 of probability concepts. For a powerful critique of Popper's whole conception of propensities, see [19]. An informative summary of all this is given by [20], pp. 448–453. His footnote 44, p. 449, also provides a synopsis of earlier antecedents of probabilities as "propensities", going back to Maimonides.

[13] It may with some justification be claimed that "Bohmian mechanics" should not be classified as an "interpretation" of quantum mechanics at all, for it is, rather, an alternative physical theory which is contrived to reproduce the experimental results predicted by quantum mechanics. We include it in our table anyway because of its significance as a gadfly challenging Copenhagen interpretations as well as in order to illustrate the remarkable differences in concepts of probability offered by empirically indistinguishable theories. Furthermore, it does retain quantum state functions (defined on configuration space) along with their evolution according to the time-dependent Schrödinger equation. In this latter respect, Bohmian mechanics differs with recent physical collapse theories, known as GRW [21], in which random collapses interrupt the continuous, linear Schrödinger evolution of quantum state functions, but in such a way as to be practically certain in measurement-type situations (where we need them) but practically impossible in circumstances prevailing at the atomic scale in which Schrödinger dynamics is empirically confirmed as accurate.

Interpretations of Probability in Quantum Mechanics 219

since these position properties evolve deterministically, probabilities are needed only because of our ignorance of the precise details of initial configurations and velocities. That is, they reflect our ignorance rather than indeterminism in nature and so are reasonably classed as "subjective" along dimension 1, what the numbers measure. To be sure, they measure relative frequencies in certain ensembles (at least approximately), but these ensembles are selected as a matter of human convenience and necessity due to the inaccessibility of exact microstates. This apparatus restores classicality of property ascriptions, avoids the quantum/classical cut problem of Copenhagen, restores determinism in the evolution of definite values of position, and avoids the measurement problem. The main price is a high degree of non-locality and related problems with extending the theory to relativistic quantum fields. (We will return to this theory/interpretation in the final section, below.)

This brings us finally to the fourth quadrant, "Subjective–Subjective". Here we encounter extreme empiricist or instrumentalist versions of Copenhagen. Probabilities measure relative frequencies in ensembles of observations, described either macro-physically or mentalistically in terms of "appearances", and they are *of* measurement outcomes so described. In the most extreme versions, one does not even attribute properties to micro-systems in eigenstates, but confines oneself to "pointer readings" (or appearances thereof). The contrast with "objective Copenhagen" discussed above would be hard to overstate. Indeed, so far from explaining observed statistics via "physical probabilistic dispositions," the subjective-subjective version renounces *any* hope of *explaining in physical terms the statistical distributions of measurement outcomes one observes*. Prediction and practical application replace that classical preoccupation (regarded as "quaint", or "old-old-European"?), and, by fiat, there are no problems of interpreting the physical significance of state functions. And there is certainly no measurement problem, for consistency can be enforced by withholding the quantum formalism from any system that appears not to obey it (i.e. appears definite in ways that the quantum formalism fails to deliver).

As indicated, new Bayesian views of quantum probability belong in this fourth quadrant as well. Quantum probabilities measure degrees of rational belief and these beliefs are of measurement outcomes (idealized in Pitowsky's "quantum gambling devices" [30]). This view shares with subjective Copenhagen the advantages of avoiding theoretical problems. But, it is to be noted, it also shares the renunciation of the goal of *explaining* observed quantum statistics. After all, quantum states can be identified as generalized probability measures. If these probability measures are then understood as giving rational betting quotients, then quantum states can hardly be called upon to *explain* observed relative frequencies in ensembles or why those quotients agree (indeed agree so well) with those relative frequencies. Of course, they had better agree, i.e. if we don't want to "lose our shirts", our bets need to conform to the long-run frequencies actually encountered.[14] But here the arrow of

---

The reader is nevertheless invited to extend the classification scheme we are presenting to cover these and other theories that have been or may be proposed.

[14] This idea underlies Lewis's "principal principle", applicable to Bayesian reasoning generally: informally, this says that degrees of belief in predicted outcomes of experiments, say, should be

220   G. Hellman

explanation is reversed, as it should be: our degrees of belief are adjusted to fit
the empirical facts, or we're not rational. But those degrees of belief cannot possibly *account* for those facts, unless you subscribe to a truly radical psychokinesis!
Thus like its subjective cousin from Denmark, this approach to quantum probability
avoids the main foundational problems and puzzles of quantum mechanics, but one
might say that it does so at the price of renouncing the enterprise of physics.

3 Where's Everett?

Conspicuous by its absence from our table of interpretations of quantum probabilities is the so-called "Everett interpretation", after Hugh Everett [22], who invented it in his Princeton doctoral dissertation in physics (1957) entitled "Theory
of the Universal Wave Function", supervised by John Archibald Wheeler. This was
an attempt to provide an alternative to the Copenhagen interpretation, avoiding its
partition of reality into observed quantum system and classical observing system
and avoiding the notorious collapse of the wave-function upon measurement. The
Everett interpretation soon underwent something of a metamorphosis (at the hands
of Bryce DeWitt) into what became known as "the many worlds interpretation", a
notorious metaphysical extravaganza in which, upon "quantum measurements", the
whole universe "splits" into multiple successor universes corresponding to different branches of the universal wave-function (of the whole physical cosmos) in turn
corresponding to eigenstates in which a "measured observable" has a definite value
(eigenvalue). In *this* theory, collapses are replaced with literal splittings of the universe into mutually causally non-interacting universes, each with its own spacetime
and physical contents. Stories can be told about why it is that no one can ever experience any such splitting. And stories can also be told about how quantum mechanical
probabilities of outcomes of measurements on *ensembles* of systems within a single
universe are in *some* sense respected. However, since collapses never occur within
a world and since values of observables do not go beyond what the eigenvalue-
eigenstate link allows, we can never say that the component individual systems of
such ensembles actually take on "measured values" of quantum observables. Instead, everything must be translated into statements of (applied) wave mechanics,
such as that a universal wave function (at a time in a suitable reference frame) is
small in a region (say, of configuration space) in which the frequency of "measurement outcomes" in an ensemble , as normally described, departs appreciably from
the predicted quantum mechanical probability of the "outcome" in question.[15] And

---

guided by what is known about "objective probabilities" of those outcomes. For discussion, see
[23].

[15] Geroch [24], in his interpretation of Everett, to be described in a moment, refers to such regions
as "precluded", and deploys this notion to replace ordinary quantum probabilities. While "precluded" itself is not a notion of Schrödinger wave mechanics *per sē*, the suggestion is that it can be
used to eliminate "probability" in any *application* of wave mechanics (whence our own reference
to "applied wave mechanics"). For critical discussion of this point, see [25].

Interpretations of Probability in Quantum Mechanics   221

the appeal of Everett's ideas to cosmologists, based on the applicability of quantum mechanics to the universe as a whole without the need to suppose any "outside observer", must surely be overwhelmed by the problems raised by fantastically many, mutually non-interacting but partially resembling universes constantly undergoing splitting—as if the task of accounting for the evolution of a single universe weren't enough! In any case, serious discussions of Everett do better to treat not a "many-worlds interpretation" at all, but rather a far simpler scheme, a *one-world version of Everett* (perhaps as he intended it) in which neither splittings nor collapses ever occur and the universe evolves strictly in accordance with Schrödinger dynamics.

Now, whereas on the "many worlds" theory, too *much* was happening, here the problem is that *too little happens*, viz. when at the conclusion of a quantum measurement we normally say that a definite outcome has occurred even though the quantum mechanical probability of that outcome assigned by the pure state of the total system involved is strictly between 0 and 1, on *one-world Everett* we cannot say this but rather must continue to describe our experience of definite outcomes with what will, mathematically, be merely a complicated component ("branch") of an extremely complicated total evolving wave-function of the universe. *It follows immediately from the eigenvalue-eigenstate link, which one-world Everett tacitly accepts, that such outcome events do not actually occur.*[16] *A fortiori*, probabilities of such occurrences do not make sense in the theory, i.e. probability functions cannot have such events in their domains of definition.

Such a view is essentially what Robert Geroch described in his [1984] paper [24] on Everett. There he makes an intriguing comparison with the theoretical situation presented by Einsteinian relativity theory: we have learned that various commonplace ideas of time and space—e.g. that we all share a unique standard of simultaneity, that "before" and "after" are absolute notions, that mass is velocity-independent, etc.—should be treated as phenomena of our *experience* to be explained rather than as corresponding to physical reality.[17] Much the same, it is suggested, should be said about our commonplace beliefs about definiteness of measurement outcomes and quantum reality. The amazing teachings of quantum physics that we must learn how to assimilate tell us that "measurement outcomes" as we ordinarily describe them don't actually occur in many, many cases, since the recording devices and events involved are in fact bound up, even if only weakly, with goings-on elsewhere in the universe (some near, some far) so that the local "systems" in which we are interested typically do not even possess pure quantum states. (They only occupy *improper* mixed states obtained from vast superpositions of states of much larger

---

[16] Thus, what we are calling "one-world Everett" is to be sharply distinguished from what Healey [26] called the "one-world version" [of the 'many-worlds' interpretation], which *does* give up the eigenvalue-eigenstate link (from left to right), thereby coming much closer to a modal interpretation.

[17] Actually, Geroch concentrates on more problematic aspects of *experience*, such as our direct awareness of a present moment of time, which find no place at all in spacetime physics. But, as Stein [25] points out, this is not a particular feature of relativistic physics, but of the science of physics in general. For the sake of argument (indeed, argument that has been made at least in conversation on this point), we have given examples of notions that relativity theory rules ill-conceived that do not raise such general (or deep) problems.

systems, perhaps extending to the whole universe, by tracing over many degrees of freedom pertaining to that larger context. On pain of contradiction, given the eigenvalue-eigenstate link (from left to right), these improper mixed states of subsystems of the universe cannot be given an *ignorance interpretation*, that is they cannot be understood as merely reflecting our uncertainty as to a particular value of the observable in question which the subsystem is supposed to possess.) Our experience of definiteness, in this sense,[18] however useful for practical purposes, is strictly illusory: accounting for such experience is indeed a challenge, but it is one for a future psycho-physics; and in any case (dare I say, "in any event") not counter-evidence to quantum mechanics itself or to the (one-world) Everett interpretation of it. How good is this analogy?

Not good, I would argue. It breaks down for the following reason: as quantum observers, situations in which we say that "it seems to us that pointers point" are themselves—as we may assume (call this "*assumption (0)*")—the result of physical processes in our brains, and so the very assumptions that the view (one-world Everett) is founded upon and certainly *not* challenging, viz. (1) the validity of quantum mechanics *without* the projection postulate, (2) its universal applicability to physical reality, and (3) the eigenvalue-eigenstate link (in both directions) will lead to a contradiction in many situations: the wave function describing (enough of) the universe, including our brains, will not be in an eigenstate of "the pointer seems to so-and-so at time t to be definite ($\otimes$ other components of a very big tensor product state...) " when it needs to be, and so a value of a quantum observable will have been attributed in violation of the eigenvalue-eigenstate link (3). (Granted it's not an *ordinary* quantum observable in any sense, but neither is "the pointer pointed up" at the end of, say, a Stern-Gerlach experiment testing for spin. Anyway, playing this game (with universality (2) as an assumption) inevitably involves us in extraordinary observables relative the ordinary practice of quantum mechanics.)

(What justifies assumption (0), that we may assume that the subjective experience of definiteness in the minds of human observers in the relevant situations that obtain after good quantum measurements may be thought of as purely physical conditions of those observers (mainly their central nervous systems), hence falling within the purview of quantum mechanics according to universality (assumption (2))? Surely, we cannot just dogmatically *assert* this physicalist view of the mental, on pain of weakening the argument. Indeed, but we are *not asserting* it; we are merely requiring that the contrary non-physicalist view of the mental *not* be assumed, that any satisfactory resolution of the quantum measurement problem within our current

---

[18] Here and below, "experience of definiteness" is used to include experience of particular outcomes or readings of experiments, not merely that one outcome *or* another (*or ...*) was obtained. Playing by the usual idealized rules stipulating an exhaustive set of mutually exclusive possibilities (corresponding to an eigenbasis of the system observable in question, where the system may include a person with experiential and belief states, etc.), a "bare theory" with Schrödinger dynamics (but no collapse or projection postulate) can claim a kind of Pyrrhic victory in "respecting definiteness" in the sense of assigning probability 1 to the relevant *disjunction* over possible outcomes, associated with the whole Hilbert space of the system (spanned by a complete set of eigenvectors of the operator for the relevant observable). See [27], and, for a detailed, critical discussion of the bare theory, [28].

Interpretations of Probability in Quantum Mechanics 223

state of knowledge must be *compatible* with a thoroughgoing physicalism regarding the mind-body question. For otherwise an appeal to any version of the Everett interpretation is simply *récherché*: if one is prepared already to treat mental experience as ontologically *non-physical*, one can simply declare by fiat that quantum mechanics, while it may apply to the physical world in its entirety, does not govern our mental experience, so that in assigning definiteness to "observables" corresponding to that experience, we are not ever violating the eigenvalue-eigenstate link, i.e. those "observables" are not really quantum mechanical anyway and so fall outside the scope of that rule. In effect, an *a priori* assumption of definiteness of mental states is used to obtain effective collapses of wave functions. Difficulties that such a view faces apart (such as how to explain the remarkable psycho-physical correlations that we observe), this approach is completely contrary to the spirit if not the letter of Everett, since in effect it leads right back to *Bohr's cut* between observed and observing systems which Everett seeks to transcend.)

Thus, assumptions (0)–(3) force us, in certain circumstances in which we claim honestly to experience definite pointings of pointer systems, to deny even that it *appears* to us that certain pointer systems definitely point![19] Following Geroch's suggestion, we presumably would say that it only *appears* to us that it is definite that it appears to us that pointers point, and that the great revolutionary new thing that Everettian quantum mechanics highlights for us is that it is this *appearance of definiteness* (of our appearances of pointers) that is illusory and requires scientific explanation. But—and it is just here that one sees most clearly why the analogy with relativity breaks down—this just pushes the problem up yet one more level, i.e. this leads to a vicious regress—which may aptly be called "*Descartes' regress*")— *whereas no regress is generated by the confrontation between relativity and experience*. At some point, in describing the situation in some way that can "save enough of the phenomena" for the experimental confirmation of ordinary quantum mechanics to make any sense at all, we need to say something about how things seem to us, i.e. that certain appearance statements are *true* (even if they are only about appearances of appearances of ...of pointers). At some level, it must be conceded in effect that we are *not deceived*. And then, you are stuck with a "revved up" version of the original measurement problem. QED[20]

It should, moreover, now be clear why (one-world) Everett finds no place in our table of possibilities for interpreting probabilities in quantum mechanics. We simply do not see how this radical view can make sense of typical quantum probabilities for lack of suitable events or outcomes that the domain of definition of probability measures would comprise and without which we cannot make sense of the empirical confirmation of the theory (supporting assumption (1) above, in the first place).

---

[19] See the note immediately preceding.

[20] The reader is invited to compare this line of argument with one in a quite similar spirit given by Stein [25] in his examination and critique of Geroch's version of the Everett interpretation. As Stein puts it, whereas on Geroch's version of Everett, a great many "classical occurrences" disappear entirely, "there is no such disappearance according to the theory of relativity". (p. 644) The former, but not the latter, should be posing Chico Marx's question in *Duck Soup*: "Who you gonna believe, me or your own eyes?" (I am grateful to Howard Stein for this quote and its source.)

224 G. Hellman

Appeals to *decoherence*— the widespread phenomenon of *practical vanishing*, in very short times in measurement-type situations, of interference terms in the evolving wave function of a larger system incorporating the *environment* of the object system of interest—do not really help; approximate collapses are not genuine collapses, and without giving up the eigenvalue-eigenstate link, definite outcomes still literally do not occur, and probabilities remain ill-defined. No wonder we did not list this interpretation in the table.[21]

4 Bohmian Mechanics and Experimental Metaphysics

As the table makes clear, Bohmian mechanics stands in the way of the conclusions we are tempted to draw from the empirical successes of quantum mechanics (including the Bell results) and of relativistic physics as well, in which "locality", at least in the sense of "parameter independence", is rooted. Our prime example of "ultimate randomness in nature" is paradigmatic. Note that this is common to the interpretations of the top row. We have already encountered strong reasons to avoid the fourth quadrant (Subjective 1 and 2). Thus, it is only Bohmian mechanics that keeps us from confinement to the top row and the conclusion of ultimate randomness as a strongly empirically supported lesson of our experience with quantum mechanics. How seriously must we take the Bohmian challenge?

As indicated in the brief summary of main features of Bohmian mechanics above, it does succeed in recovering all the statistical predictions of ordinary (non-relativistic) quantum mechanics on a basis that can be called "classical" in respect of its (theoretical, in-principle) ascription of precise values of positions and velocities at all times to particles which evolve in these variables deterministically. Probabil-

---

[21] Another, more recent branch of Everett-inspired interpretations due to Deutsch, Wallace, et al., explicitly invokes decoherence to identify privileged observables whose eigensubspaces are effectively "separated" (over very brief interaction times) from one another within a universal wave function. These approximately non-overlapping "branches" correspond to quantum experiments (on individual systems as well as ensembles) with different outcomes and associated "weights"— squared amplitudes got from coefficients of the privileged basis vectors in the wave function, behaving as probabilities in accordance with the Born rules—all of which are said to be "realized" or "equally real"; this is the Everettian twist. (See [29] and references cited therein.) In effect, in contrast with the Geroch version of Everett, the eigenvalue-eigenstate link is really being given up, and the post measurement-type interaction superposition is being treated as a mixture.

Now an adequate treatment of this approach cannot be given here, and we will rest with a pointed question: why not just stipulate, along with modal interpretations, that such states are to be understood as if they were mixtures (approximately delivered by decoherence), and proceed to take on the various problems that then arise (especially the problem of Lorentz invariance), without the additional metaphysical burdens of "many worlds"—only one of which is accessible to *our* experience—, duplicated individual systems, including persons, etc.? What work, in other words, is done by reading all the branches as actually realized ("with the appropriate probabilities," whatever that really means), rather than saying, with modal interpretations, that only the branches that we actually observe occur with the probabilities assigned, applying an ignorance interpretation of mixtures? All the work seems to be done by decoherence + the additional definiteness of the relevant properties. Why shouldn't the distinctively Everettian baggage be discarded?

Interpretations of Probability in Quantum Mechanics

ities, recall, are epistemic, reflecting our imperfect information about initial conditions (hence anthropocentric, i.e. "Subj. 1"), but they are *of* objectively possessed position properties ("Obj. 2"). On this theory, there is no place for ultimate physical randomness: that certain aspects of our world *appear* to behave randomly in an ultimate sense is really an illusion, arising from our ignorance of the precise details of quantum particle configurations; moreover, this ignorance is in a strong sense *"perpetual"*: since the velocity functions of the theory are functions of the quantum wave function (defining a ray in Hilbert space), and, since the predictions of Bohmian mechanics recover the Heisenberg "uncertainty relations", so long as the world is genuinely quantum mechanical, we could never be in a position to know the precise values of enough Bohmian hidden variables to violate the appearance of random outcomes of quantum measurements. In this manner, Bohmian mechanics is truly diabolical in character: it posits an underlying classical level but one that is always accompanied by enough quantum-mechanical statistical behavior so as always to elude detection. No experiment we can perform will distinguish this theory with its extraordinary posits from quantum mechanics as it is ordinarily practiced (if not well understood). No wonder that physicists of a positivist inclination would tend to dismiss this theory (if they ever studied it).

But it gets worse: As is evident from the equations of motion of Bohmian mechanics, position values typically depend instantaneously on values at a distance, in principle as far away as you like from a given space-time region. Indeed, it is not hard to see that, if we had precise information about enough precise positions, physical information could be transmitted superluminally, violating parameter independence (in Bell-EPR-type experiments); and outcomes of such experiments on separated or spread-out systems could be seen to depend on an absolute temporal ordering, i.e. it would make sense to say that a particular inertial frame agrees with an absolute time-ordering of events, i.e. defining a *privileged frame, but that which frame it is must remain forever undetectable.* (For a proof-sketch, see [27], pp. 159–160.) Thus, not only is physical randomness an illusion, so is special relativity with its frame-dependence of "simultaneity", "before and after" of space-like related events, and so forth. You don't have to be a positivist to find yourself recoiling from this implication!

The contrast with quantum mechanics as understood through interpretations falling in the first row of our table deserves emphasis. There "non-locality" according to various definitions also must be recognized. It seems to be a fact of life that quantum statistics present us with a kind of "holism" of complex quantum systems, violating certain forms of locality such as "outcome independence", of the form

$$\Pr(A/B\&\lambda) = \Pr(A/\lambda); \Pr(B/A\&\lambda) = \Pr(B/\lambda),$$

where 'A' and 'B' stand for local outcomes on the respective parts of a two-component system and 'λ' stands for the most complete physical state we can find for the whole system consisting of strongly correlated parts as in EPR-Bell-type systems. (Such holistic states generate joint probabilities which are not "factorizable", contrary to Reichenbach's conception of "screening off" as integral to scientific explanation of correlations.) Similarly, we cannot expect there to be separate physical states of such parts which fix the respective outcomes of (certain relevant) experiments on those parts. (Such holistic systems are in this sense "non-separable", contrary to Einstein's conception of acceptable physics of separated systems.) But precisely because these interpretations also make room for ultimate physical randomness of particular measurement outcomes, signal locality (e.g. in the form of parameter independence) is respected. Bell-type systems cannot be contrived to transmit physical information superluminally precisely because outcomes of, say, Stern-Gerlach measurements on spin of one of a strongly correlated pair of particles are beyond experimental control. This sounds anthropocentric, but it is so only in a superficial sense, as it is a limitation affecting any possible epistemic agents as well as ourselves, resulting from the inherent randomness of the events involved, despite the strong correlations among them.

What Bohmian mechanics shows is that these conclusions are not absolutely forced on us by the data alone. *Experimental metaphysics, however, does not operate in a theoretical vacuum. If* we are prepared to accept enough grossly non-local, hidden physics masked by "illusory" phenomena as effectively described by special relativity and objective-1 interpretations of quantum phenomena recognizing ultimate physical probabilities, *then* we might be able to salvage determinism-in-principle—provided Bohmian mechanics can be convincingly extended to quantum field theory. But if we require that some experimental evidence favor such hidden posits (as exact trajectories of particles and a privileged inertial frame), insisting that the case not rest *entirely* upon some theoretically appealing consequences (which, after all, are accompanied by some rather repugnant ones, as sketched), then we will be within our rights to assert that ultimate randomness is one of the surprising lessons of twentieth century physics, and, moreover, that a better solution of the measurement problem than that afforded by Bohmian mechanics must still be found.

Acknowledgment The author is grateful to Wayne Myrvold and Michel Janssen for useful comments and to audiences at the Seven Pines Symposium (May, 2006) and at the Perimeter Institute Symposium in Honor of Abner Shimony (July, 2006) for helpful discussion.

References

1. Shimony, A. [1984] "Contextual Hidden Variables Theories and Bell's Inequalities", *British Journal for Philosophy of Science* **35**: 25–45.
2. Aspect, A., Dalibard, J., and Roger, G. [1982] "Experimental Tests of Inequalities Using Variable Analyzers", *Physical Review Letters* **49**: 1804–1807.
3. Clauser, J.F., Horne, M.A., Shimony, A., and Holt, R.A. [1969] "Proposed Experiment to Test Hidden Variable Theories", *Physical Review Letters* **23**, 15: 880–883.
4. Hellman, G. [1992] "Bell-Type Inequalities in the Nonideal Case: Proof of a Conjecture of Bell", *Foundations of Physics* **22**, 6: 807–817.
5. Einstein, A., Podolski, B., and Rosen, N. [1935] "Can Quantum Mechanical Description of Physical Reality Be Considered Complete?" *Physical Review* **47**: 777–780.

Interpretations of Probability in Quantum Mechanics

6. Kosso, P. [1998] *Appearance and Reality: An Introduction to the Philosophy of Physics* (New York: Oxford University Press).
7. Jarrett, J. [1984] "On the Physical Significance of the Locality Conditions in the Bell Arguments", *Noûs* **18**, 4: 569–589.
8. Hellman, G. [1982] "Einstein and Bell: Strengthening the Case for Microphysical Randomness", *Synthese* **53**: 445–460.
9. Gleason, A.M. [1957] "Measures on the Closed Subspaces of a Hilbert Space", *Journal of Mathematics and Mechanics* **6**: 885–893.
10. Bohm, D. [1951] *Quantum Theory* (Englewood Cliffs, NJ: Prentice-Hall).
11. Healey, R.A. [1989] *The Philosophy of Quantum Mechanics: An Interactive Interpretation* (Cambridge: University of Cambridge Press).
12. Vermaas, P. [1999] *A Philosopher's Understanding of Quantum Mechanics: Possibilities and Impossibilities of a Modal Interpretation* (Cambridge: Cambridge University Press).
13. van Fraassen, B.C. [1991] *Quantum Mechanics: An Empiricist View* (Oxford: Oxford University Press).
14. Myrvold, W. [2002] "Modal Interpretations and Relativity", *Foundations of Physics* **32**: 1773–1784.
15. Berkovitz, J. and Hemmo, M. [2005] "Can Modal Interpretations of Quantum Mechanics Be Reconciled with Relativity?" *Philosophy of Science* **72**: 789–801.
16. Earman, J. and Ruetsche, L. [2005] "Relativistic Invariance and Modal Interpretations", *Philosophy of Science* **72**: 557–583.
17. Feyerabend, P. [1968, 1969] "On a Recent Critique of Complementarity", *Philosophy of Science* **35**: 309–331, **36**: 82–105.
18. Bub, J. [1975] "Popper's Propensity Interpretation of Probability and Quantum Mechanics", G. Maxwell and R. M. Anderson, eds., *Induction, Probability, and Confirmation* (Minnesota Studies in the Philosophy of Science Vol. VI, Minneapolis, MN: University of Minnesota Press), pp. 416–429.
19. Sklar, L. [1970] "Is Probability a Dispositional Property?" *Journal of Philosophy* **67**: 355–366.
20. Jammer, M. [1974] *The Philosophy of Quantum Mechanics* (New York: Wiley).
21. Ghirardi, G.C., Rimini, A., and Weber, T. [1986] "Unified Dynamics for Microscopic and Macroscopic Systems", *Physical Review D***34**: 470.
22. Everett, H. [1957] "The Theory of the Universal Wave Function", in B. DeWitt and N. Graham, eds. *The Many-Worlds Intrerpretation of Quantum Mechanics* (Princeton: Princeton University Press, 1973).
23. Earman, J. [1992] *Bayes or Bust?* (Cambridge, MA: MIT).
24. Geroch, R. [1984] "The Everett Interpretation", *Noûs* **18**, 4: 617–633.
25. Stein, H. [1984] "The Everett Interpretation of Quantum Mechanics: Many Worlds or None?" *Noûs* **18**, 4: 635–652.
26. Healey, R.A. [1984] "How Many Worlds?" *Noûs* **18**, 4: 591–616.
27. Albert, D.Z. [1992] *Quantum Mechanics and Experience* (Cambridge, MA: Harvard University Press).
28. Bub, J., Clifton, R., and Monton, B. [1998] "The Bare Theory Has No Clothes", in R. Healey and G. Hellman, eds. *Quantm Measurement: Beyond Paradox* (Minneapolis: University of Minnesota Press), pp. 32–51.
29. Wallace, D. [2003] "Everett and Structure", *Studies in the History and Philosophy of Modern Physics* **34**: 87–105.
30. Pitowsky, I. [2003] "Betting on the Outcomes of Measurements: A Bayesian Theory of Quantum Probability", *Studies in the History and Philosophy of Modern Physics* **34**: 395–414.

"No Information Without Disturbance": Quantum Limitations of Measurement

Paul Busch

Abstract In this contribution I review rigorous formulations of a variety of limitations of measurability in quantum mechanics. To this end I begin with a brief presentation of the conceptual tools of modern measurement theory. I will make precise the notion that quantum measurements necessarily alter the system under investigation and elucidate its connection with the complementarity and uncertainty principles.

1 Introduction

It is a great honor and pleasure for me to contribute to this celebration of the scientific life work and achievements of Abner Shimony, from whom I have received much inspiration, personal encouragement and the gift of friendship in a decisive period of my scientific career. When I came to know Abner more closely, I was thrilled to realize the close agreement between our quantum mechanical world views; and ever since, when contemplating foundational issues, I found myself often wonder: "What would Abner say?". I am proud to share with Abner one piece of work on an important item of "unfinished business", a paper on the insolubility of the quantum measurement problem [1], which I hope may prove useful as a stepping stone towards resolving this problem. In this contribution I will address another area of concern to Abner, one that remains even when the measurement problem is suspended: quantum limitations of measurements.

By way of introduction of terminology and notation I briefly review the basic and most general probabilistic structures of quantum mechanics, encoded in the concepts of states, effects and observables; I then recall how these objects enter the modeling of measurements (Section 2).

P. Busch
Department of Mathematics, University of York, York, UK
e-mail: pb516@york.ac.uk

W.C. Myrvold and J. Christian (eds.), *Quantum Reality, Relativistic Causality, and Closing the Epistemic Circle*, The Western Ontario Series in Philosophy of Science 73,
© Springer Science+Business Media B.V. 2009

230   P. Busch

This general framework of quantum measurement theory will then be used to obtain precise formulations and proofs of some long-disputed limitations of quantum measurements, such as the inevitability of disturbance and entanglement in a measurement, the impossibility of repeatable measurements for continuous quantities, and the incompatibility between conservation laws and the notion of repeatable sharp measurements (Section 3). In Section 4 the *"classic"* quantum limitations expressed by the complementarity and uncertainty principles are revisited. Appropriate operational measures of inaccuracy and disturbance for the formulation of quantitative trade-off relations for (joint) measurement inaccuracies and disturbances have been introduced in recent years; these will be discussed in Section 5.

I conclude with an outlook on open questions (Section 6).

2 Quantum Measurement Theory—Basic Concepts

2.1 States, Effects and Observables

Every quantum system is represented by a finite or infinite-dimensional, separable Hilbert space $\mathcal{H}$ over the complex field $\mathbb{C}$. States are described as positive operators[1] T of trace equal to one.[2] The set of states $\mathcal{S}(\mathcal{H})$ is a convex subset of the real vector space of all self-adjoint trace-class operators. The role of a quantum state is to assign a probability to the outcome of any measurement; in other words, associated with every measurement with possible outcomes ω_i, $i = 1, 2, \ldots$, are mappings $\mathcal{E}_i :$ $\mathcal{S}(\mathcal{H}) \to [0, 1]$ assigning the probabilities $p_T(\omega_i) \equiv \mathcal{E}_i(T)$. Since mixtures of states lead to the corresponding mixtures of probabilities, it follows that the mappings $\mathcal{E}_i$ are affine and hence extend uniquely to bounded positive linear mappings. Since the dual space of the trace class is isomorphic to the vector space of bounded operators, each $\mathcal{E}_i$ is of the form $\mathcal{E}_i(T) = \text{tr}\,[T E_i]$, where E_i is an operator satisfying $\mathbb{O} \leq E_i \leq \mathbb{1}$ (here $\mathbb{1}$ denotes the identity operator). Such operators are called *effects*. The set of effects will be denoted $\mathcal{E}(\mathcal{H})$. The normalization of the probability distributions p_T ($\sum_i p_T(\omega_i) = 1$) entails the condition

$$\sum_i E_i = \mathbb{1}. \qquad (1)$$

---

[1] The term operator will be taken as shorthand for "linear operator". With $A \leq B$ or equivalently $B \geq A$ we denote the usual ordering of self-adjoint operators; thus, $A \leq B$ if and only if $\langle \varphi | A\varphi \rangle \leq \langle \varphi | B\varphi \rangle$ for all $\varphi \in \mathcal{H}$. An operator A is *positive* if $A \geq \mathbb{O}$, the null operator.

[2] We remark that our notation follows closely that of the monograph [2]. The letter T was chosen there to denote a state since it is the first letter of the Finnish word for "state"; the authors of that monograph found this preferable to W, which would stand for the German word for "knowledge", or ρ, which is reminiscent of the phase space density with its classical connotations. Linguistic balance between the authors was maintained by taking Z to denote the pointer ("Zeiger") observable in a measurement scheme (see below). Naturally, $\mathcal{M}$ will stand for the English term "measurement".

"No Information Without Disturbance": Quantum Limitations of Measurement   231

The mapping $\omega_i \mapsto E_i$ together with the property (1) is a (discrete) instance of a normalized positive-operator-valued measure (POVM), the general definition being that of an operator-valued mapping $X \mapsto E(X)$ with the following properties: (i) the domain consists of all elements X of a σ-algebra Σ of subsets of an outcome space Ω; (ii) the operators $E(X)$ in the range are effects; (iii) the mapping is σ-additive (with infinite sums defined as weak limits): $E(\bigcup_i X_i) = \sum_i E(X_i)$ for any finite or countable family of mutually disjoint sets in Σ; (iv) $E(\Omega) = \mathbb{1}$. POVMs are taken as the most general representation of an *observable*. In this contribution the measurable space of outcomes (Ω, Σ) will be $(\mathbb{R}, \mathcal{B}(\mathbb{R}))$ or $(\mathbb{R}^2, \mathcal{B}(\mathbb{R}^2))$, where $\mathcal{B}(\mathbb{R}^n)$ denotes the Borel algebra of subsets of $\mathbb{R}^n$. The usual notion of observable is then recovered as the special case of a projection-valued measure (PVM) on $\mathcal{B}(\mathbb{R})$, which is nothing but the spectral measure associated with a selfadjoint operator. Observables represented by PVMs are called *sharp* observables, all other POVMs are referred to as *unsharp* observables. The extreme case of a *trivial* observable arises when all the effects in its range are *trivial*, that is, of the form $E(X) = \lambda_X \mathbb{1}$; the statistics associated with trivial effects and observables carries no information about the state.

2.2 Measurement Schemes

Measurements are physical processes and as such they are subject to the laws of physics. In quantum mechanics, a measurement performed on an isolated object is described as an interaction between this object system and an apparatus system, both being treated as quantum systems. Being a macroscopic system, the apparatus will interact with a wider environment, but it is often convenient and sufficient to subsume the degrees of freedom of this "rest of the world" into the description of the apparatus.

The quantum description of a measurement is succinctly summarized in the notion of a *measurement scheme*, i.e., a quadruple $\mathcal{M} := \langle \mathcal{H}_A, T_A, U, Z \rangle$, where $\mathcal{H}_A$ is the Hilbert space of the apparatus (or probe) system, T_A the initial apparatus state, $U = U(t_0, t_0 + \Delta t) : \mathcal{H} \otimes \mathcal{H}_A \to \mathcal{H} \otimes \mathcal{H}_A$ is the unitary operator representing the time evolution and ensuing coupling between the object system and apparatus during the period of measurement from time t_0 to $t_0 + \Delta t$. Finally, Z is the apparatus pointer observable, usually modeled as a sharp observable.

A schematic sketch of a measurement process is given in Fig. 1 which is taken from [2]. Here T and T_A denote the initial states of the object and apparatus, and $V(T \otimes T_A) := UT \otimes T_A U^*$ is the final state of the compound system after the measurement coupling has ceased. It is understood that upon reading an outcome, symbolized in the diagram with a discrete label k, the apparatus is considered to be describable in terms of a pointer eigenstate $T_{A,k}$, and this determines uniquely the associated final state T_k of the object, as will be shown below.

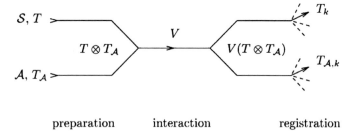

Fig. 1 Sketch of a measurement scheme. The symbols are explained in the main body of the text

The observable measured by such a scheme is determined by the pointer statistics for every object input state and is thus represented by a POVM E that is unambiguously defined by the following *probability reproducibility* condition:

$$\mathrm{tr}\left[UT\otimes T_A U^* I\otimes Z(X)\right] =: \mathrm{tr}\left[T\,E(X)\right] \equiv \mathsf{p}_T^E(X). \tag{2}$$

Here X is any element of a σ-algebra Σ of subsets of an outcome space Ω. The positivity of the operators $E(X)$ in the range of the map $X \mapsto E(X)$ and the measure properties of this map follows from the fact that the maps $X \mapsto \mathsf{p}_T^E(X)$ are probability measures for every state T.

The state T_X of the object after recording a measurement outcome in the set X is determined by the following *sequential joint* probability for a value of the pointer to be found in X and an immediately subsequent measurement of an effect B to yield a positive outcome:

$$\mathrm{tr}\left[UT\otimes T_A U^* B\otimes Z(X)\right] =: \mathrm{tr}\left[\mathcal{I}_X(T)B\right] \equiv \mathrm{tr}\left[T_X B\right] \tag{3}$$

The maps $T \mapsto \mathcal{I}_X(T) = T_X$, called (quantum) operations, are affine and trace norm-nonincreasing:

$$\mathrm{tr}\left[\mathcal{I}_X(T)\right] = \mathrm{tr}\left[T_X\right] = \mathrm{tr}\left[T\,E(X)\right] \leq \mathrm{tr}\left[T\right] = 1, \tag{4}$$

and they compose an *instrument*, that is, an operation-valued map $X \mapsto \mathcal{I}_X$. Note that these maps $\mathcal{I}_X$ extend in a unique way to linear maps on the complex vector space of trace class operators. The above equation shows that every instrument defines a unique POVM.

An important property of the operations $\mathcal{I}_X$ deriving from a measurement scheme is their *complete positivity*: for every $n \in \mathbb{N}$, the linear map defined by $T \otimes \Theta \mapsto \mathcal{I}_X(T) \otimes \Theta$ (where T is any trace class operator on $\mathcal{H}$ and Θ is any trace class operator on $\mathbb{C}^n$) is positive, that is, it sends state operators to (generally non-normalized) state operators.[3] The instrument composed of the completely positive operations is also called completely positive.

---

[3] An example of a positive state transformation that is *not* completely positive is given by $T \mapsto CTC^*$, where C is antilinear operator such as complex conjugation $\psi(x) \mapsto \psi(x)^*$ for $\psi \in L^2(\mathbb{R})$.

"No Information Without Disturbance": Quantum Limitations of Measurement   233

Every measurement scheme defines thus a unique completely positive instrument, and the latter fixes a unique POVM which represents the observable measured by the scheme. Starting from ground-breaking mathematical work of Neumark and Stinespring, the converse statement was developed in increasing generality by Ludwig and collaborators, Davies and Lewis, and Ozawa (detailed references can be found in [2]).

Theorem 1 (Fundamental Theorem of Quantum Measurement Theory). *Every observable, represented as a POVM E, admits infinitely many completely positive instruments $\mathcal{I}$ from which it arises via Eq. (2), and every completely positive instrument admits infinitely many implementations by means of a measurement scheme according to Eq. (3).*

2.3 Examples

Next I recall some model realizations of measurement schemes and completely positive instruments; these will provide valuable case studies in subsequent sections.

2.3.1 Von Neumann Model of an Unsharp Position Measurement

On the final pages of his famous book of 1932, *"Mathematische Grundlagen der Quantenmechanik"*, von Neumann introduces a mathematical model of what he describes as a measurement of the position of a particle in one spatial dimension. Both the particle and measurement probe are represented by the Hilbert spaces $\mathcal{H} = \mathcal{H}_{\mathcal{A}} = L^2(\mathbb{R})$; and the coupling

$$U = \exp(-\tfrac{i}{\hbar}\lambda Q \otimes P_{\mathcal{A}}), \tag{5}$$

generates a correlation between the observable intended to be measured, Q, and the pointer observable $Z = P_{\mathcal{A}}$.[4] To simplify the calculations, one assumes that the interaction is *impulsive*, that is, the coupling constant is large so that the duration of the interaction can be kept small enough so as to neglect the free Hamiltonians of the two systems. It is further assumed that the initial state of the probe is a pure state, $T_{\mathcal{A}} = P[\phi]$, with $\langle Q_{\mathcal{A}}\rangle_{\phi} := \langle \phi|Q_{\mathcal{A}}\phi\rangle = 0$ and finite variance $\text{Var}(Q_{\mathcal{A}}, \phi) = \langle Q_{\mathcal{A}}^2\rangle_{\phi} - \langle Q_{\mathcal{A}}\rangle_{\phi}^2$.

Von Neumann proceeded to calculate the correlation between the particle's position and the pointer observable *after* the coupling period and took this *measure of repeatability* as an indication of the quality of the measurement. Had he made the computation associated with Eq. (2) above, he would have found the actually measured observable to be a *smeared position* observable Q_e:

---

[4] The letters Q, P denote the selfadjoint canonical position and momentum operators, and their spectral measures are denoted Q, P, respectively.

$$E = Q_e : X \mapsto Q_e(X) = \chi_X * e(Q) = \int_{\mathbb{R}} \chi_X * e(q) Q(dq), \tag{6}$$

$$\text{where } e(q) = \lambda |\phi(-\lambda q)|^2.$$

Here $*$ denotes the convolution. Thus von Neumann was very close to discovering the representation of observables as POVMs! The variance $\mathrm{Var}(p_T^{Q_e})$ of the probability distribution $p_T^{Q_e}$ is

$$\mathrm{Var}(Q_e, T) = \int_{\mathbb{R}} (x - \bar{x})^2 p_T^{Q_e}(dx) = \mathrm{Var}(Q, T)^2 + \mathrm{Var}(e), \tag{7}$$

where $\bar{x} = \int_{\mathbb{R}} x p_T^{Q_e}(dx) = \mathrm{tr}\,[TQ]$. The second term in the expression for the variance, $\mathrm{Var}(e)$, indicates the unsharpness of the observable Q_e and at the same time is a measure of the inaccuracy of the measurement, that is, the separation between Q and Q_e.

The instrument induced by von Neumann's measurement scheme is given as follows:

$$\mathcal{I}^{Q_e} : X, T \mapsto \mathcal{I}_X^{Q_e}(T) = \int_X K_q T K_q^* dq, \tag{8}$$

$$\text{where } (K_q \varphi)(x) = \sqrt{\lambda} \phi\left(\lambda(q - x)\right) \varphi(x).$$

2.3.2 Ozawa's Model of a Sharp Position Measurement

It turned out much more intricate to find a measurement scheme realizing a measurement of the sharp position observable. One solution was presented by Ozawa [3, 4] who introduced the following coupling:

$$U = \exp\left[-\frac{i\pi}{3\sqrt{3}\hbar}(2Q \otimes P_A - 2P \otimes Q_A + QP - Q_A P_A)\right]$$

$$= \exp\left(-\frac{i}{\hbar}Q \otimes P_A\right) \exp\left(\frac{i}{\hbar}P \otimes Q_A\right). \tag{9}$$

Taking the pointer as $Z = Q_A$, the measured observable is Q, the sharp position, independently of the choice of initial probe state T_A. Indeed, the associated instrument is found to be

$$\mathcal{I}_X^{\mathrm{Ozawa}}(T) = \int_X \mathrm{tr}\,[TQ(dq)]\, e^{-\frac{i}{\hbar}qP}\, T_A\, e^{\frac{i}{\hbar}qP}, \tag{10}$$

so that $\mathrm{tr}\,[TE(X)] = \mathrm{tr}\,\left[\mathcal{I}_X^{\mathrm{Ozawa}}(T)\right] = \mathrm{tr}\,[TQ(X)]$ for all states T of the system.

3 Quantum Limitations on Measurability

The formalism of quantum measurements reviewed above provides a framework for the rigorous formulation of limitations on the measurability of physical quantities arising from quantum structures.

"No Information Without Disturbance": Quantum Limitations of Measurement 235

3.1 "No Information Gain Without Disturbance"

There has been much debate over the claim that according to quantum theory, every measurement necessarily "disturbs" the object system. Here is a theorem that states a precise sense in which this claim is true.

Theorem 2. *There is no instrument that leaves unchanged all states of the system unless the associated observable is trivial. More precisely: if an instrument $\mathcal{I}$ on (Ω, Σ) satisfies $\mathcal{I}_\Omega(T) = T$ for all $T \in \mathcal{S}(\mathcal{H})$, then $T \mapsto \mathrm{tr}\left[\mathcal{I}_X(T)\right] =: \lambda(X)$ is a constant map for all $X \in \Sigma$, and so the induced observable E is trivial, $E(X) = \lambda(X)\mathbb{1}$.*

The proof is quickly sketched: if $T = P[\varphi] \mapsto \mathcal{I}_\Omega(P[\varphi]) = \mathcal{I}_X(P[\varphi]) + \mathcal{I}_{\Omega \setminus X}(P[\varphi]) = P[\varphi]$, then $\mathcal{I}_X(P[\varphi]) = \lambda(X)P[\varphi]$. Due to the linearity of $\mathcal{I}_X$, the term $\lambda(X)$ is independent of φ, and the measured observable E gives probabilities independent of φ: $p_\varphi^E(X) = \mathrm{tr}\left[\mathcal{I}_X(P[\varphi])\right] = \lambda(X)$. *QED*

Hence a measurement scheme with no state change yields no information gain. We note that "disturbance" has here been interpreted as state change. This conclusion immediately leads to another question: is it possible to restrict the quality or accuracy of a measurement and thereby control the extent of the disturbance? This will be addressed in Section 5.

3.2 "No Measurement Without (Some Transient) Entanglement"

It is a general fact of quantum mechanics that interactions between two systems lead to *entanglement* between them, that is, to states which are not of product form. From this it would seem to follow that in a measurement the object system and apparatus end up necessarily in an entangled state at the end of the coupling period. The next theorem shows that this implication does not hold true without qualifications.

Theorem 3. *Let $U : \mathcal{H}_1 \otimes \mathcal{H}_2 \to \mathcal{H}_1 \otimes \mathcal{H}_2$ be a* non-entangling *unitary measurement coupling such that for a fixed vector ϕ_0 and all vectors $\varphi \in \mathcal{H}_1$, one has $U(\varphi \otimes \phi_0) = \varphi' \otimes \phi'$. Then U acts in one of the following ways:*
(a) $U(\varphi \otimes \phi_0) = V(\varphi) \otimes \phi'$, where V is an isometry;
(b) $U(\varphi \otimes \phi_0) = \varphi' \otimes W_{12}\varphi$, where $W_{12} : \mathcal{H}_1 \to \mathcal{H}_2$ is a surjective isometry and φ' is a fixed vector in $\mathcal{H}_1$

The proof is given in [5]. From this result it follows that if one aims at constructing a measurement scheme that leaves the object and apparatus in a non-entangled (separable) state after the coupling, and if this measurement is to transfer information about the initial object state φ to the apparatus, then the coupling U must act as in (b). It is therefore conceivable that after a suitable coupling interaction has been applied, the object and apparatus are left in an non-entangled state and yet complete information about the object state has been transferred to the apparatus. However, due to the continuity of the unitary dynamical evolution $t \mapsto U_t$ which comprises the coupling operator $U_{t+\Delta t}$, not all $U_{t'}$ with $t < t' < t + \Delta t$ can be of the non-entangling form (b), since that operator is not continuously connected with the identity operator U_0 at $t = 0$. It follows that some intermittent entanglement must build up during the interval $[t, t + \Delta t]$.

In order to extend this proof to measurement schemes for which the initial apparatus state T_A is not pure, it is necessary to sharpen the no-entanglement condition of the theorem to hold for any vector in $\mathcal{H}_A$ whose projection operator can arise as a convex component of T_A. These vectors are known to be given exactly by those in the range of $T_A^{1/2}$ [6]. The following theorem, also proven in [5], can then be applied to take a step towards extending the above discussion to mixed apparatus states.

Theorem 4. *Let $U : \mathcal{H}_1 \otimes \mathcal{H}_2 \rightarrow \mathcal{H}_1 \otimes \mathcal{H}_2$ be a unitary mapping such that for all vectors $\varphi \in \mathcal{H}_1$, $\phi \in \mathcal{H}_2$, the image of $\mathcal{H}_1 \otimes \mathcal{H}_2$ under U is of the form $U(\varphi \otimes \phi) = \varphi' \otimes \phi'$. Then U is one of the following:*
(A) $U = V \otimes W$ where $V : \mathcal{H}_1 \rightarrow \mathcal{H}_1$ and $W : \mathcal{H}_2 \rightarrow \mathcal{H}_2$ are unitary;
(B) $U(\varphi \otimes \phi) = V_{21}\phi \otimes W_{12}\varphi$, where $V_{21} : \mathcal{H}_2 \rightarrow \mathcal{H}_1$ and $W_{12} : \mathcal{H}_1 \rightarrow \mathcal{H}_2$ are surjective isometries.
The latter case can only occur if $\mathcal{H}_1$ and $\mathcal{H}_2$ are Hilbert spaces of equal dimensions.

It is not hard to construct a measurement scheme with a non-entangling coupling of the form (B) for *any* object observable E. This can be achieved by making the object interact with another system of the same type onto which the state of the original system is identically copied.

Example 1. Let $\mathcal{H}_1 = \mathcal{H}_2 = \mathcal{H}$. Let $E : \Sigma \rightarrow \mathcal{E}(\mathcal{H})$ be a POVM in $\mathcal{H}$. Define $U(\varphi \otimes \phi) = \phi \otimes \varphi$. Then we have

$$\langle U\varphi \otimes \phi | I \otimes E(X) U\varphi \otimes \phi \rangle = \langle \varphi | E(X) \varphi \rangle. \qquad (11)$$

3.3 "No Repeatable Measurement for Continuous Observables"

3.3.1 Repeatability and Ideality

A measurement and its associated instrument are called *repeatable* if the probability for obtaining the same result upon immediate repetition of the measurement is equal to one:

$$\text{tr}[\mathcal{I}_X(\mathcal{I}_X(T))] = \text{tr}[\mathcal{I}_X(T)] \quad \text{for all } X \in \Sigma, \, T \in \mathcal{S}(\mathcal{H}). \qquad (12)$$

A measurement of a discrete observable and its associated instrument is called *ideal* if it does not change any eigenstate; thus, if the state T is such that a particular outcome is certain to occur, then an ideal instrument does not alter the state:

$$\text{for all } T, k, \quad \text{if tr}[TP_k] = 1 \text{ then } \mathcal{I}_k(T) = T. \qquad (13)$$

Examples of repeatable measurements are the von Neumann and Lüders measurements which will be defined next.

"No Information Without Disturbance": Quantum Limitations of Measurement   237

Let A be an observable with discrete spectrum and associated spectral decomposition $A = \sum_k a_k P_k$. We allow the eigenvalues to have multiplicity greater than one, so that the spectral projections can be decomposed into a sum of orthogonal rank-1 projections: $P_k = \sum_\ell P[\varphi_{k\ell}]$. Then a *von Neumann measurement* is a measurement whose associated instrument has the form

$$\mathcal{I}_k^{\text{vN}}(T) = \sum_\ell P[\varphi_{k\ell}] T P[\varphi_{k\ell}]. \tag{14}$$

A *Lüders measurement* is a measurement whose associated instrument is given by:

$$\mathcal{I}_k^{\text{L}}(T) = P_k T P_k. \tag{15}$$

Note that Lüders measurement are ideal but von Neumann measurements are not ideal if at least one eigenvalue is degenerate. The ideal measurements are uniquely characterized by the form of their instruments [2]:

Theorem 5. *Any ideal measurement of a discrete sharp observable is a Lüders measurement.*

In particular, it follows that every ideal measurement is repeatable. A much deeper result is the following, conjectured by Davies and Lewis in 1970 [7] and proven by M. Ozawa in 1984 [8]. An observable E on (Ω, Σ) is *discrete* if there is a countable subset of N of Ω such that $E(N) = \mathbb{1}$.

Theorem 6. *If a measurement of an observable E is repeatable then E is discrete.*

I discuss briefly the implications of these results. First observe that the existence of ideal measurements enables the applicability of the famous reality criterion of Einstein et al. [9]:

> If, without in any way disturbing a system, we can predict with certainty (i.e., with probability equal to unity) the value of a physical quantity, then there exists an element of physical reality corresponding to that physical quantity.

Since ideal measurements are repeatable, the associated observables must be discrete. Hence the EPR criterion can only be applied to discrete observables or discrete coarse-grainings of continuous observables.

3.3.2 Approximate Repeatability

While strict repeatability is impossible for continuous observables such as position (or momentum), there do exist instruments for position (say) that are approximately repeatable in the following sense. Let $\delta > 0$, and for any (Borel) subset X of $\mathbb{R}$ let X_δ denote the set of all points which have a distance of not more than δ from some point in X. (Since $X_\delta = \bigcup_{x \in X}[x - \delta, x + \delta]$, this set X_δ is a Borel set.) An instrument $\mathcal{I}$ on $\mathcal{B}(\mathbb{R})$ is *δ-repeatable* if for all states T and all $X \in \mathcal{B}(\mathbb{R})$,

$$\text{tr}\left[\mathcal{I}_{X_\delta}(\mathcal{I}_X(T))\right] = \text{tr}\left[\mathcal{I}_X(T)\right]. \tag{16}$$

An example is given by Ozawa's instrument of a sharp position measurement, Eq. (10) if the probe state T_A is chosen such that its position distribution $p_{T_A}^{Q_A}$ is concentrated within $[-\delta, \delta]$.

The same form of instrument can also be defined for an unsharp position observable Q_e,

$$\mathcal{I}_X^{Q_e}(T) = \int_X \operatorname{tr}[TQ_e(dq)]\, e^{-iqP} T_A e^{iqP}, \tag{17}$$

and if T is chosen as before, one can find $d > 0$ such that

$$\operatorname{tr}\left[\mathcal{I}_{X_d}^{Q_e}(\mathcal{I}_X^{Q_e}(T))\right] \geq (1-\varepsilon)\operatorname{tr}\left[\mathcal{I}_X^{Q_e}(T)\right]. \tag{18}$$

Instruments with this property can be called $(d, 1-\varepsilon)$-repeatable. A detailed proof can be found in [10], and connections with the intrinsic unsharpness of the observable Q_e have recently been studied in [11].

3.3.3 Approximate Ideality

Ideality is a form of nondisturbance, but it is restricted to the eigenstates of the measured observable: if the quantity being measured has a definite value, then such measurements do not change the state. But any state other than an eigenstate will be disturbed: it will be transformed into one of the eigenstates due to the repeatability property of an ideal measurement.

The tight link between ideality and repeatability is relaxed if unsharp observables are considered: these still allow a notion of approximate ideality, but that does not imply approximate repeatability. I illustrate the last statement by means of the *generalized Lüders instrument* associated with a discrete observable $E : \omega_i \mapsto E_i$:

$$\mathcal{I}_i^L(T) = E_i^{1/2} T E_i^{1/2}. \tag{19}$$

The operations $\mathcal{I}_i^L$ have the following property:

$$\text{if } \operatorname{tr}[TE_i] \geq 1 - \varepsilon \text{ then } \operatorname{tr}\left[\mathcal{I}_i^L(T)E_i\right] \geq (1-\varepsilon)\operatorname{tr}[TE_i]. \tag{20}$$

That is, they do not decrease the probability. Further, it can be shown that for all states T for which $\operatorname{tr}[TE_i] \geq 1 - \varepsilon$, the (trace norm) difference between the states T and $\mathcal{I}_i^L(T)$ is of the order $\varepsilon^{1/2}$; this is the sense in which the generalized Lüders instruments are approximately ideal. Approximately ideal measurements enable a weakening of the EPR criterion applicable to unsharp or continuous observables, thus yielding a notion of *unsharp reality* [12].

It is not hard to construct examples of effects (with some eigenvalues small) such that the associated Lüders operation does not increase the small probability represented by that eigenvalue since the corresponding eigenstate is left unchanged. This shows that repeatability does not hold even in an approximate sense. Thus unsharp observables sometimes admit measurements that are less invasive than measurements of sharp observables.

"No Information Without Disturbance": Quantum Limitations of Measurement   239

The notion of a Lüders measurement was introduced by G. Lüders in 1951 [13] (english translation in [14]) who showed that such measurements can be used to test the compatibility of sharp observables.

Theorem 7 (Lüders Theorem). *Let $A = \sum_k a_k P_k$ and B be two (discrete) observable. The following are equivalent:*
(a) For all states T, $\mathrm{tr}\left[\sum_k P_k T P_k B\right] = \mathrm{tr}[TB]$;
(b) $AB = BA$.

The statement also holds if the observable B is not discrete or bounded; in that case statements (a) and (b) can be rephrased by replacing B with all spectral projections of B. This theorem has been used in relativistic quantum theory to motivate the "local commutativity" condition by virtue of the postulate that measurements in one spacetime region should not lead to observable effects in another, spacelike separated region.

According to the Lüders theorem, any observable B not commuting with A is sensitive to a Lüders measurement being performed on A. In other words, a Lüders measurement of A disturbs the distributions of B in some states if B does not commute with A. If A, B are allowed to be unsharp observables, the corresponding statement is no longer true in general but requires stronger assumptions [15].

Theorem 8. *Let $E : \omega_i \mapsto E_i$ be a discrete observable and B an effect. The following are equivalent if one of the assumptions (I) or (II) or (III) stated below holds:*
(a') For all states T, $\mathrm{tr}\left[\sum_k E_k^{1/2} T E_k^{1/2} B\right] = \mathrm{tr}[TB]$;
(b') $E_k B = BE_k$ for all k.
The assumptions are:
(I) E is a simple observable with only two effects $E_1, E_2 = \mathbb{1} - E_1$.
(II) B has a discrete spectrum of eigenvalues that can be numbered in decreasing or increasing order.
(III) Condition (a') is also stipulated for the effect B^2.

That *some* additional assumptions are necessary has been demonstrated by means of a counter example in [16]. There a discrete unsharp observable E and effect B not commuting with E were found such that the generalized Lüders instrument of E does not disturb the statistics of B.

3.4 Measurement Limitations due to Conservation Laws

There is an obvious limitation on measurability due to the fact that the physical realization of a measurement scheme depends on the interactions available in nature. In particular, the Hamiltonian of any physical system has to satisfy the symmetry requirements associated with the fundamental conservation laws. This measurement limitation is reviewed in Abner Shimony's contribution, so that here some complementary points and comments will be sufficient.

An early demonstration of the impact of the existence of additive conserved quantities on the measurability of a physical quantity was given by Wigner in 1952 [17]. Wigner showed that repeatable measurements of the x-component of a spin-1/2 system are impossible due to the conservation of the z-component of the total angular momentum of the system and the apparatus. The conclusion was generalized by other authors to the statement that a repeatable measurement of a discrete quantity is impossible if there is a (bounded) additive conserved quantity of the object plus apparatus system that does not commute with the quantity to be measured.

Wigner's resolution was to show that a successful measurement can be realized with an angular-momentum-conserving interaction and with an arbitrarily high success probability if the apparatus is sufficiently large. Thus he allowed for an additional measurement "outcome" that indicated "no information" about the spin. The outcomes associated with "spin up" and "spin down" were shown to be reproduced with probabilities that came arbitrarily closely to the ideal quantum mechanical probabilities. In [18, Sec. IV.3] it was shown that this resolution amounts to describing the measurement by means of a POVM with three possible outcomes and associated effects $E_+, E_-, E_?$, where the effects $E_\pm = (1 - \varepsilon)P_\pm^{s_x}$, i.e., they are "close to" the spectral projections of s_x if $0 < \varepsilon \le 1$, and the effect $E_? = \varepsilon \mathbb{1}$ is a multiple of $\mathbb{1}$. It can be shown that ε can be made very small if the size of the measuring system is large.

These considerations show that it is a matter of principle that measurements of spin can never be perfectly accurate as a consequence of the additive conservation law for total angular momentum. The necessary inaccuracy is appropriately described by a POVM of the kind described above. However, the common description of a sharp spin measurement is found to be an admissible idealization; the error made by breaking (ignoring) the fundamental rotation symmetry of the measurement Hamiltonian is negligible due to the fact that the measuring system is very large.

It seems to be a difficult problem to decide whether a limitation of measurability arises also in cases where the observable to be measured and the conserved quantity are unbounded and have continuous spectra. This question was raised by Shimony and Stein in 1979 [19]. The most general result at that time was the following (expressed in the notation of the present paper):

Theorem 9. *If a sharp observable E admits a repeatable measurement, and if $L \otimes \mathbb{1} + \mathbb{1} \otimes L_A$ is a bounded selfadjoint operator representing a conserved quantity for the combined object and apparatus system, then E commutes with L.*

Since repeatable measurements exist only for discrete observables (Theorem 6), the above statement is only applicable to object observables with discrete spectra. Hence it does not apply to measurements of position.

Ozawa [20] presented what seems to be a counter example, using a coupling that is manifestly translation invariant. However, this model constitutes an unsharp position measurement which becomes a sharp measurement only if the initial state of the apparatus is allowed to be a non-normalizable state (that is, not a Hilbert

"No Information Without Disturbance": Quantum Limitations of Measurement   241

space vector or state operator).[5] A proof that a sharp position measurement (without repeatability, but with some additional physically reasonable assumptions) cannot be reconciled with momentum conservation was given in [21]. A general proof is still outstanding.

Here we use another modification of the von Neumann model to demonstrate that momentum conservation is compatible with unsharp position measurements where the inaccuracy can be made arbitrarily small [18, Sec. 4.3]. Note that the total momentum $P + P_A$ commutes with the coupling

$$U = \exp\left(-i\tfrac{\lambda}{2}\left[(Q - Q_A)P_A + P_A(Q - Q_A)\right]\right). \tag{21}$$

The pointer is again taken to be $Z = Q_A$. Then the measured observable is the smeared position $Q_e = e * Q$, where $e(q) = \left(e^{\lambda} - 1\right)\left|\phi\left(-(e^{\lambda} - 1)q\right)\right|^2$.

One can argue that the clash between the conservation law and position measurement has been shifted and reappears when the measurement of Q_A is considered. However, if momentum conservation is taken into account in the measurement of the pointer, it would turn out that the pointer itself is only measured approximately, that is, an unsharp pointer $Q_{A,h}$ is actually measured, which then yields the measured observable as Q_{e*h}.

The lesson of the current subsection is this: to the extent that the limitation on measurability due to additive conservation laws holds as a general theorem, it shows that the notion of a sharp measurement of the most important quantum observables is an idealization which can be realized only approximately *as a matter of principle*; yet the quality of the approximation can be extremely good due to the macroscopic nature of the measuring apparatus.

To conclude this section, it is worth remarking that the quantum limitations of measurements described here are valid independently of the view that one may take on the measurement problem. This is the case because these limitations follow from consideration of the total state of system and apparatus as it arises in the course of its unitary evolution.

4 Complementarity and Uncertainty

The *"classic"* expressions of quantum limitations of preparations and measurements are codified in the complementarity and uncertainty principles, formulated by Bohr and Heisenberg 80 years ago.

This section offers a "taster" for two recent extensive reviews on the complementarity principle, [22], and the uncertainty principle, [23], which together develop a novel coherent account of these two principles. In a nutshell, complementarity states a strict exclusion of certain pairs of operations whereas the uncertainty principle

---

[5] The same observation applies to the von Neumann measurement model of which Ozawa's model is a modification.

242   P. Busch

shows a way of "softening" complementarity into a graded, quantitative relationship, in the form of a trade-off between the accuracies with which these two options
can be realized together approximately. This interpretation is compatible with, if not
envisaged in, the following passage of Bohr's published text of his famous Como
lecture of 1927 [24].

> In the language of the relativity theory, the content of the relations (2) [the uncertainty rela-
> tions] may be summarized in the statement that according to the quantum theory a general
> reciprocal relation exists between the maximum sharpness of definition of the space-time
> and energy-momentum vectors associated with the individuals. This circumstance may be
> regarded as a simple symbolical expression for the complementary nature of the space-time
> description and claims of causality. At the same time, however, the general character of this
> relation makes it possible to a certain extent to reconcile the conservation laws with the
> space-time co-ordination of observations, the idea of a coincidence of well-defined events
> in a space-time point being replaced by that of *unsharply* defined individuals within finite
> space-time regions.

Bohr summarizes here his idea of complementarity as the falling-apart in quantum
physics of the notions of observation, which leads to *space-time description*, and
state definition, linked with *conservation laws* and *causal description*; he regarded
the possibility of combining space-time description and causal description as an
idealization that was admissible in classical physics. Note also the reference to *unsharpness* (the emphasis in the quotation is ours), which seems to constitute the first
formulation of an intuitive notion of *unsharp reality* (and the first occurrence of this
teutonic addition to the English language).

4.1 The Complementarity Principle

In a widely accepted formulation, the *Complementarity Principle* is the statement
that there are pairs of observables which stand in the relationship of complementarity. That relationship comes in two variants, stating the mutual exclusivity of
preparations or *measurements* of certain pairs of observables. In quantum mechanics there are pairs of observables the eigenvector basis systems of which are mutually unbiased. This means that the system is in an eigenstate of one observable, so
that the value of that observable can be predicted with certainty, the values of the
other observable are uniformly distributed. This feature is an instance of *preparation complementarity*, and it has been called *value complementarity*. *Measurement
complementarity* of observables with mutually unbiased eigenbases can be characterized by the following property: any attempt to obtain simultaneous information
about both observables by first measuring one and then the other is bound to fail
since the first measurement completely destroys any information about the other
observable; that is to say, the second measurement gives no information about the
state prior to the first measurement. This will be illustrated in an example below.
We conclude that the "principle" of complementarity, as formalized here, is in fact
a consequence of the quantum mechanical formalism.

"No Information Without Disturbance": Quantum Limitations of Measurement   243

Examples of complementary pairs of observables are spin-1/2 observables such as s_x, s_z, and the canonically conjugate position and momentum observables Q, P of a free particle. A unified formalization of preparation and measurement complementarity can be given in terms of the spectral projections of these observables ($P_\pm^x, P_\pm^z$ for s_x, s_z), and $Q(X), P(Y)$ for Q, P:

$$
\begin{aligned}
P_k^x \wedge P_\ell^z = \mathbb{O} \quad &\text{for } k, \ell = +, -; \\
Q(X) \wedge P(Y) = \mathbb{O} \quad &\text{for bounded intervals } X, Y.
\end{aligned}
\tag{22}
$$

The symbol $\wedge$ represents the lattice-theoretic infimum of two projections, that is, for example, $Q(X) \wedge P(Y)$ is the projection onto the closed subspace which is the intersection of the ranges of $Q(X)$ and $P(Y)$. These relations entail, in particular, that complementary pairs of observables do not possess joint probability distributions associated with a state T in the usual way: for example, there is no POVM $G : \mathcal{B}(\mathbb{R}^2) \to \mathcal{E}(\mathcal{H})$ such that $G(X \times \mathbb{R}) = Q(X)$ and $G(\mathbb{R} \times Y) = P(Y)$ for all $X, Y \in \mathcal{B}(\mathbb{R})$. In fact, if these marginality relations were satisfied for all bounded intervals X, Y, then one must have $G(X \times Y) \leq Q(X)$ and $G(X \times Y) \leq P(Y)$, and this implies that any vector in the range of $G(X \times Y)$ must also be in the ranges of $Q(X)$ and $P(Y)$, hence $G(X \times Y) = \mathbb{O}$.

Example 2 (Complementarity for measurement sequences (1)). Let A, B be observables in $\mathbb{C}^n$, $n \geq 2$, with mutually unbiased eigenbases $\varphi_1, \varphi_2, \dots, \varphi_n$ and $\psi_1, \psi_2, \dots, \psi_n$, respectively. (Hence A, B are value complementary.) Let $\mathcal{I}^A$ be the repeatable (von Neumann-Lüders) instrument associated with A: $\mathcal{I}_k^A(T) := \langle \varphi_k | T \varphi_k \rangle | \varphi_k \rangle \langle \varphi_k |$. Let $\mathcal{I}_\mathbb{R}^A := \sum_k \mathcal{I}_k^A$ be the nonselective measurement operation, then the probability for a B measurement following the A measurement is $p_{\mathcal{I}_\mathbb{R}^A(T)}^B(\ell) = 1/n$, which is independent of T. This can be expressed by saying that the observable effectively measured in this process is not B but the trivial POVM whose effects are $E_\ell = \frac{1}{n}\mathbb{1}$.

Example 3 (Complementarity for measurement sequences (2)). Consider a measurement of position Q followed by a measurement of momentum P. Let $\mathcal{I}^Q$ be the instrument representing the position measurement. Then the following defines a joint probability distribution:

$$
\mathrm{tr}\left[\mathcal{I}_X^Q(T) P(Y) \right] = p_T(X \times Y) =: \mathrm{tr}\left[T G(X \times Y) \right], \quad X, Y \in \mathcal{B}(\mathbb{R}).
\tag{23}
$$

The marginal observables are sharp position and a "distorted momentum" observable, $G(X \times \mathbb{R}) = Q(X)$ and $G(\mathbb{R} \times Y) = \widetilde{P}(Y)$. Since one of these marginal observables is a sharp observable, it follows that the effects of the other marginal observable commute with the sharp observable. But Q is a maximal observable, and so the effects $\widetilde{P}(Y)$ are in fact functions of the position operator. The attempted momentum measurement only defines an effectively measured observable which contains a "shadow" of the information of the first position measurement. Hence a sharp measurement of position destroys all prior information about momentum (and vice versa).

244   P. Busch

The following defines a completely positive instrument $\mathcal{I}^Q$ which renders the effective observable defined by a subsequent momentum measurement trivial: let T_x be the continuous family of positive operators of trace one, generated by $T_x := U_x T_0 U_x^{-1}$, where U_x are unitary operators that commute with momentum P. Then put

$$\mathcal{I}_X^Q(T) := \int_X T_x \operatorname{tr}[TQ(dx)] . \tag{24}$$

The associated measured observable is indeed the sharp observable Q since $\operatorname{tr}\left[\mathcal{I}_X^Q(T)\right] = \operatorname{tr}[TQ(X)]$. Then the distorted momentum observable $\widetilde{P}$ defined above is found to be:

$$\begin{aligned}
\operatorname{tr}\left[T\widetilde{P}(Y)\right] &:= \operatorname{tr}\left[\mathcal{I}_\mathbb{R}^Q(T)P(Y)\right] = \int_\mathbb{R} \operatorname{tr}[T_x P(Y)] \operatorname{tr}[TQ(dx)] \\
&= \int_\mathbb{R} \operatorname{tr}\left[T_0 U_x^{-1} P(Y) U_x\right] \operatorname{tr}[TQ(dx)] \\
&= \int_\mathbb{R} \operatorname{tr}[T_0 P(Y)] \operatorname{tr}[TQ(dx)] = \operatorname{tr}[T_0 P(Y)] . \tag{25}
\end{aligned}$$

Thus $\widetilde{P}$ is a trivial observable. Note that in this calculation Q could have been replaced by any observable as the first-measured observable. However, if the instrument (24) is required to be approximately repeatable, then T_0 must have a position distribution concentrated around the origin 0, and U_x must ensure that T_x has a position distribution concentrated around the point x; this is achieved if U_x is chosen to be $exp(\frac{i}{\hbar}xP)$. Notice that this form is in fact realized in the Ozawa instrument for a sharp position measurement, Eq. (10). While we have not shown that this form is necessary, this consideration suggests that for approximately repeatable position measurements a subsequent momentum measurement leads to a (nearly) trivial observable as the distorted momentum.

4.2 The Uncertainty Principle

Following [22], we propose that the term *uncertainty principle* refers to the broad statement that there are pairs of observables for which a trade-off relationship pertains for the degrees of sharpness of the preparation or measurement of their values, such that a simultaneous or sequential determination of the values requires a nonzero amount of unsharpness (latitude, inaccuracy, disturbance). This gives rise to *three* variants of uncertainty relations, exemplified here for position and momentum: first there is the well-known inequality for the widths of the probability distributions of position and momentum in any quantum state that can be expressed in terms of the standard deviations,

$$\Delta(Q, T)\Delta(P, T) \geq \tfrac{1}{2}\hbar. \tag{26}$$

"No Information Without Disturbance": Quantum Limitations of Measurement 245

Second, one may consider a trade-off relation for the inaccuracies in any attempted *joint measurement* of position and momentum,

$$\delta(\widetilde{Q}, Q)\, \delta(\widetilde{P}, P) \geq C\hbar, \tag{27}$$

where the inaccuracies are to be defined appropriately as measures of the differences between the sharp position and momentum observables Q, P and their approximations $\widetilde{Q}, \widetilde{P}$, respectively, which are to be measured jointly. Finally, there is a trade-off between the accuracy of an approximate measurement of position (momentum) and a necessary disturbance of the momentum (position) distribution:

$$\delta(\widetilde{Q}, Q)\, D(\widetilde{P}, P) \geq C\hbar, \quad \delta(\widetilde{P}, P)\, D(\widetilde{Q}, Q) \geq C\hbar, \tag{28}$$

where $D(\widetilde{Q}, Q)$ and $D(\widetilde{P}, P)$ denote appropriate measures of the disturbance of position and momentum, respectively.

Suitable measures of inaccuracy and disturbance which make the last two measurement uncertainty relations precise will be presented in Section 5. It thus turns out that similar to the complementarity principle, the uncertainty principle in its three manifestations is also a formal consequence of the noncommutativity of the observables in question. The term "principle" may still be used to highlight the fact that the uncertainty relations reflect an important nonclassical feature of quantum mechanics.

4.3 Complementarity Versus Uncertainty?

The reviews [22, 23] propose a resolution of a long-standing controversy over the relationship, relative roles and interplay of the complementarity and uncertainty principles. This resolution will be briefly summarized here. As indicated in the introductory quote from Bohr (1928), the traditional view describes the uncertainty relations as a formal expression of the complementarity principle. However, as a quick survey of the research and textbook literature on quantum mechanics shows, this view has met with a considerable degree of uneasiness by many. Some authors consistently avoid any reference to complementarity while others play down the significance of the uncertainty relations, denying them the status of a principle which they reserve for complementarity.

Yet, in recent years there has been a shift of perspective which was indeed anticipated in the same quote of Bohr: complementarity is seen as a statement of the *impossibility* of jointly performing certain pairs of preparation or measurement procedures, whereas the role of the uncertainty principle is to quantify the degree to which an approximate reconciliation of these mutually exclusive options becomes a *possibility*. It seems that in this way a more balanced assessment has been achieved: compared to the view that emphasized complementarity over uncertainty, the positive role of the uncertainty relations as enabling joint determinations and joint

246   P. Busch

measurements is now highlighted more prominently; and even though it is true (as shown in [22]) that the uncertainty relations entail the complementarity relations in a suitable limit sense, it is still appropriate to point out the strict mutual exclusivity of sharp value assignments which, after all, is the reason for the quest for an approximate reconciliation in the form of simultaneous but unsharp value assignments.

The principles of complementarity and uncertainty are extreme manifestations of the existence of noncommuting pairs of observables and of superpositions of states, which both entail fundamental limitations of the possibilities of preparing or measuring simultaneous sharp values of observables that do not commute. These limitations are consequences of a famous theorem of von Neumann which we summarize here as follows.

Theorem 10. *Let A and B be two sharp observables represented as selfadjoint operators. The following are equivalent:*
(a) A and B possess a joint spectral representation (possibility of preparing joint sharp values).
(b) A and B possess a joint observable that defines joint probabilities for them (jointly measurability).
(c) $AB = BA$.

The reason for the long-standing debate over the superiority of either the complementarity *principle* or the uncertainty *principle* seems to lie in the fact that the *features* of complementarity and uncertainty are formally intertwined in Hilbert space quantum mechanics. It is only in the context of theoretical frameworks more abstract and general than quantum or classical theories that the logical relationships between complementarity and uncertainty postulates can be investigated; in such a generalized setting these postulates can in fact be used as principles within a set of axioms from which the Hilbert space framework of quantum mechanics can be deduced. As an example, we note the work of Lahti together with the late Bugajski [25] who used appropriate formalizations of complementarity and the existence of von Neumann-Lüders measurements in the so-called convexity framework to derive Hilbert space quantum theory.

5 Inaccuracy and Disturbance in Quantum Measurements

It remains to show how the above programmatic statement of the uncertainty principle for joint and sequential measurements can be made precise by appropriate measures of inaccuracy and disturbance. Such measures are also applicable in the analysis of the other quantum limitations of measurability discussed in Section 3.

First I will introduce the idea of an approximate joint measurement of two noncommuting quantities and present an operational definition of measurement error applicable to continuous observables such as position and momentum; the error measures for these observables obey a trade-off relation valid in any approximate joint measurements. Then I will show that a trade-off relation between the accuracy of a measurement and the disturbance of the distributions of an observable not commuting with the measured observable can be considered as an instance of a trade-off relation between the inaccuracies in an approximate joint measurement of two noncommuting observables.

5.1 Approximate Joint Measurements

A necessary criterion for the joint measurability of two observables is the existence of a joint probability distribution for every state T in the usual quantum mechanical form. Von Neumann's theorem entails that two noncommuting sharp observables such as position and momentum do not possess joint distributions (for all states). Hence these observables are not jointly measurable. However, for the joint measurability of pairs of *unsharp* observables, commutativity is *not* a necessary requirement. This suggests the following consideration: it should be possible to find two jointly measurable observables M_1, M_2 on $\mathcal{B}(\mathbb{R})$ which are *approximations*, in a suitable sense, of position Q and momentum P, respectively. Then a measurement of a joint observable M on $\mathcal{B}(\mathbb{R}^2)$ of M_1, M_2 will be accepted as an *approximate joint measurement* of Q, P if the deviations of M_1 from Q and of M_2 from P are finite in some appropriate measure. This constellation is shown in Fig. 2.

Two tasks need to be addressed in order to complete the above program. First, one needs to introduce suitable operational measures of inaccuracy, that is, of the deviation between two observables defined on the same outcome space (Ω, Σ). Second, since we are interested in good joint approximations of noncommuting pairs of observables, the optimal approximators M_1, M_2 must be expected to be noncommuting and hence unsharp observables in order to be jointly measurable; therefore, the problem arises to quantify the necessary degree of unsharpness required for the joint measurability given the finite "distance" of M_1, M_2 from two noncommuting observables.

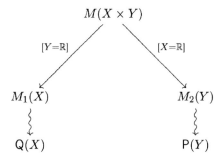

Fig. 2 Idea of a joint approximate measurement of position Q and momentum P, by means of an observable M on $\mathcal{B}(\mathbb{R}^2)$ whose marginals M_1 and M_2 are approximations of Q and P, respectively

248   P. Busch

The definition of such measures of inaccuracy and unsharpness will in general depend on the type of outcome space. A variety of approaches for the case $(\mathbb{R}, \mathcal{B}(\mathbb{R}))$ are analyzed in [23] and compared in detail in [26], and the case of discrete (qubit) observables is investigated in [27]. Here I will give a brief survey of notions applicable to the position-momentum case.

5.1.1 Standard Error

The only known measure that is universally applicable to different types of outcome spaces (barring questions of domains of unbounded operators) is a quantity that may be called *standard error* as it is defined in terms of the first and second moments of the relevant operator measures, similar to the standard deviation. This seems to be the only measure of inaccuracy or error that has been in use in the literature over an extended period. Examples of its application in the formulation of uncertainty relations for joint measurements are the works of Appleby [28, 29], Hall [30], and Ozawa (e.g., [31, 32]).

For an observable E on $\mathcal{B}(\mathbb{R})$, let $E[k] := \int x^k E(dx)$ denote the kth moment operator of E (defined on its natural domain $D(E[k]) := \{\varphi \in \mathcal{H} : |\int x^k \langle \psi | E(dx) \varphi \rangle| < \infty$ for all $\psi \in \mathcal{H}\}$ [33]). Assume $\mathcal{M}$ is a measurement scheme defining an observable E on $\mathcal{B}(\mathbb{R})$ which is intended to approximate the sharp position Q. Then a suggestive choice of measure of inaccuracy is

$$\varepsilon(Z, Q; T) := \mathrm{tr} \left[UT \otimes P[\phi] U^* (\mathbb{1} \otimes Z[1] - Q \otimes \mathbb{1})^2 \right]^{1/2}. \tag{29}$$

This can be expressed in terms of the actually measured observable E:

$$\varepsilon(E, Q; T) := \left(\mathrm{tr} \left[T(E[1] - Q)^2 \right] + \mathrm{tr} \left[T(E[2] - E[1]^2) \right] \right)^{1/2}. \tag{30}$$

The inaccuracy in a momentum measurement is defined similarly. Ozawa proved the following universal uncertainty relation for the marginals M_1, M_2 of an observable M on $\mathcal{B}(\mathbb{R}^2)$:

$$\varepsilon(M_1, Q; T)\varepsilon(M_2, P; T) + \varepsilon(M_1, Q; T)\Delta(P, T) + \Delta(Q, T)\varepsilon(M_2, P; T) \geq \frac{1}{2}\hbar. \tag{31}$$

He noted that the first product term can be zero (this happens in Ozawa's model of a sharp position measurement introduced above), and considers this to be a demonstration that the Heisenberg uncertainty principle for joint measurements of position and momentum and that for inaccuracy vs disturbance does not have the common form with a state-independent lower bound.

However, this way of reasoning ignores two crucial deficiencies in the definition of $\varepsilon(E, Q; T)$ as a measure of inaccuracy. First, the above uncertainty relation is not a statement solely about measurement inaccuracies since it depends on the preparation of the system. An appropriate definition of measurement inaccuracy should give an estimate of error which can be obtained without reference to the state of

"No Information Without Disturbance": Quantum Limitations of Measurement 249

the measured object (which usually is unknown in a measurement). This point was observed by Appleby in 1998 who introduced what we propose to call the *(global) standard error*:

$$\varepsilon(E,Q) := \sup_{T \in \mathcal{S}(\mathcal{H})} \left(\operatorname{tr}\left[T(E[1] - Q)^2 \right] + \operatorname{tr}\left[T(E[2] - E[1]^2) \right] \right)^{1/2}. \tag{32}$$

This quantity gives rise to a universal trade-off relation for joint measurement errors.

Theorem 11. *Let M be an observable on $\mathcal{B}(\mathbb{R}^2)$. Its marginals M_1, M_2 obey the following:*

$$\varepsilon(M_1, Q)\varepsilon(M_2, P) \geq \tfrac{1}{2}\hbar. \tag{33}$$

This result was proven by Appleby [29].

The second deficiency of the definition of $\varepsilon(E,Q;T)$—and also of $\varepsilon(E,Q)$—lies in the fact that this quantity cannot be estimated in terms of the measurements of E and Q under consideration unless the operators $E[1]$ and Q commute so that they can be jointly measured to determine the expectation of the operator $(E[1] - Q)^2$. If $E[1]$ and Q do not commute then normally the squared difference operator does not commute with either of them and a third, quite different measurement is required to find its expectation value. This is to say that the standard error is not *operationally significant*, in general.

An interesting but very special subclass of measurements where this deficiency does not arise is the family of *unbiased* measurements, for which $E[1] = Q$. In this case the standard error is given solely by the second term in Eq. (30), which is actually an operational measure of the intrinsic *noise* or unsharpness of the approximator E of Q (see below).

5.1.2 A Distance Between Observables on $\mathcal{B}(\mathbb{R})$

In 2004, Werner [34] introduced a distance $d(E,F)$ between two observables E and F on $\mathcal{B}(\mathbb{R})$ which is sensitive to the distance of the bulks of probability distributions p_T^E and p_T^F, and he derived an uncertainty relation for position and momentum. Some definitions are required in order to present this result.

For any bounded continuous function g on $\mathbb{R}$, one can define the operator $L(g,E) := \int_{\mathbb{R}} g(x)E(dx)$. The definition of $d(E,F)$ makes use of the set of (Lipshitz) functions $\Lambda := \{g : \mathbb{R} \to \mathbb{R} : g \text{ bounded}, |g(x) - g(y)| \leq |x - y|\}$. Werner's distance then is given as follows:

$$d(E,F) := \sup \left\{ \|L(g,E) - L(g,F)\| : g \in \Lambda \right\} \tag{34}$$

Werner's joint measurement uncertainty relation is stated as follows [34].

Theorem 12. *Let M_1, M_2 be marginals of an observable M on $\mathcal{B}(\mathbb{R}^2)$. The distances $d(M_1, Q)$ and $d(M_2, P)$ obey the inequality*

$$d(M_1, Q)\, d(M_2, P) \geq C\hbar. \tag{35}$$

250   P. Busch

Here the optimal constant C is determined via $C\hbar = E_0^2/(4ab)$, where E_0 is the lowest (positive) eigenvalue of the operator $a|Q| + b|P|$ for some $a,b > 0$. Its value is given by $C \approx 0.304745$.

This result constitutes the first universal joint measurement inaccuracy relation for operationally significant measures of inaccuracy. Moreover, the proof techniques used turn out to be applicable for quite different definitions of inaccuracy (see [26, 35]). The distance $d(E,F)$ is geometrically appealing and constitutes a natural choice due to its connection with the so-called Monge metric on the space of probability measures on $\mathcal{B}(\mathbb{R})$, as explained in [34]. However, from an experimenter's perspective, it may be considered less appealing to be asked to estimate $d(E,F)$ by measuring differences of expectation values for $L(g,E)$ and $L(g,F)$, where g runs through the set Λ of Lipshitz functions.

5.1.3 Error Bar Width

A measure of measurement inaccuracy that would appear natural to an experimenter is the width of error bars, which is estimated in a process of calibration: the measurement scheme to be calibrated is fed with systems prepared with fairly sharply defined values of (say) the position observable. For each value, one estimates the spread of output values which gives a measure of the *error bar width*. If this measure is found to be bounded across all input values, the measurement will be considered to constitute a good approximation of the position observable to be measured. This consideration is captured in the following definitions.

Let M_1 be an observable on $\mathcal{B}(\mathbb{R})$ which is to approximate Q. Let $J_{q;\delta} := [q - \delta/2, q + \delta/2]$. By $\mathcal{W}_{\varepsilon_1,\delta}(M_1,Q)$ I denote the *inaccuracy*, defined as the smallest interval width w such that whenever the value of Q is certain to lie within an interval $J_{q;\delta}$, then the output distribution $p_\varphi^{M_1}$ is concentrated to within $1 - \varepsilon_1$ in $J_{q;w}$:

$$\mathcal{W}_{\varepsilon_1,\delta}(M_1,Q) := \inf\{w \,|\, \text{for all } q \in \mathbb{R}, \, \psi \in \mathcal{H},$$
$$\text{if } p_\psi^Q(J_{q;\delta}) = 1 \text{ then } p_\psi^{M_1}(J_{q;w}) \geq 1 - \varepsilon_1\}. \tag{36}$$

The inaccuracy describes the range within which the input values can be inferred from the output distributions, with confidence level $1 - \varepsilon_1$, given initial localizations within δ. The inaccuracy is an increasing function of δ, so that one can define the *error bar width* of M_1 relative to Q:

$$\mathcal{W}_{\varepsilon_1}(M_1,Q) := \inf_\delta \mathcal{W}_{\varepsilon_1,\delta}(M_1,Q) = \lim_{\delta \to 0} \mathcal{W}_{\varepsilon_1,\delta}(M_1,Q). \tag{37}$$

If $\mathcal{W}_{\varepsilon_1}(M_1,Q)$ is finite for all $\varepsilon_1 \in (0,\frac{1}{2})$, we will say that M_1 approximates Q in the sense of *finite error bars*. Similar definitions apply to approximations M_2 of momentum P, yielding $\mathcal{W}_{\varepsilon_2,\delta}(M_2,P)$ and $\mathcal{W}_{\varepsilon_2}(M_2,P)$.

It is interesting to note that the finiteness of either $\varepsilon(M_1,Q)$ or $d(M_1,Q)$ implies the finiteness of $\mathcal{W}_{\varepsilon_1}(M_1,Q)$ [26]. Therefore, among the three measures of

"No Information Without Disturbance": Quantum Limitations of Measurement 251

inaccuracy introduced above, the condition of finite error bars gives the most general criterion for selecting "good" approximations of Q and P.

The following uncertainty relation for error bar widths is proven in [35].

Theorem 13. *Let M be an observable on $\mathcal{B}(\mathbb{R}^2)$. The marginals M_1, M_2 obey the following trade-off relation (for $0 < \varepsilon_1, \varepsilon_2 < \frac{1}{2}$):*

$$W_{\varepsilon_1}(M_1, Q)\, W_{\varepsilon_2}(M_2, P) \geq 2\pi\hbar\, (1 - \varepsilon_1 - \varepsilon_2)^2 . \tag{38}$$

5.1.4 Unsharpness

A measure of the intrinsic unsharpness of an observable E on $\mathcal{B}(\mathbb{R})$ is given by the *resolution width* (at confidence level $1 - \varepsilon$), as defined in [11]:

$$\gamma_\varepsilon(E) := \inf\{w > 0 \,|\, \forall x \in \mathbb{R}\, \exists \rho \in S : \rho^E([x - \tfrac{w}{2}, x + \tfrac{w}{2}]) \geq 1 - \varepsilon\}. \tag{39}$$

This measure describes the smallest interval width for which probability of no less than $1 - \varepsilon$ can be achieved, irrespectively of where the interval is located. For a sharp observable E on $\mathcal{B}(\mathbb{R})$ with support equal to $\mathbb{R}$ (so that $E(X) \neq O$ for any open interval), the resolution width is $\gamma_\varepsilon(E) = 0$ for all $\varepsilon \in (0,1)$.

I put forth the conjecture (being studied in [26]) that the resolution widths in approximate joint measurements of position and momentum obey the following trade-off relation.

Conjecture. Let M be an approximate joint observable for Q, P in the sense of finite error bars. Then the resolution widths of M_1 and M_2 obey the following inequality:

$$\gamma_{\varepsilon_1}(M_1)\, \gamma_{\varepsilon_2}(M_2) \geq 2\pi\hbar\, (1 - \varepsilon_1 - \varepsilon_2)^2 . \tag{40}$$

The special case where M is a covariant phase space observable has already been proven in [11].

5.2 Inaccuracy-Disturbance Trade-Off

We have seen that a momentum measurement following a sharp position measurement defines an observable that carries no information about the momentum distributions of the states prior to the position measurement. A sharp measurement of position thus destroys completely the momentum information contained in the initial state. The question arises whether the disturbance of momentum can be diminished if the position is measured approximately rather than sharply.

This possibility was already envisaged by Heisenberg in his discussion of thought experiments illustrating the uncertainty relations [36, 37]. For example, in the case of a particle passing through a slit he noted that due to the diffraction at the slit, an initially sharp momentum distribution is distorted into a broader distribution whose width Δp is of the order $\hbar/\delta x$, where δx is the width of the slit. The width Δp is a measure of the change, or disturbance, of the momentum distribution, and δx can be interpreted as the inaccuracy of the position determination effected by the slit. Further, one may also consider the recording of the location at which the particle hits the screen as a geometric determination of the (direction) of its momentum, the inaccuracy δp of which is given by the width Δp of the distribution obtained after many repetitions of the experiment. In this way the passage through the slit followed by the recording at the screen constitutes an approximate joint measurement of the position and momentum of the particle at the moment of its passage through the slit; see Fig. 3.

Generalizing this idea of making an approximate joint measurement by way of a sequence of approximate measurements, we consider the schemes of Figs. 4 and 5). Here M_1 is either the sharp position Q or an unsharp position observable Q_e measured first, followed by a sharp momentum observable, whose measurement is to be followed by a sharp momentum measurement. The observable M_2 effectively mea-

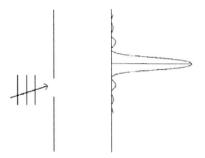

Fig. 3 Slit experiment as an approximate (sequential) joint measurement of position and momentum

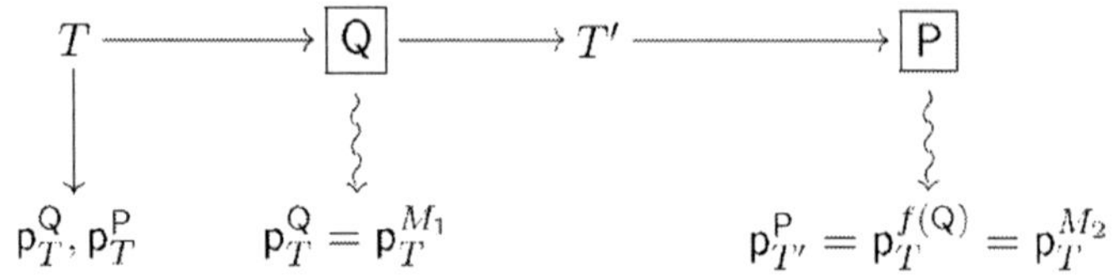

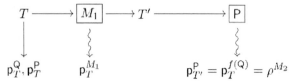

Fig. 4 Sharp position measurement followed by a sharp momentum measurement. The two marginals $M_1 = Q$ and M_2 commute and have a unique joint observable M

Fig. 5 Approximate joint measurement of position and momentum defined by an unsharp position measurement followed by a sharp momentum measurement. The marginals are $M_1 = Q_e$ and $M_2 = P_f$ where e, f are probability distributions which are related as described in the main text

"No Information Without Disturbance": Quantum Limitations of Measurement   253

sured by this momentum measurement is defined via $\mathsf{p}_T^{M_2} := \mathsf{p}_{T'}^{\mathsf{P}}$, for all initial states T, where T' is the state after the position measurement. Thus M_2 is the "distorted" momentum observable. Collecting the probabilities for finding an outcome in a set X for the first measurement and an outcome in Y for the second measurement defines a probability measure for each state T via $X \times Y \mapsto \mathsf{p}_T(X \times Y)$. Hence there is a unique joint observable M for M_1 and M_2 determined by the given measurement scheme [23].

In the first case, since the marginal $M_1 = \mathsf{Q}$ is sharp, M_2 commutes with Q and is therefore *not* a good approximation of the momentum observable P. However, in the second case, $M_1 = \mathsf{Q}_e$, it is known [38] that the second marginal observable M_2 is a smeared momentum observable, $M_2 = \mathsf{P}_f$, if the first, unsharp position measurement is such that the induced instrument is the von Neumann instrument (8). The inaccuracy distributions are then related as follows (cf. Eq. (6)):

$$e(q) = \lambda |\phi(-\lambda q)|^2, \quad f(p) = \tfrac{1}{\lambda} |\widetilde{\phi}(-\tfrac{1}{\lambda} p)|^2. \tag{41}$$

Here $\widetilde{\phi}$ is the Fourier transform of ϕ, from which it follows that the standard deviations of the distributions e, f obey the uncertainty relation:

$$\Delta(e)\Delta(f) \geq \tfrac{1}{2}\hbar. \tag{42}$$

Note that $\Delta(e)$, $\Delta(f)$ are measures of how well the sharp observables Q, P are approximated by M_1, M_2, respectively. Thus they are measures of measurement inaccuracy, and at the same time $\Delta(f)$ quantifies the disturbance of the momentum distribution due to the position measurement.

These considerations show that an operational definition disturbance of the momentum distribution due to a position measurement is obtained by considering the sequential joint measurement composed of first measuring position and then momentum. The inaccuracy of the second measurement, that is, any measure of the separation between P and M_2, is also a measure of the momentum disturbance. Consequently, all the joint measurement inaccuracy relations discussed above apply to sequential joint measurements of position and momentum, and in this case they constitute rigorous versions of the long-sought-after inaccuracy-vs-disturbance trade-off relations.

6 Conclusion

Using the apparatus of modern quantum measurement theory, I have reviewed rigorous formulations of some well-known quantum limitations of measurements: the inevitability of disturbance and (transient) entanglement; the impossibility of repeatable measurements for continuous quantities, the restrictions on measurements arising from the presence of an additive conserved quantity, and the necessarily approximate and unsharp nature of joint measurements of noncommuting quantities.

In each case, a strict no-go theorem is complemented with a positive result describing conditions for an approximate realization of the impossible goal: repeatability can be approximated arbitrarily well for continuous sharp observables, also in the presence of a conservation law. It was found that ideal measurements of sharp observables are necessarily repeatable, but in the case of unsharp observables, approximate ideality can be achieved without forcing approximate repeatability. Thus, unsharp measurements may be less invasive than sharp measurements.

The impossibility of joint sharp measurements of complementary pairs of observables can be modulated into the possibility of *approximate* joint measurements of such observables, *provided* the inaccuracies are allowed to obey a universal Heisenberg uncertainty relation. Likewise, the complete destruction of momentum information by a sharp position measurement can be avoided if an *unsharp* position measurement is performed. The trade-off between the information gain in the approximate measurement of one observable and the disturbance of (the distribution of) its complementary partner observable was found to be an instance of the joint-measurement uncertainty relation.

These results, some of which were made precise in very recent investigations, open up a range of interesting new questions and tasks. In particular, it will be important to find operational measures of inaccuracy that are applicable to all types of observables, whether bounded or unbounded, discrete or continuous. This would probably enable a formulation of a universal form of joint measurement uncertainty relation for arbitrary pairs of (noncommuting) observables, thus generalizing the relations presented here for the special case of complementary pairs of continuous observables such as position and momentum.

Acknowledgement This work was carried out during my visiting appointment at the Perimeter Institute (2005–2007). Hospitality and support by PI are gratefully acknowledged.

References

1. P. Busch and A. Shimony. "Insolubility of the quantum measurement problem for unsharp observables." *Stud. Hist. Phil. Mod. Phys.*, **27**:397–404, 1996.
2. P. Busch, P.J. Lahti, and P. Mittelstaedt. *The Quantum Theory of Measurement.* Springer, Berlin, second revised edition, 1996.
3. M. Ozawa. "Measurement breaking the standard quantum limit for free-mass position." *Phys. Rev. Lett.*, **60**:385–388, 1988.
4. M. Ozawa. "Position measuring interactions and the Heisenberg uncertainty principle." *Phys. Lett. A*, **299**:1–7, 2002.
5. P. Busch. "The role of entanglement in quantum measurement and information processing." *Int. J. Theor. Phys.*, **42**(5):937–941, 2003.
6. N. Hadjisavvas. "Properties of mixtures of non-orthogonal states." *Lett. Math. Phys.*, **5**:327–332, 1981.
7. E.B. Davies and J.T. Lewis. "An operational approach to quantum probability." *Comm. Math. Phys.*, **17**:239–260, 1970.

"No Information Without Disturbance": Quantum Limitations of Measurement 255

8. M. Ozawa. "Quantum measuring processes of continuous observables." *J. Math. Phys.*, **25**: 79–87, 1984.
9. A. Einstein, B. Podolsky, and N. Rosen. "Can quantum-mechanical description of physical reality be considered complete?" *Phys. Rev.*, **47**:777–780, 1935.
10. P. Busch and P. Lahti. "Some remarks on unsharp quantum measurements, quantum nondemolition, and all that." *Ann. Phys.*, **47**:369–382, 1990.
11. C. Carmeli, T. Heinonen, and A. Toigo. "Intrinsic unsharpness and approximate repeatability of quantum measurements." *J. Phys. A*, **40**:1303–1323, 2007.
12. P. Busch. "Can quantum theoretical reality be considered sharp?" In P. Mittelstaedt and E.W. Stachow (eds.), *Recent Developments in Quantum Logic*. Bibliographisches Institut, Mannheim, pp. 81–101, 1985.
13. G. Lüders. "Über die Zustandsänderung durch den Meßprozeß." *Ann. Physik*, **8**:322–328, 1951.
14. G. Lüders. "Concerning the state-change due to the measurement process." *Ann. Phys. (Leipzig)*, **15**(9):663–670, 2006.
15. P. Busch and J. Singh. "Lüders theorem for unsharp quantum measurements." *Phys. Lett. A*, **249**:10–12, 1998.
16. A. Arias, A. Gheondea, and S. Gudder. "Fixed points of quantum operations." *J. Math. Phys.*, **43**(12):5872–5881, 2002.
17. E.P. Wigner. "Die Messung quantenmechanischer Operatoren." *Z. Phys.*, **133**:101–108, 1952.
18. P. Busch, M. Grabowski, and P.J. Lahti. *Operational Quantum Physics*. Springer, Berlin, second corrected printing, 1997.
19. A. Shimony and H. Stein. "A problem in Hilbert space theory arising from the quantum theory of measurement." *Am. Math. Mon.*, **86**:292–293, 1979.
20. M. Ozawa. "Does a conservation law limit position measurements?" *Phys. Rev. Lett.*, **67**(15): 1956–1959, 1991.
21. P. Busch. "Momentum conservation forbids sharp localisation." *J. Phys. A: Math. Gen.*, 18: 3351–3354, 1985.
22. P. Busch and C. Shilladay. "Complementarity and uncertainty in Mach–Zehnder interferometry and beyond." *Phys. Rep.*, **435**:1–31, 2006.
23. P. Busch, T. Heinonen, and P.J. Lahti. "Heisenberg's uncertainty principle." *Phys. Rep.*, **452**:155–176, 2007.
24. N. Bohr. "The quantum postulate and the recent development of atomic theory." *Nature*, **121**: 580–590, 1928.
25. P.J. Lahti and S. Bugajski. "Fundamental principles of quantum theory. II. From a convexity scheme to the DHB theory." *Int. J. Theor. Phys.*, **24**:1051–1980, 1985.
26. P. Busch and D.B. Pearson. "Inaccuracy and unsharpness in approximate joint measurements of position and momentum." In preparation, 2008.
27. P. Busch and T. Heinosaari. "Approximate joint measurements of qubit observables." *Quantum Inf. & Comput.* **8**, 797–818, 2008.
28. D.M. Appleby. "Concept of experimental accuracy and simultaneous measurements of position and momentum." *Int. J. Theor. Phys.*, **37**:1491–1509, 1998.
29. D.M. Appleby. "Error principle." *Int. J. Theor. Phys.*, **37**:2557–2572, 1998.
30. M.J.W. Hall. "Prior information: How to circumvent the standard joint-measurement uncertainty relation." *Phys. Rev. A*, **69**:052113/1–12, 2004.
31. M. Ozawa. "Universally valid reformulation of the Heisenberg uncertainty principle on noise and disturbance in measurement." *Phys. Rev. A*, **67**:042105, 2003.
32. M. Ozawa. "Uncertainty relations for noise and disturbance in generalized quantum measurements." *Ann. Phys. (N.Y.)*, **311**:350–416, 2004.
33. A. Dvurečenskij, P. Lahti, and K. Ylinen. "Positive operator measures determined by their momentum sequences." *Rep. Math. Phys.*, **45**:139–146, 2000.
34. R.F. Werner. "The uncertainty relation for joint measurement of position and momentum." *Qu. Inf. Comp.*, **4**:546–562, 2004.

35. P. Busch and D.B. Pearson. "Universal joint-measurement uncertainty relation for error bars." *J. Math. Phys.*, **48**:082103, 2007.
36. W. Heisenberg. "Über den anschaulichen Inhalt der quantentheoretischen Kinematik und Mechanik." *Z. Phys.*, **43**:172–198, 1927.
37. W. Heisenberg. *The Physical Principles of the Quantum Theory*. University of Chicago Press, Chicago, 1930.
38. E.B. Davies. "On the repeated measurements of continuous observables in quantum mechanics." *J. Funct. Anal.*, **6**:318–346, 1970.

How Stands Collapse II

Philip Pearle

Abstract I review 10 problems associated with the dynamical wave function collapse program, which were described in the first of these two papers. Five of these, the *interaction, preferred basis, trigger, symmetry* and *superluminal* problems, were shown there to have been resolved. In this volume in honor of Abner Shimony, I discuss the five remaining problems, *tails, conservation law, experimental, relativity, legitimization*. Particular emphasis is given to the tails problem, first raised by Abner. The discussion of legitimization contains a new argument, that the energy density of the fluctuating field which causes collapse should exert a gravitational force. This force can be repulsive, since this energy density can be negative. Speculative illustrations of cosmological implications are offered.

1 Introduction and Recapitulation

All things in the world come from being. And being comes from non-being.

The Way of Lao Tzu

In 1977, a graduate student at the University of Edinburgh's department of Sociology of Science named Bill Harvey (presently Deputy Director of the Scottish Education Funding Council) was doing his PhD thesis, and wrote to physicists working in the field of foundations of quantum theory, including myself, to ask if he could visit and ask questions. After my interview, which took place at Hamilton College, Bill, two colleagues and I went out to dinner and, as we drove back, I asked him what his PhD thesis was about. He said: "Social deviance."

In the first of these papers [1], hereafter referred to as paper I, as well as in a previous festschritt for Abner Shimony [2], I presented some personal history, my route to becoming a social deviant. Closet deviance, shared by a small (but growing,

---

P. Pearle
Hamilton College, Clinton, NY 13323, USA
e-mail: ppearle@hamilton.edu

W.C. Myrvold and J. Christian (eds.), *Quantum Reality, Relativistic Causality, and Closing the Epistemic Circle,* The Western Ontario Series in Philosophy of Science 73,
© Springer Science+Business Media B.V. 2009

I hope) group of physicists, is the belief that standard quantum theory, handed down on Mount Copenhagen, while a most marvelous set of laws, has conceptual flaws. Outright deviance is the temerity to try and do something about it.

(Parenthetically, Abner Shimony, whom I first met in Wendell Furry's office at Harvard around 40 years ago, has over these years been supportive of my apostasy. Since Abner is both a physicist and philosopher, he is at most half a deviant, since what is deviant in physics is normal in philosophy).

The flaws are encapsulated in the inadequate answer given by standard quantum theory to what has been called "the measurement problem," but which I prefer to call "the reality problem":

For a closed system of any kind, given a state vector and the Hamiltonian, specify the evolving realizable states and their probabilities of realization.

That is, there is no well-defined procedure within standard quantum theory for, at any time, plucking out from the state vector the possible states which describe what we see around us. At best, in a restricted set of situations, namely measurement situations by human beings, which are a small subset of the full set of situations in the universe created by nature, one can apply procedures that work FAPP ("For All Practical Purposes," a useful acronym coined by John Bell, in his pungent critique of standard quantum theory [3]). These procedures require additional, ad hoc (which means "for this case only") information: *this* is the apparatus, *that* is the environment, etc.

Paper I [1] described the Continuous Spontaneous Localization (CSL) dynamical wave function collapse theory [4,5]. It consists of two equations. A *dynamical equation* describes how the state vector evolves under the joint influence of the Hamiltonian and an operator depending upon an arbitrarily chosen fluctuating scalar field $w(\mathbf{x}, \mathbf{t})$. A *probability rule* equation gives the probability that this $w(\mathbf{x}, \mathbf{t})$ is realized in nature. Then, the answer given by CSL to the measurement/reality problem is simply:

Given any $w(\mathbf{x}, \mathbf{t})$, a state vector evolving according to the dynamical equation is a realizable state, and the probability rule gives its probability of realization.

The claim of CSL is "what you see (in nature) is what you get (from the theory)." Among other considerations, in this paper it will be argued that this works well.

1.1 CSL Lite

Que será, será, whatever will be, will be...
Jay Livingstone and Ray Evans, sung by Doris Day

In order that this paper be self contained, some of paper I's discussion of CSL will be repeated here, First comes "CSL lite," a simplified formulation which illustrates essential features. An initial state vector

$$|\psi, 0\rangle = \sum_{n=1}^{N} c_n |a_n\rangle \tag{1}$$

How Stands Collapse II 259

(the $|a_n\rangle$ are eigenstates of an operator A with nondegenerate eigenvalues a_n) evolves according to the *dynamical equation*

$$|\psi,t\rangle_w \equiv e^{\frac{1}{4\lambda}\int_0^t dt'[w(t')-2\lambda A]^2}|\psi,0\rangle$$

$$= \sum_{n=1}^{N} c_n|a_n\rangle e^{-\frac{1}{4\lambda}\int_0^t dt'[w(t')-2\lambda a_n]^2}. \qquad (2)$$

In Eq. (2), $w(t)$ is a sample random function of white noise type, and λ characterizes the collapse rate. The state vector given by (2) is not normalized to 1, so one must remember to normalize it when calculating expectation values, the density matrix, etc.

The probability associated to $|\psi,t\rangle_w$ is given by the *probability rule*

$$P_w(t)Dw \equiv_w\langle\psi,t|\psi,t\rangle_wDw = \sum_{n=1}^{N} |c_n|^2 e^{-\frac{1}{2\lambda}\int_0^t dt'[w(t')-2\lambda a_n]^2}Dw. \qquad (3)$$

To see that the integrated probability is 1, discretize the time integral in Eq. (3), so that it appears as a product of gaussians and, using

$$Dw \equiv \frac{dw(0)}{\sqrt{2\pi\lambda/\Delta t}} \frac{dw(\Delta t)}{\sqrt{2\pi\lambda/\Delta t}} \cdots \frac{dw(t)}{\sqrt{2\pi\lambda/\Delta t}},$$

integrate over all $dw(n\Delta t)$ from $-\infty$ to ∞.

Here is a proof (not given in paper I, where the result was just cited) that, as $t \to \infty$, Eqs. (2) and (3) describe collapse to one of the eigenstates $|a_m\rangle$ with probability $|c_m|^2$.

Consider first the special class of $w(t)$, labeled $w_a(t)$, which have the asymptotic behavior

$$\lim_{T\to\infty} (2\lambda T)^{-1} \int_0^T dt w_a(t) \to a,$$

where a is a constant. Write $w_a(t) = w_0(t) + 2\lambda a$, and define

$$(2\lambda T)^{-1} \int_0^T dt w_0(t) \equiv \varepsilon(T),$$

so $\lim_{T\to\infty} \varepsilon(T) \to 0$. Then Eq. (3) may be written

$$P_w(t) = \sum_{n=1}^{N} |c_n|^2 e^{-\frac{1}{2\lambda}\int_0^t dt' w_0^2(t')} e^{-2\lambda t(a-a_n)[2\varepsilon(t)+(a-a_n)]}. \qquad (4)$$

If $a \neq a_n$ for any n, the probability density (4) vanishes for $t \to \infty$, since it is a sum of terms which vanish as $\exp-2\lambda t(a-a_n)^2$. The (normalized) state vector corresponding to such a $w_a(t)$, as given by Eq. (2), is generally not a collapsed state, but its asymptotic probability of occurrence is zero.

If $a = a_m$, Eqs. (2) and (3) respectively become

$$|\psi,t\rangle_w = e^{-\frac{1}{4\lambda}\int_0^t dt' w_0^2(t')}\left[c_m|a_m\rangle + \sum_{n\neq m}^N c_n|a_n\rangle e^{-\lambda t(a_m-a_n)[2\varepsilon(t)+(a_m-a_n)]}\right]$$

$$\rightarrow e^{-\frac{1}{4\lambda}\int_0^\infty dt' w_0^2(t')}c_m|a_m\rangle \tag{5}$$

$$P_w(t) = e^{-\frac{1}{2\lambda}\int_0^t dt' w_0^2(t')}\left[|c_m|^2 + \sum_{n\neq m}^N |c_n|^2 e^{-2\lambda t(a_m-a_n)[2\varepsilon(t)+(a_m-a_n)]}\right]$$

$$\rightarrow |c_m|^2 e^{-\frac{1}{2\lambda}\int_0^\infty dt' w_0^2(t')}. \tag{6}$$

Eq. (5) shows that collapse to $|a_m\rangle$ occurs for any $w_{a_m}(t)$. When Eq. (6) is integrated over all possible $w_{a_m}(t)$, (i.e., over all possible $w_0(t)$), the total associated probability is $|c_m|^2$.

There are other possibilities for $w(t)$ other than the $w_a(t)$, namely the cases for which $T^{-1}\int_0^T dt w(t)$ it has no asymptotic limit. However, since the probability for the $w_{a_m}(t)$'s totals to 1, these possibilities have measure 0. End of proof.

The density matrix constructed from (2), (3) is

$$\rho = \int P_w(t) Dw \frac{|\psi,t\rangle_w\,_w\langle\psi,t|}{_w\langle\psi,t|\psi,t\rangle_w} = \sum_{n,m=1}^N c_n c_m^* |a_n\rangle\langle a_m| e^{-(\lambda t/2)(a_n-a_m)^2}. \tag{7}$$

Thus, the off-diagonal elements decay at a rate determined by the squared differences of eigenvalues.

For many mutually commuting operators A_k, and with a possibly time-dependent Hamiltonian $H(t)$ to boot, the evolution (2) becomes

$$|\psi,t\rangle_w \equiv \mathcal{T}e^{-\int_0^t dt'\{iH(t')+\frac{1}{4\lambda}\sum_k[w_k(t')-2\lambda A_k]^2\}}|\psi,0\rangle, \tag{8}$$

where $\mathcal{T}$ is the time-ordering operator. With $H = 0$, the probability $\sim_w\langle\psi,t|\psi,t\rangle_w$ is asymptotically non-vanishing only when $w_k(t)$ has its asymptotic value equal to 2λ multiplied by an eigenvalue of A_k, for each k. The collapse is to the eigenstate labeled by these joint eigenvalues.

1.2 CSL

For full-blown CSL, the index k corresponds to spatial position $\mathbf{x}$: $w_k(t) \rightarrow w(\mathbf{x},t)$ is considered to be a physical scalar field. The commuting operators $A_k \rightarrow A(\mathbf{x})$ are taken to be (proportional to) the mass density operator $M(\mathbf{x})$ "smeared" over a region of length a around x. Thus, the *dynamical equation* is

$$|\psi,t\rangle_w \equiv \mathcal{T}e^{-\int_0^t dt'\{iH(t')+\frac{1}{4\lambda}\int d\mathbf{x}[w(\mathbf{x},t')-2\lambda A(\mathbf{x})]^2\}}|\psi,0\rangle, \tag{9}$$

$$A(\mathbf{x}) \equiv \frac{1}{m_0 (\pi a^2)^{3/4}} \int d\mathbf{z} e^{-\frac{1}{2a^2}(\mathbf{x}-\mathbf{z})^2} M(\mathbf{z}). \tag{10}$$

In Eq. (9), m_0 is taken to be the proton's mass, and the choices $\lambda \approx 10^{-16} \mathrm{s}^{-1}$, $a \approx 10^{-5} \mathrm{cm}$, the values suggested by Ghirardi, Rimini and Weber for their Spontaneous Localization (SL) theory ([6]) are taken, although the present experimental situation allows a good deal of latitude [7, 8]. The *probability rule* is, as before,

$$P_w(t)Dw =_w\langle \psi, t | \psi, t \rangle_w \prod_{\mathbf{x},t=0}^{t} \frac{dw(\mathbf{x},t)}{\sqrt{2\pi\lambda/\Delta\mathbf{x}\Delta t}}. \tag{11}$$

Thus, for a state which initially is a superposition of states corresponding to different mass density distributions, ideally (i.e., if one neglects the Hamiltonian evolution, and waits for an infinite time) one state survives under the CSL dynamics. The greater the mass density distribution differences between the states, the more rapid is the collapse rate. When describing the collapse competition between macroscopically distinguishable states, the Hamiltonian evolution can have little effect when it is slow compared to the collapse rate, or when it does not materially affect the mass distribution.

2 Problem's Progress

Paper I discusses a framework for dynamical collapse models begun in the 70's [9, 10]. I listed nine problems which were evident then. Then, SL came along, a well-defined model of instantaneous collapse, which provides a resolution of four problems, but raised one more. CSL, which was stimulated by the earlier work and by SL, provides a (somewhat different) resolution of these five problems. The five problems and their resolutions are:

Interaction problem: what should be the interaction which gives rise to collapse? This is specified in Eqs. (9) and (10).

Preferred basis problem: what are the states toward which collapse tends? They are eigenstates of the (smeared) mass density operator (10).

Trigger problem: how can it be ensured that the collapse mechanism is "off " for microscopically distinguishable states, but "on" for macroscopically distinguishable states? This is resolved in CSL, as in SL, by having the collapse always "on." In CSL, the collapse rate is slow in the microscopic case because the mass density differences are small, and fast in the macroscopic case because the mass density differences are large.

Symmetry problem: how to make the collapse mechanism preserve the exchange symmetry properties of fermionic and bosonic wave functions, which was a problem of SL [11]? This is ensured by the symmetry preserving mass density operator in Eq. (10).

Superluminal problem: how can it be ensured that the collapse dynamics does not allow superluminal communication? Gisin [10] pointed out a necessary condition.

262   P. Pearle

It is that the density matrix $\rho(t)$, evolving from an initial density matrix matrix $\rho(0)$ which can be composed from pure state vectors in various ways, only depend upon $\rho(0)$ and not upon this composition. It is straightforward to see this is satisfied in CSL, since the density matrix, from Eqs. (9) and (11), is

$$\rho(t) \equiv \int Dw P_w(t) \frac{|\psi,t\rangle_w \, _w\langle\psi,t|}{_w\langle\psi,t|\psi,t\rangle_w}$$

$$= \mathcal{T}e^{-\int_0^t dt' \{iH_L(t')-iH_R(t')+\frac{\lambda}{2}\int d\mathbf{x}[A_L(\mathbf{x})-A_R(\mathbf{x})]^2\}}\rho(0) \qquad (12)$$

(the subscripts L or R mean that the operators are to appear to the left or right of $\rho(0)$, and $\mathcal{T}$ time-reverse orders operators to the right). The other necessary ingredient is that the interaction not be long-range. The gravitational and electrostatic interactions are non-local but not long-range. In a relativistic theory, of course, these interactions are local, transmitted with speed c. In a non-relativistic theory, where particles interact via a non-local potential, the best one can expect is the prevention of long-range communication. In CSL, the interaction is via the gaussian-smeared local mass density operator (10), so it is non-local, but it is not long-range.

In the remainder of this paper I shall discuss five problems which remained after the advent of CSL, the *tails, experimental, conservation law, relativity* and *legitimization* problems. They shall be defined when encountered. I shall spend most time on the tails problem, because it was first raised by Abner.

3 Tails Problem

> *With a little bit, with a little bit, ...*
>
>   *My Fair Lady*, A. J. Lerner and F. Loewe

In November 1980, Abner kindly invited me to stay at his home in Wellesley. We discussed various aspects of my dynamical collapse program. In the course of the discussion, Abner expressed the point of view that, in a collapse situation involving macroscopically distinguishable alternatives, one cannot justify saying a definite outcome has occurred if the amplitude of the outcome state is not precisely 1 (i.e., if the amplitudes of the rest of the states—the "tails"—are not precisely zero, no matter how small they are). Outcomes are observed to occur in a finite time, and the framework for collapse models I had developed allowed different models, ones where the tails vanish in a finite time or in an infinite time. When I was looking for a physical principle to enable selection of one model over another, I bought Abner's argument and seized upon this to make a choice ([9], 1985). However, Gisin [10] had a better physical principle, avoidance of the superluminal problem. He proposed a model in which the superluminal problem is avoided, but for which the collapse time is infinite. I showed ([9], 1986) that, generally, solution of the superluminal problem comes with infinite collapse time. So, CSL entails the tails problem.

At a conference in Amherst in June 1990, which was the last time many of us saw John Bell, I remarked in an open session at the end of the conference that I had

How Stands Collapse II   263

previously phrased the tails situation in CSL, quite poetically I had thought, as "a
little bit of what might have been is always present with what is," at which point
John frowned. But, I went on, I had learned from him not to say this, for one should
not express a new theory in an old theory's language, at which he beamed.

John died on October 1, 1990. At a memorial session at the end of that month,
Abner, GianCarlo and I gave talks [12, 13] about dynamical collapse, which had
been championed by John as a conceptually clear alternative to standard quantum
theory. Abner's talk was entitled "Desiderata for a Modified Quantum Mechanics."
A number of his desiderata involved the tails issue, raising the question as to whether
CSL is indeed conceptually clear, in particular:

> ... it should not permit excessive indefiniteness of the outcome, where "excessive" is defined
> by considerations of sensory discrimination ... it does not tolerate "tails" which are so broad
> that different parts of the range of the variable can be discriminated by the senses, even if
> very low probability amplitude is assigned to the tail.

A decade ago, in a festschritt for Abner, GianCarlo and Tullio Weber [14] and
I [15] gave responses to Abner's position (as did Sohatra Sarkar [16], who adopted
it)—see also the lucid paper of Albert and Loewer [17]. The problem, in a collapse
theory with tails, is to provide a well-defined criterion for the existence of possessed properties of macroscopic variables which coincides with the evidence of, in
Abner's words, "sensory discrimination."

3.1 Smeared Mass Density Criterion

Ghirardi and co-worker's response is based upon the smeared mass density (SMD)
whose operator is $A(\mathbf{x})$ (Eq. (10)). For a state $|\psi\rangle$, their criterion for the SMD at $\mathbf{x}$ to
have a possessed value (or, in their language, "accessible" value) is when the ratio
$\mathcal{R}(\mathbf{x})$ of variance of $A(\mathbf{x})$ to $\langle\psi|A(\mathbf{x})|\psi\rangle^2$ satisfies $\mathcal{R}(\mathbf{x}) \ll 1$: then one identifies
the possessed value of the SMD with $\langle\psi|A(\mathbf{x})|\psi\rangle$.

In measurement situations, because of CSL dynamics, the possessed SMD value
criterion very rapidly becomes consistent with our own observations of SMD, for
macroscopic objects. For microscopic objects, e.g., in regions where only a few particles are cavorting, the SMD does not have a possessed value but, as Abner stressed,
the point of the criterion is to serve to compare the theory with our macroscopic
experience.

However, as Ghirardi et al. point out, for a macroscopic object in a superposition of two locations, after a short time undergoing CSL evolution, $\mathcal{R}(\mathbf{x}) \gg 1$ in
the region where the object in the tail is located, so the SMD does not have a possessed value there: one would wish for the value 0. (This presumes there is no air
in the region; when air at STP is present, the SMD possesses a value in agreement
with experience, the air density). Nonetheless, although the criterion fails there,
$\langle\psi|A(\mathbf{x})|\psi\rangle \ll m_0/a^3$ in that region, which is consistent with the experienced value
0. Another place where the criterion fails is in neither location, where there is no
mass density, since $\mathcal{R}(\mathbf{x}) = 0/0$.

One would like the criterion for the SMD to be possessed to include these cases, since zero mass density is, in principle, a macroscopic observable. Although the authors do not give one, it is easy to obtain: the SMD possesses the value $\langle\psi|A(\mathbf{x})|\psi\rangle$ if either $\mathcal{R}(\mathbf{x}) \ll 1$ OR $\mathcal{R}(\mathbf{x} \gg 1$ but $\langle\psi|A(\mathbf{x})|\psi\rangle \ll m_0/a^3$ OR $\langle\psi|A(\mathbf{x})|\psi\rangle = 0$. There still is an ambiguity as to how small is $\ll 1$, which I shall try to make precise later, in the context of my own response to Abner's challenge.

As I wrote to GianCarlo and Abner, I regard this as an elegant answer to the question: "What is the *minimum* structure which will allow one to attribute *macroscopic* reality?" I addressed, and will address here a different question: "What is the *maximum* structure which will allow one to attribute reality, *both macroscopic and microscopic?*"

3.2 Qualified Possessed Value Criterion

Rather than reprise my previous argument, I wish to take this opportunity to make it more simple and general. My point of view is that a collapse theory is different from standard quantum theory and, as I said to John Bell in Amherst, therefore requires a new language, conceptual as well as terminological.

The second sentence of Abner's desideratum quoted above utilizes some words and concepts which, while appropriate for standard quantum theory, are inappropriate for CSL: at the end of this discussion, I shall be more specific. But, in the first sentence, Abner was absolutely right: a conceptually sound collapse theory with tails must allow an interpretation which provides no "indefiniteness of the outcome" and that what is crucial to characterize the definite outcome, are "considerations of sensory discrimination."

The new language I propose devolves upon the meaning of the words *correspond* and *possess* which, to emphasize their importance, I shall irritatingly continue to italicize. For expository reasons, I shall first review the use of these words in classical and standard quantum physics, before addressing their use in a dynamical collapse theory.

3.2.1 Classical Theory Language

In classical physics, to a physical state of a system *corresponds* its "mathematical descriptor" (e.g., a vector in phase space for a mechanical system) and, *corresponding* to either, every variable *possesses* a value.

When one is in ignorance about the physical state, then every variable *possesses*, not a value but, rather, a probability distribution of values. However, these *possessed* entities *correspond* to one's state of ignorance of the physical state of the system, not to the (unknown, but existing) physical state of the system.

How Stands Collapse II   265

3.2.2 Standard Quantum Theory Language

With the advent of quantum phenomena, physicists (especially Bohr) tried to maintain as much classical language as possible. But something had to give. What gave is the correspondence of the physical state of the system to the mathematical descriptor, the state vector.

For a microsystem, the notion of *possessed* value of a variable is preserved by the so-called "eigenstate-eigenvalue link": a variable has a *possessed* value only if the operator *corresponding* to the variable has the state vector as an eigenstate, and then the *possessed* value is the eigenvalue. But, generally, only for very few state vectors can a useful variable can be found which has a *possessed* value. Even for a system of modest complexity, for the overwhelming majority of state vectors which describe it, variables which have *possessed* values are of limited interest, e.g., the projection operator on the state itself.

For a macrosystem, the precisely applied eigenstate-eigenvalue link does not work. For example, for such variables as the center of mass position of a meter needle, the location of the ink in a symbol on a computer printout, or the excited state of a radiating pixel on a computer screen, a reasonable state vector *corresponding* to an observed physical state is not an eigenstate of the *corresponding* operator. However, if the wave function (the projection of the state vector on a basis vector of the operator) in some sense has a narrow range, one may try to adopt some criterion to assign a "near" *possessed* value to the variable, a value within the range [13, 15, 17].

By a preparation or a measurement, i.e., a judicious coupling of a microsystem to a macrosystem, one can force a microsystem to change its physical state to a more desirable one. Initially, the physical state *corresponds* to microscopic variables which *do not* have *possessed* values and macroscopic variables which *do* have (near) *possessed* values. Afterwards, the physical state *corresponds* to microsystem and macrosystem variables, which both *do* have *possessed* values or near *possessed* values.

The problem, of course, is that the state vector corresponding to the physical state is not produced by the theory. Schrödinger's equation evolves the initial state vector into a state vector where neither microscopic nor macroscopic variables have *possessed* values. One might regard the evolved state vector as a sum (superposition) of state vectors, each of which corresponds to a different possible physical state.

Because the evolved state vector is not the descriptor of the state of the evolved physical system, there are various positions taken, within the framework of standard quantum theory to make sense of this situation.

One position is to try to maintain the *correspondence* between the state of the physical system and the state vector by introducing the collapse postulate. To try to select the possible physical states, the collapse results, out of the superposition, there may be pressed into service a (near) *possessed* value criterion for the macroscopic variables, or properties of a distinguished part of the physical system, the "environment," may be relied upon. However, these criteria are ad hoc: for each different situation they require different knowledge outside the theory. Sometimes

266   P. Pearle

selections made can be quite arbitrary e.g., when the superposition of states is a continuum [18]. Indeed, the collapse postulate itself is also ill-defined [19] with regard to when and under what circumstances to apply it.

Another position is to regard quantum theory solely as a theory of measurement [20], and the state vector as a calculational tool. Thus, Heisenberg considered the state vector of a microsystem to be the repository of "potentia," the capability to describe potential outcomes of future experiments. Schrödinger [21], in discussing this position (with which he was not comfortable), called the state vector which evolves after a measurement the "expectation catalog," in the sense that it tells one what to expect. To pluck out the macroscopically distinguishable alternatives from the catalog, again one utilizes the (near) *possessed* value criterion, informed by the experimental situation. The ambiguity of when to apply it is of no concern: it is any time after the measurement is completed. The circumstances of application are limited to experimental situations: although what that means is ill-defined, that is also of no concern to people who take such a pragmatic view of the purpose of quantum theory.

Suppose one takes this position, or adopts the ensemble interpretation, the position that it is an ensemble of physical states which *corresponds* to the state vector [19, 20]. One thus gives up the idea of the *correspondence* of the state of the physical system to the state vector. If one also believes, as did Bohr, that standard quantum theory cannot be improved upon, one thereby gives up the possibility of the physical system's state having any kind of mathematical descriptor. Bell was moved to say that adoption of this position "is to betray the great enterprise" [3]. At the least, it certainly is a great break with classical physics ideas.

A position which does not make that break is the "histories" program [23]. Here, the state of a physical system *corresponds* to a mathematical construct different from the state vector, the so-called "decoherence functional." Utilizing standard quantum theory structures, the hope is to have the decoherence functional correspond to variables which occasionally have *possessed* values.

In all these cases, one is ignorant of the outcome of an experiment. Thus, just as in classical physics, *corresponding* to one's state of ignorance of the physical state, a viable variable *possesses* a probability distribution of values.

For these positions, how is a tails situation treated? Suppose a state vector evolving in a measurement situation becomes a superposition of two states whose ratio of amplitudes is enormous. Suppose also that the values *possessed* by a macroscopic variable characterizing these states, in Abner's words, "can be discriminated by the senses even if very low probability is assigned to the tail." This state vector is interpreted as describing a two-outcome measurement, albeit one outcome is much less likely than the other. (In the histories scheme, which does not use a state vector, a similar interpretation arises.) When this situation arises in CSL, one needs a different conclusion, that this state vector describes a one-outcome experiment. This requires a new language.

How Stands Collapse II

3.2.3 Dynamical Collapse Theory Language

CSL retains the classical notion that the physical state of a system *corresponds* to the state vector. *Corresponding* to a random field $w(\mathbf{x},t)$ whose probability of occurrence (11) is non-negligible, the dynamics always evolves a realizable state. Therefore, one is freed from requiring the (near) eigenstate-eigenvalue link criterion for the purpose of selecting the realizable states. I suggest that the eigenstate-eigenvalue link criterion be subsumed by a broader concept. It must be emphasized that this new conceptual structure is only applicable for a theory which hands you macroscopically sensible realizable states, not superpositions of such states.

In the new language, corresponding to a quantum state, *every variable possesses a distribution of values*, defined as follows.

If the normalized state is $|\psi\rangle$, consider a variable *corresponding* to the operator B, with eigenvalues b. Denote the eigenvectors $|b,c\rangle$, where c represents eigenvalues of other operators C which commute with B, all comprising a complete set. The variable's *possessed* distribution is defined to be $Tr_c|\langle b,c|\psi\rangle|^2$ (Tr_c represents the trace operation over C's eigenstates). One may generalize this to say that the set of variables corresponding to the complete set of commuting operators *possesses* a joint distribution $|\langle b,c|\psi\rangle|^2$.

What does it mean to say that a variable *possesses* a distribution? I am never sure what it means to ask what something means[1], except that it is a request for more discourse.

I choose to call this a distribution, *not* a probability distribution, even though it has all the properties of a probability distribution. This is because, in classical physics, a probability distribution is what *corresponds* to a state of ignorance, and that is not the case here. What is it a distribution of, if not probability? Following [15], one may give the name "stuff" to a distribution's numerical magnitude at each value of the variable, as a generalization of Bell's quasi-biblical characterization [3], "In the beginning, Schrödinger tried to interpret his wavefunction as giving somehow the density of the stuff of which the world was made."

One is encouraged to think of each variable's stuff distribution as something that is physically real. The notion allows retention of the classical idea that, for a physical state, *every* variable *possesses* an entity. What is different from classical ideas is that the entity is not a number. One may think of this difference as an important part of what distinguishes the quantum world picture from the classical world picture.

But, the distribution notion also differs from standard quantum theory, where one is precluded from thinking of simultaneous values of complementary variables.

---

[1] This brings to mind the "shaggy dog" story of a man who is driven to find the meaning of life. After incredible hardship over many years (which takes a long time to recount, but which will be forgone here) he reaches a high mountain in India and obtains an audience with a Guru whom he has been told is the wisest man on earth and who can answer his question. He asks "Oh sir, what is the meaning of life," and the Guru answers serenely, "My son, life is a bridge." At this the man jumps up, visibly upset, and recounts the incredible hardships he has endured (again taking a long time to recount, but which will be forgone here) ending with, "And, after all that, what I get is this lousy answer, that life is a bridge? To which the Guru also jumps up, visibly upset, and says "Y-y-you mean, life isn't a bridge?"

In the present view, simultaneously, every variable *possesses* its stuff distribution. Complementarity here means that variables whose operators don't commute do not *possess* joint distributions, but they do jointly *possess* distributions.

Here are a few simple examples.

If B is the position operator of a particular particle, one may think of the associated position-stuff as representing something real flowing in space. If the particle undergoes a two-slit interference experiment, something real is going through both slits and interfering. Likewise, for the particle's momentum operator, real momentum-stuff also flows in momentum space. The "something real" can be stuff for any variable represented by an operator function of position and momentum, and all these are *possessed* simultaneously.

If B is the operator representing spin in the $\hat{n}$ direction of a spin-1/2 particle, one may think of the $\hat{n}$-spin variable as *possessing* something real, $\hat{n}$-spin-stuff *corresponding* to both values $+\hbar/2$ and $-\hbar/2$, in varying amounts. Just as in classical physics where a spinning object has a projection of angular momentum on each direction, and all those values are simultaneously *possessed*, the particle state *corresponds* to variables for all directions, all of whose spin-stuff distributions are simultaneously *possessed*. There is one direction, $\hat{m}$, in which the $\hat{m}$-spin-stuff distribution has magnitude 1 at value $+\hbar/2$ and magnitude 0 at value $-\hbar/2$. In this case, one can use the language that the $\hat{m}$-spin *possesses* the value $+\hbar/2$.

3.2.4 Qualified Possessed Value

A criterion is needed for when it is appropriate to promote a macroscopic variable's *possessed* stuff distribution to a *possessed* value. This must be done in order to compare the theory with observation, since observers insist that macroscopic variables *possess* values. We shall follow Abner's insightful recourse to "sensory discrimination," as well as take sustenance from a remark in a recent article on the Federal Reserve in The New Yorker [24]: "As social scientists have long recognized, we prefer confident statements of fact to probabilistic statements... ." Here are two probabilistic considerations.

The first consideration is that an observer's quotation of a *possessed* value of a macroscopic variable, such as location, velocity, rotation, trajectory, color, brightness, length, hardness, ..., is not sufficient. It should contain an error bar. Such a qualification can readily be supplied, although usually it is not. Thus, CSL need only present a *possessed* value prediction within the observer's supplied error bar, to favorably compare with the observer's value.

The second consideration is that, when an observer makes a "confident statement" about the *possessed* value of a macroscopic variable (plus error bar), it needs to be qualified in another way. If this is to be compared with the theory, there is the implication that anyone who observes this variable will quote the same value. This is a prediction, an assertion about the observations of other observers in similar circumstances, and so it requires qualification by providing a measure of the confidence one may give to the assertion or, alternatively, to its falsification.

How Stands Collapse II   269

For example, one might confidently say that all observers will see that lamp is on the table, all observers will see that board's thickness is 0.75 ± 0.01", all observers who toss 100 coins will see them not all come up heads, all observers who spill water on the floor will not see it jump back up into the glass, all observers will see that a particular star is in the heavens, all observers of me today can see me tomorrow, etc. However, each statement is not absolutely sure, and each should be qualified by giving the probability of its falsification, although sometimes that is not so easy to estimate.

In summary, a statement about an observed variable, should be characterized by three numbers, a possessed value, the error bar associated with that value, and the probability the statement of value plus error bar is false. We shall use the latter two numbers in conjunction with a macroscopic variable's stuff distribution, to obtain a criterion for assigning a *possessed* value to the macroscopic variable, for comparison with the first number.

From the theory, for the state vector of interest, take the stuff-distribution *possessed* by the macroscopic variable of interest, graphed as stuff versus variable value. From the observation, take the error bar and slide it along the variable value axis until the maximum amount of stuff lies within the error bar. (If the variable has a continuous range of values, and Δ is the error bar, this condition is simply

$$\frac{\partial}{\partial b} \int_{b-\frac{1}{2}\Delta}^{b+\frac{1}{2}\Delta} db' Tr_c |\langle b',c|\psi\rangle|^2 = Tr_c |\langle b+\frac{1}{2}\Delta,c|\psi\rangle|^2 - Tr_c |\langle b-\frac{1}{2}\Delta,c|\psi\rangle|^2 = 0.$$

If the amount of stuff outside the error bar is less than the probability of falsification, then the criterion is met, and we shall say that the macroscopic variable has a *qualified possessed value*.

That value is found, first, by dividing the variable's distribution by the amount of stuff within the error bar. The resulting "renormalized" distribution is restricted to the error bar range, so that the renormalized amount of stuff within the error bar $= 1$. The *qualified possessed value* is defined as the mean value of the variable calculated with this renormalized distribution. This *qualified possessed value* is what is to be compared with the observed *possessed* value, in order to test the validity of the theory.

(An alternative is to simply use the variable's unrenormalized distribution to calculate the mean, and call this the variable's *qualified possessed value*, if it lies within the error bar. However, even if the tail amplitude is very small, the variable's value at the tail could be so large that it makes a significant contribution to the mean, putting it outside the error bar, which is why this alternative might not produce a *qualified possessed value* which agrees with observation, in circumstances where it ought).

3.2.5 Comparison with Observation

Consider a simple example, a wave packet modeled by a sphere of mass density 1g/cc and radius 10^{-4} cm. Suppose the variable of interest is the center of mass position of the sphere. According to CSL [7], its center of mass wave packet achieves an equilibrium width of $\approx 10^{-8}$ cm in about 0.6 s, due to the competition between spreading caused by the Schrödinger evolution and contracting caused by the collapse evolution. Suppose the wave packet has that equilibrium width.

Suppose somehow the particle is put into a superposition of two states of equal amplitude, where the centers of mass are further apart than the radius. According to CSL, the collapse rate $R \approx \lambda \times$ (number of nucleons within a volume a^3) $\times$ (number of nucleons within the sphere). Thus, $R \approx 10^{-16} \times 6 \cdot 10^8 \times 2.5 \cdot 10^{12} = 1.5 \cdot 10^5 \text{s}^{-1}$. Since the tail's squared amplitude $\sim \exp{-Rt}$, when 1 ms has passed, this is $\approx \exp{-150}$ (and is overwhelmingly likely to be rapidly going down). Therefore, after 1 ms, the typical state describes a center of mass stuff distribution which consists of a packet *corresponding* to squared amplitude ≈ 1 and width $\approx 10^{-8}$ cm at one location and squared amplitude $\leq \exp{-150}$ and width $\approx 10^{-8}$ cm at the other location.

We wish to know whether a *qualified possessed value* of the center of mass exists, according to the criterion and, if so, if it agrees with what an observer would say. An observer sees the sphere at one location. "See," is meant literally: observers use optical light. We thus can conservatively assign a light wavelength-restricted error bar of $\approx 10^{-5}$ cm.

Moreover, we believe that, in all of human history, all observers in like circumstances would see the same thing. However, we cannot be absolutely sure of this belief—it hasn't been tested, and can't be. This suggests that the measure of falsification is not larger than the following stringent estimate. If all the homo sapiens who have ever lived, an upper estimate of $\approx 10^{11}$ people, were each to spend their whole lives (upper estimate of 100 years $\approx 3 \cdot 10^{12}$ ms) doing nothing else but observing such a sphere every millisec ($<<$ human perception time of ≈ 100 ms), that in such circumstances only one person once might report seeing something else. This amounts to a probability of falsification of $\approx 1/[10^{11} \times 3 \cdot 10^{12}] = 3 \cdot 10^{-24} \approx \exp{-54}$.

The *qualified possessed value* criterion is met. The error bar of 10^{-5}cm is much larger than the 10^{-8}cm spread of the center of mass wave packet. Essentially all the stuff at the location *corresponding* to squared amplitude ≈ 1 can be considered to be within the error bar e.g., if the center of mass wave function is $\sim \exp{-r^2/(10^{-8})^2}$, this has the value $\exp{-10^6}$ at $r = 10^{-5}$cm. Therefore, the amount of stuff outside the error bar is $\exp{-150}$, solely due to the tail. It is much less than the probability of falsification: $\exp{-150} << \exp{-54}$. Thus, the theory's assignment of *possessed* center of mass location agrees with the observer's assignment. For a larger object than a mote of dust, it would be satisfied even more easily.

More generally, CSL can be applied to the state of an arbitrarily large fraction of the universe (idealized as isolated), in principle even up to the universe itself. The physical system should be describable by macroscopic variables with *possessed* values all over space, even if no observer is there. For the picture given by CSL, it is helpful [15] to probe the *corresponding* state vector with operators representing density variables of every sort: density of various elementary particle types (i.e., proton, neutron, electron, photon, etc.), density of bound state types, (i.e., nucleii or atoms), density of mass, momentum, velocity, angular momentum, energy,..., inte-

How Stands Collapse II   271

grated over a conveniently sized volume. For each spatial location of the volume, each such variable will *possess* its distribution. Because the CSL collapse mechanism rapidly collapses to states where macroscopic objects are well localized, as one moves the probe volume over the space, one recognizes locations where the variable's distribution exhibits the behavior discussed above, a narrow width packet of total squared amplitude very close to 1, and a small tail. Thus, one can assign *qualified possessed values* to these variables, and so build up a picture of the macroscopic structure of the system described by the state vector.

3.2.6 Desideratum Revisited

I believe the second sentence in Abner's desideratum,

> ... it does not tolerate "tails" which are so broad that different parts of the range of the variable can be discriminated by the senses, even if very low probability amplitude is assigned to the tail.,

in referring to the nature and amplitude of a tail state, uses language appropriate for a quantum theory of measurement, but inappropriate for CSL, which is a quantum theory of reality.

Consider the example of a state vector which is a superposition of two macroscopically distinguishable states, a "dominant" state with squared amplitude $1 - \varepsilon$ and an orthogonal tail state of extremely small squared amplitude ε. According to standard quantum theory, if somehow a measurement of this state can be made in the future (for it is possible in principle, but generally not in practice, to measure a superposition of macroscopically distinguishable states), ε is the probability that the result will correspond to the tail state. Since repeated measurements do not *always* yield the dominant state, in a theory where 100% reproduceability of measurement results is the criterion for assigning values to variables, one cannot say that the state vector corresponds to the dominant state.

In CSL, the tail state and its squared amplitude represent something rather different than a possible outcome of a future measurement and its probability. The tail state represents an unobservably small amount of stuff which allows describing the state vector by (qualified) possessed values assigned to macroscopic variables, consistent with the dominant state.

The role of a tail state's squared amplitude in CSL is best understood by considering the gambler's ruin game analogy to the collapse process. This was described in paper I but, for completeness, here is a brief recapitulation, in the context of our example. Two gamblers correspond to the two states. They toss a coin, which corresponds to the fluctuating field. They exchange money, depending upon the toss outcome, and their net worth fluctuations correspond to fluctuations of the squared amplitudes. A result is that a gambler who possesses a fraction ε of the total money has the probability ε of eventually winning all the money. In particular, even if ε is extremely small, so one of the gamblers has almost lost all his money, it still is possible that a highly improbable sequence of coin tosses favorable to that gambler can occur, which completely reverses the two gambler's fortunes.

Analogously, for our example, this means that the dominant state and the tail state have the probability ε of spontaneously changing places, what I call a "flip." What does this imply about the picture of nature provided by the theory?

It means that there is a highly improbable possibility that nature, "on a whim" (i.e., by choosing an appropriate field $w(\mathbf{x}, t)$ for a sufficient time interval), can change the universe to a different universe. In either universe, macroscopic objects have (qualified) possessed values of macroscopic variables.

Note that such a flip is not triggered by a "measurement" by anybody: it is something that can happen spontaneously, at any time. But, consider a flip, by nature's whim, occurring right after a measurement with two possible outcomes, where the state vector is as described above. Before the flip, the universe contains an observer who is sure that result 1 has occurred, and the (qualified) possessed values of macroscopic variables all concur. After the flip, the universe contains an observer who is sure that result 2 has occurred, and the (qualified) possessed values of macroscopic variables all concur.

To summarize, in the quantum theory of measurement, because one only has the eigenstate–eigenvalue link as a tool for assigning reality status, one must conclude that a state vector with a tail *cannot* be assigned a reality status consistent with the dominant state. In CSL, where the dynamics and the (qualified) possessed value criterion are what allows assigning reality status, one concludes that the state vector with a tail *can* be assigned a reality status consistent with the dominant state. There is no problem here, before or after the flip, with assigning a reality status and reconciling an observer's observations with the theory.

Then, what, in CSL, corresponds to the difficulty faced by the quantum theory of measurement? The difficulty belongs, not to an observer within the universe, but to some hypothetical being outside the universe (a theoretical physicist?) who keeps track of its state vector. This being cannot say with 100% certainty that the realistic universe with a certain history may not at some future time be replaced by another realistic universe with a somewhat different history. Observers within the universe will be oblivious to this (highly improbable) possibility. And, the theory describes their observations.

Although I have argued here against Abner's position, I find impressive his insight, a quarter of a century ago, that the tails issue is key to an understanding of important interpretational implications of a dynamical collapse theory.

4 Experimental Problem

CSL is a different theory than standard quantum theory, and so makes different predictions in certain situations. The problem is to find and perform experiments which test these predictions, with the ultimate goal of either refuting or confirming CSL vis-a-vis standard quantum theory.

Perhaps the quintessential experimental test involves interference [25, 26]. Suppose an object undergoes a two slit interference experiment. According to CSL,

How Stands Collapse II 273

once there are two spatially separated packets which describe the center of mass exiting the separated slits, they play the gambler's ruin game and their amplitudes will fluctuate. Thus, when the packets are brought together once more and their interference is observed, the pattern which results from repeated measurements is predicted to have less contrast (be "washed out") as compared to the prediction of standard quantum theory. Indeed, if the packets are separate long enough so that one packet is dominant, the interference pattern essentially disappears.

The largest objects so far undergoing interference experiments are C^{60} and C^{70}(fullerene or buckyball) [27]. These experiments involved a diffraction grating, so one may visualize a superposition of wave packets emerging from each slit and thereafter all pairs of packets simultaneously compete in the gambler's ruin game. The off-diagonal elements of the density matrix between two such packet states decay just as do those for two-slit interference. The decay factor can be obtained from Eqs. (10) and (12) (with the slit size and therefore the packet size less than a): it is $\exp -\lambda t n^2$, where n is the number of nucleons in the molecule ($n = 720$ for C^{60}). The time of flight of a C^{60} was about 0.05 s and, if one takes the agreement of the observed diffraction pattern with standard quantum theory's prediction to be of 1% accuracy, this places the limit $\lambda \times .05 \times 720^2 < .01$, or $\lambda^{-1} > 10^6$. A recent proposal [28] to test dynamical collapse, involving the superposition of a mirror in states undisplaced and displaced, has the capability of pushing this limit to $\lambda^{-1} > 10^{10}$ [29]. Thus, at present, interference experiments have only had a mild impact on CSL.

The only experiments which, so far, have had an important impact upon CSL, look for "spontaneous" increase in particle energy. It is these experiments which have strongly suggested that a viable CSL must have the mass-density proportional coupling given in Eq. (10).

Because collapse narrows wave packets, this leads to momentum increase by the uncertainty principle, and therefore energy increase, of all particles. According to Eqs. (10) and (12), independently of the potential, the average rate of increase of energy is

$$\frac{d\overline{E}_A}{dt} = \sum_k \frac{3\lambda \alpha_k^2 n_k \hbar^2}{4 m_k a^2}, \tag{13}$$

for any state describing n_k particles of type k and mass m_k ($\alpha_k \equiv (m_k/m_0)$). However, the SL model, and CSL following it, initially assumed that all particles had the same collapse rate, so that $\alpha_k = 1$.

More generally, assume that $\alpha_p = 1$ for the proton and α_k is unknown for other particles. Eq. (13) is an average: CSL predicts that, occasionally, a particle can get a large excitation, which could be detected if a large enough number of particles is observed for a long enough time.

One can find the probability/sec of a transition from an initial bound state to a final state, from Eq. (12) expanded in a series in the size of the bound state divided by a. With the effect of the center of mass wavefunction integrated out, denoting the initial bound state $|\psi_0\rangle$ and the final state $|\psi_f\rangle$ (bound or free), where these states are eigenstates of the center of mass operator with eigenvalue 0, the transition rate is [30]

$$\frac{dP}{dt} = \frac{\lambda}{2a^2} |\langle \psi_f | \sum_{j,k} \alpha_k \mathbf{r}_{jk} | \psi_0 \rangle|^2 + o(\text{size}/a)^4, \tag{14}$$

where $\mathbf{r}_{jk}$ is the position operator of the jth particle of kth type. Interestingly, if $\alpha_k \sim m_k$, the matrix element of the center of mass operator appears in (14), which vanishes. Then, dP/dt depends upon the much smaller $o(\text{size}/a)^4$ term.

For this reason, experiments which put an upper limit on spontaneous excitation from bound states of atoms or nucleii can constrain the ratios of α_k's to be close to the ratios of masses.

An experiment, which looks for unexplained radiation appearing within a $\approx 1/4$ kg slab of germanium [31] over a period of about a year, has been applied to a putative CSL ionization of a Ge atom by ejection of a 1s electron [32]. Such an excitation should yield a pulse of radiation, 11.1 keV from photons emitted by the other electrons in the atom as they cascade down to the new ground state plus the kinetic energy of the ejected electron deposited in the slab. The probability to ionize the atom is calculated and compared with the experimental upper limit on pulses above 11.1 keV. The result at present is $0 \leq \alpha_e/\alpha_N \leq 13 m_e/m_N$, where the subscripts e, N refer to the electron and nucleon (proton and neutron parameters are assumed identical).

In the Sudbury Neutrino Observatory experiment [33], solar neutrinos can collide with deuterium in a sphere 12 m in diameter. The result is dissociation of deuterium. (Thereafter, the released neutron, thermalized by collisions, bonds with a deuterium nucleus to form tritium, releasing a 6.25 MeV gamma which then Compton scatters from electrons which emit Cerenkov radiation detected by photodetectors bounding the sphere). The experiment took data for ≈ 254 days, and the observed number of deuterium nucleii was $\approx 5 \times 10^{31}$. The predicted result, using the standard solar model with neutrino oscillations and the neutrino-deuterium dissociation cross-section, agreed well with the experimental result, within an error range. Taking this error range as representing an upper limit to CSL excitation of the deuteron, the result is $\alpha_n/\alpha_p = m_n/m_p \pm 4 \times 10^{-3}$ [34] (note, $4 \times 10^{-3} \approx 3(m_n - m_p)/(m_n + m_p)$).

These results make plausible the use of the mass density as the discriminating operator in CSL, $\alpha_k = m_k/m_0$. The rate of energy increase (13) is thus quite small, e.g., over the 13.7×10^9 year age of the universe, with the SL values for λ and a, a single particle acquires energy $E \approx 1.3 \cdot 10^{-16} m_k c^2$.

Steve Adler [8] has discussed a number of experiments which could reveal CSL collapse behavior, were λ to be substantially larger, than the SL value, say by a factor of 10^6 or more. I know of only one experimental proposal at present [7], which appears to be currently technically feasible, which could test CSL with the SL value of λ.

The idea is that a small sphere will undergo random walk due to CSL [35]. The expansion of the center of mass wave packet due to Schrödinger dynamics is counteracted by the contraction of the wave packet due to CSL dynamics, which results in an equilibrium size for the wave packet. However, since a collapse contraction can occur anywhere within the wave packet, the center of the packet jiggles about.

How Stands Collapse II   275

Actually, the proposal is rather to observe the random rotation of a small disc: the mechanism is similar to that discussed above. The disc, charged and made of metal, could be suspended and maintained on edge in a Paul trap (an oscillating quadrupole electric field) or, as suggested by Alain Aspect (private communication), a dielectric disc suspended by laser tweezers might be feasible.

It is a consequence of (12) that the ensemble average rms angular deflection of the disc is $\Delta\Theta_{CSL} \approx (\hbar/ma^2)(\lambda ft^3/12)^{1/2}$ (f is a form factor of order 1, depending on the disc dimensions). For a disc of radius 2×10^{-5} cm and thickness 0.5×10^{-5} cm, $\Delta\Theta_{CSL}$ diffuses through 2π rad in about 70 s. For comparison, according to standard quantum theory, $\Delta\Theta_{QM} \approx 8\hbar t/\pi m R^2$ which, in 70 s, is about 100 times less than $\Delta\Theta_{CSL}$. For example, at an achieved low pressure of 5×10^{-17} Torr at liquid helium temperature [36], the mean time between gas molecule collisions with the disc is about 45 min, allowing for even a diffusion of the magnitude of $\Delta\Theta_{QM}$ to be observable.

I hope that someone interested in testing fundamental physics will undertake this experiment.

5 Conservation Law Problem

The problem here is that the collapse process appears to violate the conservation laws. For example, as discussed in the previous section, particles gain energy from the narrowing of wave functions by collapse. The resolution is that the conservation laws *are* satisfied when not only the particle contributions, but also the contributions of the $w(\mathbf{x},t)$ field to the conserved quantities are taken into account. The easiest way to see this is to take a detour which is interesting in its own right.

The detour is to discuss a way to quantize the $w(\mathbf{x},t)$ field, and obtain an ordinary Hamiltonian evolution which is mathematically completely equivalent to CSL [37–40]. For this reason (and because I like the alliteration) I call it a "Completely Quantized Collapse" (CQC) model although, as will be seen, strictly speaking, this is not what is usually considered as a collapse model. But, then, it is easy to identify the space-time and rotation generators as conserved quantities, as is usual in a Galilean-invariant quantum theory, and then extract from them the contributions of the classical $w(\mathbf{x},t)$ field in CSL.

5.1 CQC

Define the quantum fields

$$W(x) \equiv \frac{\lambda^{1/2}}{(2\pi)^2} \int d^4k [e^{ik\cdot x}b(k) + e^{-ik\cdot x}b^\dagger(k)], \tag{15}$$

$$\Pi(x) \equiv \frac{i}{2\lambda^{1/2}(2\pi)^2} \int d^4k [-e^{ik\cdot x}b(k) + e^{-ik\cdot x}b^\dagger(k)], \tag{16}$$

where x is a four vector, $k \cdot x \equiv \omega t - \mathbf{k} \cdot \mathbf{x}$ and $[b(k), b^\dagger(k')] = \delta^4(k - k')$. It is readily verified that $[W(x), W(x')] = 0$, $[\Pi(x), \Pi(x')] = 0$ (the negative energy contribution to these commutators cancels the positive energy contribution) and $[W(x), \Pi(x')] = i\delta^4(x - x')$.

Thus, although $W(x)$ is a quantum field, its value can be simultaneously specified at all space-time events, just like a classical field. At the space-time event x, a basis of eigenstates of $W(x)$ can be constructed: $W(x)|w\rangle_x = w|w\rangle_x$, where $-\infty < w < \infty$. Using these, a basis $|w(x)\rangle \equiv \prod_x |w\rangle_x$ of eigenstates of the operator $W(x)$ at all events can be constructed, where the eigenstate $|w(x)\rangle$ can have any eigenvalue at any x, and so is labeled by a white noise "function" $w(x)$. (For later use, define $|w(x)\rangle_{(a,b)} \equiv \prod_{x,t=a}^{t=b} |w\rangle_x$, with $|w(x)\rangle = |w(x)\rangle_{(-\infty,\infty)}$).

If the "vacuum" state $|0\rangle$ is defined by $b(k)|0\rangle = 0$, it follows from (15), (16) that

$$\langle w(x)| [W(x) + 2i\lambda \Pi(x)]|0\rangle = [w(x) + 2\lambda \frac{\delta}{\delta w(x)}]\langle w(x)|0\rangle,$$

so

$$\langle w(x)|0\rangle = \exp^{-\frac{1}{4\lambda} \int_{-\infty}^{\infty} d^4 x w^2(x)}, \tag{17}$$

with the notation $\int_a^b d^4 x \equiv \int_a^b dt \int_{-\infty}^{\infty} d\mathbf{x}$.

If $|\psi, 0\rangle|0\rangle$ is the initial state, where $|\psi, 0\rangle$ is the initial particle state, define the evolution in the interaction picture as

$$|\Psi, t\rangle \equiv T e^{-2i\lambda \int_0^t d^4 x' A(x') \Pi(x')} |0\rangle |\psi, 0\rangle \tag{18}$$

so, from (17), (18),

$$\langle w(x)|\Psi, t\rangle = C(t) T e^{-\frac{1}{4\lambda} \int_0^t d^4 x' [w(x') - 2\lambda A(x')]^2} |\psi, 0\rangle, \tag{19}$$

where $C(t) = \exp -(4\lambda)^{-1} [\int_{-\infty}^0 + \int_t^\infty] d^4 x' w^2(x')]$ and $A(x)$ is the Heisenberg picture operator, $A(x) \equiv \exp(iH_A t) A(\mathbf{x}) \exp(-iH_A t)$ (H_A is the particle Hamiltonian).

The expression in Eq. (19), apart from the factor $C(t)$, is the CSL interaction picture statevector $|\psi, t\rangle_w$, corresponding to the Schrödinger picture Eq. (9). Thus it follows from (19) that $|\Psi, t\rangle$ may be written as

$$|\Psi, t\rangle = |\chi\rangle \int Dw_{(0,t)} |w(x)\rangle_{0,t} |\psi, t\rangle_w \tag{20}$$

where $|\chi\rangle \equiv \int Dw_{(-\infty,0)} Dw_{(t,\infty)} C(t) |w(x)\rangle_{-\infty,0} |w(x)\rangle_{t,\infty}$ and $Dw_{(0,t)}$ is as defined in Eq. (11).

Equations (17)–(19) show that this interaction may be thought of as having the form of a sequence of brief von Neumann measurements, where each point of space-time contains a measurement "pointer." A "pointer" $w(x)$, labeled by x, has an initial wave function $\exp -(4\lambda)^{-1} d^4 x w^2(x)$, a very broad gaussian. The pointers at all $\mathbf{x}$ with common time t are idle until time t, when the brief (duration dt) entanglement interaction occurs (Eq. (18) with the integral over t removed), and they are once again idle. Each measurement is quite inaccurate, as its variance is $\sim (d^4 x)^{-1}$.

How Stands Collapse II

The resulting wave function Eq. (19) describes the state of all pointers having made measurements over the interval $(0,t)$, with $C(t)$ describing the pointers labeled by $t < \infty$ which will never make measurements, while the pointers labeled by $t > 0$ stand waiting to make measurements.

I call $|\Psi,t\rangle$, given by Eq. (20), the "ensemble vector." It is the "sum" of the (non-orthogonal) CSL states, each multiplied by a unique pointer state, the mutually orthogonal eigenstates of the quantized $w(\mathbf{x},t)$ field. Therefore, the product states are mutually orthogonal, do not mutually interfere, and they may be unambiguously identified. One may think of the ensemble vector as representing a precisely defined example of Schrodinger's "expectation catalog," a "horizontal listing" (i.e., the sum) of the real states of nature, identifiable with the "vertical listing" of the same states given by CSL.

The difficulty in making standard quantum theory provide a precise description of the real states of nature, compared with the success of collapse models, was succinctly characterized by John Bell as "AND *is not* OR." But, with CQC, "AND *is* OR." CQC provides a successful model for any interpretation of standard quantum theory, Environmental Decoherence (the w-field is the environment), Consistent Histories, Many Worlds, Modal Interpretations, ... Key is that, as the particle states evolve, they are generally not orthogonal, but CQC "tags" them with eigenstates $|w(x)\rangle$ which are orthogonal, allowing the eigenstate–eigenvalue link to be successfully employed. Also crucial is that the particle states *can* be regarded as realizable, sensible states of nature, as they are CSL states.

A possible benefit of CQC is that it is formulated in standard quantum theory terms, albeit with the strange W-field. This may make it easier to connect the collapse mechanism with physical mechanisms proposed for other purposes, which are formulated in standard quantum theory terms (see Section 7).

5.2 Conservation of Energy

The free $W(x)$-field time-translation generator is its energy operator:

$$H_w \equiv \int_{-\infty}^{\infty} d^4 k \, \omega b^{\dagger}(k) b(k) = \int_{-\infty}^{\infty} d^4 x \dot{W}(x) \Pi(x). \tag{21}$$

(the order of $\dot{W}(x)$ and $\Pi(x)$ can be reversed). In the Schrödinger picture, the Hamiltonian

$$H = H_w + H_A + 2\lambda \int d\mathbf{x} A(\mathbf{x}) \Pi(\mathbf{x},0) \tag{22}$$

is the time-translation generator, and is conserved. Because the Hamiltonian is translation and rotation invariant, the momentum and angular momentum operators are likewise conserved (e.g., the momentum operator is $-\int_{-\infty}^{\infty} d^4 x \nabla \mathbf{W}(x) \Pi(x) + \mathbf{P}_A$). Conservation of energy can be expressed in terms of the constancy of the moment-generating function,

$$\langle \Psi, t | e^{-i\beta H} | \Psi, t \rangle = \langle \Psi, t | \Psi, t + \beta \rangle = \langle \psi, 0 | \langle 0 | e^{-i\beta H} | 0 \rangle | \psi, 0 \rangle \qquad (23)$$

$$= \langle \psi, 0 | \langle 0 | e^{-i\beta(H_w + H_A)} \mathcal{T} e^{-i2\lambda \int_0^\beta d^4 x A(x)\Pi(x)} | 0 \rangle | \psi, 0 \rangle$$

$$= \langle \psi, 0 | e^{-i\beta H_A} \mathcal{T} e^{-\frac{\lambda}{2} \int_0^{|\beta|} d^4 x A^2(x)} | \psi, 0 \rangle$$

$$= \langle \psi, 0 | e^{-\left[i\beta H_A + |\beta| \frac{\lambda}{2} \int dx A^2(x)\right]} | \psi, 0 \rangle. \qquad (24)$$

Its fourier transform, $\mathcal{P}(E) \equiv (2\pi)^{-1} \int d\beta \exp i\beta E \langle \Psi, t | \exp -i\beta H | \Psi, t \rangle$, is the probability distribution of the energy:

$$\mathcal{P}(E) = \frac{1}{\pi} \langle \psi, 0 | \frac{1}{(E - H_A + i(\lambda/2) \int dx A^2(\mathbf{x}))} \cdot (\lambda/2) \int d\mathbf{x} A^2(\mathbf{x})$$

$$\cdot \frac{1}{(E - H_A - i(\lambda/2) \int dx A^2(\mathbf{x}))} | \psi, 0 \rangle, \qquad (25)$$

which is, roughly speaking, like the form $\pi^{-1} \langle \psi, 0 | c/[(E - \overline{H}_A)^2 + c^2] | \psi, 0 \rangle$, where $\overline{H}_A \equiv \langle \psi, 0 | H_A | \psi, 0 \rangle$. In the limit $\lambda \to 0$, (25) reduces to $\langle \psi, 0 | \delta(E - H_A) | \psi, 0 \rangle$. For $\lambda \neq 0$, the interaction spreads the distribution: $\overline{E} = \overline{H}_A$, $\overline{E^2} = \infty$.

Similarly, expressions can be written for the probability distribution of E_w, E_A, E_I or any sum of two of these, which generally vary with time since these are not constants of the motion. For example, it follows from (19) that the mean energies are:

$$\langle \Psi, t | H_A | \Psi, t \rangle = \langle \psi, 0 | \mathcal{T}_r e^{-\frac{\lambda}{2} \int_0^t dx [A_L(x) - A_R(x)]^2} H_A | \psi, 0 \rangle, \qquad (26)$$

$$\langle \Psi, t | H_w | \Psi, t \rangle = \langle \psi, 0 | \int_0^t \mathcal{T}_r e^{-\frac{\lambda}{2} \int_0^{t'} dx [A_L(x) - A_R(x)]^2} \cdot \int dx' [A(x'), [A(x'), H_A]] | \psi, 0 \rangle, \qquad (27)$$

$$\langle \Psi, t | H_I | \Psi, t \rangle = 0. \qquad (28)$$

where $\mathcal{T}_r$ is the time-reversal ordering operator (A_L's are time-reversed, A_R's are time-ordered). Taking the time derivative of (26), (27) shows that $d\overline{H}_A/dt = -d\overline{H}_w/dt$: in particular, in CSL, the mean particle kinetic energy increase (13) resulting from (26) is compensated by the mean w-field energy decrease.

When collapse has occurred, e.g., following a measurement, the ensemble vector (20) can be written as a sum of macroscopically distinguishable states:

$$|\Psi, t\rangle = \sum_n |\chi\rangle \int_{\Omega_n} Dw_{(0,t)} |w_n(x)\rangle_{(0,t)} |\psi, t\rangle_{w_n} \equiv \sum_n |\Psi, t\rangle_n, \qquad (29)$$

where $|\psi, t\rangle_{w_n}$ is a CSL state corresponding to the nth outcome engendered by the field $w_n(x)$, and Ω_n is the set of such fields. Here, not only $_{(0,t)}\langle w_m(x) | w_n(x)\rangle_{(0,t)} = 0$ for $m \neq n$, but also the CSL states are orthogonal (modulo tails), $_{w_m}\langle \psi, t | \psi, t\rangle_{w_n} \approx 0$. This is because "macroscopically distinguishable states" means that the mass density distributions of the CSL states have non-overlapping wave functions (except for tails) in some spatial region(s).

How Stands Collapse II   279

In this case, energy expressions may be written as the sum of contributions of the separate CSL outcome states. The product of powers of the energy operators H_A, H_w, H_I, acting on $|\Psi, t\rangle_n$, is essentially orthogonal (that is, up to tails contributions) to states $|\Psi, t\rangle_m$, where $m \neq n$. This is because none of these operators affects the non-overlapping nature of the mass density distribution wave functions. H_w doesn't act on $|\psi, t\rangle_n$. H_A is the integral over the energy density operator, so it only changes the wave function of the state where the mass density is non-zero. H_I behaves similarly as it depends upon the integral of the mass density operator. Thus, for any operator Q formed from these energy operators,

$$\langle \Psi, t | Q | \Psi, t \rangle \approx \sum_n \langle \Psi, t | Q | \Psi, t \rangle_n.$$

In this manner, generating functions and probabilities can be expressed as the sum of the separate contributions of the CSL states.

Mention should be made of a recent interesting work by Angelo Bassi, Emiliano Ippoliti and Bassano Vacchini [41], who consider a single free particle. The collapse engendering operator is the position, but modified by adding to it a small term proportional to momentum. The result is that the energy does not increase indefinitely, but reaches an asymptote, in analogy to the behavior of a particle reaching equilibrium with a thermal bath. The hope is to eventually model the bath and obtain, as discussed here, energy conservation when the particle and bath are both considered.

6 Relativity Problem

The problem is to make a relativistic quantum field theory which describes collapse. Although a good deal of effort has been expended upon it [42–47], there is not a satisfactory theory at present.

The difficulty is that, while the collapse behavior seems to work just fine, the collapse interaction produces too many particles out of the vacuum, amounting to infinite energy per second per volume.

6.1 With White Noise

By replacing $A(\mathbf{x})$ in Eqs. (9) and (12) by a Heisenberg picture quantum field operator $\Phi(x)$ which is a relativistic scalar, replacing $\int_0^t dt' H(t')$ by a space-time integral over the usual quantum field theory interaction density $V_I(x)$ and performing the space-time integral over the region between space-like hypersurfaces σ_0, σ, one obtains interaction picture state vector and density matrix evolution equations which are manifestly covariant:

$$|\psi, \sigma\rangle_w \equiv \mathcal{T} e^{-\int_{\sigma_0}^{\sigma} d^4x \{iV_I(x) + \frac{1}{4\lambda}[w(x) - 2\lambda\Phi(x)]^2\}} |\psi, \sigma_0\rangle, \tag{30}$$

$$\rho(\sigma) = Te^{-\int_{\sigma_0}^{\sigma} d^4x\{i[V_{IL}(x)-V_{IR}(x)]+\frac{\lambda}{2}[\Phi_L(x)-\Phi_R(x)]^2\}}\rho(\sigma_0). \tag{31}$$

The probability density in (11) is essentially unchanged, with t replaced by σ.

Suppose $\Phi(x)$ is a scalar quantum field. If $V_I(x) = g\Phi(x) : \overline{\Psi}(x)\Psi(x) :$, where $\Psi(x)$ is a Dirac fermion quantum field representing some particle type of mass M, then the scalar field "dresses" the particle field, distributing itself around the particle mass density. Thus, a superposition representing different particle mass distributions will also be a superposition of different scalar field spatial distributions, and collapse will occur to one or another of these.

To see what goes wrong, it is easiest to work in what I like to call the "collapse interaction picture," where $\Phi(x)$ is the Heisenberg picture scalar field: this eliminates $V_I(x)$'s explicit presence in Eqs. (30) and (31). In a reference frame where (σ_0, σ) are constant time hyperplanes $(0, t)$, consider the average energy for an initial density matrix $|\phi\rangle\langle\phi|$:

$$\begin{aligned}
\overline{H}(t) &= \text{Tr}\{HTe^{-\frac{\lambda}{2}\int_{\sigma_0}^{\sigma} d^4x[\Phi_L(x)-\Phi_R(x)]^2}|\phi\rangle\langle\phi|\} \\
&= \langle\phi|\{H - \frac{\lambda}{2}\int_0^t d^4x[\Phi(x)[\Phi(x),H]] + ...|\}|\phi\rangle \\
&= \langle\phi|H|\phi\rangle - \frac{i\lambda}{2}\int_0^t d^4x\langle\phi|\{[\Phi(x),\dot{\Phi}(x)]|\}|\phi\rangle \\
&= \langle\phi|H|\phi\rangle + \frac{\lambda}{2}\int_0^t d^4x\delta(0) \\
&= \langle\phi|H|\phi\rangle + \frac{\lambda t}{2}V\frac{1}{(2\pi)^3}\int d\mathbf{k}.
\end{aligned} \tag{32}$$

In Eq. (32), $\int d\mathbf{x} = V$ is the volume of space, $\delta(0) = (2\pi)^{-3}\int d\mathbf{k}\exp i\mathbf{k}\cdot\mathbf{0}$ is the sum over modes of the vacuum and is the 0th component of the four-vector $(2\pi)^{-3}\int(d\mathbf{k}/E)\{E,\mathbf{k}\}$. Although the energy increase/second-vol per mode is small, the vacuum gains infinite energy/second-vol because the vacuum has an infinite number of modes.

The reason the vacuum is excited can be seen by writing Eq. (30) in fourier transform form, mentioned in Eq. (20) of paper I:

$$|\psi,t\rangle_w = \int D\eta e^{-\lambda\int_{\sigma_0}^{\sigma} d^4x\eta^2(x)}e^{i\int_{\sigma_0}^{\sigma} d^4x\eta(x)w(x)} \cdot Te^{-i\int_{\sigma_0}^{\sigma} d^4x\{V_I(x)+2\lambda\eta(x)\Phi(x)\}}|\psi,0\rangle. \tag{33}$$

This can be regarded as an ensemble average over a classical white noise field $\eta(x)$ (the first term in (33) is the white noise gaussian probability distribution). The average is a superposition of unitary evolutions. The collapse evolution is due to the "interaction Hamiltonian" density $\eta(x)\Phi(x)$. Since $\eta(x)$ is a classical white noise field, it contains all frequencies and wave numbers in equal amounts. As a result, because of its interaction with $\Phi(x)$, it excites Φ-particles of all allowed frequencies and wavelengths out of the vacuum.

How Stands Collapse II
281

Indeed, if any mode of the vacuum is excited, for a relativistically invariant theory, all modes must be excited, since that mode looks like another mode in another, equivalent, reference frame.

6.2 Gaussian Noise

To try to remove the vacuum excitation, it is worth considering a noise field that is not white noise, and therefore doesn't have all frequencies and wavelengths [37, 48, 49]. A generalization of Eqs. (30) and (31) is

$$|\psi, \sigma\rangle_w \equiv T e^{-i\int_{\sigma_0}^{\sigma} d^4 x V_I(x)}$$
$$\cdot e^{-\frac{1}{4\lambda}\int\int_{\sigma_0}^{\sigma} d^4 x d^4 x' [w(x) - 2\lambda\Phi(x)]G(x-x')[w(x') - 2\lambda\Phi(x')]}|\psi, \sigma_0\rangle, \quad (34)$$

$$\rho(\sigma) = T e^{-i\int_{\sigma_0}^{\sigma} d^4 x\{V_{IL}(x) - V_{IR}(x)\}}$$
$$\cdot e^{-\frac{\lambda}{2}\int\int_{\sigma_0}^{\sigma} d^4 x d^4 x' [\Phi_L(x) - \Phi_R(x)]G(x-x')[\Phi_L(x') - \Phi_R(x')]}\rho(\sigma_0), \quad (35)$$

with

$$G(x - x') = \frac{1}{(2\pi)^4}\int d^4 k e^{ik \cdot (x - x')}\tilde{G}(p^2), \quad (36)$$

where $\tilde{G}(p^2) \geq 0$: if $\tilde{G}(p^2) = 1$, this reduces to the white noise case.

CSL, although non-relativistic, can be written in this form. Put the expression for $A(\mathbf{x})$ from Eq. (10) into Eq. (9), as well as replace $w(\mathbf{x}, t)$ by

$$w(\mathbf{x}, t) \equiv (\pi a^2)^{-3/4}\int d\mathbf{z} e^{-\frac{1}{2a^2}(\mathbf{x} - \mathbf{z})^2}w'(\mathbf{z}, t),$$

and perform the integral over $\mathbf{x}$ in the exponent. The result is

$$|\psi, t\rangle_w \equiv T e^{-i\int_0^t dt' H(t')} \cdot e^{-\frac{1}{4\lambda}\int\int_0^t dz dz' [w'(z) - \frac{2\lambda}{m_0}M(z)]G(z-z')[w'(z') - \frac{2\lambda}{m_0}M(z')]}|\psi, 0\rangle \quad (37)$$

where

$$G(z - z') = \delta(t - t')e^{-\frac{1}{4a^2}(\mathbf{z} - \mathbf{z}')^2}. \quad (38)$$

6.3 Tachyonic Noise

6.3.1 Φ = Free Scalar Field

To see how this flexibility can help with the energy creation problem, reconsider the calculation of $\overline{H}(t)$ given in (32), with the density matrix (35), with (σ_0, σ) replaced by $(-T/2, T/2)$ as $T \to \infty$, and with $\Phi(x)$ a free scalar field of mass m ($V_I(x) = 0$):

282   P. Pearle

$$\bar{H}(t) = \langle \phi | \{ H - \frac{\lambda}{2} \int \int_{-T/2}^{T/2} d^4x d^4x' G(x-x') \mathcal{T}_r \{ [\Phi(x)[\Phi(x'),H]] + ... \} | \phi \rangle$$

$$= \langle \phi | H | \phi \rangle + \frac{\lambda T}{2} V \tilde{G}(m^2) \frac{1}{(2\pi)^3} \int d\mathbf{k} \qquad (39)$$

($\mathcal{T}_r$ is the time-reversed-ordering operator). So, if $\tilde{G}(m^2) = 0$, there is no energy creation from the vacuum in this case. But, then, nothing else happens either!

This can be seen in terms of Feynman diagrams for the density matrix. Write the density matrix (35) in fourier transform form:

$$\rho(\frac{T}{2}) = \int D\eta e^{-2\lambda \int \int_{-T/2}^{T/2} d^4x d^4x' \eta(x) G^{-1}(x-x') \eta(x')}$$

$$\cdot \mathcal{T} e^{-i\int_{-T/2}^{T/2} d^4x \{V_I(x)+\eta(x)2\lambda\Phi(x)\}} \rho(-\frac{T}{2}) \mathcal{T}_r e^{i\int_{-T/2}^{T/2} d^4x \{V_I(x)+\eta(x)2\lambda\Phi(x)\}}. \qquad (40)$$

The last line of (40) is a unitary transformation, so it can be expanded in a power series, and Wick's theorem used to replace a time-ordered product of operators by a product of positive and negative frequency normal ordered operators and Feynman propagators. Then, $\int D\eta$ can be performed, resulting in $\int \int_{-T/2}^{T/2} d^4x d^4x' \eta(x)\eta(x') \to$ $\int \int_{-T/2}^{T/2} d^4x d^4x' G(x-x')$: a term containing an even number of $\eta(x)$ factors becomes a sum of terms with all possible pairings of $\eta(x)$'s replaced by G's. (A term with an odd number of $\eta(x)$ factors vanishes). When the integrals over x are performed, the result is the momentum space expression for the sum of Feynman diagrams. $\tilde{G}(p^2)$ plays the role of the Feynman propagator for the η field.

Return to the case of the free Φ field (i.e., $V_I(x) = 0$). Before and after integration over η, every normal-ordered positive or negative frequency Φ operator appears in an integral,

$$\int_{-T/2}^{T/2} d^4x \eta(x)\Phi^\pm(x) \to \int_{-T/2}^{T/2} d^4x G(x'-x)\Phi^\pm(x) = \tilde{G}(m^2)\Phi^\pm(x') = 0,$$

i.e., G and Φ are orthogonal if $\tilde{G}(m^2) = 0$. Thus, the operators disappear from (40). Then, $\rho(\frac{T}{2}) = C\rho(-\frac{T}{2})$: when the trace is taken, this implies the c-number $C = 1$, i.e., there is no evolution.

To see how the energy creation disappears, look at the first order in λ Feynman graph which describes creation of a Φ-particle from the vacuum, and is responsible for the energy increase given by Eq. (39). Represent the Φ field by $\langle$ and the η propagator by ___. To lowest order (terms quadratic in η), the relevant diagram is $\langle\langle$. The Φ particle created out of the vacuum appears to the left and right sides of the initial density matrix $\rho(-\frac{T}{2}) = |0\rangle\langle 0|$. The η propagator crosses from one side to the other. Because the 4-momentum p is conserved (it goes in at the right and out at the left), the diagram is proportional to $\tilde{G}(p^2) = \tilde{G}(m^2)$, with no contribution if $\tilde{G}(m^2) = 0$.

How Stands Collapse II   283

6.3.2 Φ = Interacting Scalar Field

With $V_I(x) \neq 0$, there can be particle creation out of the vacuum to first order in λ. The relevant diagram is $\)($ with a fermion-antifermion pair $\vee$ tacked on to the end of each $\)$ ($\)$ attached at both ends then represents a Φ-particle propagator). If p_1 and p_2 are the outgoing fermion 4-momenta, the diagram is proportional to $\tilde{G}([(p_1 + p_2]^2)$. Vanishing of the contribution of this diagram requires G to vanish for the range of its argument $(2M)^2 \leq p^2 < \infty$. If M can be arbitrarily small, then $\tilde{G}(p^2)$ *must vanish for all time-like p*. Thus, if we take $\tilde{G}(p^2) = 0$ for $0 \leq p^2 < \infty$, there is no particle creation from the vacuum to first order in λ: a space-like (i.e., tachyonic) 4-momentum (for which $\tilde{G}(p^2)$ does not vanish) cannot equal a time-like 4-momentum (of the outgoing fermions).

So, the time-like 4-momenta of $\tilde{G}(p^2)$ are responsible for the energy creation from the vacuum to first order in λ. In the next subsection we shall see that it is the space-like 4-momenta of $\tilde{G}(p^2)$ which are responsible for collapse.

First note that, for non-vanishing diagrams describing collapse, any $\)$ attached to a $\_$ must be a Φ-particle propagator since, if it represents a free Φ-particle, the diagram's contribution $\sim \tilde{G}(m^2) = 0$. But, then, this diagram segment's contribution is

$$\sim \frac{1}{p^2 - m^2 + i\varepsilon}\tilde{G}(p^2) = [\mathcal{P}\frac{1}{p^2 - m^2} - i\pi\delta(p^2 - m^2)]\tilde{G}(p^2) = \mathcal{P}\frac{1}{p^2 - m^2}\tilde{G}(p^2).$$

Thus, the Φ-particle propagator can be absorbed into the η propagator: $\mathcal{P}[p^2 - m^2]^{-1}\tilde{G}(p^2) \equiv \tilde{G}'(p^2)$. In Feynman diagrams, this means that the Φ-particle propagator line can be replaced by a point: for example, the diagram described in the second sentence of this section, $\)($ with $\vee$ tacked on to the end of each $\)$, can be replaced by $\underline{\vee\vee}$ (which, of course, vanishes).

6.3.3 Φ = Fermion Density

Therefore, we may just consider the model with collapse directly toward fermion density eigenstates, putting $\Phi(x) =: \overline{\Psi}(x)\Psi(x):$ (and setting V_I as the usual interaction Hamiltonian for the fermion field with e.g., photons, mesons, ...) into Eqs. (34, 35, 40).

In the non-relativistic limit, (36) becomes

$$G(x - x') \to \lim_{c \to \infty} \frac{1}{(2\pi)^4} \int dE d\mathbf{p} e^{iE(t-t') - i\mathbf{p}\cdot(\mathbf{x}-\mathbf{x}')}\tilde{G}[\left(\frac{E}{c}\right)^2 - \mathbf{p}^2]$$

$$= \delta(t - t')\frac{1}{(2\pi)^3} \int d\mathbf{p} e^{-i\mathbf{p}\cdot(\mathbf{x}-\mathbf{x}')}\tilde{G}(-\mathbf{p}^2). \tag{41}$$

With the choice

$$\tilde{G}(p^2) \equiv (4\pi a^2)^{3/2}\Theta(-p^2)e^{a^2 p^2} \to (4\pi a^2)^{3/2}e^{-a^2\mathbf{p}^2}$$

(Θ is the step function), (41) is identical to the CSL form (38). Another interesting choice is the spectrum $\tilde{G}(p^2) \equiv \delta(p^2 + \mu^2)$ of a tachyon of mass $\mu \approx \hbar/ac \approx 2eV$. Then, (41) becomes $G(x - x') \to (2\pi)^{-2}\delta(t - t')\sin\mu|x - x'|/|x - x'|/$, which is a perfectly good substitute for the gaussian smearing function.

Indeed, with one of these choices, if one regards the non-relativistic limit of $: \overline{\Psi}(x)\Psi(x) :$ as allowing one to neglect its pair creation and annihilation terms, the remainder would be the operator $M(x)/m_0$, the sum of the number operators for fermion and anti-fermion. Then (35) would become (37): the model would reduce to CSL in the non-relativistic limit. Unfortunately, one cannot neglect these terms.

So, while there is no vacuum production of particles to order λ, alas, in the relativistic model, there is production to order λ^2. The expansion of (40) to fourth order in η produces the vacuum excitation diagram $\vee\!\!-\!\!\vee$: two space-like four-momenta of the two η propagators can add up to the timelike four-momentum of the excited fermion pair.

Thus, I have given up trying to make a satisfactory relativistic collapse model. A reason I have gone over this failure in such detail is that it might perhaps stimulate someone to succeed in this endeavor. Another reason is that, if this failure persists, it helps motivate my fall-back position, the "Quasi-relativistic" model sketched below [50].

6.4 Quasi-Relativistic Collapse Model

In this model, which has no particle creation from the vacuum, the state vector and density matrix evolution equations are Eqs. (30) and (31), with

$$
\begin{aligned}
\Phi(x) &\equiv (4\pi a^2)^{3/4} e^{-\frac{a^2}{2}\Box} [\overline{\Psi}^+(x)\Psi^-(x) + \Psi^+(x)\overline{\Psi}^-(x)] \\
&= \frac{1}{2^{1/2}(\pi a^2)^{5/4}} \int db^0 d\mathbf{b} \, e^{-\frac{1}{2a^2}[(b^0)^2 + \mathbf{b}^2]} \\
&\quad \cdot [\overline{\Psi}^+\Psi^- + \Psi^+\overline{\Psi}^-](t + ib^0, \mathbf{x} + \mathbf{b}),
\end{aligned} \tag{42}
$$

where $\Box$ is the D'Alembertian. This should be compared with the CSL expression for A in Eq. (10), written in terms of the particle annihilation and creation operators $\xi(t, \mathbf{x}), \xi^\dagger(t, \mathbf{x})$:

$$
\begin{aligned}
A(x) &\equiv (4\pi a^2)^{3/4} e^{-\frac{a^2}{2}\nabla^2} \xi^\dagger(t, \mathbf{x})\xi(t, \mathbf{x}) \\
&= \frac{1}{(\pi a^2)^{3/4}} \int d\mathbf{b} \, e^{-\frac{1}{2a^2}\mathbf{b}^2} \xi^\dagger\xi(t, \mathbf{x} + \mathbf{b}),
\end{aligned} \tag{43}
$$

to which (42) reduces in the non-relativistic limit (when the anti-particle term $\Psi^+\overline{\Psi}^-$ is discarded, and the spin degrees of freedom are ignored). Of course, to agree with CSL, when more than one fermion type is considered, there should be a sum of terms with coefficients proportional to their masses.

How Stands Collapse II   285

The first expression in (42) is manifestly a Lorentz scalar, but the model given by Eqs. (30, 31, 42) is not Lorentz invariant. This is because, while $\overline{\Psi}\Psi$ does commute with itself at space-like separations, $[\overline{\Psi}^{+}\Psi^{-} + \Psi^{+}\overline{\Psi}^{-}]$ does not. Therefore, the time-ordering operation in one Lorentz frame is not the time-ordering operation in another Lorentz frame. However, it can be shown that, for $-(x-x')^2 > a^2$, the commutator $[\Phi(x), \Phi(x')] \sim \exp-[a/(\hbar/Mc)]$ which for nucleons is $\approx \exp-10^9$, i.e., it "almost" commutes. It is in this sense that the model is quasi-relativistic.

Since there is a preferred reference frame in the model, the one in which time-ordering prevails, it is natural to take it as the co-moving frame in the universe. Since the earth is not far from the co-moving frame, and the non-relativistic limit of the model is CSL, it so far agrees with experiment. It would be worthwhile exploring whether there are feasible experiments predicted by the model which would show deviations from relativistic invariance, e.g., experiments with apparatus moving rapidly with respect to the preferred frame.

A number of theoretical proposals [35, 51–53] have suggested that collapse is related to gravity. This idea has been buttressed, in the context of CSL, by the experimental evidence for coupling of the fluctuating field $w(\mathbf{x}, t)$ to the mass density operator. Therefore, there is a positive aspect to a model which is most naturally specialized to the co-moving frame, in that it additionally suggests a cosmological connection for collapse (see the next section).

It may also be observed that the relativistic collapse models, which do produce satisfactory collapse behavior in the midst of unsatisfactory excitation, require the causes of collapse and the space-time locations of the regions where the wave function collapses (rapidly diminishes or grows) to be reference frame dependent [42, 43, 54]. This is not a problem, since the causes and locations of collapse cannot be observed. But, this differs from the situation in standard quantum field theory, where the amplitudes for particles being in a space-time region do not change when the reference frame changes. It may then be considered a benefit of this quasi-relativisitic model that it also possesses this same behavior of standard quantum field theory, since the causes and locations associated with collapse are those of the preferred frame.

7 Legitimization Problem

When, over 35 years ago, as described in paper I, I had the idea of introducing a randomly fluctuating quantity to cause wave function collapse, I thought, because there are so many things in nature which fluctuate randomly, that when the theory was better developed, it would become clear what thing in nature to identify with that randomly fluctuating quantity. Perhaps ironically, this problem of legitimizing the phenomenological CSL collapse description by trying it in a natural way to established physics remains almost untouched [55].

Although, as mentioned in the previous section, various authors, as well as the experimental evidence supporting coupling of the collapse-inducing fluctuating field

286 P. Pearle

to mass density, have suggested a connection between collapse and gravity, it is fair to say that the legitimization problem is still in its infancy. No convincing connection (for example, identification of metric fluctuations, dark matter or dark energy with $w(\mathbf{x},t)$) has yet emerged. But, I shall give here a new argument that the w-field energy density must have a gravitational interaction with ordinary matter, and a perhaps less-convincing argument, that the w-field energy density could be cosmologically significant.

7.1 Gravitational Considerations

What happens to the w-field energy once it is created, either in small amounts as in measurement situation collapses, or in large amounts as will be suggested below? Suppose we do not alter the CQC Hamiltonian (22). Then this energy just sits where it was created, and has no other effect on matter. The picture given in Section 5.1 is that the w-field in an infinitesimal space-time volume is like a pointer making a measurement, which briefly interacts and therefore changes during the measurement, but is unchanged before and after, and its associated energy density has the same behavior.

But, here is an argument that the CQC Hamiltonian (22) must be altered, so that the w-field energy density exerts a gravitational force on matter. Consider the equation of quasi-classical general relativity, $G^{\mu,\nu} = -8\pi G \langle \Psi | T^{\mu,\nu} | \Psi \rangle$, i.e., $G^{\mu,\nu}$ is classical, but the classical stress tensor is replaced by the quantum expectation value of the stress tensor operator. Of course, the latter must obey the conservation laws if the equation to be consistent. However, due to the collapse interaction, the expectation value of the particle energy-momentum is not, by itself, conserved. As discussed in Section 5.2, it is the expectation value of the sum of particle energy-momentum and w-field energy-momentum which is conserved. Therefore, $T^{\mu,\nu} = T_A^{\mu,\nu} + T_w^{\mu,\nu}$: the expectation value of the sum of particle and w-field stress tensor operators must be utilized.

In the non-relativistic limit, $G^{0,0} = -8\pi G \langle \Psi | T^{0,0} | \Psi \rangle$ reduces to $\nabla^2 \phi = 4\pi G \langle \Psi | T^{0,0} | \Psi \rangle$. Thus, the w-field energy density acts just like matter's energy density in creating a gravitational potential, except that the w-field energy density can be negative or positive.

Therefore, when modeling the local behavior of the w-field in CQC, and wishing to take into account its gravitational behavior, one ought to modify the CQC Hamiltonian (22), adding a term representing the gravitational interaction of the w-field energy density with the matter energy density:

$$ H_G \equiv -G \int_{-\infty}^{\infty} dx \dot{W}(x) \Pi(x) \int d\mathbf{z} \frac{1}{|\mathbf{x} - \mathbf{z}|} \mathcal{H}_M(\mathbf{z}). $$

With this addition, although the w-field energy, once created in a volume, still sits in that volume as if nailed in space, it now has an effect on matter, which is repelled/attracted by a region containing negative/positive w-field energy.

How Stands Collapse II   287

One could consider further alterations in the local CQC Hamiltonian, to make the
w-field energy density dynamic, for example, to treat it like a fluid. Then, it would
be gravitationally attracted by matter, or repelled/attracted by itself, if the w-field
energy density is negative/positive. One might add a positive constant w-field energy
to the Hamiltonian, so that the w-field energy, although decreased by the collapse
interaction, remains positive. We shall not consider such modifications here.

To reiterate, the argument here is that compatibility with general relativity requires a gravitational force exerted upon matter by the w-field.

7.2 Cosmological Creation of Negative w-Field Energy

It was discussed in Section 5.2 that, as the mean energy of matter increases due to
collapse, the mean w-field energy goes negative by an equal amount. Thus, if there
is an amount of negative w-field energy which is of cosmological significance, it
would repel matter, and contribute to the observed cosmic acceleration [57].

But, as pointed out in Section 4, the mean amount of kinetic energy (13) gained
by a particle of mass m over the age of the universe is very small,

$$E \approx \frac{\lambda a_0^2}{\lambda_0 a^2} 1.3 \times 10^{-16} mc^2$$

(λ_0, a_0 are the SL values of λ and a). A factor of 10^{16} increase in λ/a^2 which makes
this energy comparable to mc^2 would violate already established experimental lim-
its, e.g., on "spontaneous" energy production in atoms or nucleii. Thus, w-field energy created by the collapses accompanying the dynamical evolution of the particles
in the universe is not of cosmological significance.

However, it is in the spirit of models of the beginning of the universe to imagine that the universe started in a vacuum state, and that it was briefly governed by
a Hamiltonian which describes production of particles from the vacuum. We now
illustrate, by a simple model, that negative w-field energy of a cosmologically significant amount could be generated in such a scenario. Suppose that, even under such
circumstances, the CSL collapse equations apply. If collapse went on then, as we
suppose it does now, the universe would have been in a superposition of the vacuum
state and states with various numbers of particles in various configurations, and the
collapse mechanism would have been responsible for choosing the configuration of
our present universe.

This model can also be utilized to describe continuous production of particles as
the universe evolves, as in the steady state cosmology. However, we shall not make
that application here.

In this simple model, only scalar particles of mass m are produced, and the
Hamiltonian is

$$H_A = \int_V d\mathbf{x}\{m\xi^\dagger(\mathbf{x})\xi(\mathbf{x}) + g[\xi(\mathbf{x}) + \xi^\dagger(\mathbf{x})]\}. \tag{44}$$

where $\xi(\mathbf{x})$ is the annihilation operator for a scalar particle at $\mathbf{x}$, V is the volume of the early universe and g is a coupling constant. With initial state $|\psi,0\rangle = |0\rangle$, with $A(x) = \exp(iH_A t)A(\mathbf{x})\exp(-iH_A t)$ and $A(\mathbf{x})$ given in Eq. (10), we obtain from (26), (44):

$$\overline{N}(t) \equiv \langle \Psi,t| \int_{V_1} d\mathbf{x}\, \xi^\dagger(\mathbf{x})\xi(\mathbf{x})|\Psi,t\rangle$$

$$= \frac{g^2 V_1}{m^2 + (\lambda/2)^2}\left\{\lambda t - 2(\cos\theta - e^{-\frac{\lambda t}{2}}\cos(\theta + mt))\right\}, \qquad (45)$$

$$\overline{Q}(t) \equiv \langle \Psi,t| \int_{V_1} d\mathbf{x}\, \frac{1}{2}[\xi^\dagger(\mathbf{x}) + \xi(\mathbf{x})]|\Psi,t\rangle$$

$$= \frac{-gmV_1}{m^2 + (\lambda/2)^2}\left\{1 - e^{-\frac{\lambda t}{2}}(\cos mt) + \frac{\lambda}{2m}\sin mt\right\}, \qquad (46)$$

$$\overline{H}_A(t) \equiv \langle \Psi,t|H_A|\Psi,t\rangle = m\overline{N}(t) + 2g\overline{Q}(t) \qquad (47)$$

where $\lambda = \lambda_0(m/m_0)^2$, $V_1 \subseteq V$, and $\theta \equiv 2\tan^{-1}(2m/\lambda)$. One can check that the $\lambda \to 0$ limit of these equations is the usual oscillatory quantum mechanical result (since $\theta = \pi/2$, $m\overline{N}(t) = -2g\overline{Q}(t)$, and so $\overline{H}_A(t) = 0$). Also, all expressions $\to 0$ as $\lambda \to \infty$, i.e., in that case the universe remains in the vacuum state due to "watched pot" or "Zeno's paradox" behavior (the collapse occurs so fast that there is no chance for the vacuum state to evolve).

The interesting thing is that the coefficient of the linear increase in $\overline{N}(t)$ is $\sim g^2\lambda$: the Hamiltonian, acting by itself, generates and annihilates particles, but without linear growth. It is the collapse dynamics which, favoring creation over annihilation, is ultimately responsible for creating the matter in the universe, according to this model.

Because $d\overline{H}_w/dt = -d\overline{H}_A/dt$, the mean w-field energy $\overline{H}_w(t)$ goes linearly negative. Moreover, if $\mathcal{H}_w(\mathbf{x},t)$ and $\mathcal{H}_A(\mathbf{x},t)$ are the w-field and particle energy densities at time t, it can be shown that $d\overline{\mathcal{H}}_w(\mathbf{x},t)/dt = -d\overline{\mathcal{H}}_A(\mathbf{x},t)/dt$. Since the initial value of each is zero, we have that $\overline{\mathcal{H}}_w(\mathbf{x},t) = -\overline{\mathcal{H}}_A(\mathbf{x},t)$, where the average is over the ensemble of possible universes, one of which became ours, due to collapse. In each universe, the particle and w-field energy densities vary from place to place: in particular, the w-field energy density can be negative or positive.

We can say something interesting about the total w-field and particle energies, H_w and H_A in any one universe. Suppose we divide a particular universe into N equal volumes ΔV, and calculate the mean of the sum $S \equiv (w\text{-field energy} + \text{particle energy}$ in kth volume)$/\Delta V$ over that universe, i.e., the mean of $S = \sum_{k=1}^{N}(E_k/\Delta V)/N = (H_w + H_A)/V$. The probability of E_k is independent of k, and the mean of E_k is zero, so the mean of S is zero. By the law of large numbers, as $N \to \infty$ (which can be achieved by letting ΔV become infinitesimal), S achieves zero variance. Thus, each universe satisfies $H_w = -H_A$.

This may be thought of as a crude model for the reheating after inflation which produces matter. It should not be taken too seriously: for one thing, one would prefer

How Stands Collapse II   289

drawing conclusions from the collapse mechanism applied to the field accompanying an accepted inflationary model. But, it suggests that it is possible for the w-field energy to be of cosmological significance, that regions of both positive and negative w-field energy would then be present, the former attracting matter, the latter repelling matter. If the collapse interaction is not limited to ordinary matter, but includes dark matter, then it suggests that there is a negative amount of w-field energy in the universe equal in magnitude to the mass–energy of all matter.

7.3 Some Cosmological Considerations

Astronomical observation and theory, which lead to what is called the "standard model," are woven together in a tight web, so it is rather presumptuous to inject the w-field into the mix, especially since the suggestion described at the end of the last subsection is not very detailed. However, it may stimulate further scrutiny to return to semi-classical gravity, model the quantum expectation values of the matter and w-field energy densities in our universe by classical distributions $\rho_m(\mathbf{x},t)$, $\rho_w(\mathbf{x},t)$ and discuss a few ways in which the w-field might play a role in affecting the evolution of the universe, with regard to both fluctuations about the mean behavior and the mean behavior itself.

With regard to fluctuations, following the suggestion of the model in Section 7.2, we suppose, after the period of particle production in the early universe, that the w-field energy density ρ_w is fixed in space, varies from place to place on the scale of a, can be positive or negative, and is initially overlain by mass density ρ_m. The negative w-field energy density should repel the mass density nearby, the positive w-field energy density should attract it, and so the scenario of matter density fluctuations in the early universe could be affected. One might speculate that the presently observed voids between galaxies could initially have been sites of negative w-field energy density, perhaps initially of scale a which expanded with the universe, that the sites of positive w-field energy density could have helped seed initial galactic gravitational collapse and could play a role similar to that of the CDM, etc. If the w-field negative energy density (perhaps equal in magnitude to the matter mass–energy, estimated at $\approx \rho_c/4$, where $\rho_c \equiv 3H_0^2/8\pi G$ is the critical mass density which makes the universe flat) is spread fairly uniformly throughout the universe, its gravitational repulsive effect on matter would not seem to have much of an effect on the behavior of formed galaxies, because the density of matter in galaxies is so much greater than ρ_c.

As is well known, the mean behavior of the universe is described by the Friedmann–Robertson–Walker general relativistic homogeneous isotropic cosmological model, which gives rise to two equations. One of these can be taken to be the conservation of energy equation relating the universe's scale factor R, the energy density ρ, and the pressure p:

$$\frac{d}{dR}(\rho R^3) = -3pR^2. \tag{48}$$

This equation holds for ρ_w by itself ($p_w = 0$) which, glued to space, evolves only due to the expansion of the universe after the period of particle creation has ended, $\rho_w \sim R^{-3}(t)$. Also, $\rho_m \sim R^{-3}(t)$ as the matter pressure is negligible. Let us neglect the radiation density and pressure, and assume a cosmological constant $\rho_\Lambda = -p_\Lambda$ which also satisfies (48).

The second equation, the evolution equation for the scale factor $R(t)$, and its current consequence are

$$\frac{\dot{R}^2}{R^2} = \frac{8\pi G}{3}\rho - \frac{k}{R^2} \equiv H_0^2\left[\frac{\rho}{\rho_c} - \frac{k}{R^2 H_0^2}\right] \implies 1 = \Omega_m + \Omega_w + \Omega_\Lambda + \Omega_k \tag{49}$$

where R_0 is the present scale factor, $H_0 \equiv \dot{R}_0/R_0$ is Hubble's constant, $k = 1, 0, -1$ depending respectively upon whether the universe is closed, flat or open, $\Omega_m \equiv \rho_{0m}/\rho_c$ etc., and $\Omega_k \equiv -k/H_0^2 R_0^2$. From (48), (49) also follows the useful expression

$$-\frac{\ddot{R}}{H_0^2 R} = \frac{\rho}{2\rho_c} + \frac{3p}{2\rho_c} \implies q_0 = \frac{1}{2}(\Omega_m + \Omega_w) - \Omega_\Lambda, \tag{50}$$

where the deceleration parameter is $q_0 \equiv -\ddot{R}_0 R_0/\dot{R}_0^2$.

The matter mass density and w-field energy density affect equations (49), (50) only through their sum. If we suppose that the w-field collapse interaction generates not only the ordinary matter in the universe, but the CDM as well, then $\Omega_m + \Omega_w = 0$. The consequent result from (49), $\Omega_\Lambda + \Omega_k = 1$, appears to be within 1σ of the microwave radiation background data [56], assuming a flat universe, $\Omega_k = 0$. But, when combined with the result from (50), $q_0 = -\Omega_\Lambda = -1$, while qualitatively consistent with the observed cosmic acceleration, appears to be 3σ from the Hubble plot data [57]. However, these analyses assume certain prior constraints, and analyzing the prior $\Omega_m + \Omega_w = 0$ has not received priority.

It is likely that the simple scenario given here will conflict with various astronomical observations and constraints. There are variants of the model which could be explored to resolve such conflicts, e.g., the parameters λ, a could vary with time, the w-field energy could be made dynamic, its magnitude could be smaller than the magnitude of the matter energy (e.g., $\approx 20\%$ of it because the collapse interaction could only occur for ordinary matter and not dark matter), its magnitude could be larger than that of the matter energy (e.g., because collapse could be governed by energy density rather than mass density, and so could occur for light as well as matter), it could play a role in the generation of dark energy, or even be dark energy, etc. The purpose of this discussion is to illustrate the hope that progress may be made in legitimizing the phenomenological CSL collapse dynamics by connecting it to the still mysterious contents of the universe.

References

1. P. Pearle, "How Stands Collapse I," *J. Phys. A: Math. Theor.* **40**, 3189 (2007).
2. P. Pearle, *Experimental Metaphysics: Quantum Mechanical Studies for Abner Shimony*, R. S. Cohen, M. Horne and J. Stachel, eds. (Kluwer, Dordrecht, 1997), p. 143.
3. J. S. Bell in *Sixty-Two Years of Uncertainty*, A. I. Miller, ed. (Plenum, New York, 1990), p. 17.
4. P. Pearle, *Phys. Rev. A* **39**, 2277 (1989).
5. G. C. Ghirardi, P. Pearle and A. Rimini, *Phys. Rev. A* **42**, 78 (1990).
6. G. C. Ghirardi, A. Rimini and T. Weber, *Phys. Rev. D* **34**, 470 (1986); *Phys. Rev. D* **36**, 3287 (1987).
7. B. Collett and P. Pearle, *Found. Phys.* **33**, 1495 (2003).
8. S. L. Adler, "Lower and Upper Bounds on CSL Parameters from Latent Image Formation and IGM Heating": quant-ph/0605072, *J. Phys. A: Math. Theor.* **40**, 2935 (2007).
9. P. Pearle, *Phys. Rev. D* **13**, 857 (1976); *Int'l. J. Theor. Phys.* **18**, 489 (1979); *Found. Phys.* **12**, 249 (1982); in S. Diner, D. Fargue, G. Lochat and F. Selleri, eds., *The Wave-Particle Dualism*. (D. Reidel, Dordrecht, 1984); *Phys. Rev. D* **29**, 235 (1984); *Phys. Rev. Lett.* **53**, 1775 (1984); J. Stat. Phys. **41**, 719 (1985); P. Pearle, *Phys. Rev.* D**33**, 2240 (1986); in D. Greenberger, ed., *New Techniques and Ideas in Quantum Measurement Theory*, p. 457 (New York Academy of Sciences, Vol. 480, 1986), p. 539.
10. N. Gisin, *Phys. Rev. Lett.* **52**, 1657 (1984); *Phys. Rev. Lett.* **53**, 1776 (1984).
11. C. Dove and E. J. Squires, *Found. Phys.* **25**, 1267 (1995) found a hitting process which preserves symmetry. See also R. Tumulka, *Proc. Roy. Soc.* **A462**, 1897 (2006).
12. A. Shimony in *PSA 1990*, Vol. 2, A. Fine, M. Forbes and L. Wessels, eds. (Philosophy of Science Association, East Lansing, 1991), p. 17.
13. G. C. Ghirardi and P. Pearle in *PSA 1990*, Vol. 2, A. Fine, M. Forbes and L. Wessels, eds. (Philosophy of Science Association, East Lansing, 1991), pp. 19, 35.
14. G. C. Ghirardi and T. Weber in R. S. Cohen, M. Horne and J. Stachel, eds., *Potentiality, Entanglement and Passion-at-a-Distance, Quantum Mechanical Studies for Abner Shimony, Vol. 2.* (Kluwer, Dordrecht, 1997), p. 89. See also G. C. Ghirardi, R. Grassi and F. Benatti, *Found. Phys.* **20**, 1271 (1990).
15. P. Pearle in *Experimental Metaphysics, Quantum Mechanical Studies for Abner Shimony, Vol. 1*, R. S. Cohen, M. Horne and J. Stachel, eds. (Kluwer, Dordrecht, 1997), p. 143.
16. S. Sarkar in *Experimental Metaphysics, Quantum Mechanical Studies for Abner Shimony, Vol. 1*, R. S. Cohen, M. Horne and J. Stachel, eds. (Kluwer, Dordrecht, 1997), p. 157.
17. D. Z. Albert and B. Loewer in R. Clifton, ed., *Perspectives on Quantum Reality* (Kluwer, Dordrecht, 1996), p. 81.
18. H. P. Stapp, *Can. J. Phys.* **80**, 1043 (2002) and quant-ph 0110148.
19. P. Pearle, *Phys. Rev.* **35**, 742 (1967).
20. C. A. Fuchs and A. Peres, *Phys. Today*, **70**, 3 (2000).
21. E. Schrödinger, "Die gegenwärtige Situation in der Quantenmechanik", *Naturwissenschaften* **23**: pp. 807–812; 823–828; 844–849 (1935). See the translation by J. D. Trimmer in J. A. Wheeler and W. H. Zurek, eds., *Quantum Theory and Measurement* (Princeton University Press, New Jersey, 1983).
22. L. Ballentine, *Quantum mechanics. A Modern Development* (World Scientific, Singapore, 1998).
23. R. Omnès, *Quantum Philosophy* (Princeton University Press, Princeton, 1999), and references therein to work by R. B. Griffiths, M. Gell-Mann and J. Hartle.
24. J. Surowiecki, *New Yorker*, p. 40 (July 10 &17, 2006).
25. P. Pearle, *Phys. Rev. D* **29**, 235 (1984).
26. A. Zeilinger in *Quantum Concepts in Space and Time*, R. Penrose and C. J. Isham eds, (Clarendon, Oxford, 1986), p. 16.
27. O. Nairz, M. Arndt and A. Zeilinger, *Am. J. Phys.* **71**, 319 (2003).
28. W. Marshall, C. Simon, R. Penrose and D. Bouwmeester, *Phys. Rev. Lett.* **91**, 130401 (2003).
29. A. Bassi, E. Ippoliti and S. l. Adler, *Phys. Rev. Lett.* **94**, 030401 (2005).

292   P. Pearle

30. P. Pearle and E. Squires, *Phys. Rev. Lett.* **73**, 1 (1994).
31. E. Garcia, *Phys. Rev. D* **51**, 1458 (1995).
32. B. Collett, P. Pearle, F. Avignone and S. Nussinov, *Found. Phys.* **25**, 1399 (1995): P. Pearle, J. Ring, J. I. Collar and F. T. Avignone, *Found. Phys.* **29**, 465 (1999).
33. SNO collaboration, *Phys. Rev. Lett.* **92**, 181301 (2004).
34. G. Jones, P. Pearle and J. Ring, *Found. Phys.* **34**, 1467 (2004).
35. F. Karolyhazy, *Nuovo Cimento* **42A**, 1506 (1966) presented a theory of collapse engendered by phase decoherence induced by metric fluctuations. As is the case with CSL, random walk of a small object is predicted by this theory, and Karolyhazy suggested testing it by looking for such motion.
36. G. Gabrielse et al., *Phys. Rev. Lett.* **65**, 1317 (1990).
37. P. Pearle, *Phys. Rev. A* **48**, 913 (1993).
38. P. Pearle in D. H. Feng and B. L. Hu, eds., *Quantum Classical Correspondence: Proceedings of the 4th Drexel Conference on Quantum Nonintegrability* (International Press, Singapore, 1997), p. 69.
39. P. Pearle in F. Pettruccione and H. P. Breuer, eds., *Open Systems and Measurement in Relativistic Quantum Theory* (Springer, Heidelberg, 1999).
40. P. Pearle, *Phys. Rev. A* **72**, 022112 (2005), p.195.
41. A. Bassi, E. Ippoliti and B. Vacchini, *J. Phys. A: Math-Gen* **38**, 8017 (2005).
42. P. Pearle in A.I. Miller, ed., *Sixty-Two Years of Uncertainty* (Plenum, New York, 1990), p. 193.
43. G. Ghirardi, R. Grassi and P. Pearle, *Found. Phys.* **20**, 1271 (1990).
44. P. Cvitanovich, I. Percival and A. Wirzba eds., *Quantum Chaos-Quantum Measurement*, (Kluwer, Dordrecht, 1992), p. 283.
45. N. Dowrick, *Path Integrals and the GRW Model* (preprint, Oxford, 1993).
46. P. Pearle, *Phys. Rev. A* **59**, 80 (1999).
47. O. Nicrosini and A. Rimini, *Found. Phys.* **33**, 1061 (2003).
48. P. Pearle in *Stochastic Evolution of Quantum States in Open Systems and in Measurement Processes*, L. Diosi and B. Lukacs, eds. (World Scientific, Singapore, 1994), p. 79.
49. P. Pearle in R. Clifton, ed., *Perspectives on Quantum Reality* (Kluwer, Dordrecht, 1996), p. 93.
50. P. Pearle, *Phys. Rev. A* **71**, 032101 (2005).
51. L. Diosi, *Phys. Rev. A* **40**, 1165 (1989). See also the discussion and modification in G. C. Ghirardi, R. Grassi and A. Rimini, *Phys. Rev. A* **42**, 1057 (1990).
52. R. Penrose, *Gen. Rel. Grav.* **28**, 581 (1990).
53. P. Pearle and E. Squires, *Found. Phys.* **26**, 291 (1996).
54. Y. Aharonov and D. Z. Albert, *Phys. Rev. D* **29**, 228 (1984).
55. There is a connection to an interesting and imaginative piece of unestablished physics. S. L. Adler, *Quantum theory as an emergent phenomenon* (Cambridge Univversity Press, Cambridge 2004), argues that the formalism of quantum theory can be derived from statistical mechanical equilibrium behavior of a classical dynamics of certain matrix variables, and that the fluctuations from equilibrium can give rise to CSL-type collapse behavior. For a review which summarizes this argument, see P. Pearle, *Studies in History and Philosophy of Modern Physics* **36**, 716 (2005).
56. A. G. Riess et al., *Astr. J.* **116**, 1009 (1998).
57. D. N. Spergel et al., *Ap. J. Suppl.* **170**, 377 (2007).

Is There a Relation Between the Breakdown of the Superposition Principle and an Indeterminacy in the Structure of the Einsteinian Space-Time?

Andor Frenkel

Abstract It has been shown that a small, assessable amount of quantum indeterminacy in the space-time structure leads to the destruction of the coherence of the wave functions (of the Ψ's) of macroscopic bodies, whereas the coherence of the Ψ's of the microparticles remains nearly perfect. Assuming that whenever the coherence gets lost, a breakdown of the superposition principle takes place whether observed or not, it has been possible to formulate the rule for the breakdown such that, due to an instantaneous, stochastic contraction of Ψ, the loss of the coherence is counterbalanced. After each breakdown Ψ undergoes a Schrödinger time evolution until the next breakdown. The successive expansion–contraction cycles keep the indeterminacy of the position of the center of mass (c.m.) of a macroscopic body microscopically small, whereas the indeterminacy of the position of a microparticle may be, and often is macroscopic. The mechanism of the observation-independent decay of a superposition of droplet-tracks in a cloud chamber is also presented.

1 Introduction

I would like to thank Wayne Myrvold, Joy Christian and Kate Gillespie for having organized this colorful, high-spirited conference. Also, I am grateful to Abner for more than 30 years of friendship, and for his help in my work through encouragement and benevolent criticism.

Turning to my topic, let me recall that while the breakdown of the superposition principle is an essential element of the orthodox (von Neumann) theory of quantum measurement, there is no mathematical, quantitative criterion in this theory to tell when, under what circumstances does the superposition principle break down. One

---

A. Frenkel
Research Institute for Particle and Nuclear Physics, Budapest, Hungary
e-mail: frenkel@rmki.kfki.ha

W.C. Myrvold and J. Christian (eds.), *Quantum Reality, Relativistic Causality, and Closing the Epistemic Circle*, The Western Ontario Series in Philosophy of Science 73,
© Springer Science+Business Media B.V. 2009

of the first, perhaps the very first theoretical construct providing such a criterion has been proposed by Károlyházy about 40 years ago [1, 2]. First, he noticed that the incompatibility between the sharply determined structure of the Einsteinian space-time on the one hand, and the quantum mechanical indeterminacies of the positions and of the momenta of the bodies determining that structure on the other hand, can be lifted if one associates a small amount of indeterminacy with the space-time structure. The order of magnitude of this structural indeterminacy could be estimated relying on basic formulas of the two theories (of General Relativity and of Quantum Mechanics) themselves.

Next, Károlyházy has shown that, in its turn, the indeterminacy of the space-time structure induces indeterminacies in the relative phases of the Schrödinger wave function of any physical system propagating on the somewhat "hazy" space-time. For some simple but important generic systems the order of magnitude of the relative phase indeterminacies could be calculated. This made possible, for these systems, to characterize quantitatively the stage at which the coherence gets destroyed. Assuming that at that stage the superposition principle breaks down independently of observation (or of measurement), one obtains sensible results without having introduced any free parameter. Namely, in the case of bodies empirically perceived as macroscopic, the stage for the breakdown is reached, whereas in the case of the feather-weight microparticles it is not even approached.

In Sections 2 and 3 the order of magnitude of the structural indeterminacy in the length of a time interval and in the synchronization of moments of time is given. In Section 4 the indeterminacy (more precisely the spread) of the difference in the values of two synchronized moments of time is expressed in terms of an expectation value in the ground state of space-time. With the help of this expression, in Section 5 the general formula for the spread in the relative phases of the wave function of any isolated physical system is obtained in a simpler form than in [2]. In Sections 6 and 7 the notions of the coherence cell and of the cell length are recalled, and with their help the mass regions of microscopic and macroscopic behavior in the case of homogeneous, spherical solid bodies are quantitatively characterized. In Section 8 it is shown that the transition between these regions takes place at $M^{tr} \approx 10^{-14}$ g, the mass of dust particles and of colloidal grains. In Section 9 the notions of the expansion–contraction cycles and of the cycle period are exposed. Section 10 deals with the "anomalous Brownian motion" [2] of a body on the hazy space-time, and references to papers discussing the prospects of the observation of this anomaly are given. In Section 11 the competition between the localizing effect of the surroundings and of the expansion–contraction cycles is shortly described. In Section 12 the "submacroscopic decay" [2–4] of the superposition of droplet-tracks in a cloud chamber is discussed. Section 13 contains a comment on the preferred basis problem. In the concluding section it is emphasized that quantum measurements and observations should also be described by the unified dynamics exposed in the talk.

2 Lower Bound of the Indeterminacy in the Length of a Time Interval

Examining the indeterminacy of the structure of the Einsteinian space-time induced by various quantum mechanical bodies, Károlyházy noticed the existence of a lower bound of this indeterminacy. The order of magnitude of that bound could be conveniently expressed (see the Appendix) in the form of a relation between the length T of a time interval, and the value $\Delta_L T$ of the smallest possible indeterminacy of T (L stands for lower bound):

$$\Delta_L T \approx T_P^{2/3} T^{1/3} \tag{1}$$

where

$$T_P = 5.4 \times 10^{-44} \text{ s} \tag{2}$$

is the Planck time. The symbol "$\approx$" in (1) means "equal, up to a numerical factor between 0.1 and 10". In the absence of the unified theory of General Relativity (GR) and Quantum Mechanics (QM), this factor could not be calculated precisely. Its looseness does not influence the general picture presented in this talk.

Two restrictions should be put on the value of T:

(i) When deducing relation (1), formulas of non-relativistic QM have been used. Accordingly, (1) can be applied only to time intervals taken along $|v| \ll c$ worldlines (in a reference frame in which the 2.7° K background radiation is isotropic).

(ii) When $T \lesssim T_P$, the very concept of space-time becomes questionable. Therefore for the T's in (1) the restriction

$$T \gg T_P \tag{3}$$

should hold. It is needed to keep well out from those very small space-time domains inside which the laws of physics are not known. The spatial extension of such a domain is, of course, the Planck length

$$\Lambda = cT_P = 1.6 \times 10^{-33} \text{ cm.} \tag{4}$$

Notice that $\Delta_L T$ is an absolute lower bound. For a given value of T, $\Delta_L T$ cannot decrease at the expense of the increase of the indeterminacy of an other quantity, like e.g. Δx at the expense of Δp or vice versa. Notice also how small $\Delta_L T$ is. For $T = 1$ s

$$\Delta_L T \approx 10^{-29} \text{ s} \tag{5}$$

only. Still, as we shall see, such tiny indeterminacies are large enough to destroy the coherence of the wave functions of macroscopic bodies, and small enough to leave the coherence of the Ψ of a microparticle nearly perfect.

It should be emphasized that the relation between $\Delta_L T$ and T involves only the universal constant

$$T_P = \sqrt{\frac{G\hbar}{c^5}}, \tag{6}$$

where G is the constant of gravitation. The parameters (mass, velocity, etc.) of the quantum object used in the deduction of relation (1) dropped out from the final formula. Now, a space-time relation independent of any particular property of matter can be, perhaps even must be attributed to the space-time itself, in particular also to the empty space-time, which in the future unified {GR + QM} theory should be the ground state of space-time. Relation (1) says that the compatibility of GR with QM demands, as a minimum, to admit that the structure of the empty space-time has a slight indeterminacy, therefore it cannot be exactly Minkowskian. We know that in the quantum world the expectation concerning the simplicity of physical ground states (or "vacuums") had to be given up. Examples are the zero point energy of the oscillator, the Dirac sea, the BCS energy gap of superconductors, the vacuum of the elektroweak quantum field theory. A deeper understanding of the slightly indeterminate, non-Minkowskian character of the empty space-time probably cannot be reached without a genuine unification of GR and QM.

3 Lower Bound of the Indeterminacy in the Synchronization of Moments of Time

The existence of the lower bound of the indeterminacy in the length of time intervals along $|v| \ll c$ worldlines involves the existence of a lower bound of the indeterminacy in the synchronization of the moments of time between two such worldlines [3]. To see this consider, first in Minkowskian space-time, two $|v| = 0$ worldlines W_1 and W_2 at a distance r from each other (Fig. 1). A light signal emitted on W_1 arrives back to W_1 from W_2 after a time $2T$, where

$$T = \frac{r}{c}. \tag{7}$$

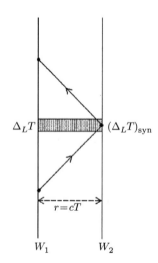

Fig. 1 Synchronization indeterminacy generated by $\Delta_L T$

Breakdown of Superposition Principle due to Space-Time Indeterminacy 297

Since the time interval $2T$ along W_1 has the indeterminacy

$$\Delta_L(2T) \approx T_P^{2/3}(2T)^{1/3} \approx 2^{1/3}\Delta_L T \approx \Delta_L T, \tag{8}$$

the moment of arrival of the signal to W_2 cannot be synchronized with a corresponding moment on W_1 better than with the indeterminacy

$$(\Delta_L T)_{\text{syn}} \approx T_P^{2/3}\left(\frac{r}{c}\right)^{1/3} = \frac{\Lambda^{2/3}r^{1/3}}{c}. \tag{9}$$

Similarly to $\Delta_L T$, this lower bound in the synchronization is an attribute of the space-time structure, therefore it cannot be beaten by clocks, however exact. Notice that, again similarly to $\Delta_L T$, $(\Delta_L T)_{\text{syn}}$ is very small. The relative lower bound

$$\frac{(\Delta_L T)_{\text{syn}}}{T} = \frac{(\Delta_L T)_{\text{syn}}}{r/c} \approx \left(\frac{\Lambda}{r}\right)^{2/3}, \quad (r \gg \Lambda) \tag{10}$$

tends to zero when r tends to infinity, and e.g. for $r = 10$ m ($T \approx 10^{-8}$ s) it is already as small as 10^{-24}. The relative indeterminacy in the time-keeping of the best atomic clocks is much larger.

4 Tentative Expression of $(\Delta_L T)_{\text{syn}}$ Through Vacuum Expectation Values

As exposed in a recent paper of mine [5], in the future unified {GR + QM} theory the indeterminacies $\Delta_L T$ and $(\Delta_L T)_{\text{syn}}$ will presumably appear as spreads of time intervals in the ground state of space-time. In what follows we shall need only the expression for $(\Delta_L T)_{\text{syn}}$. To obtain it, let us consider $t = \text{constant}$ hyperplanes in Minkowskian space-time. For any pair of points x, x' on such a plane we have, of course,

$$t_{x'} = t_x = t, \tag{11}$$

that is

$$t_{x'} - t_x = 0. \tag{12}$$

In the space-time where (9) holds, the value of the difference $t_{x'} - t_x$ is slightly indeterminate. It seems reasonable to assume that in the ground state of space-time (for short in the vacuum state) the expectation value of this difference is equal to the Minkowskian zero value, while the value of its squared spread is equal to $(\Delta_L T)^2_{\text{syn}}$.

Quantities having indeterminacies in their values can be represented by operators. Instead of (12) we should therefore write

$$\langle \hat{t}(t,x') - \hat{t}(t,x) \rangle = 0, \tag{13}$$

where $\langle \ldots \rangle$ denotes the vacuum expectation value of an operator, in our case of the difference of a local time operator $\widehat{t}$ taken at the world points (t,x'), (t,x) [5]. However, hoping that this will not lead to misunderstanding, here we shall use the lighter notation

$$\langle t_{x'} - t_x \rangle = 0. \tag{14}$$

Taking into account (9) and (14), we find

$$
\begin{aligned}
(\Delta_L T)^2_{\text{syn}} &= \langle (t_{x'} - t_x - \langle t_{x'} - t_x \rangle)^2 \rangle \\
&= \langle (t_{x'} - t_x)^2 \rangle \approx \frac{\Lambda^{4/3} |x' - x|^{2/3}}{c^2}.
\end{aligned} \tag{15}
$$

5 Indeterminacies in the Relative Phases of a Quantum State

In regular, non-relativistic QM the time evolution of an isolated physical system of N microparticles with masses

$$M_1, M_2, \ldots, M_N \tag{16}$$

can be described by the Schrödinger wave function

$$\Psi(x,t) = e^{-\frac{i}{\hbar}Ht}\Psi(x,0). \tag{17}$$

Since the time evolution on a Minkowskian space-time is deterministic, Ψ, and therefore its relative phase between any pair of points

$$x = (x_1, \ldots, x_N) \tag{18}$$

and x' of the $3N$-dimensional configuration space, is sharply determined. The wave function of any isolated system, no matter how complicated, is and remains perfectly coherent.

How does the indeterminacy in the synchronization affect the relative phases if Ψ propagates on the hazy space-time? At this point one has to remember [2,3] that on the r.h.s. of (17) the rest energy phase factor

$$e^{-\frac{i}{\hbar}\sum\limits_{\ell=1}^{N} M_\ell c^2 t} := e^{i\phi(t)} \tag{19}$$

has been omitted, because, being independent of x and of $p = -i\hbar(\nabla_1, \ldots, \nabla_N)$, it drops out from all the observables. However, if there is an indeterminacy in the synchronization, then

$$\phi(t) \to \phi(t,x) = -\frac{c^2}{\hbar} \sum_{\ell=1}^{N} M_\ell t_{x_\ell}, \tag{20}$$

Breakdown of Superposition Principle due to Space-Time Indeterminacy 299

because the coordinate of the particle with mass M_ℓ is x_ℓ. The relative rest energy phase between two points x, x' becomes

$$\phi(t,x,x') = -\frac{c^2}{\hbar} \sum_{\ell=1}^{N} M_\ell \left(t_{x'_\ell} - t_{x_\ell}\right). \tag{21}$$

Here a remark is in order. It is not clear what kind of a substitution should be made for t in the phase factor

$$e^{-\frac{i}{\hbar}Ht}, \tag{22}$$

but, at any rate, in the non-relativistic regime the absolute values of the matrix elements of the Hamiltonian H are much smaller than the rest energy, therefore the latter gives the leading contribution.

Relying on our formulas (14), (15) and (21), the spread $\Delta\widehat{\phi}(t,x,x')$ of the relative phase can be easily calculated. From (14) and (21) one sees that

$$\langle \phi(t,x,x') \rangle = 0, \tag{23}$$

therefore

$$\Delta_\phi^2(t,x,x') = \langle (\phi(t,x,x'))^2 \rangle$$
$$= \frac{c^4}{\hbar^2} \sum_{i,\ell=1}^{N} M_i M_\ell \left\langle \left(t_{x'_i} - t_{x_i}\right)\left(t_{x'_\ell} - t_{x_\ell}\right)\right\rangle. \tag{24}$$

The product under the r.h.s. bracket is identically equal to the algebraic sum

$$\frac{1}{2}\left[\left(t_{x'_i} - t_{x_\ell}\right)^2 + \left(t_{x_i} - t_{x'_\ell}\right)^2 - \left(t_{x_i} - t_{x_\ell}\right)^2 - \left(t_{x'_i} - t_{x'_\ell}\right)^2\right] \tag{25}$$

of full squares. Furthermore, due to the $i \leftrightarrow \ell$ symmetry of the product $M_i M_\ell$, the two first terms of (25) give equal contributions to the sum. Taking into account the expression for $\langle (t_x - t_{x'})^2 \rangle$ in (15) one finds

$$\Delta_\phi(x,x') \approx \Lambda^{4/3} \frac{c^2}{\hbar^2} \sum_{i,\ell=1}^{N} M_i M_\ell \left(|x'_i - x_\ell|^{2/3} - \frac{1}{2}|x_i - x_\ell|^{2/3} - \frac{1}{2}|x'_i - x'_\ell|^{2/3}\right). \tag{26}$$

The time argument of Δ_ϕ has been dropped, because Δ_ϕ turned out to be time independent.

We see that Δ_ϕ increases with the masses and with the number of the particles constituting the system, and for a given system Δ_ϕ increases with the distances $|x'_i - x_\ell|$, that is with the separation between the points x, x' in the configuration space. These are encouraging properties from the point of view of the expected loss of the coherence between "macroscopically distinct" components of the wave function of a massive body.

The formula for Δ_ϕ deduced by Károlyházy in the early seventies ([2], see also [6, 7]) looks very different from (26). Namely, it is a momentum space integral of a weighted difference of the Fourier transform of the mass distributions in the two configurations x and x'. It was a surprise to me that the simpler expression (26) is not just approximately, but exactly equal to the original one.

6 The Coherence Cell and the Cell Length of a Single Microparticle

Let us look at Δ_ϕ in the case of a single microparticle. Then in (26) $N = 1$, so that i and ℓ are restricted to $i = \ell = 1$. In the parenthesis only the first term survives. Dropping the index 1 of M_1, x_1 and x'_1, we find

$$\Delta_\phi(x,x') \approx \Lambda^{2/3} \frac{cM}{\hbar} |x' - x|^{2/3} = \frac{\Lambda^{2/3} a^{1/3}}{L}, \tag{27}$$

where

$$L = \frac{\hbar}{cM} \tag{28}$$

is the Compton wavelength of the particle, and

$$a = |x' - x| \tag{29}$$

is the distance between two points $x = (x)$ and x' of the configuration space. From (27) we see that for a given particle Δ_ϕ depends only on a. Namely, $\Delta_\phi = 0$ if $a = 0$, and it increases monotonically with a. This means that the coherence of a normalized wave function is never perfect, because the support of such a wave function cannot be concentrated to a single point, and $\Delta_\phi > 0$ for any pair of points $x' \neq x$. However, it may be that the coherence is and remains nearly perfect even if Ψ extends over a very large domain of the configuration space, and this is what happens in the case of a microparticle.

Indeed, let us consider a sphere of diameter a in the configuration space (x) of a microparticle, e.g. of a neutron. Due to its monotonicity, Δ_ϕ is maximal at the diametrically opposed pairs of points on the surface of the sphere. The Compton wavelength of the neutron being 1.8×10^{-14} cm, we find that for a sphere of diameter as large as that of the Earth ($\approx 10^9$ cm)

$$\Delta_\phi(a_{\text{Earth}}) \approx \frac{(10^{-33})^{2/3}}{10^{-14}} (10^9)^{1/3} = 10^{-5} \ll \pi. \tag{30}$$

This means that the Ψ of a neutron would be still far from losing its coherence even if it expanded over a domain as large as the Globe. The message is: there is no hope to see the neutron interference to die out, no matter how large a monocrystal one would use in the interferometer.

A remark concerning the expression "the domain occupied by the wave function" seems to be in order here. As a rule, Ψ is non-zero (almost) everywhere, so that, strictly speaking, it occupies the whole configuration space. In the present context "the domain occupied by Ψ" refers to the smallest domain Ω in which the weight

$$w_\Omega = \int_\Omega dx |\Psi(x,t)|^2 \tag{31}$$

Breakdown of Superposition Principle due to Space-Time Indeterminacy 301

of Ψ is close to the maximal weight 1, e.g. in which

$$w_\Omega = 1 - 10^{-4}.\tag{32}$$

There is an obvious, but tolerable arbitrariness in the choice of the closeness of w_Ω to 1, similar to that in the choice of the point beyond which exponentially falling tails are considered negligible.

Let me now recall the important notions of the coherence cell and of the cell length.

Between two points x, x' of the configuration space the coherence is completely lost if

$$\Delta_\phi(x,x') \geq \pi;\tag{33}$$

a maximal domain Ω_c of the configuration space in which

$$\Delta_\phi(x,x') \leq \pi \quad \text{for all } x,x' \in \Omega_c,\tag{34}$$

i.e. in the interior of which the coherence is still not lost, has been called "coherence cell" in [2]. Thus, a coherence cell of a microparticle is a sphere of diameter a_c, where a_c is determined by the relation

$$\Delta_\phi(a_c) = \pi.\tag{35}$$

From (27) one finds that

$$\Delta_\phi(a_c) \approx \frac{\Lambda^{2/3} a_c^{1/3}}{L},\tag{36}$$

and, dropping a factor $\pi^3 \approx 30$, (35) gives

$$a_c \approx \left(\frac{L}{\Lambda}\right)^2 L.\tag{37}$$

For the neutron

$$a_c^{ne} \approx \left(\frac{10^{-14}}{10^{-33}}\right)^2 \cdot 10^{-14} = 10^{24} \text{ cm.}\tag{38}$$

The diameter of our Galaxy, with its spherical halo included, is only 10^{21} cm. The wave function of a neutron, and of any other microparticle, occupies only a tiny part of a coherence cell. Ψ is therefore always almost perfectly coherent, no breakdown of the superposition principle occurs. The same situation prevails for any microsystem, i.e. for any system composed of a moderate number of microparticles, free or bound.

The diameter of the coherence cell of a microparticle, characterizing the size of the cell, has been called "coherence width" in [2]. Since this term is often used in other contexts, the name "cell length" seems preferable. Of course, in the case of complex systems—as a matter of fact, already in the case of a system composed of two free microparticles—the coherence cell is not spherical, and several cell lengths may be needed to characterize the coherence cell.

7 The Coherence Cell and the Cell Length of a Homogeneous, Spherical Solid Body; Microscopic and Macroscopic Behavior

At first sight one would think that not much can be said about the indeterminacy of the relative phases of the Ψ of a system consisting of $N \approx 10^{23}$ microparticles. Surprisingly enough, one can say a lot. In the case of a (macroscopically) homogeneous, spherical solid body the coherence cell turns out to be again spherical, and the cell length a_c (the diameter of the cell) can be estimated. The calculations, carried out in two different ways, can be found in [2,6] and in [5].

A salient point hinting at the relative simplicity of the situation, is that the wave function of an isolated solid body expands practically only in the center of mass coordinate subspace. In the $3(N-1)$ dimensional subspace of the relative coordinates there is no appreciable expansion, because the system is tightly bound. Therefore, in very good approximation, in $\Delta_\phi(x,x')$ only the center of mass coordinates $x_{\text{c.m.}}$ and $x'_{\text{c.m.}}$ play a role. Furthermore, if the body is homogeneous and spherical, then the coherence cell turns out to be spherical in the c.m. subspace, and Δ_ϕ depends only on the distance

$$a = \left| x'_{\text{c.m.}} - x_{\text{c.m.}} \right| \tag{39}$$

between the centers of mass in the two configurations x, x'. The calculation shows that the expression for the cell length a_c (the value of a at which Δ_ϕ reaches the value $\pi \approx 1$) depends now on whether a_c turns out to be larger or smaller than the diameter $2R$ of the body in question. Namely,

$$a_c \approx \left(\frac{L}{\Lambda} \right)^2 L \qquad \text{if } a_c \geq 2R, \tag{40}$$

$$a_c \approx \left(\frac{2R}{\Lambda} \right)^{2/3} L \qquad \text{if } a_c \leq 2R, \tag{41}$$

where

$$L = \frac{\hbar}{cM} \tag{42}$$

is the Compton wavelength corresponding to the full mass M of the body. In [2] Károlyházy considered these relations to be "the most important results" of his thesis. Their physical meaning is easy to grasp if one looks at the extreme cases $a_c \gg 2R$ and $a_c \ll 2R$.

If $a_c \gg 2R$, then the wave function of the body can, without losing its coherence, expand in the c.m. coordinate subspace over a domain much larger than the geometrical size $2R$ of the body. A large indeterminacy in the position of the center of mass is a basic characteristic of microbehavior. Notice also that in the whole region $a_c \geq 2R$ the formula for a_c is the same as for the microparticles. (The detailed calculation shows that this is true up to small corrections hidden here in the "$\approx$" symbol.)

Let us now look at the case $a_c \ll 2R$. Then Ψ begins to develop incoherent components as soon as it expands over a domain of linear size larger than a_c, still much

Breakdown of Superposition Principle due to Space-Time Indeterminacy   303

smaller than the geometrical size of the body. At that stage a stochastic contraction of Ψ to one of the coherence cells takes place. So, the indeterminacy of the position of the c.m. remains of the order of $a_c \ll 2R$. Small indeterminacy in the position of a body is a characteristic of macrobehavior. Also, it is noteworthy that in the whole region $a_c \leq 2R$ the formula for a_c depends not only on M, but also on R. It should be noted that (40) and (41) are valid only up to about $R = 1$ m, because for larger bodies the vibrational degrees of freedom contribute appreciably to Δ_ϕ [2].

8 The Transition Between Microscopic and Macroscopic Behavior

It is clear from the preceding section that for spherical, homogeneous solid bodies the transition between micro- and macrobehavior takes place in the region where

$$a_c \approx 2R, \tag{43}$$

that is, where the indeterminacy in the position of the c.m. equals the geometrical size of the body. Taking into account that

$$L = \frac{\hbar}{Mc} = \frac{3\hbar}{4\pi\rho\varrho R^3 c}, \tag{44}$$

for water density $\varrho = 1$ g/cm^3 one finds from (40), or equivalently from (41), that the transition values are

$$a_c^{\text{tr}} \approx 2R^{\text{tr}} \approx 2.5 \cdot 10^{-5} \text{ cm}, \tag{45}$$

$$M^{\text{tr}} = \frac{4\pi}{3}\varrho(R^{\text{tr}})^3 \approx 10^{-14} \text{ g}. \tag{46}$$

This is the mass region of dust particles and of colloidal grains. (The reason for having kept factors of the order of unity when estimating the transition values is that $(R^{\text{tr}})^{10}$ appears in the equation, and e.g. $2^{10} = 1024$.)

The idea that the indeterminacy of the space-time structure may play a role in the transition from micro to macrobehavior, has been raised, among others, by Feynman [8]. The idea has not been pursued, probably because it was thought that if gravitation is at work, then the transition should take place in the region of the Planck mass

$$M_P = \frac{\hbar}{c\Lambda} = 2.2 \times 10^{-5} \text{ g}. \tag{47}$$

Now, even bodies of considerably smaller mass are known to behave macroscopically, therefore M_P cannot represent the transition region.

However, as we have seen, only the amount of the indeterminacy in the space-time structure is fixed by M_P (or, what is the same, by $T_P = \hbar/M_P c^2$). The degree of the loss of the coherence is determined by Δ_ϕ, which depends not only on M_P, but also on the masses constituting the system considered. This leads to

$$M^{\text{tr}} \approx 10^{-14} \text{ g} \ll M_P, \tag{48}$$

a plausible value.

9 Expansion–Contraction Cycles; the Cycle Period τ_c

The observation-independent rule for the breakdown of the superposition principle can be formulated in several ways. The rule proposed by Károlyházy in his thesis [2] says that the breakdown occurs as soon as Ψ occupies at least two non-overlapping coherence cells, and it consists in the instantaneous, stochastic contraction of Ψ to one of these cells (and not to a single point of a cell), with probability proportional to the weight of Ψ in that cell just before the breakdown. The rule for the calculation of the contracted Ψ is also given in [2,7].

In the case of a spherical, homogeneous solid body, after a contraction Ψ expands in the c.m. coordinate subspace as dictated by the Schrödinger equation. When the expansion reaches the linear size $2a_c$, a new contraction throws back Ψ to a single cell, and so on. In the generic case of a Gaussian $\Psi_{\text{c.m.}}$, the time τ_c needed for the doubling of the width from a_c to $2a_c$ is

$$\tau_c \approx \frac{M a_c^2}{\hbar}. \tag{49}$$

In Table 1 a_c and the cycle period τ_c are given for a microparticle, and for bodies in the micro-, the transition, and the macro-region. The astronomical values of a_c and τ_c of a microparticle are due to the strong dependence of a_c on M in the micro-region $a_c \geq 2R$. Indeed, from (40) one sees that a_c is inversely proportional to the third power of M. The same reason explains the spectacular growth of a_c from the transition value 10^{-5} cm to 100 m while M decreases only from 10^{-14} to 10^{-17} g. In the macro-region $a_c \leq 2R$ the dependence of a_c on M is much milder. Namely, according to (41)

$$a_c \sim R^{2/3} L \sim M^{2/9} \cdot M^{-1} \sim M^{-7/9}; \tag{50}$$

Table 1 Characteristic values of a_c and of τ_c

	neutron	mini grain	dust	marble
M (g)	10^{-24}	10^{-17}	10^{-14}	1
$2R$ (cm)	0	10^{-6}	10^{-5}	1
a_c (cm)	10^{24}	10^4	10^{-5}	10^{-16}
τ_c (s)	10^{51}	10^{18}	10^3	10^{-5}

Breakdown of Superposition Principle due to Space-Time Indeterminacy 305

still, when we arrive at $M = 1$ g, $a_c \approx 10^{-16}$ cm only. Thus, the localization of the c.m. of a marble of 1 g is almost pointlike due to the frequent, 10^5 s contractions.

In the Károlyházy cycles the contractions are jumps from the size $2a_c$ to a_c, and they follow each other at finite time intervals τ_c. Pearle [9], Gisin [10, 11] and Diósi [12] proposed procedures in which the breakdowns are infinitesimal, and they are repeated at infinitesimally small time intervals. The applicability of such a procedure to the Károlyházy model has been shown in [13], the main topic of that paper being the comparison of the K model with the model of Ghirardi et al. [14].

10 Anomalous Brownian Motion

The momentum (more precisely the momentum distribution) of the center of mass of an isolated body is conserved during the Schrödinger time evolution. However, at each contraction the indeterminacy in the position of the c.m. shrinks from the linear size $2a_c$ to a_c, and this involves a change

$$\Delta p \approx \frac{\hbar}{\Delta x_{\text{c.m.}}} \approx \frac{\hbar}{a_c} \tag{51}$$

in the momentum of the c.m. The direction of this momentum kick being random, the center of mass deviates from the straight Newtonian (or Ehrenfestian) trajectory, producing a zig-zaging "anomalous Brownian motion" (ABM) [2,6]. For an isolated marble of 1 g the elongation of the ABM would reach the respectable value of 10^{-5} cm within about an hour. Notice that under pure Schrödinger evolution during this time the width of a Gaussian wave packet would increase from the initial size $a_c \approx 10^{-16}$ cm only to 10^{-7} cm. Here the prediction of the K model differs not only from the classical, but also from the orthodox quantum mechanical prediction.

The ABM is fuelled by the kinetic energy

$$\varepsilon = \frac{(\Delta p)^2}{2M} \approx \frac{\hbar^2}{Ma_c^2} \approx \frac{\hbar}{\tau_c} \tag{52}$$

produced at each contraction. For our marble $\varepsilon \approx 10^{-22}$ erg $= 10^{-29}$ J. Since there are about 10^5 contractions per second, during a year an isolated marble would store only 10^{-11} erg. Of course, under normal conditions the surroundings would convert this tiny energy into heat long before an appreciable elongation could appear. On the prospects of observing the ABM under special laboratory conditions see [2,6].

The energy (52) is furnished by the hazy space-time. Whether this is an energy production slightly violating the energy conservation law, or this energy gets back to the space-time structure, is again a question which probably can be answered only in the framework of the future unified theory of GR and QM.

11 Taking into Account the Surroundings

The localizing effect of various natural environments (air, radiations) on quantum objects of various size has been estimated by Joos and Zeh [15]. The comparison of their results with those of the K model shows [13] that for a marble of 1 g the effect of the expansion–contraction cycles is much stronger than that of the surroundings. On the other hand, in the transition mass region the localization by the surroundings (e.g. by the air molecules) is more effective than the 10^{-5} cm wide localization due to the cycles. Notice, however, that while according to the K model the contractions do not allow the localization of the c.m. of a dust particle (of a marble) to be worse than about 10^{-5} cm (10^{-16} cm), the surroundings, while keeping the localization within certain limits in the branches of a superposition, does not prevent the development of branches macroscopically separated from each other. This is a basic difference between decoherence theories with no breakdown of the superposition principle, and theories in which the breakdown takes place [2, 9–12, 14, 16].

12 Submacroscopic Decay of a Superposition of Droplet-Tracks in a Cloud Chamber

The description of the expansion–contraction cycles of a spherical, homogeneous body is simple because Δ_ϕ is determined by a single degree of freedom (the c.m. coordinate of the body).

In the case of a gas, the many quasi-independent degrees of freedom of the gas molecules all contribute to Δ_ϕ, and the situation becomes complicated. Still, realistic results have been obtained in [2] (see also [3, 7]).

Let us consider a cloud chamber filled with vapor. The walls of the chamber are taken into account only as boundary conditions: $\Psi = 0$ on the walls. We shall look at the fate of a state in which $\Psi(x,t)$ is a superposition of two components of comparable weights. In both components there is a water droplet surrounded by the vapor molecules. The components differ in the positions of the droplet.

Imagine, first, a superposition with a droplet in the two branches, but without the vapor. Then, if the droplet is massive enough, the two components of the superposition will belong to different coherence cells, and a random contraction will choose one of the components. If the droplet is not massive enough, then no contraction occurs, the state remains a superposition.

However, if the vapor is there, the thermal motion of its molecules leads to a "submacroscopic" decay of the superposition, even if the droplet is not massive enough to trigger its own expansion–contraction cycle. Namely, due to the difference in the positions of the droplet in the two branches, in the configuration space of the vapor molecules such subdomains develop between which $\Delta_\phi > \pi$, but inside each of which $\Delta_\phi < \pi$. The ensuing random contraction to one of these subdomains does not select at once one of the branches of the superposition. Instead, it modifies randomly the coefficients c_1 and c_2 of the superposition by a small, calculable

Breakdown of Superposition Principle due to Space-Time Indeterminacy 307

amount. As has been shown in [2, 7], the repetition of this procedure amounts to the well-known "win or lose" game between the coefficients. The ith coefficient wins the game with probability $|c_i|^2$, in agreement with the prediction of the orthodox QM. The smaller the droplet, the longer the game lasts. A rough estimate shows [2] that with a droplet composed of 10^6 molecules in the two components of the superposition, the game—that is the exponential decay of the superposition—is practically over in 10^{-7} s, if initially $|c_1| = |c_2| = 1/\sqrt{2}$, and if the separation between the positions of the droplet in the two branches is macroscopic (e.g. 1 cm). Let me point out that both breakdown processes (the expansion–contraction cycles and the submacroscopic decay) satisfy Gisin's criterion [11] of non-superluminal signalling.

13 Comment on the Preferred Basis Problem

In the von Neumann measurement theory the breakdown projects the wave function onto an eigenstate of an operator of some observable. In the K model Ψ is projected not onto an eigenstate, but on a coherence cell. The latter is determined by the dimensionless condition $\Delta_\phi \leq \pi$, which does not choose a basis. We have seen that in the case of a homogeneous, spherical solid body the coherence cell can be conveniently described in the c.m. coordinate subspace, there it is simply a sphere. But this does not mean that the position is a preferred basis. The condition $\Delta_\phi \leq \pi$ could be expressed also e.g. in the c.m. momentum subspace, but the shape of the coherence cell would be more complicated. Thus, in the K model there is no preferred basis, but it is advantageous to choose suitable variables. In the case of the droplet-tracks these are the c.m. coordinates of the droplet in the two branches, and appropriately chosen momentum space coordinates of the vapor molecules [2].

14 Conclusion

According to the orthodox view, in order to extract observable results from QM, one needs at least one agent which does not obey the laws of QM, but behaves "classically".

However, if one allows for a small indeterminacy of the space-time structure, involving also a limitation of the validity of the superposition principle, then the division of the world into objects with purely quantum mechanical or purely classical behavior is not needed. The time evolution of any physical system obeys a unified dynamics, in which the deterministic Schrödinger time evolution is blended with stochastic contractions of the wave function. In principle the Ψ of a microparticle is also subject to contraction, but in practice this never happens. For a solid macroobject the contractions occur frequently, and they would keep the localization of its c.m. very tight even if the body would be completely isolated, and if nobody would observe it.

308   A. Frenkel

The measurement processes and the observations should also fit into the framework of the unified dynamics. In most of the cases troublesome superpositions are killed by one of the two mechanisms described above, independently of a measurement or of observation. This is certainly the case for the superposition of a dying and a healthy Schrödinger cat. In other cases superpositions may decay in the nervous system—a system with many quasi-independent degrees of freedom—while travelling towards the cortex. And one cannot exclude a priori that in some cases an entangled state of a certain domain of the cortex with an outer object develops, even if we do not know what do we feel then.

Thank you for your attention.

Appendix

In this appendix a deduction of relation (1)

$$\Delta_L T \approx T_P^{2/3} T^{1/3} \tag{A1}$$

will be presented, following a line of thought exposed in [2]. Although many of the equalities below are only order of magnitude estimates, for simplicity the equality sign will be used everywhere.

Let us consider the indeterminacy ΔT of the length T of a time interval in the case of a quantum body of mass M, moving with velocity $v \ll c$, forgetting for a moment about General Relativity. If the indeterminacy in the position of the center of mass is Δx, then the indeterminacy in the moment of time of the passage of the c.m. near a point P is

$$\Delta T = \frac{\Delta x}{v}. \tag{A2}$$

Δx should be chosen such that its order of magnitude does not change during the travelling time T. The relation

$$T = \frac{M(\Delta x)^2}{\hbar} \tag{A3}$$

will give a good Δx, because at the end of the travel time the initial Δx becomes only $2\Delta x$. With

$$\Delta x = \frac{\hbar}{\Delta p} \tag{A4}$$

one finds

$$\Delta T = \frac{\hbar}{v \Delta p} = \frac{\hbar}{\Delta K}, \tag{A5}$$

where K is the kinetic energy of the body. (It is easy to verify that from the natural requirement $\Delta T \ll T$ it follows that $\Delta v \ll v$.)

Now let General Relativity act. If the body were at rest, the Minkowskian time interval T would be changed in the vicinity of the body into

Breakdown of Superposition Principle due to Space-Time Indeterminacy
309

$$T' = \left(1 - \frac{r_s}{2R}\right) T, \tag{A6}$$

where R is the linear size of the body, and

$$r_s = \frac{2GM}{c^2} \tag{A7}$$

is its Schwartzshcild radius. If the body moves with velocity $v \ll c$, then in (A7) one should put

$$M \longrightarrow M + \frac{K}{c^2}. \tag{A8}$$

(The correctness of this simple-minded substitution is corroborated by a $1/c^4$ order post-Newtonian calculation. I am indebted to John Stachel for this comment.)

The salient point made by Károlyházy is that although $K \ll Mc^2$ in the non-relativistic regime, ΔK is not zero, and it gives an important contribution to $\Delta T'$:

$$\Delta T' = \Delta T + \frac{G \Delta K}{c^4 R} T. \tag{A9}$$

Indeed, since ΔT and ΔK work against each other, $\Delta T'$ has a minimum. Inserting (A5) into (A9) one gets

$$\Delta T' = \Delta T + \frac{G \hbar}{c^4 R \Delta T} T. \tag{A10}$$

In order to make $\Delta T'$ as small as possible, one should make R as large as possible. Károlyházy argued [2] that since the c.m. of the body reaches distinguishable positions during the time interval ΔT, material parts from which even a light signal cannot reach the c.m. during the time ΔT, cannot belong to the body. Accordingly,

$$R_{\max} = c \Delta T. \tag{A11}$$

Inserting (A11) into (A10) one finds

$$\Delta T' = \Delta T + \frac{T_P^2 T}{(\Delta T)^2}. \tag{A12}$$

The minimal value of $\Delta T'$ as a function of ΔT is

$$(\Delta T')_{\min} = \left(2^{1/3} + \frac{1}{2^{2/3}}\right) T_P^{2/3} T^{1/3}, \tag{A13}$$

and this minimum is reached when

$$(\Delta T) = 2^{1/3} T_P^{2/3} T^{1/3}, \tag{A14}$$

so that the order of magnitude of the lower bound of the indeterminacy of the length of the time interval T is indeed $T_P^{2/3} T^{1/3}$.

References

1. F. Károlyházy, *Nuovo Cimento* **42**, 390, 1966.
2. F. Károlyházy, *Gravitation and Quantum Mechanics of Macroscopic Bodies* (Thesis in Hungarian), Magyar Fizikai Folyóirat **22**, 23, 1974; for a review in English see [4] and [7].
3. F. Károlyházy, in M. Ferrero and A. van der Merwe, eds., *Fundamental Problems in Quantum Mechanics*, Kluwer Academic, Dordrecht, 1995.
4. A. Frenkel, in R. S. Cohen, M. Horne and J. Stachel, eds., *Experimental Metaphysics: Quantum Mechanical Studies for Abner Shimony*, Vol. 1, Kluwer Academic, Dordrecht, 1997.
5. A. Frenkel, *Found. Phys.* **32**, 751, 2002.
6. F. Károlyházy, A. Frenkel and B. Lukács, in A. Shimony and H. Feschbach, eds., *Physics as Natural Philosophy: Essays in Honor of László Tisza*, MIT, Cambridge, 1982.
7. F. Károlyházy, in A. Miller, ed., *Sixty-Two Years of Uncertainty*, Plenum, New York, 1990.
8. R. P. Feynman, *Lectures on Gravitation*, California Institute of Technology, Berkeley, 1962.
9. P. Pearle, *Phys. Rev. D* **13**, 857, 1976; Int. J. Theor. Phys. **18**, 489, 1979; *Phys. Rev. D* **29**, 235, 1984.
10. N. Gisin, *Phys. Rev. Lett.* **52**, 1657, 1984.
11. N. Gisin, *Helv. Phys. Acta* **62**, 363, 1989.
12. L. Diósi, *Phys Lett. A* **120**, 377, 1987; *Phys. Rev. A* **40**, 1165, 1989.
13. A. Frenkel, *Found. Phys.* **20**, 159, 1990.
14. G. C. Ghirardi, A. Rimini and T. Weber, *Phys. Rev. D* **34**, 470, 1986; G. C. Ghirardi, R. Grassi and F. Benatti, *Found. Phys.* **25**, 5, 1995.
15. E. Joos and H. D. Zeh, *Zeit. Phys. B* **59**, 223, 1985.
16. R. Penrose, *Gen. Rel. Grav.* **28**, 581, 1996.

Indistinguishability or Stochastic Dependence?

D. Costantini and U. Garibaldi

Abstract Our approach casts some light upon the probabilistic difference existing among elementary particles, bosons, fermions and classical ones. In order to describe the behaviors of elementary particles we refers to the different values of a parameter governing the stochastic (in)dependence. The equilibrium probability distributions we attain are defined on the same finite and discrete set of vectors.

1 Introduction

Abner Shimony played a very important role in our scientific development. After having devoted ourselves to the foundation of statistics/and solide state physics respectively, in the middle of the 80's we (the authors of the present paper) met in Genova and began to study the foundations of statistical mechanics. Our first paper in this direction [1] was devoted to the (in)distinguishability of classical particles. This paper had been prepared by both of us but, for reasons that are to too long to be recalled here, only one of us signed it. Broadening our field of research, we moved on to elementary particle statistics, namely to Maxwell-Boltzmann, Bose-Einstein, and Fermi-Dirac statistics [2, 3]. Our work aimed at giving a characterization of elementary particles statistics based on the notions of exchangeability and invariance, a probability condition generalizing Carnap's λ-principle [4, 5]. Essentially, what we did was to treat these statistics as probability distributions and to deduce them

---

D. Costantini
Health Physics's Laboratory, Department of Physics, Università di Genova, via Dodecaneso 33, 16146 Ganova, Italy
e-mail: costantinistudio@libero.it

U. Garibaldi
IMEM-CNR, c/o Depeartment of Physics, Università di Genova, via Dodecaneso 33, 16146 Ganova, Italy
e-mail: garibaldi@fisica.unige.it

W.C. Myrvold and J. Christian (eds.), *Quantum Reality, Relativistic Causality, and Closing*   311
the Epistemic Circle, The Western Ontario Series in Philosophy of Science 73,
© Springer Science+Business Media B.V. 2009

312   D. Costantini and U. Garibaldi

by assuming the validity of exchangeability and invariance [6]. Although we were acquainted with the fact that those statistics are equilibrium distributions, we faced them as abstract probability distributions. That is, by means of the following formula (1) we calculated the probability of the descriptions of the elements (complexions) and then by the multiplicities we arrived at those of the descriptions of the cells (occupation vectors), i.e. to the core of the elementary particles statistics. To be more precise, we reached the equilibrium distribution of the occupation vectors of n elements on d cells of the same energy, the difference among the various types of particles being entirely ascribed to different value of λ.

Discussing the talk that we presented at the conference "Probability, dynamics and causality" held in Luino in the spring of 1995 [7], Abner remarked that our research was purely logical. As a consequence our deduction of the three elementary particles statistics was intrinsically static. Moreover with great lucidity he expressed the conviction that an equilibrium distribution must result from time evolution, generated by interparticle collisions, which conserve the total energy of the system, not the energy of a single particle. Thus it is not to be deduced as an abstract theorem regarding the probabilities of individuals constrained to move among cells of the same energy level. Briefly: a satisfactory reconstruction of an equilibrium distribution must be dynamical, and dynamics must be rooted on collisions.

Abner's criticism struck us profoundly. It compelled us to leave the static way we had followed until then. We began to think how to deduce the elementary particle statistics taking time into account. The first consequence of this change of perspective was published two years later [8]. The more recent results in this direction appeared in [9,10]. This is the reason why it seems appropriate to us to give an account of our results in this "dynamical" direction at a Conference in honor of Abner.

Before beginning we would like to recall an occasion when we first showed Abner the example we are going to present in Section 6. The story goes as follows. The Springer-Verlag Italia had decided to translate Abner's wonderful novel into Italian "Tibaldo and the hole in the calender". After the book was printed, the best place to present the Italian translation was surely Bologna, the city in which Tibaldo lived. The presentation of the Italian translation was arranged for late autumn 2000. For the event we invited Abner and his son Jonathan (who illustrated the book) to Bologna. In the few days Abner spent in this town we also organized a brief workshop at the Faculty of Statistics of the University of Bologna. On that occasion we gave a talk on the example to be discussed in Section 6. Unfortunately, there is no published record of that workshop so that the proceedings of this Conference seemed to us the best opportunity to introduce this example.

2 Indistinguishability

Recently the problem of indistinguishability has been once more tackled by Saunders [11]. The question posed by this author is: Why does indistinguishability, in quantum mechanics but not in classical mechanics, force a change in

Indistinguishability or Stochastic Dependence? 313

Statistics? Or: What can explain the difference between classical and quantum statistics? Saunders gave the following answer:

> The structure of their state spaces: in the quantum case the measure is discrete, the sum over states, but in the classical case it is continuous. This makes a difference when one passes to the quotient space under permutations, as we should for particles intrinsically alike. [11]

Thus the equilibrium measure on classical phase space is continuous, whilst it is discrete on Hilbert space.

In order to show a mistake usually made when discussing classical and non classical probability, we focus on a very simple example made by Saunders to stress the difference he points out. He says

> therein lies the difference with classical theory (and the reason why, for two quantum coins, the probabilities $\{H,T\}$ $\{H,H\}$ and $\{T,T\}$ are all the same). Arriving at a quantity $(C\tau)^N/N!$, rather than $(C+N-1)!/N!(C-1)!$, is not an option. [11]

In other words, in the case of throwing the two imaginary quantum coins one must assume the uniform distribution on the three outcomes H,T (one head and one tail), H,H (two heads) and T,T (two tails). On the contrary, when we are throwing two customary classical coins we must assume a uniform probability distribution on the four outcomes H,H (head in the first and in the second) H,T (head in the first and tail in the second), T,H (tail in the first and head in the second) and T,T (tail in the first and in the second). We do not enter into the details of the difference between these two uniform distributions because we shall consider them in the following. We only note: Laplace [12], producing the first scientific work on statistical inference, used a "quantum" probability distribution thus imagining an infinity of "quantum" cards in order to derive his rule of succession. Broad [13] in reconstructing the celebrated Laplace's rule in the case of a bag containing a finite number of counters also worked with a "quantum" probability distribution, as a matter of fact he started hypothesizing an uniform probability on all the $\dfrac{(C+N-1)!}{N!(C-1)!}$ possible "quantum" composition of the bag. It is completely superfluous to note that both Laplace and Broad have had nothing to do with quantum objects.

Saunders' explanation suggests essentially that the difference between classical and quantum statistics is mainly due to the difference in phase space volume. His argument refers to the different nature of elementary particles and it follows the usual old pattern. It is a paraphrase of the well-known argument: Classical particles have trajectories while quantum particles have no trajectories. Such an answer completely obscures the probabilistic side of the question. This has been stressed by A. Bach [14, 15] and us on many occasions.. Regarding classical indistinguishability, Saunders refers to Bach. But he also says

> If there is to be a departure from classical statistics [...], it will require the existence of a fundamental unit of phase space volume, with the dimensions of action [11]

adding in note that this consideration also applies to our attempt, performed in our paper of 1987, to explain quantum statistics by making use of a classical probability function. To this respect, in [1, 2] our argument was essentially that both classical and quantum particles may be described with exchangeable probability-functions. Hence as a consequence of the introduction of an exchangeable probability function, classical and quantum particles are assumed to be probabilistically indistinguishable. But now we are able to go much further. More precisely, we can give in fully probability terms a deduction of the equilibrium probability distribution for the elements of an abstract system. Specializing this distribution we reach equilibrium distributions for classical particles, bosons and fermions. Moreover in a form differing from that of Gentile [25], we may deduce some parastatistics too. These kinds of parastatistics are one parameter equilibrium distributions arising from parameter values different from those of the three well-known statistics. These equilibrium distributions can be applied to econophysics, a new discipline that bridges physics and economics [16].

3 A Result on Random Variables

We begin by recalling a result to be used in what follows. Let $X_1, X_2, ..., X_n, ...$ be a sequence of random variables whose range is $\{1, ..., g\}$, $n_j \equiv \#\{X_i = j; i = 1, 2, ..., n\}$ the occupation number of j in the evidence, that is in the first n random variables, $\mathbf{n} \equiv (n_1, ..., n_g), \sum_{j=1}^{g} n_j = n$, the corresponding occupation vector, and $P(.|.)$ an exchangeable and invariant probability-function. For the definitions of exchangeability and invariance we refer to [6]. It can be proved that, if $\mathbf{x} \equiv X_1 = j_1, ..., X_n = j_n$, then for all j

$$P(X_{n+1} = j|\mathbf{x}) = \frac{\lambda_j + n_j}{\lambda + n}, \tag{1}$$

where $\lambda = \sum_{j=1}^{g} \lambda_j$ and $\frac{\lambda_j}{\lambda} \equiv P(X_1 = j) \equiv p_j$ is the initial probability. $\{\lambda_1, ..., \lambda_g\}$ or equivalently $\{\lambda, p_1, ..., p_{g-1}\}$ is a set of g free parameters to be fixed. From the equation $\lambda \equiv \dfrac{P(X_2 = j|X_1 = h)}{p_j - P(X_2 = j|X_1 = h)}, j \neq h$, we see that: if $\lambda > 0$, the probability-function $P(.|.)$ is positively stochastic dependent; if $\lambda < 0$, the probability-function $P(.|.)$ is negatively stochastic dependent; if $\lambda \to \infty$, the probability-function $P(.|.)$ is stochastic independent. Hence the value of (1) depends on the initial distribution $\mathbf{p} \equiv (p_1, ..., p_g), \sum_{j=1}^{g} p_j = 1$, and the stochastic (in)dependence fixed by λ.

4 Destruction and Creation Probabilities

Now we consider a system S amounting to n *elements* and g *cells* $1, ..., g$. We shall call the number of elements of S the *size* of S. The system-state is the *cells description* or *cell occupation vector* $\mathbf{n} \equiv (n_1, ..., n_j, ..., n_g)$ that may be seen as the set of

Indistinguishability or Stochastic Dependence?

all *element descriptions (complexions)* $\mathbf{x} \equiv X_1 = j_1, ..., X_i = j_i, ..., X_n = j_n$, for all i, $j_i \in \{1, ..., g\}$, whose occupation numbers are $n_1, ..., n_g$. We shall denote by $\mathcal{X}^{(g,n)}$ and $\mathcal{N}^{(g,n)}$ respectively the set of all elements descriptions and the set of all cells descriptions (cell occupation vectors) of S.

When $\mathbf{n}$ is the system-state of S, we look for the probabilities of destroying and creating elements in cells. During these operations the size S shrinks or enlarges, as elements are supposed to be extracted from or added to S. To this end we assume the validity of the condition:

C (*general condition*). Destruction and creation probabilities are exchangeable and invariant, that is they follow (1).

The free parameters in (1) are fixed by the following supplementary conditions:

DC (*destruction condition*). If the starting state is $\mathbf{n}$, for the destruction sequence $D_1, D_2, ..., D_i, i \leq n$, the parameters of (1) take the values $\lambda_j^d = -n_j$ for all j.

CC (*creation condition*). If the starting state is $\mathbf{n}$, for the creation sequence $C_1, C_2, ..., C_i, ...,$ the parameters of (1) take the values $\lambda_j^c = \alpha_j + n_j$ for all j that are available.

We see immediately that the destruction probability is completely determined by the starting state. It does not depend on the type of particle and does not suffer any other constraint. The destruction sequence is finite, and proves to be a random sampling without replacement.

On the contrary this is not the case for **CC**. This latter condition depends on the initial state and on the set $\alpha = \{\alpha_j, ..., \alpha_g\}$ (the initial creation weights) which varies according to the type of particles being created.

In the system we are considering creations may occur only after destructions, and their number must be equal to that of destructions. In other words, sequences of destructions and creations do not change the size of the system. In what follows we shall consider two destructions immediately followed by two creations. According to **C**, **DC** and **CC** these probabilities are

$$P(D_1 = k, D_2 = l; \mathbf{n}) = \frac{n_k}{n} \cdot \frac{n_l}{n-1}, \tag{2}$$

$$P(C_1 = m, C_2 = o; \mathbf{n}_{kl}, \alpha) = \frac{\alpha_m + n_m}{\alpha + n - 2} \cdot \frac{\alpha_o + cn_o}{\alpha + n - 1} \tag{3}$$

where the starting state for creation $\mathbf{n}_{kl} \equiv (n_1, ..., n_k - 1, ..., n_l - 1, ..., n_g)$ is written in terms of the initial state $\mathbf{n}$ of the double operation.

In all the following applications the initial creation weigths $\{\alpha_j, ..., \alpha_g\}$ are supposed to be uniform, so that it is convenient to pose $\alpha_j = \frac{1}{c}$ for all j. Thus (3) can be usefully rewritten as

$$P(C_1 = m, C_2 = o; \mathbf{n}_{kl}, c) = \frac{1 + cn_m}{g + c(n-2)} \cdot \frac{1 + cn_o}{g + c(n-1)}. \tag{4}$$

The $\mathbf{n}$ after the semicolon ";" in (2) and $\mathbf{n}_{kl}$ and c after the semicolon ";" in (3) recall that the free parameters depend on the initial state and the type of particle to which destructions and creations occur.[1] If c is positive, the creation sequence is not limited; if c is negative (in this case its absolute value must be the reciprocal of an integer, see [2]), the creation sequence is limited to $g|c|$. A main constraint, typical of elastic collisions, comes from the conservation of the total energy, as we will see immediately. Indeed (3) is a normalized probability function if all cells can be reached by the double creation. If some couple of cells were forbidden, then the denominator is the sum of the probability of all admitted creations:

$$P(C_1 = m, C_2 = o; \mathbf{n}_{kl}, c) = \frac{(1 + cn_m)(1 + cn_o)}{\sum_{i,j}(1 + cn_i)(1 + cn_j)}. \tag{5}$$

Now we have put forward all probabilistic conditions that can be used to describe elastic binary collisions i.e. collisions of one particle with another one, conserving the total energy. We denote by $\varepsilon(j)$ the energy of the cell j. If an elastic binary collision takes place, then a particle may gain energy whilst the other may lose the same amount of energy. Then for the four cells m, o, k and l involved in such a collision, $\varepsilon(m) + \varepsilon(o) = \varepsilon(k) + \varepsilon(l)$ hold.[2] An elastic binary collision changes the system-state from $\mathbf{n}$ to $\mathbf{n}_{kl}^{mo} \equiv (n_1, ..., n_k - 1, ..., n_l - 1, ..., n_m + 1, ..., n_o + 1, ..., n_g)$. We suppose that k, l, m and o are all different. By (2) and (3) the transition probability we are interested in is

$$P(\mathbf{n}_{kl}^{mo}|\mathbf{n}) = P(\mathbf{n}_{kl}|\mathbf{n})P(\mathbf{n}_{kl}^{mo}|\mathbf{n}_{kl}) = A_{kl}^{mo}(\mathbf{n})n_k n_l (1 + cn_m)(1 + cn_o), \tag{6}$$

where $A_{kl}^{mo}(\mathbf{n})$ takes into account both denominators of (2) and (5). For the following it is useful to write also the inverse transition, from $\mathbf{n}_{kl}^{mo}$ to $\mathbf{n}$:

$$P(\mathbf{n}|\mathbf{n}_{kl}^{mo}) = P(\mathbf{n}_{kl}|\mathbf{n}_{kl}^{mo})P(\mathbf{n}|\mathbf{n}_{kl})$$
$$= A_{mo}^{kl}(\mathbf{n}_{kl}^{mo})(n_m + 1)(n_o + 1)(1 + c(n_k - 1))(1 + c(n_l - 1)). \tag{7}$$

We note that the constants of proportionality in (6) and (7) are equal, i.e. $A_{mo}^{kl}(\mathbf{n}_{kl}^{mo}) = A_{kl}^{mo}(\mathbf{n})$ (see [6]).

5 Stochastic Dynamics

Now we consider the stochastic dynamics of $\mathcal{S}$, that is the sequence of random variables

$$\mathbf{S}(0), \mathbf{S}(1), ..., \mathbf{S}(t), \mathbf{S}(t+1), ... \tag{8}$$

---

[1] For the sake of simplicity we considered destructions and creations occurring in all different cells. To take the general case into account one must introduce Kronecker's functions for repeated indices (see [6]).

[2] We suppose that all microstates of the same energy communicate via binary collisions. If this were not the case, we could consider ternary or even r-ary collisions, till the ergodic set fills the whole energy shell. The form of the equilibrium distribution does not depend on r. The sole effect is that of enlarging its domain, improving communication within the energy shell, and of increasing the rate of approach to equilibrium.

Indistinguishability or Stochastic Dependence?   317

whose range is the set of possible occupation vectors, that from now on will be called microstates. If $P(\mathbf{S}(t+1) = \mathbf{n}'|\mathbf{S}(t) = \mathbf{n})$ is a function of $\mathbf{n}$ and $\mathbf{n}'$ the sequence is a homogeneous Markov chain. The transition matrix that describes a binary elastic collision is (6). Let us suppose that for the initial state $P(\mathbf{S}(0) = \mathbf{n}_0) = 1$ holds. It is apparent that due to repeated transitions (collisions) the probability mass spreads on all microstates that are reachable from $\mathbf{n}_0$ *via* binary elastic collisions. All these microstates have the same energy as the initial one, say $E_0 = E(\mathbf{n}_0)$.[3] A probability distribution $\pi(\mathbf{n})$ defined on the set of microstates is an equilibrium probability distribution for S when, whatever may be the initial state $\mathbf{n}_0$, $\lim_{t \to \infty} P(\mathbf{S}(t) = \mathbf{n}|\mathbf{S}(0) = \mathbf{n}_0)$ exists and neither depends upon time nor upon the initial state of the system. Instead a probability distribution on the states of S is invariant (stationary) if it does not change with time. If a set of states is ergodic (that is all states communicate with each other reversibly), there exists a stationary distribution for this set. But the stationary distribution and the equilibrium distribution coincide only if the set is aperiodic. As our set is aperiodic, we have to search the stationary distribution $\pi(\mathbf{n})$ of the Markov chain whose transition probability is (6).

The probabilistic time evolution of S is given by the Chapman–Kolmogorov equation, that is

$$P(\mathbf{S}(t+1) = \mathbf{n}) = \sum_{\mathbf{n}'} P(\mathbf{S}(t+1) = \mathbf{n}|\mathbf{S}(t) = \mathbf{n}')P(\mathbf{S}(t) = \mathbf{n}'), t = 0, 1, 2, \ldots,$$

that might be written as

$$P(\mathbf{S}(t+1) = \mathbf{n}) - P(\mathbf{S}(t) = \mathbf{n}) =$$

$$\sum_{\mathbf{n}'} P(\mathbf{S}(t+1) = \mathbf{n}|\mathbf{S}(t) = \mathbf{n}')P(\mathbf{S}(t) = \mathbf{n}') - \sum_{\mathbf{n}'} P(\mathbf{S}(t+1) = \mathbf{n}'|\mathbf{S}(t) = \mathbf{n})P(\mathbf{S}(t) = \mathbf{n}).$$

This is a discrete "Master Equation". When for any pair $\mathbf{n}' \neq \mathbf{n}$, the equality

$$P(\mathbf{S}(t+1) = \mathbf{n}|\mathbf{S}(t) = \mathbf{n}')P(\mathbf{S}(t) = \mathbf{n}') = P(\mathbf{S}(t+1) = \mathbf{n}'|\mathbf{S}(t) = \mathbf{n})P(\mathbf{S}(t) = \mathbf{n}) \quad (9)$$

holds, then

$$P(\mathbf{S}(t+1) = \mathbf{n}) = P(\mathbf{S}(t) = \mathbf{n}) = \pi(\mathbf{n}).$$

This equality asserts that the distribution $\pi(\mathbf{n})$ does not change over time. Hence a distribution satisfying (9) is invariant. The set of equations (9) expresses the detailed balance between pairs of occupation vectors belonging to the same ergodic set. Roughly speaking, the meaning of (9) is that the probability flux from $\mathbf{n}$ to $\mathbf{n}'$ equals the flux from $\mathbf{n}'$ to $\mathbf{n}$.

---

[3] Here we suppose that all microstates belonging to the same energy E_0 (that is the energy shell) may reach each other via binary elastic collisions. If it is not the case, see the previous footnote.

6 Equilibrium

For binary collision such that $\varepsilon(m)+\varepsilon(o)=\varepsilon(k)+\varepsilon(l)$, considering (6) and (7), the equations expressing the detailed balance ensure that the equilibrium distribution is reached if a probability $\pi_c(\cdot)$ exists on all possible microstates such that

$$\frac{\pi_c(\mathbf{n}_{kl}^{mo})}{\pi_c(\mathbf{n})} = \frac{n_k n_l(1+cn_m)(1+cn_o)}{(n_m+1)(n_o+1)(1+c(n_k-1))(1+c(n_l-1))}. \tag{10}$$

This c-distribution exists, and the quantum cases are trivial. For $c = +1$, the left side of (10) is equal to 1. This means that all microstates reached by elastic binary collision have the same probability, that is the equilibrium probability distribution $\pi_1(\mathbf{n})$ is uniform on the energy shell. This is the Bose–Einstein statistics. For $c = -1$, the left side of (10) is equal to 1 if $n_k = n_l = 1$ and $n_m = n_o = 0$. This means that all microstates satisfying the Pauli exclusion principle reached by elastic binary collisions have the same probability. That is the equilibrium probability distribution $\pi_{-1}(\mathbf{n})$ is uniform on all microstates of the energy shell which satisfy Pauli's principle. This is the Fermi–Dirac statistics. For $c = 0$, $\dfrac{\pi_0(\mathbf{n}_{kl}^{mo})}{\pi_0(\mathbf{n})} = \dfrac{n_k n_l}{(n_m+1)(n_o+1)}$.

This equality is satisfied by a $\pi_0(\mathbf{n})$ proportional to $\dfrac{n!}{\prod_{j=1}^{g} n_j!}$. This means that the equilibrium probability distribution allots the same probability to all elements description (complexions) belonging to the energy shell. This is the Maxwell–Boltzmann statistics.

Hence the homogeneous Markov chain whose transition probability is governed by **C**, **CD** and **CC** leads the system towards the following equilibrium probability distributions

$$\pi_c(\mathbf{n}) = \begin{cases} \text{const} & \text{for } E(\mathbf{n})=E(\mathbf{n}_0) & \text{if } c = +1 \\ \text{const} & \text{for } E(\mathbf{n})=E(\mathbf{n}_0),\, n_j = 0,1 & \text{if } c = -1 \\ \text{const} \times (\prod_{j=1}^{g} n_j!)^{-1} & \text{for } E(\mathbf{n})=E(\mathbf{n}_0) & \text{if } c = 0. \end{cases} \tag{11}$$

where the set $\{\mathbf{n}: E(\mathbf{n}) = E(\mathbf{n_0})\}$ is the energy shell of the initial state $\mathbf{n}_0$.

Values of c different from those considered in (11) have no interest in elementary particle physics. These values eventually characterize parastatistics. Nothing prevents our approach from being used fruitfully with new values of c or, dropping the equality of all p_j, for arbitrary values of λ_j. This happens in econophysics where parastatistics may be profitably used [16, 17].

The three distributions (11) are equilibrium distributions of a family of Markov chains that mimic binary elastic collisions between elements, trigged by a sole parameter, c. We note that they are usually assumed as postulates (if we follow Tolman [18], to our knowledge still the best account of the foundations of Statistical

Indistinguishability or Stochastic Dependence? 319

Mechanics). With regard to the different interpretations of the probability in the usual vindication *versus* the probabilistic dynamics here presented, we refer to [19].

We now consider the level occupation vector (the macrostate) $\mathbf{N}$, whose ith component is the sum of the occupation vectors of all g_i cells belonging to the ith energy level ε_i. Its Markov transition matrix, derived by (6), is given by

$$P(\mathbf{N}_{ij}^{kl}|\mathbf{N}) = P(\mathbf{N}_{ij}|\mathbf{n})P(\mathbf{N}_{ij}^{kl}|\mathbf{N}_{ij}) = A_{ij}^{kl}(\mathbf{N})N_iN_j(g_k+cN_k)(g_l+cN_l) , \qquad (12)$$

with the constraint $\varepsilon_k + \varepsilon_l = \varepsilon_i + \varepsilon_j$. The equilibrium distribution $\pi_c(\mathbf{N})$ is proportional to $\prod_i \dfrac{c^{N_i}(g_i/c)^{[N_i]}}{N_i!}$, that is

$$\pi_c(\mathbf{N}) \sim \prod_i \frac{c^{N_i}(g_i/c)^{[N_i]}}{N_i!}, \qquad (13)$$

that specializes in the well-known combinatorial factors $\prod_i \dbinom{N_i+g_i-1}{N_i}, \prod_i \dfrac{g_i^{N_i}}{N_i!}$ and $\prod_i \dbinom{g_i}{N_i}$ for $c = 1, 0, -1$. In the thermodynamic limit the most probable macrostate is given by

$$N_i^* = \frac{g_i}{e^{\beta\varepsilon_i-\nu} - c}. \qquad (14)$$

Remember that to achieve (14) it is necessary to assume Stirling's approximations of factorials, that is $g_i, N_i \gg 1$.[4] The same form can be deduced for a quite different probability notion, the expectation value of N_i. In fact by considering the average of the balance equation:

$$E[N_iN_j(g_k+cN_k)(g_l+cN_l)] = E[N_kN_l(g_i+cN_i)(g_j+cN_j)], \qquad (15)$$

and supposing that the mean of the product is equal to the product of the means, (15) is satisfied if for any level $\dfrac{E[N_i]}{g_i+cE[N_i]} = e^{-\beta\varepsilon_i+\nu}$, that is if

$$E[N_i] = \frac{g_i}{e^{\beta\varepsilon_i-\nu} - c}. \qquad (16)$$

For systems whose equilibrium probability has a strong maximum around $\mathbf{N}^*$ we have $E[N_i] \equiv \sum_{\mathbf{N}} N_i P(\mathbf{N}) \to N_i^*$. As the covariance of two level occupation numbers is a very rapidly decreasing function of the number of levels, it happens that (16) approximately holds also for very small systems, where N_i^* is meaningless.

---

[4] Note that for quantum particles in the case of no degeneracy (that is $g_i = 1$) all level occupation vectors are equiprobable, and (14) has no meaning, notwithstanding the system is large. This is the deep reason for grouping energy levels into classes (see [20]), so that the corresponding macrostate is a much coarser description than the exact one presented here .

7 c-Thermodynamics of Small Systems

We stress that our derivation is exact. An interesting consequence of this is that the equilibrium probability distribution (13) does not depend upon the size of S. Obviously when the size is that of the systems usually taken into account, i.e of the order of 10^{23} elements, the thermodynamics is the usual one, that is based on (14). But two improvements are possible: the first is to consider c for values different from $0, \pm 1$, that is abandoning the comfortable realm of uniform equilibrium distributions on some domains. This improvement has probably little use in Physics, but can be interesting for systems of elements whose correlation is whatever, and whose motion is constrained by some conservation law. A general conceptual advance should be reached for the notion of entropy, usually tied to the number of available microstates, and so bounded to uniform distributions. A second improvement is the possibility to treat exactly the probabilistic c-thermodynamics of very small systems. In this section we shall give an example of such a thermodynamics.

7.1 An Elastic Dipole

Let us consider two particles of mass m in a cubic box of side L. The correlation between them is described by c. Consider the stationary wave functions. There is one fundamental state whose energy is $\varepsilon_1 = 3\frac{h^2}{8mL^2}$, $g_1 = 1$, where h denotes Planck's constant. Three states whose energy is $\varepsilon_2 = 6\frac{h^2}{8mL^2}$, $g_2 = 3$. Three states whose energy is $\varepsilon_3 = 9\frac{h^2}{8mL^2}$, $g_3 = 3$. Let the two particles are in some cell of the first excited level. We limit our attention to macrostates. Let the initial macrostate be $(0, 2, 0) \equiv A$. After the first collision we have two possibilities: either the system stays in A, or it will be found in $(1, 0, 1) \equiv B$ (see Fig. 1)

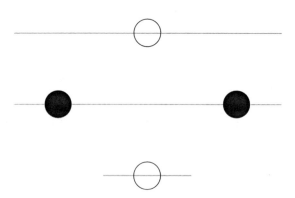

Fig. 1 The elastic dipole. The width of the levels is proportional to g_i; the macrostate A is black, B is white

Indistinguishability or Stochastic Dependence? 321

From (12) $P(B|A) \sim 2g_1g_3$, while $P(A|B) \sim g_2(g_2+c)$, and the equilibrium probability distribution is such that

$$\frac{\pi(A)}{\pi(B)} = \frac{P(A|B)}{P(B|A)} = \frac{g_2(g_2+c)}{2g_1g_3} = \frac{3+c}{2} \tag{17}$$

Hence the odds for A are respectively 2, 1.5 and 1 in the thee cases $c = 1, 0$ and -1.[5] The state wherein particles are close together is preferred proportionally to c.

Now suppose that $\pi(A)$ and $\pi(B)$ are estimated by the Maximum Entropy Principle [21]. Following Jaynes, the cell entropy associated to a macrostate is $S(\mathbf{N}) = \sum S(N_i)$, and $S(N_i) = -\sum_{\mathbf{n}^{(i)} \in N_i} P(\mathbf{n}^{(i)}|N_i) \ln P(\mathbf{n}^{(i)}|N_i)$, where $\{\mathbf{n}^{(i)} \in N_i\}$ denotes the set of all occupation numbers of a level containing N_i particles. This quantity is easy to calculate only in the quantum cases due to the fact that $P(\mathbf{n}^{(i)}|N_i)$ is uniform, and one gets $S_{BE}(\mathbf{N}) = \sum_i \ln \binom{N_i + g_i - 1}{N_i}$, and $S_{FD}(\mathbf{N}) = \sum_i \ln \binom{g_i}{N_i}$. It is easy to see that, posing $P(\mathbf{N}) \sim e^{S(\mathbf{N})}$, we obtain the result (13) in the case $c = \pm 1$. As long as we deal with uniform distributions it is right that $\pi(\mathbf{N}) \sim e^{S(\mathbf{N})}$. But in general $P(\mathbf{n}^{(i)}|N_i)$ is a g_i- generalized Polya distribution, whose probabilistic entropy is hard to calculate. The case $c = 0$ it is a multinomial distribution, and its probabilistic entropy is different from $\ln \frac{g_i^{N_i}}{N_i!}$ forseen by (13).[6]

This example is so simple that we calculate exactly the following table:

	$c = 1$	$c = 0$	$c = -1$
$S(A)$	$\ln 6$	1.74	$\ln 3$
$S(B)$	$\ln 3$	$\ln 3$	$\ln 3$
$exp[S(A) - S(B)]$	2	1.9	1
$\frac{\pi(A)}{\pi(B)}$	2	1.5	1

It is obvious that the "true" probability is that provided by the probabilistic dynamics. In fact we show the results of a computer simulation consisting in 10,000 collisions, starting from B. Here $v(A)$ vs $v(B)$ are the frequencies for the system to stay in A vs B

	$c = 1$	$c = 0$	$c = -1$
$v(A)$	6,656	5,992	4,941
$v(B)$	3,344	4,008	5,059
$\frac{v(A)}{v(B)}$	1.99	1.495	.98
$\frac{\pi(A)}{\pi(B)}$	2	1.5	1

---

[5] If the case $c = -1$ is referred to a fermion, one could object that the degeneracies have to be multiplied by 2 due to the spin of the particle. In this case $\frac{g_2(g_2+c)}{2g_1g_3} = \frac{6+c}{4} = \frac{5}{4}$.

[6] Note that for $c = 0$ it is easy to calculate the entropy of particles, that is $\ln g_i^{N_i}$, as particles are uniformly distributed. The lacking factor $N_i!^{-1}$ can be added without danger because in most physical applications of the Mawell–Boltzmann statistics N_i is considered very small with respect to g_i.

For $c = \pm 1$ to maximize $P(\mathbf{N})$ is equivalent to chose the macrostate whose entropy is maximum. In fact, in absence of any dynamics, it may be reasonable to suppose that will be achieved the macrostate whose entropy is maximum.

As a consequence for $c \neq \pm 1$ the most probable macrostate does not coincide with the macrostate which maximizes the entropy of microstates. And the thermodynamical entropy is bounded to the probability of the macrostate, not to its entropy.

7.2 No Most Probable Macrostate

Let us consider a system of equispaced non-degenerate energy levels, that is $\varepsilon_i = i\varepsilon$, $g_i = 1$. Suppose that the initial state is such that $n = 30$ elements are put in the third level. They suffer binary elastic collisions, and allow the correlation parameter $c = 1$. Figure 2 shows the result of a simple computer simulation: the grey dots represent the mean occupation numbers of the first, second,..., 40th level after $s = 10,000$ collisions.

The conservation of the energy $E = 90\varepsilon$ implies that the maximum reachable level corresponds to $i = 61$. The line (continuous for graphical opportunity) represents the function $\frac{1}{e^{\beta i - \nu} - 1}$, where β and ν are obtained as numerical solutions of the system $\sum_{i=1}^{61} \frac{1}{e^{\beta i - \nu} - 1} = 30$, $\sum_{i=1}^{61} \frac{i}{e^{\beta i - \nu} - 1} = 90$. It is apparent that the two curves are close to each other, so that we can conclude that (16) is not so bad even in the case of only 30 elements. We can reasonably suppose that when (16) begins to hold, it makes sense to introduce intensive variables like the temperature and the chemical potential, so that the usual apparatus of thermodynamics "emerges" form the probabilistic dynamics.

The example has been chosen for the following reason. If $c = 1$ all microstates are equiprobable, which is also the case for all macrostates if $g_i = 1$. In this peculiar

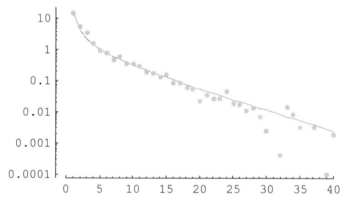

Fig. 2 Simulated (gray dots) and theoretical (continuous) mean occupation numbers of the energy levels; $n = 30, c = 1, s = 10,000$

Indistinguishability or Stochastic Dependence? 323

case there is no most probable macrostate, and (14) is meaningless. This not only because the system is small, but also due to the essential uniformity of the equilibrium distribution.[7]

8 Conclusion

Our approach casts some light upon the difference among elementary particles. The difference lies in the properties of the probability-function used to describe their behaviour. The discrete or continuous nature of the phase-space does not matter. The equilibrium probability distributions we attained, i.e. (11), are defined on the same finite and discrete set of vectors, that is, the set of all vectors which can be attained by elastic collisions starting from the initial state. The difference between classical and quantum statistics is explained by the values of c used in the transition probabilities. In the various cases we consider different creation probabilities which allot different probability values to elements descriptions (complexions) of the same system. This has already been seen by Ehrenfest [22], who has twice played an influential role. Firstly, when he (together with Tatiana) introduced the celebrated urn model [23], whose broad generalization is at the base of our attempt. Secondly, there is the far less known paper written five years later, where he pointed out that the essential traits (*Zuege*) of the hypothesis of *Lichtquanten* were to be located in the different probabilities allotted to the complexions. In other words, in order to reach a uniform distribution on the set of cells' descriptions it is necessary to allot different probability values to the elements' descriptions. Twenty years later substantial progress in this approach has been made another neglected work, due to Brillouin [24], who tried to unify the three statistics using a classical probability space. The c-Thermodynamics is promising for economic and genetic applications, while the thermodynamics of small systems, apart from its foundational interest, is possibly fruitful in the realm of nanoscience.

References

1. Costantini, D. (1987) "Symmetry and the indistinguishability of classical particles", *Physics Letters A* **123**, 433–436.
2. Costantini, D. and Garibaldi, U. (1988) "Elementary particles statistics reconsidered", *Physics Letters A* **134**, 161–164.
3. Costantini, D. and Garibaldi, U. (1990) "A unified theory for elementary particles statistics", *Il Nuovo Cimento* **105**, 371–380.
4. Carnap, R. (1952) *The Continuum of Inductive Methods*, The University of Chicago Press, Chicago.
5. Carnap, R. (1971) "A basic system of inductive logic, Part 2", *Studies in Inductive Logic and Probability, Volume II* (Jeffrey ed.). University of California Press, Berkely and LosAngeles.

---

[7] Observe that the case $c = 0$, where all descriptions of the elements are equiprobable, leads to the well-known Einstein's crystal.

6. Costantini, D. and Garibaldi, U. (1989) "Classical and quantum statistics as finite random processes", *Foundations of Physics* **19**, 743–754.
7. Costantini, D. and Garibaldi, U. (1997) "Predictive laws of association in statistics and physics", *Probability, Dynamics and Causality* (Costantini and Galavotti eds.), Kluwer Academic Publishers, Dordrecht.
8. Costantini, D. and Garibaldi, U. (1998) "A probability foundation of elementary particle statistics. Part II", *Studies in the Historical Philosophy of Modern Physics* **29**, 37–57.
9. Costantini, D. and Garibaldi, U. (2000) "A purely probabilistic representation for the dynamics of a gas of particles", *Foundations of Physics* **30**, 81–999.
10. Costantini, D. and Garibaldi, U. (2004) "The Ehrenfest fleas: from model to theory", *Synthese* **139**, 107–142.
11. Saunders, S. (2006) "On the explanation for quantum statistics", *Studies in History and Philosophy of Modern Physics* **37**, 192–211.
12. Laplace, P. S. (1774) "Mémoire sur la probabilité des causes par les évènements", *Mémoires de l'Académie royale des science presentés par divers savan* **6**, 621–656.
13. Broad, C. D. (1927/8) "The principles of problematic induction", *Proceedings of the Aristotelian Society* **28**, 1–46.
14. Bach, A. (1988) "The concept of indistinguishable particles in classical and quantum physics", *Foundations of Physics* **18,** 639–649.
15. Bach, A. (1997) *Indistinguishability Classical Particles*. Berlin: Springer.
16. Costantini, D. and Garibaldi, U. (2006) "Parastatistics in Economics?" in *The foundations of Quantum Mechanics* (Garola, Rossi and Sozzo eds.), Word ScientificPublishing Co, Singapore.
17. Garibaldi, U., Penco, M. A., and Viarengo, P. (2003) "An exact physical approach for market participation models", in *Heterogeneous Agents, Interactions and Economic Performance* (Cowan and Jonard eds.) Lecture Notes in Economics and Mathematical Systems Series, 91–103.
18. Tolman, R.C. (1938) *The Principles of Statistical Mechanics*, Oxford University Press, London.
19. Costantini, D. and Garibaldi, U. (2006) "Liouville Equation and Markov Chains: Epistemological and ontological probabilities", in *Quantum Mechanics* (Bassi, Duer, Weber, and Zanghì eds.), Aip, Melville, NewYork, pp. 68–76.
20. ter Haar, D. (1954) *Elements of Statistical Mechanics*, Rinehart & Company, New York.
21. Jaynes, E. T. (1983) *Papers on Probability, Statistics and Statistical Physics* (Rosenkranz ed.) Reidel, Dordrecht.
22. Ehrenfest, P. (1911) "Welche Zuege der Lichtquantenhypothese spielen in der Theorie der Waermestrahlung eine wesentliche Rolle?" *Annalen der Physik* **36**, 91–118.
23. Ehrenfest, P. and Ehrenfest, T. (1907) "Über zwei bekannte Einwände gegen Boltzmanns *H*-Theorem", *Physikalische Zeitschrift* **8**, 311–316.
24. Brillouin, L. (1927) "Comparaison des différentes statistiques appliqué aux problèmes des quanta", *Annales de Physique (Paris)* **VII**, 315–331.
25. Gentile, G. Jr. (1940) "Osservazioni sopra le statistiche intermedie", *Il Nuovo Cimento*, *a.XVII*, 10.

Part V
Relativity

Plane Geometry in Spacetime

N. David Mermin

Abstract Minkowski's spacetime diagrams are extracted directly from Einstein's 1905 postulates, using only some very elementary plane geometry.

I have spent a significant part of my career looking at familiar results from unfamiliar perspectives. This has been partly because I very much enjoy teaching courses about aspects of physics to nonscientists, for whom the conventional approach is almost always an impenetrable thicket. But it's also because I have a lot of trouble understanding physics myself. If I don't think it through in my own terms, more often than not I can't make much sense of it.

The fruits of this self-indulgent approach have elicited a broad spectrum of reactions. The worst was a proposal review from an extremely distinguished senior physicist who anonymously complained that my work was characterized by a surprising lack of originality. (The only encouraging word in the entire report was "surprising".) Less depressing was "What made you write a paper about *that*?" from a cosmologist friend and "I *still* don't understand what bothers you", from a beloved postdoctoral advisor. Moving in the positive direction through nervous silence, one passes by "Nice!" in varying degrees of intensity and sincerity until one finally reaches those who have truely appreciated what I was trying to do. Of these two stand out from the others in the warmth of their encouragement, the penetration of their constructive criticisms, and the high regard in which I hold them quite aside from our direct interactions. One was the late Ed Purcell. The other was, and remains to this day, Abner Shimony.

So I know that Abner won't mind my revisiting a subject that has been with us in pretty much its current form for 98 years, and since this meeting is for Abner, the rest of you will just have to put up with it. Even Abner, however, may be tempted to catch up on lost sleep, because much of my talk can be found in Chapter 10 of a new book I sent him last year. Although I can hope that he has not read it, knowing Abner

---

N.D. Mermin
Laboratory of Atomic and Solid State Physics, Cornell University, Ithaca, NY 14853-2501, USA
e-mail: ndm4@cornell.edu

W.C. Myrvold and J. Christian (eds.), *Quantum Reality, Relativistic Causality, and Closing the Epistemic Circle,* The Western Ontario Series in Philosophy of Science 73,
© Springer Science+Business Media B.V. 2009

I very much fear that he has. Nevertheless, Abner, pay attention. While I was putting this talk together my muse—curiously enough, also named Abner Shimony—kept raising questions that hadn't occurred to me, cleverly objecting to obvious points, and otherwise helping me to come up with a better formulation than the one you've already read.

My subject is how to construct Minkowski's space-time diagrams (1908) straight from Einstein's postulates (1905) without the intermediate of the Lorentz transformation or the invariant interval. I also assume no familiarity with graphs or coordinate systems, and in this respect my approach is to the conventional approach as the plane geometry of Euclid is to the analytical geometry of Descartes.

As this point of view evolved I found myself led into some entertaining bywaters, which various relativists I consulted seemed not to have visited. In particular, if you do chose to catch up on some lost sleep, keep an ear open for the notion of the *light rectangle* which I commend to all as a genuinely useful unfamiliar relativistic tool.

Here are Einstein's two postulates. Purcell once complained that I had quoted an execrable translation of them, so I give below my own translation. But because my knowledge of German is limited to Schubert lieder and the operas of Richard Strauss, I also reproduce the original texts.

1. In electrodynamics, as well as in mechanics, no properties of phenomena correspond to the concept of absolute rest. ... *dem Begriffe der absoluten Ruhe nicht nur in der Mechanik, sondern auch in der Elektrodynamik keine Eigenschaften der Erscheinungen entsprechen.*
2. Light always propagates in empty space with a definite velocity c, independent of the state of motion of the emitting body. ... *sich das Licht im leeren Raume stets mit einer bestimmten, von Bewegungszustande des emittierenden Körpers unabhängigen Geschwindigkeit V fortpflanze.*

My Abner-muse kept pointing out that I was slipping in other postulates down the road, so I decided to present up front a third postulate. Having formulated it, I found it sounded familiar, so I looked up Einstein (1905) and sure enough, he states it himself a few pages after the two famous ones:

3. If a clock at A runs synchronously with clocks at B and C, then the clocks at B and C also run synchronously relative to each other. *Wenn die Uhr in A sowohl mit der Uhr in B als auch mit der Uhr in C synchron läuft, so laufen auch die Uhren in B und C synchron relativ zueinander.*

The third postulate comes in an obviously equivalent version:

3'. If event A is simultaneous with event B and event C, then events B and C are also simultaneous.

I will combine this with its equivalent under the interchange of space and time:

3''. If an event A happens in the same place as event B and event C, then the events B and C also happen in the same place.

While Postulate 3 can probably be extracted from Postulates 1 and 2 in the case of a single spatial dimension, it has an independent status in two or more spatial dimensions and constructing this talk led me to appreciate why Einstein had made what I had once regarded as an unnecessarily pedantic remark. No pedant he.

Plane Geometry in Spacetime

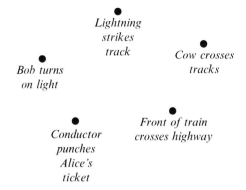

Fig. 1 Some events of interest to Alice

An *event*, of course, is something that happens at a definite place and time: a point in spacetime. It is, as Einstein (1905) also wisely points out, an idealization and, I would have said, a classical (i.e. pre-quantum) idealization, in that it is an unambiguous phenomenon whose spatial and temporal extent we are willing to treat as inconsequential.

Alice wants to make a plane diagram depicting events that happen at various times and places in one spatial dimension—e.g. along a long straight railroad track. Figure 1 gives an example of such a diagram:

Alice organizes events in her diagram according to the time at which they take place. Postulate 3 permits her to place all the events to which she assigns the same time on a single straight line, which we shall call an *equitemp*. (I would have preferred calling it an *isochron* but an internet search revealed "isochron" to be a standard term in radioactive dating and dendrochronology. "Equitemp" only produced a few furnaces and kettles.)

Equitemps associated with different times must be parallel, for if they were not then the event represented by their point of intersection would happen at two different times, contradicting the idea that an event is localized in time. Following the usual conventions of mapmakers (Fig. 2) Alices takes the distance between two equitemps in her diagram to be proportional to the time between the two sets of events they represent.

Alice still has the freedom of sliding points along her equitemps and she takes advantage of it to organize events in her diagram according to the location at which they happen, requiring (as Postulate 3 again permits) all events to which she assigns the same place to lie on a single straight line, which we shall call an *equiloc*. (Here, the more elegant "isotop" is clearly out of the question.)

Equilocs associated with different positions must be parallel, or the event represented by their point of intersection would happen in two different places, contradicting the idea that an event is localized in space. Alice again follows mapmakers (Fig. 3) in taking the distance between equilocs in her diagram to be proportional to the spatial separation of the two sets of events they represent.

The intersection of an equiloc and equitemp specifies a definite place and time and therefore corresponds to a unique event, so equilocs cannot be parallel to

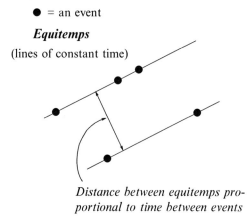

Fig. 2 Some of Alice's equitemps

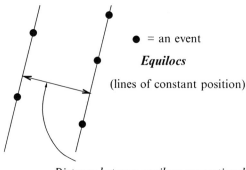

Fig. 3 Some of Alice's equilocs

equitemps, but aside from that restriction the angle Θ between Alice's equilocs and equitemps is free for her to chose as she wishes. Picking it to be a right angle obscures many important features of the diagrams, so we shall keep it general.

It is convenient for Alice to redefine the foot (ft). I remind our Canadian hosts that the foot is an obscure unit of distance, still used south of the border and in a few other backward countries. It is defined by 1 ft = 0.3048 m. Alice prefers a foot (f) that is just a little more than 1% shorter than the foot (ft): 1f = 0.299792458 m. Since the meter is now defined so that the speed of light is 299,792,458 m/s, the speed of light is exactly 1 f/ns. Once the United States abandons this archaic practice (along, one hopes, with a few other barbarisms) I confidently expect the foot (f) to reemerge on the international scene as the light nanosecond.

It is also highly convenient for Alice to relate the spatial and temporal scales in her diagrams so that equilocs representing events 1 f apart are same distance λ apart in a diagram as equitemps representing events 1 ns apart. The scale factor λ, with dimensions centimeter (of diagram) per nanosecond (of time), or centimeter (of diagram) per foot (of distance), is strictly analogous to the mapmakers scale factor of, for example, centimeter (of map) per kilometer (of distance).

It is conventional to orient such a diagram so that equilocs are more vertical than horizontal and equitemps are more horizontal than vertical. It is also useful to introduce an alternative scale factor μ, defined to be the distance along an equiloc of two events 1 ns apart in time or, equivalently, the distance along an equitemp of two events 1 f apart in space. All this is illustrated in Figs. 4 and 5.

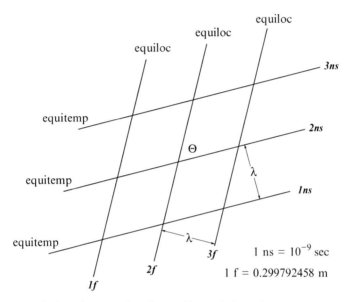

Fig. 4 Some of Alice's equitemps and equilocs and her scale factor λ

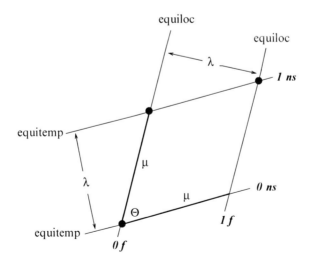

Fig. 5 The scale factor μ

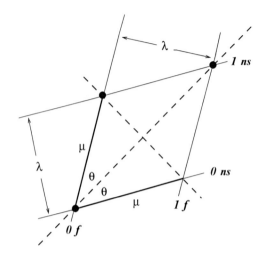

Fig. 6 Photon trajectories (dashed lines). $\theta = 1/2\,\Theta$

Equilocs and equitemps are thus characterized by two independent parameters, which can be any two of λ, μ, and Θ (the three are related by $\lambda = \mu \sin\Theta$). Note (Fig. 5) that the area of a *unit rhombus* whose opposite sides represent events 1 f and 1 ns apart is given by $\lambda\mu = \mu^2 \sin\Theta$.

Of particular importance in Alice's diagrams are *photon trajectories*, which are lines (conventionally dashed) that contain all events in the history of something moving at the speed of 1 f/ns. Photon trajectories are the diagonals of a rhombus whose sides are equitemps and equilocs, and therefore (as Fig. 6 makes clear) photon trajectories associated with motion in opposite directions cross at right angles, and they bisect the angle $\Theta = 2\theta$ between equitemps and equilocs. Putting it another way, equilocs and equitemps are symmetrically disposed about the photon trajectories.

Bob, who moves along the tracks uniformly with respect to Alice at v_{BA} feet per nanosecond (f/ns) would like to describe the same events as Alice has described. He could, of course, make a diagram of his own, but since Alice has already put down points representing the events on a piece of paper, Bob can try to impose on those points (without moving them on the page) his own equitemps and equilocs.

It turns out that these are also straight lines. This is particularly evident for Bob's equilocs, since an equiloc for Bob is just the trajectory of something that moves at velocity v_{BA} according to Alice—i.e. the heavy straight line depicted in Fig. 7. Note that Bob's velocity according to Alice is the ratio of Alice's time between the two events indicated on Bob's equiloc to her distance between the same two events: $v_{BA} = \mu_A g / \mu_A h = g/h$.

Up to now Einstein's first two postulates have played no role. Relativity enters the story for the first time when we ask for the orientation of Bob's equitemps in Alice's diagram. This is determined by Einstein's famous gedanken experiment that determines simultaneous events on a moving train (Fig. 8.). If Bob rides along the

Plane Geometry in Spacetime

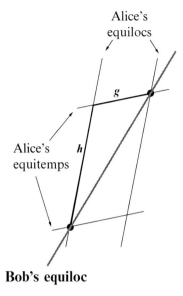

Fig. 7 One of Bob's equilocs in Alice's diagram

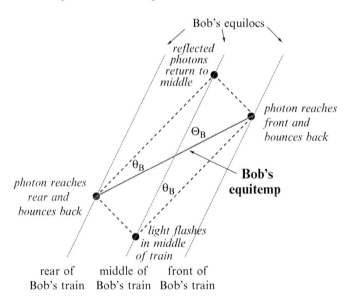

Fig. 8 Einstein's train: Determining the orientation of Bob's equitemps in Alice's diagram

tracks on a train moving uniformly with velocity v_{BA} and if two oppositely moving photons originate from a single flash at the center of the train, then the arrival of the photons at the two ends of the train constitute simultaneous events for Bob, who takes the train to be stationary. This immediately determines the orientation of Bob's equitemps in Alice's diagram (and that Bob's equitemps, like his equilocs, are straight lines).

The easiest way to see immediately how to characterize the orientation of Bob's equitemps in Alice's diagram is to continue the experiment after the photons arrive at the two ends of the train, letting them encounter mirrors and bounce back towards the middle. Since according to Bob they bounce off the mirrors at the same time (and he takes the train to be stationary), they will also arrive back at the middle at the same time. Their trajectories therefore form a rectangle—the first of many such "light rectangles" we shall be encountering. One of the diagonals of the rectangle is Bob's equitemp, and the other is his equiloc associated with events at the center of the train. It follows immediately from the symmetry of the rectangle that Bob's equitemp is tilted downward from a photon trajectory through the same angle $\theta_B = \frac{1}{2}\Theta_B$ as his equiloc is tilted upward.

So when Bob imposes his own equitemps and equilocs on Alice's diagram he finds that they are straight lines that make same angle $\theta_B = \frac{1}{2}\Theta_B$ with photon trajectories. But Alice originally made her diagram by imposing the rules that her own equitemps and equilocs should be straight lines that make same angle $\theta_A = \frac{1}{2}\Theta_A$ with the photon trajectories. There is thus no way to tell who made the diagram first and who later added their own equitemps and equilocs (Fig. 9). This is a strikingly direct demonstration that the principle of relativity is indeed entirely compatible with the principle of the constancy of the velocity of light.

It remains only to establish the relation between Bob's scale factor λ_B and Alice's scale factor λ_A. But it is worth pausing to note that the single most important quantitative result of special relativity is already at hand, even before we know Bob's scale factor. This is the quantitative statement of the relativity of simultaneity, which follows directly from the symmetry of both Bob's and Alice's equitemps and equilocs under reflection in 45° photon lines.

The events P and R in Fig. 10 lie on an equiloc of Bob, so according to Bob they happen in the same place. They lie on two different Alice-equilocs, however, and according to Alice they happen $\mu_A g$ feet apart. According to Alice they also

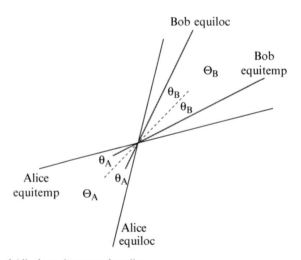

Fig. 9 Bob's and Alice's equitemps and equilocs

Plane Geometry in Spacetime   335

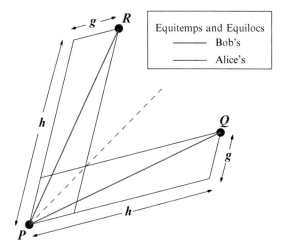

Fig. 10 The quantitative measure of the relativity of simultaneity

happen $\mu_A h$ nanoseconds apart, so although Bob says they happen in the same place Alice disagrees. She says that the distance D_A between them in feet is g/h times the time T_A between them in nanoseconds. Since $g/h = v_{BA}$ f/ns, Alice concludes that $D_A = v_{BA} T_A$: the distance between them in f is the time between them in ns multiplied by Bob's speed in f/ns. A banal conclusion well known to Galileo, except for the choice of units.

The reflection Q of the event R in the dashed photon line lies on the same Bob-equitemp with P and therefore occurs at the same time, according to Bob. Alice disagrees. The time T_A between P and Q is $\mu_A g$ nanoseconds and the distance between them is $\mu_A h$ feet. So Alice concludes that when Bob, who moves at v_{BA} f/ns, says two events are simultaneous, then if the events are D_A feet apart then they are also T_A nanoseconds apart where $T_A = v_{BA} D_A$. This is the quantitative statement of the relativity of simultaneity, in its clearest form. It is, however, surprisingly unfamiliar to some professional relativists, one of whom, as a referee, insisted that I had left out a factor of $\sqrt{1 - v_{BA}^2}$ and had to be convinced of his error.

To determine the relation between the scales used by Alice and Bob it is useful first to introduce some nomenclature. The first bit of terminology is familiar (Fig. 11). Two events are said to be *spacelike* separated if they lie on *somebody's* (Alice's, Bob's, Carol's, Dick's,...) equitemp, so for that person their separation is entirely spatial. And they are said to be *timelike* separated if they lie on somebody's equiloc, so for that person their separation is entirely temporal. Two events that lie on a single photon trajectory are said to be *lightlike* separated.

The second bit of terminology is probably unfamiliar (Fig. 12). Two spacelike or timelike separated events determine a unique rectangle of photon lines with the events at diagonally opposite vertices. We call this rectangle the *light rectangle* determined by the events. We have already encountered such a light rectangle in the Einstein-train experiment (Fig. 8), and they play a central role in what follows.

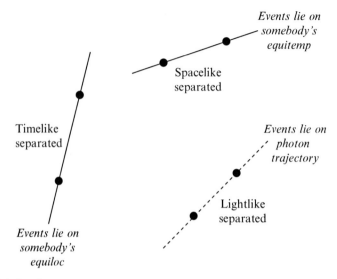

Fig. 11 Relations between events

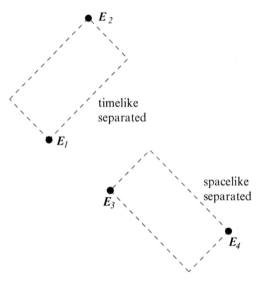

Fig. 12 Light rectangles

A light rectangle determined by two events 1 ns apart on the same Alice-equiloc (or 1 f apart on the same Alice-equitemp) is called a *unit light rectangle* for Alice (Fig. 13). The area Ω_0 of Alice's unit light rectangles is just half the product of her two scale factors: $\Omega_0 = \frac{1}{2}\lambda_A \mu_a$. This is established in Fig. 14, which shows that a unit light rectangle (connecting spacelike separated events) can be sliced along the equiloc connecting the events into two congruent triangles, four of which can be

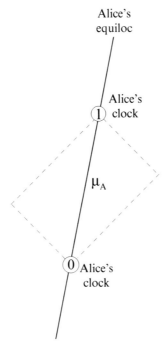

Fig. 13 Alice's unit light rectangle

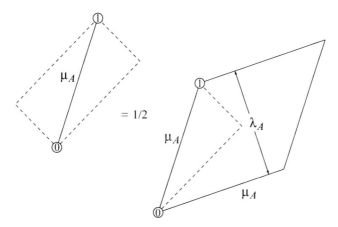

Fig. 14 Area of Alice's unit light rectangle, in terms of the area of her unit rhombus

reassembled into a rhombus whose sides are equilocs and equitemps separated by 1 f and 1 ns, and whose area is therefore just the product of the two scale factors λ and μ.

Armed with this terminology we can give a very simple geometrical specification of the connection between Alice's and Bob's scale factors by appealing to a second gedanken experiment. This one might be called the reciprocity of the Doppler effect.

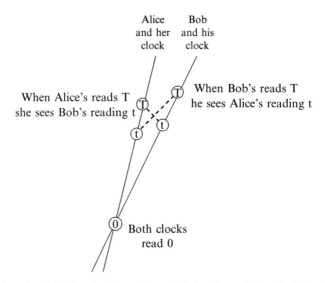

Fig. 15 Reciprocity of the Doppler effect. Alice and Bob each watch the other's clock

If Alice and Bob each carry a clock and each looks at the other's clock, Einstein's first two principles require that each should *see* the other's running at a rate that differs from the rate of their own clock in exactly the same way.

This is shown in the diagram of Fig. 15. The two solid lines are the space-time trajectories of Alice and Bob (and therefore an Alice-equiloc and a Bob-equiloc). When Alice and Bob are together in the event at the bottom of Fig. 15 they set their coincident clocks to 0. (Both clocks are shown as a single circle with a 0 inside it.) Subsequently their trajectories diverge and their clocks advance. When the clock of each reads T, each looks at the clock of the other and sees it reading t—i.e. the photons arriving from Alice's clock at the moment Bob's clock reads T, were emitted from Alice's clock at the moment it read t, and vice-versa. That each sees the same t when his or her own clock reads the same T is required by the principle of relativity and the principle of the constancy of the speed of light.

This is enough to determine the ratio of their scale factors, since the length of the line between his clocks reading 0 and T on Bob's equiloc is $\mu_B T$ while the corresponding length on Alice's equiloc is $\mu_A T$. Hence the length ratio of their two line segments is just the ratio of their scale factors. It might appear that some intricate trigonometry would be required to extract this ratio from the figure, but a simple trick gives it a very elementary geometric interpretation. The trick is to note that the two photon trajectories in the figure form segments of sides of two light rectangles, one determined by the events in which Alice's clock reads T and 0, and the other by the corresponding readings of Bob's clock. The complete light rectangles are drawn in Fig. 16.

Notice that the long side B of Bob's light rectangle exceeds the long side A of Alice's by just the ratio T/t of the two readings of Bob's clock. But this is compensated for by the fact that the short side a of Alice's light rectangle exceeds the short

Plane Geometry in Spacetime

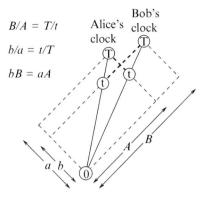

Fig. 16 The photon lines in Fig. 15 have been expanded into two light rectangles

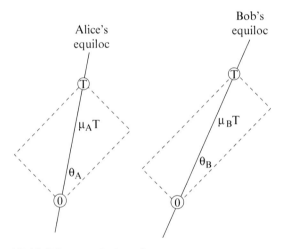

Fig. 17 Alice's and Bob's light rectangles have the same area

side b of Bob's by the ratio T/t of the two readings of *her* clock. Since these ratios are the same, *the two light rectangles have the same area*. This is shown in Fig. 17, where the two light rectangles and equilocs in Fig. 16 have been slid apart from one another.

Specializing to the case $T = 1$ we conclude that Bob's and Alice's unit light rectangles must have same area. Since the area of a unit light rectangle is given by $\Omega_0 = \frac{1}{2}\mu\lambda$, we have a simple analytic expression for this (even simpler) geometric condition: the product $\mu\lambda$ of scale factors must be the same for everyone: $\mu_A\lambda_A = \mu_B\lambda_B = \mu_C\lambda_C = \cdots$.

The area Ω of the light rectangle determined by a pair of events has an important significance of its own, which becomes immediately evident if one notes that for timelike separated events Ω/Ω_0 is the square of the time between the events in the frame in which the events happen at the same place, while for spacelike separated events Ω/Ω_0 is the square of the distance between the events in the frame in which the events happen at the same time (Fig. 18).

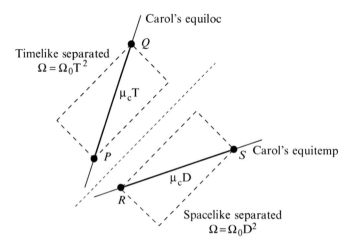

Fig. 18 Carol's is the frame in which events P and Q happen in the same place (or in which events R and S happen at the same time)

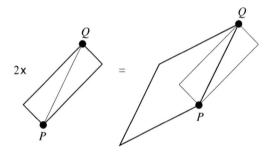

Fig. 19 Interval is also proportional to the area of the rhombus determined by the events

But this is precisely the definition of the squared interval I^2 between pairs of timelike or spacelike separated events. We conclude that the squared interval between any pair of events is just the area Ω of the light rectangle determined by those events, expressed in units of the (frame-independent) area Ω_0 of the unit light rectangle: $I^2 = \Omega/\Omega_0$. (This also works for lightlike separated events, since the light rectangle determined by such a pair degenerates to a line—the area goes to zero.)

Can we see geometrically why the squared interval is given by $I^2 = |T^2 - D^2|$, where T and D are the time and distance between the events (according to anybody—Alice, Bob, Carol, Dick,...)? This can be seen at a glance if one replaces light rectangles by the rhombi of twice the area shown in Fig. 19.

Figure 20 shows two events P and Q and the Alice-equilocs and Alice-equitemps on which the events lie. Alice's time T between P and Q is the same as her time between P and R, so T^2, the time between P and R in the frame in which they happen at the same place is proportional to the area of the rhombus they determine. For the same reason D^2, the square of Alice's distance between P and Q is also

Plane Geometry in Spacetime

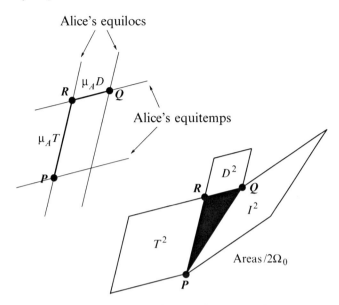

Fig. 20 Is the area of the rhombus determined by P and Q equal to the difference in areas of the rhombi determined by R and P and by Q and R?

proportional to the area of the rhombus they determine. And the squared interval I^2 between P and Q is proportional to the area of the rhombus they determine. Why should the area of the PQ rhombus be equal to the difference in areas of the PR and QR rhombi?

This is made evident in Fig. 21. Part (a) reproduces the construction of Fig. 20. Part (b) combines two copies of the black triangle with the PQ and QR rhombi, while part (c) combines the two triangles with the PR rhombus alone. But the two quadrilaterals in (b) and (c) are easily verified to be identical in size and shape, so they therefore have the same area. (One can also, of course, reach the same conclusion using light rectangles, but this is the more beautiful way to see it.)

Here is an important application of light rectangles in *three* spatial dimensions. There is a wonderful way to measure the interval between two events P and Q using only light signals and a single clock. Alice moves uniformly with her clock, and both of them are both present at event P. Bob is present at event Q. When P happens Alice's clock reads T_0. When Q happens, Bob *sees* Alice's clock reading T_1. When Alice *sees* Q happen, her clock reads T_2. The squared interval between P and Q is given by $I_{PQ}^2 = |(T_1 - T_0)(T_2 - T_0)|$.

Light rectangles provide an elegant proof of this. Since a point (event Q) and a line (Alice's trajectory) determine a plane, and since all the relevant photon trajectories lie in that plane, we can capture the situation fully with a 2-dimensional space-time diagram.

In Fig. 22 the events P and Q are timelike separated. (Note, by the way, how transparently the diagram conveys the procedure compared with the verbal description

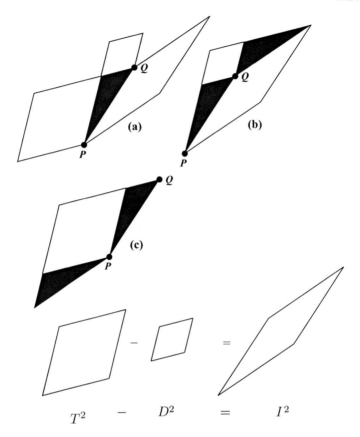

Fig. 21 Why the squared interval is $|T^2 - D^2|$

I gave above.) Alice's small light rectangle determined by her clock reading T_1 and T_0 is scaled down (in both dimensions) by a factor f from her big light rectangle determined by her clock reading T_2 and T_0 The light rectangle determined by P and Q has an area Ω_{PQ} that is smaller than the area Ω_{20} of Alice's big light rectangle by a factor f, but larger than the area Ω_{10} of Alice's small light rectangle by a factor $1/f$. So its area is the geometric mean of the areas of Alice's two light rectangles: $\Omega_{PQ}^2 = \Omega_{20}\Omega_{10}$. Since $I_{PQ}^2 = \Omega_{PQ}/\Omega_0$, $(T_2 - T_0)^2 = \Omega_{20}/\Omega_0$, and $(T_1 - T_0)^2 = \Omega_{10}/\Omega_0$, we have established the desired relation. (You can make the argument even simpler, but less symmetric, by noting right away that the factor f is just $\frac{T_1-T_0}{T_2-T_0}$.)

Exactly the same argument works for spacelike separated events, so I give the accompanying Fig. 23 without further comment.

Finally, there is the question of how to extend the whole procedure to cases where more than a single spatial dimension comes nontrivially into play. The figures, of course, become harder to draw on a page and their geometric transparency is

Plane Geometry in Spacetime

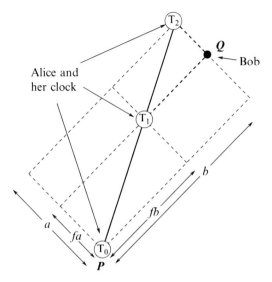

Fig. 22 Measuring the interval between timelike separated events

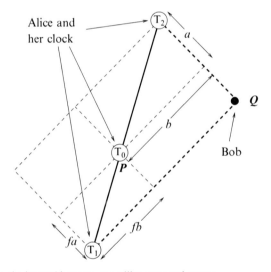

Fig. 23 Measuring the interval between spacelike separated events

considerably diminished, but there is an important point to be made. Let Alice add a second spatial dimension perpendicular to the plane of the planar diagrams we have been examining. Every two-dimensional slice of her new three-dimensional diagram parallel to the original diagram has, of course, exactly the same structure. Her equitemps expand into equitemporal planes, perpendicular to her slices. Her equilocs remain lines, parallel to her earlier equilocs, but now lying in any of the slices. Everything is as it was before for motion within a slice, but now she can

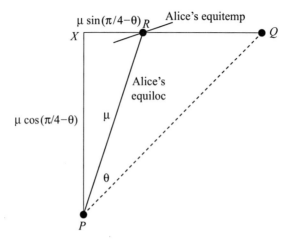

Fig. 24 A plane slice of Alice's 3-dimensional diagram in which the line from P to R is an Alice-equiloc. According to Alice events P and R are 1 ns apart

describe motion with a component perpendicular to the slices. Bob, Carol, and Dick are no longer constrained to move within one of Alice's slices but for reasons of spatial isotropy, their own slices ought to retain all the features we found before. In particular oppositely moving photon trajectories in their slices should also be perpendicular and thus their photon trajectories through any point should lie on the same right circular cone as Alice's.

Alice, on the other hand, must agree that photons move a foot in a nanosecond, even when they move out of the plane of her diagram, and it is this combination of spatial isotropy and the constancy of the speed of light, that determines Alice's scale factor for her slices.

Figure 24 shows a single one of Alice's slices—a plane diagram of just the kind we have been considering. The points P and R on Alice's equiloc represent events one ns apart, so the length of the line segment PR is just Alice's scale factor μ. A photon trajectory through P and Q is also shown. In the full three dimensional diagram the trajectories of other photons through P will fill up the right circular cone given by rotating the line PQ about the vertical line PX. The lengths of the two sides of the triangle PXR in terms of the angle $\theta = \frac{1}{2}\Theta$ between Alice's equiloc and the photon trajectory are as indicated.

We wish to determine σ, the distance in Alice's three-dimensional diagram between slices containing events separated by 1 f in the new perpendicular direction. We can do this by considering a second photon originating at P whose motion is only in the direction perpendicular to Alice's slices. Its trajectory must therefore lie on the equilocs given by displacing the equiloc PR in the perpendicular direction. After 1 ns the second photon will be represented by the point on its trajectory that intersects the slice of Alice's diagram a distance σ from the slice in Fig. 24. That point must represent an event 1 ns after the event represented by P, and it therefore lies on the equitemporal plane perpendicular to the slices passing through R.

Plane Geometry in Spacetime

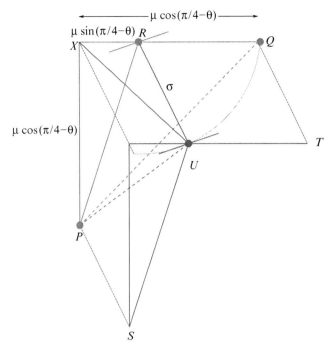

Fig. 25 Alice's scale factor σ

This whole three-dimensional situation is pictured in Fig. 25. The plane slice containing the points P, Q, and R is reproduced in the figure (along with the equiloc and photon trajectory it contains). The second plane slice, a distance σ away, contains points S, T, and U, which are just the displacements of P, Q, and R in the perpendicular direction. The line RU lies within an equitemporal plane. The trajectory of the second photon goes from P to U and lies in the plane $PRUS$. Since the events P and U in the history of the second photon are a nanosecond apart, the event U is a foot away from the event R, so the distance in the diagram from point R to point U is indeed the perpendicular scale factor σ.

Both photon trajectories must lie on the right circular cone through P (whose intersection with the plane $PRUT$ is indicated by the circular arc through Q and U). Since PX is the axis of that cone, the lengths of the lines XQ and XU must be the same. The length of the XQ is $\mu \cos(\pi/4 - \theta)$, while XU is the hypotenuse of a right triangle with sides σ and $\mu \sin(\pi/4 - \theta)$. Therefore $\sigma^2 + \mu^2 \sin^2(\pi/4 - \theta)$ $= \mu^2 \cos^2(\pi/4 - \theta)$, so $\sigma^2 = \mu^2 \cos(\pi/2 - 2\theta) = \mu^2 \sin(2\theta) = \mu^2 \sin\Theta = \mu\lambda$.

Thus the out-of-plane scale factor σ is just the geometric mean of the in-plane scale factors μ and λ. Since the product $\mu\lambda$ is the same for all observers, we have produced what has to be the world's most difficult demonstration that "$y' = y, z' = z$". I regard this as a fitting punishment for one who began this essay by boasting of his ability to make complicated matters simple.

Some Philosophical Remarks

Although I remember firmly declaring at the age of 25 that I planned to spend my declining years as a philosopher, and although I cannot imagine an occasion more appropriate for such meditations, I will be extremely brief and may well deny ever having said what follows. I emerge from these exercises thinking that:

1. The raw material of our experience consists of events.
2. Events, by virtue of being directly accessible to our experience, have an unavoidably classical character.
3. Space and time and spacetime are not properties of the world we live in, but concepts we have invented to help us organize (classical) events.
4. Notions like dimension or interval, or, to look beyond the story I tell here, curvature or geodesics, are not properties of the world we live in, but of the abstract geometric constructions we have invented to help us make sense of events.
5. When I hear that spacetime becomes a foam at the Planck scale, I don't reach for my gun (because I haven't any) but I do wonder what this has to do with (classical) events.
6. Alice knows not to confuse her diagrams with the events they help her describe. Do we?
7. Is the difficulty in reconciling quantum mechanics with general relativity better regarded as another aspect of the "measurement problem"?

Acknowledgments My last grant from the U. S. National Science Foundation (PHY-0098429) concluded on June 30, 2006, the day before I became Professor Emeritus at Cornell. I would like to thank the NSF for supporting me continuously for exactly 50 years, starting with one of their new predoctoral fellowships as I started graduate school in the fall of 1956.

Abner Shimony has not supported me for nearly as long, but I am grateful to him for good ideas and wonderful letters and conversations over the past quarter of a century.

References and Comments on Related Work

The earliest reference I have found to measuring the interval with a single clock is the unpublished 1959 Princeton senior thesis of Robert F. Marzke, as described by Marzke and John A. Wheeler in *Gravitation and Relativity*, Ed. Hong-Yee Chiu and William F. Hoffmann, Benjamin, New York, 1964, pps. 40–64. I described this approach to Minkowski's diagrams in "An Introduction to Space-Time Diagrams," *Am. J. Phys.* **65**, 476–486 (1997) and "Spacetime Intervals as Light Rectangles," *Am. J. Phys.* **66**, 1077–1080 (1998). Another version appeared as "From Einstein's Postulates to Spacetime Geometry," in the Einstein centenary issue of *Annalen der Physik* **14**, 103–114 (2005). A greatly expanded version with more applications can be found in my book *It's About Time: Understanding Einstein's Relativity* Princeton University Press, 2005. Dieter Brill and Ted Jacobson independently arrived at some very similar constructions in "Spacetime and Euclidean

Plane Geometry in Spacetime 347

Geometry," http://arxiv.org/abs/gr-qc/0407022. I learned from Brill and Jacobson of some very early work along the same lines: Edwin B. Wilson and Gilbert N. Lewis, "The Spacetime Manifold of Relativity:The Non-Euclidean Geometry of Mechanics and Electromagnetics," *Proc. Amer. Acad. Boston* **48**, 389–507 (1913). I also learned from them that some wonderful animated geometric demonstrations (that inspired the one I give above) that $I^2 = |T^2 - D^2|$ (and also of the ordinary Pythagorean theorem) can be found at the web site of Dierck-Ekkehard Liebscher, http://www.aip.de/~lie/. Liebscher's book, *The Geometry of Time*, Wiley-VCH, Weinheim, 2005, delves much deeper into this approach. And, of course, Einstein (1905) is "Zur Elektrodynamik bewegter Köpper", *Annalen der Physik* **17**, 891–921.

The Transient *nows*

Steven F. Savitt

Abstract It is often claimed that features of the spacetime of special relativity are inimical to the passage of time. In opposition to this view, I show how the passage of time is to be understood in Minkowski spacetime. A (local, specious) present is construed as an open set in the Alexandroff topology and the passage of time is a succession of presents along a timelike curve. Temporal becoming is a local, rather than a global, phenomenon.

I offer some motivations for the view I propose, and I consider five objections that might be raised against it. For instance, one general objection to the notion of the "flow" or passing of time is that one can not answer the natural question 'How fast does time "flow" or pass?' I claim that Minkowski spacetime provides a natural answer to this question.

In his "Intellectual Autobiography" the philosopher Rudolf Carnap [1, p. 37] described a conversation he had with Albert Einstein at the Institute for Advanced Study in the early 1950s:

> Once Einstein said that the problem of the Now worried him seriously. He explained that the experience of the Now means something special for man, something essentially different from the past and the future, but that this important difference does not and cannot occur within physics. That this experience cannot be grasped by science seemed to him a matter of painful but inevitable resignation.

Other distinguished scientists have had similar qualms. Hermann Weyl [2, p. 116] wrote:

> The objective world simply is, it does not happen. Only to the gaze of my consciousness, crawling upward along the life line of my body, does a section of this world come to life as a fleeting image in space which continuously changes in time.

S.F. Savitt
Department of Philosophy, The University of British Columbia, Vancouver, British Columbia, Canada V6T 1Z1
e-mail: Savitt@interchange.ubc.ca

W.C. Myrvold and J. Christian (eds.), *Quantum Reality, Relativistic Causality, and Closing the Epistemic Circle,* The Western Ontario Series in Philosophy of Science 73,
© Springer Science+Business Media B.V. 2009

Weyl is taken to be claiming that the "objective world" lacks passage, temporal becoming, or transience, a phenomenon that is merely subjective. P. C. W. Davies [3, p. 21] endorses this view forthrightly:

> The four dimensional space-time of physics makes no provision whatever for either a 'present moment' or a 'movement' of time .... Rather than thinking in terms of a succession of experiences by a particular particle, we must instead deal with its entire world line in four dimensions; in the words of H. Weyl 'the objective world simply *is*, it does not *happen*.'

Finally Carlo Rovelli wrote, "[T]he notion of *present*, of the 'now', is completely absent from the description of the world in physical terms" [4]. These views, expressed by distinguished scientists, are philosophical views and so open to philosophical examination. What I hope to do in this paper is show that there is a viable alternative picture to these views, a picture that includes, in some sense, a *now* and the passage of time. I cannot, like Einstein, talk about what can be grasped by 'science' or by 'physics'. I hope I can talk, coherently and persuasively and philosophically, about one small bit of physics, the special theory of relativity, and the treatment of time in Minkowski spacetime.

1 Closing the Circle

It is useful to begin with Carnap's response [1, pp. 37–38] to Einstein's problems with the Now:

> I remarked that all that occurs objectively can be described in science; on the one hand the temporal sequence of events is described in physics; and, on the other hand, the peculiarities of man's experience with respect to time, including his different attitude towards past, present, and future, can be described and (in principle) explained in psychology. But Einstein thought that these scientific descriptions cannot possibly satisfy our human needs .... I definitely had the impression that Einstein's thinking on this point involved a lack of distinction between experience and knowledge. Since science in principle can say all that can be said, there is no unanswerable question left. But though there is no theoretical question left, there is still the common human emotional experience, which is sometimes disturbing for special psychological reasons.

Carnap was the quintessential anti-metaphysician, and it is very tempting, for those of us who are similarly inclined, to agree with the sentiment that after all that Carnap says can be objectively described is described ("the temporal sequence of events" on the one hand, and "the peculiarities of man's experience with respect to time" on the other), then there is nothing left to be said. What I have been led to see by reading Abner Shimony is that there *is* something important left out, something left to be said. What is left out is an account of the relation between these two.

That, at least, is my way of construing the implications of his program of "closing the circle", a way of doing philosophy (or, at least, metaphysics and epistemology) that Shimony traces back to Aristotle and that lets him, in his version of it, stake out a position in opposition to the later Putnam and van Fraassen in the battle over realism. Here's his capsule description [5, p. 40]:

> The program [of closing the circle] envisages the identification of the knowing subject (or, more generally, the experiencing subject) with a natural system that interacts with other natural systems. In other words, the program regards the first person and an appropriate third person as the same entity. From the subjective standpoint the knowing subject is at the center of the cognitive universe, and from the objective standpoint it is an unimportant system in a corner of the universe.

This program guides his discussion of realism. I suggest that the program can be adapted to the philosophy of time as well, in the form of the following slogan: *Philosophy of time should aim at an integrated picture of the experiencing subject with its felt time in an experienced universe with its spatiotemporal structure.* This rationale underlies and shapes the picture of time in Minkowski spacetime that I will sketch in the rest of this paper.[1]

Shimony's summary statement of the nature of closing the circle is followed by a particularly elegant diagnosis of what goes wrong in an unbalanced discussion:

> If either the subjective or the objective aspect of the knowing subject is played down, or if the substantial identity of their two aspects is neglected, then the problem ... is flattened, or—to use quasimathematical language—it is projected into a subspace of smaller dimensionality than it deserves. [5, p. 40]

This last remark exactly captures my feeling as to what is wrong with contemporary discussions of time. Many analytic philosophers concentrate single-mindedly on the subjective side of time and find nothing more to it than the Now. Others, like the three scientists I quoted at the beginning of this talk, look first to the spacetime of physics and cannot find a Now at all. I think we will only begin to do justice to time (and all I can hope to do here is to begin to do justice to it) if we look at *both* sides *and* at the hitherto missing connection between them—if, that is, we try to close the circle.

2 What Transience is Not

The first step towards finding relativistic notions of the Now and transience (the passage, flow, or lapsing of time) must be elucidation of the classical notions. The classical Now is straightforward and not controversial. A classical Now is a global

---

[1] I would be remiss, however, if I did not mention that my views are everywhere shaped by those of two additional thinkers. The first is Howard Stein, a friend of Shimony's, whose papers on time in relativity theory are beacons of light in dark waters, and C. D. Broad, a British philosopher of the generation before them.

I am also indebted to Richard Arthur, Craig Callender, Carl Hoefer, and Wayne Myrvold for helpful discussions and suggestions as the paper evolved. Arthur was a pioneer in directing attention to what I now call Alexandroff presents. In addition, I have received individual support from the Social Sciences and Humanities Research Council, but participation in the TaU cluster, which SSHRC supports, has provided valuable opportunities to engage with others on the topics of this paper.

352   S.F. Savitt

hypersurface of simultaneous events. Any given event, if idealized as having no duration, is contained in precisely one such global hypersurface, its Now. These global hypersurfaces foliate spacetime into equivalence classes that are mutually exclusive but exhaustive Nows.

Transience is the successive occurrence of these global hypersurfaces. As Kurt Gödel [6, p. 558] put it succinctly:

> The existence of an objective lapse of time ... means (or, at least, is equivalent to the fact) that reality consists of an infinity of layers of "now" which come into existence successively.

If the "objective" lapsing of time requires frame- or observer-independent hyperplanes of simultaneity, then Gödel is right that there is no lapsing of time in Minkowski spacetime.

Gödel then presents the well-known argument that this global version of transience runs into insurmountable difficulties if transferred to a special relativistic setting:

> But, if simultaneity is something relative in the sense just explained,[2] reality cannot be split up into such layers in an objectively determined way. Each observer has his own set of "nows," and none of these various systems of layers can claim the prerogative of representing the objective lapse of time. (ibid.)

There is a variant of this argument that appears in recent paper by Dennis Dieks that I find very suggestive. Dieks argues (1) that the experiences of observers are of such short duration and occupy such a small amount of space that they can, without loss, be idealized as point-like, (2) amongst these experiences are those that convince us that time flows or passes, and (3) given the upper limit of speed of propagation of causal signals, so that no event spacelike separated from a given event can influence it causally, it follows that (4) the human experiences that suggest at any event e in the history of an observer that time flows are invariant under different choices of global hypersurface containing e. [7, Section 1]

The moral I draw from Dieks' argument is, at this point, conditional. *If* there is such a thing in special relativity theory as the passage of time and *if* it is to relate to the experiences of creatures like us in spacetime, then global hypersurfaces are irrelevant to it. This conclusion is deeply shocking to common sense metaphysics, since global Nows permeate pre-relativistic thinking about time. It is difficult to think about time without them, yet David Mermin has reminded us recently that special relativity is a radical theory [8, p. xii]. "That no inherent meaning can be assigned to the simultaneity of distant events is the single most important lesson to be learned from relativity." Time, we must learn, is *not* spread through space.

So far my argument is conditional and negative. If there is such a thing as transience countenanced in Minkowski spacetime, it won't be the successive occurrence

---

[2] "The very starting point of special relativity theory consists in the discovery of a new and very astonishing property of time, namely the relativity of simultaneity, which to a large extent implies that of succession. The assertion that the events A and B are simultaneous (and, for a large class of pairs of events, also the assertion that A happened before B) loses its objective meaning, in so far as another observer, with the same claim to correctness, can assert that A and B are not simultaneous (or that B happened before A)." [6, p. 557].

The Transient *nows* 353

of global hypersurfaces. Transience, if such there be, would have to a local rather than a global notion. But is there such a thing, relativistically, as transience? It is time, I submit, to turn to poetry.

3 Time Goes, You Say?

Consider the following clever couplet from Henry Austin Dobson [9]:

Time goes, you say? Ah no!
Alas, Time stays, *we* go...

I hope that you have the same two-fold reaction to this verse that I do. First, one's attention is drawn from the transience of time to the transience of self, as Dobson intended. But then I hope you ask yourself, "Is there really a difference here? Isn't this a cheat, some poetical slight-of-hand or misdirection?"

I think it is. I believe that there is no difference between our going and time's going, no difference (that is) between ordered events or objects moving in or through time and time's moving along or by ordered but static events or objects, no difference (that is, again) between on the one hand future events "approaching" us, becoming present, and then "receding" from us into the past or, on the other hand, our leaving the past ever further behind as we "move" into the future. The "motion" is relative to whichever of the two, time or objects, one chooses to think of as "static", and one may choose either.

I do not think that it useful to try to understand the passage of time as a kind of motion, since motion has to be understood as change of position through time. Nevertheless, I think the point of the previous paragraph is general. *If* there is a way to make sense of the passage of time (and it is a notoriously difficult idea to make sense of in any terms), then in whatever terms prove successful there will be only a verbal difference between (on the one hand) speaking of time's passing or "going" and our remaining still and (on the other hand) speaking of our going or progressing through a "static" time.

This point may seem innocuous, but when combined with the idea of the last section that special relativity constrains one to construe transience locally rather than globally, the resulting point of view is anything but commonplace. Let us reconsider in this new light, for instance, the supposedly anti-passage remark of Weyl cited earlier in the paper:

> The objective world simply is, it does not happen. Only to the gaze of my consciousness, crawling upward along the life line of my body, [emphasis added] does a section of this world come to life as a fleeting image in space which continuously changes in time.

Is there a difference between my consciousness "crawling" along my world line (on the one hand), as opposed to my consciousness being static but time itself "passing" along my world line? If not (as we have agreed), then Weyl cannot simply be read as a partisan of a static universe, as Davies does. Weyl can also be read as a proponent of the local passage of time.

354   S.F. Savitt

Perhaps I am taking advantage of an unfortunate turn of phrase to misconstrue Weyl. Perhaps. But note that in one other famous remark of Weyl's on time the same phrasing recurs.

> However deep the chasm may be that separates the intuitive nature of space from that of time in our experience, nothing of this qualitative difference enters into the objective world which physics endeavours to crystallize out of direct experience. It is a four-dimensional continuum which is neither "time" nor "space". Only the consciousness *that passes on in one portion of this world* [emphasis added] experiences the detached piece which comes to meet it and passes behind it, as history, that is, as a process that is going forward in time and takes place in space. [10, p. 217]

Suppose, then, that Weyl can be read as backhandedly legitimizing the local passage of time. Is there a local structure that can support such a notion of transience? If the program of closing the circle is to be our guide and if that program starts with the experiencing subject, then we must immediately note that this subject, as a physical system in spacetime, is represented by a timelike curve. The program of closing the circle suggests, then, that we look to timelike curves. If classical transience is the successive occurrence of global Nows, perhaps special relativistic transience is the successive occurrence of local *nows* along a timelike curve. But what can a local *now* be?

4 Interlude

Before proceeding further, it will be useful to be a bit more explicit about a few matters. For instance, I will consider *the special theory of relativity* to be the theory developed in the early chapters, especially chapters 4 and 5, of Hartle [11].[3] The setting of the theory is a four-dimensional real vector space, $\mathbf{R}^4$, along with a metric to endow it with geometric structure. We can indicate points in $\mathbf{R}^4$ by using four coordinates, suggestively labeled (t,x,y,z), and we can define a spacetime "distance" function on pairs of points (t^0, x^0, y^0, z^0) and (t^1, x^1, y^1, z^1) in $\mathbf{R}^4$ as

$$(\Delta S)^2 = -(\Delta t)^2 + (\Delta x)^2 + (\Delta y)^2 + (\Delta z)^2, \qquad (1)$$

where $\Delta t = t^1 - t^0$, etc. Hartle shows that, given the Principle of Relativity ("Identical experiments carried out in different inertial frames give identical results."), the quantity $(\Delta S)^2$ does not change when one switches from the coordinates of one inertial frame to those of another. Since the directed line segment from (t^0, x^0, y^0, z^0) to (t^1, x^1, y^1, z^1) is a vector, one can use the invariant quantity of spacetime distance to divide vectors at (t^0, x^0, y^0, z^0) (or any other point in the spacetime, for that matter) into three kinds:

1. Those for which $\Delta S^2 > 0$, the *spacelike* vectors
2. Those for which $\Delta S^2 = 0$, the *null* vectors
3. Those for which $\Delta S^2 < 0$, the *timelike* vectors

---

[3] In some places I will appeal to Naber [12] for mathematical notions that do not appear in Hartle's presentation.

The Transient *nows* 355

Equation (1) is often written in the infinitesimal form:

$$ds^2 = -dt^2 + dx^2 + dy^2 + dz^2. \tag{2}$$

When discussing timelike vectors and time, it is convenient to multiply Eq. (2) by -1, in order to change the negative quantities. The result is written

$$d\tau^2 = dt^2 - dx^2 - dy^2 - dz^2. \tag{3}$$

The quantity 'τ' is known as *proper time*.

We will call the set $\mathbf{R}^4$ with a (Lorentz) metric like Eq. (2) *Minkowski spacetime*, M. If we choose a point $p \in$ M, the set of points whose spacetime distance from $p = 0$ comprise the (exterior of) the *null cone* at p. A curve through a set of points that is inside the null cone of any point that it goes through is a *timelike curve*. The histories or careers of ordinary objects (like us), whose rest mass is greater than 0, are to be represented by timelike curves (world lines) in M.

Timelike curves or world lines can be parameterized by proper time, τ. We can define proper time lengths between two points A and B on a timelike line, τ_{AB} as:

$$\tau_{AB} = \int_A^B d\tau = \int_A^B [dt^2 - (dx^2 + dy^2 + dz^2)]^{1/2}. \tag{4}$$

If we choose some point on the timelike line and assign it proper time 0, then we can define the proper time function along the timelike line by:

$$\tau_A = \tau_{0A} \tag{5}$$

From Eq. (4) one can easily derive a useful relation between proper time τ and coordinate time t:

$$\tau_{AB} = \int_{\tau_A}^{\tau_B} dt [1 - \vec{V}^2(t)]^{1/2}, \tag{6}$$

where $\vec{V}$ is a three-dimensional velocity vector.

In order to do physics properly and relativistically in M we must have four-dimensional quantities or *four-vectors*. For instance, the four-velocity $\mathbf{u}$ of a moving object is:

$$u = (dt/d\tau, dx/d\tau, dy/d\tau, dz/d\tau), \tag{7}$$

where boldface type is used to distinguish four-vectors. One can then go on to define four-acceleration and the relativistic analogs of Newton's laws in order to do special relativistic dynamics.

It will be useful to note here one further fact. Suppose that we write

$$\mathbf{u} \cdot \mathbf{u} = -\frac{dt^2}{d\tau^2} + \frac{dx^2}{d\tau^2} + \frac{dy^2}{d\tau^2} + \frac{dz^2}{d\tau^2} = \frac{ds^2}{d\tau^2} = -1. \tag{8}$$

Then we see that, for any massive particle or object, its velocity four-vector is a timelike vector of unit length, since the length of a timelike four-vector is $\sqrt{-u \cdot u}$.

356   S.F. Savitt

Since speed is the length of the velocity vector, we have the odd result that all massive objects have the same speed in spacetime, 1.

5 Now, *now*

To return to the main line of argument, we now know what a timelike line is. In addition, in my characterization of transience as the *successive* occurrence of events on a timelike world line, I am supposing that we know what it means for one event to occur *after* another. I am, therefore, supposing not only that our timelike curves occur in a temporally orientable manifold, but also that one of the orientations has been chosen as future (and the other as past). I do not know how this orientation is selected. Perhaps the choice is based on some asymmetry amongst the fundamental laws of physics, but that is a (deep) problem for another day. I will assume that an orientation is given.[4]

Recall the way I understand "closing the circle" in thinking about time:

> Philosophy of time should aim at an integrated picture of the experiencing subject with its felt time in an experienced universe with its spatiotemporal structure.

If we begin with the experiencing subject, we notice that its present is not point-like; it is extended. We experience the whole of a spoken sentence or a musical phrase, for example, as occurring now.

This phenomenon has traditionally been referred to as *the specious present*, but has recently come to be called *the psychological present*. The temporal extent of this temporally extended present no doubt varies both inter- and intra-personally. An eminent psychologist has told me that the *specious* or *psychological* present is variously estimated to last from 0.5 to 3 s, and I will fix on a middling value of 1 s for the sake of convenience in this discussion.[5] Extending the present opens up new possibilities, as we shall see.[6]

Suppose that we parameterize the timelike curves in Minkowski spacetime M with proper time (as characterized in the preceding section). Suppose also that we choose two events, e_1 and e_2 (with e_1 earlier than e_2) that are 1 s apart on a timelike curve, λ. Call the set of events in the intersection of the interior of the future light cone of e_1 with the interior of the past light cone of e_2 *ALEX* (e_1, e_2).[7] Here is a picture of this set of events, taken from [14, p. 156].

---

[4] Tim Maudlin [13] defends passage in Minkowski spacetime, but he takes the notion as basic or primitive and uses it to select an orientation as future.

[5] There is a recent account of the specious present in [15, chapter 4]. This chapter contains useful references, but I think that the account of the specious present in chapter 35 of Broad [16] is not adequately represented.

[6] I would like to emphasize from the outset that considering an extended present makes neither the present, as I characterize it below, nor the passage of time subjective. Once τ_{AB} is fixed, the present for such an interval is an invariant open subset of Minkowski spacetime.

[7] These open sets are elements of the Alexandroff topology for M discussed in [14, Section 3]. Winnie calls them *Alexandroff intervals*, and I will call them *Alexandroff presents*.

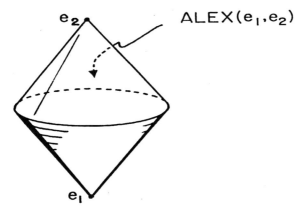

I propose that ALEX (e_1, e_2) is *the present for the interval from e_1 to e_2 along λ*.[8] If the passage of time or transience pre-relativistically is a succession of world-instants or Nows, then its relativistic successor concept must be the succession of presents (in the sense of ALEX (e_1, e_2)) along a timelike curve λ.

I would like to point out three advantages of this proposal. The first advantage applies not just in Minkowski spacetime but in any general relativistic spacetime that is *stably causal*—i.e., one that admits a global time function t: M $\rightarrow$ R, which in turn means that for all distinct events p, q $\in$ M, if $p \ll q$ (that is, if there is a smooth, future-directed timelike curve from p to q), then $t(\text{p}) < t(\text{q})$. In such a spacetime if $e' \in$ ALEX (e_1, e_2), then $t(e_1) < t(e') < t(e_2)$. If one defines a present or *now* based on an interval [e_1, e_2], then one surely would want it to meet this condition.

One reasonable demand on a scientific successor concept to a previous scientific concept or even to a folk concept is that it explains why the earlier concept is as useful or salient as it is. The explanation on offer for our commonsense or Newtonian notion of the present as the universe-at-an-instant is inspired by section V of [18]. Stein there observes that in a psychological present "light travels a spatial distance that bears a very large ratio to the spatial extent of our bodies or of ordinary objects." (p. 161) In one second, for instance, light travels in *vacuo* about 300,000 km, meaning that ALEX (e_1, e_2), with the small proper time duration of 1 s, is 300,000 km across at its widest. In the course of human conceptual development, it would be no more surprising that we developed the idea that this brief, fat structure was unbounded than that we developed the idea that the surface of the earth was flat.

Nor is it surprising that we developed the idea that we share a common present. Suppose that you walk past me at a reasonable pace of 4 km/h, that we call our meeting e, and that we compare the volumes of your present and my present, assuming they are symmetric about e (that is, each present extends 0.5 s to the future

---

[8] Throughout this discussion I represent conscious beings or "observers" as timelike world lines rather than world tubes. Relaxing this idealization will likely greatly complicate the definitions, if such relaxation is indeed possible. It is hoped, however, that the idealization itself is no more inexact than Einstein's [17, Section 1] extension of simultaneity to events "in the immediate proximity" or "in the immediate neighborhood" of a clock but not to a second clock not so located with respect to the first.

358   S.F. Savitt

and to the past along our two world lines. Then our two presents agree—that is, include the same events—up to about one half of one millionth of 1% ($\sim 5 \times 10^{-9}$).[9]

6 Reservations and Rejoinders

The first reservation that is apt to occur to one is that there is something arbitrary about the claim that the Alexandroff present is the special relativistic successor of the classical world-at-a-time. We can generate other structures by adding or deleting from an Alexandroff present. Why not them?

Note, though, that if we consider supersets of the closure of ALEX (e_1, e_2), then we must give up the first advantage that I claimed for it above. Call a superset of the closure of ALEX (e_1, e_2) that is a putative present *PRES* (e_1, e_2). In a spacetime that is oriented and stably causal, it would no longer follow that if $e' \in$ PRES (e_1, e_2), then $t(e_1) < t(e') < t(e_2)$. I think such structures would lack a feature that a *now* ought to have.

If we consider subsets of ALEX (e_1, e_2), then the one set available that is defined invariantly is just the set of events in the intersection of ALEX (e_1, e_2) with the world line λ which contains e_1 and e_2. I believe that this is the structure thought of as *now* in [5, 7], but conceiving of the relativistic *now* this way forfeits the second and third advantages that I highlighted above.

Nevertheless, this view raises a second question or challenge. As we noted above in Section 2, Dieks [7, Sections 1–3] argues that the passage of time must be construed locally. Then in Sections 4–5 he sketches of view of temporal becoming or transience that seems more austere than the one offered here. The existence of an event is just its happening,[10] and these happenings are constrained in the temporal dimension only by the partial ordering familiar from the special theory of relativity. There is no *now* and, especially, no *moving now*, since that (in his view) would require "the *addition* of something to the four-dimensional continuum" [7, p. 21].

What has to be added, according to Dieks, is "a moving very narrow 'window' through which a small portion of the continuum is made visible (or 'real')." (ibid.) This metaphysics is indeed suspect, even in the classical case, as I argue elsewhere [19]. But the *nows* as I construe them are well-defined subsets of Minkowski spacetime and are no addition to it. The *nows* along a timelike worldline can be totally ordered, given the simplifying assumption of a fixed length for the psychological present, and the *nows* in spacetime can be partially ordered by extending in an obvious way the usual partial ordering for the events of which they are comprised. Transience, along a timelike line, is just the succession of *nows* according to this ordering, and Dieks is committed to this sort of "motion" no less than I. Of course one cannot think of all the events in one *now* as "co-occurring" since many pairs of events in a *now*—such as any two distinct points along the world line λ—are timelike separated.

---

[9] My thanks to Alexandre Korolev for this calculation.

[10] On this point we agree, and he approvingly cites me [20] as a recent proponent of this view.

The Transient *nows* 359

The Shimony/Dieks view does call one's attention to the importance of timelike curves, along which we characterize transience. But this new focus in turn calls up the third reservation one might have about my proposal (in fact, about mine and the Shimony/Dieks proposal as well).

My notion of transience is local and metaphysically very austere. It may avoid the sorts of objection that are brought against metaphysically stronger (or, as I would say, more baroque) conceptions of passage, but it might be that in order to avoid these objections it falls foul of a problem at the other extreme. The problem is that one could parameterize a spacelike curve with, say, proper length. Is my notion of transience not so weak that I am, then, committed to a parallel notion of spatial becoming, change of length along a spacelike curve? And is this not a *reductio*? There is no spatial becoming. Space does not pass, and in this way space differs from time.

Before I try my hand at a rejoinder, I'd like to point to a line of thought that doesn't do the trick. It is simply to emphasize that the timelike curves along which Alexandroff presents occur successively are *time*like. Therefore, one might think, change in this dimension is trivially change (or passage) of time. But thus far, the 'time' part of timelike has not been given any *temporal* content. Timelike vectors are distinguished in Section 4 in the usual purely formal way in terms of the metric, and timelike curves are those that have timelike tangent vectors. It's true that the signature of the metric is (1,3) and so has one distinguished dimension, but it is difficult to see how pointing to this fact will help with objection raised here.

Nevertheless, I do think there is a cogent response, and I was led to it by thinking about closing the circle—about finding a connection between our experience of time (as crude clocks) and timelike worldlines. The connection is typically left implicit in discussions of special relativity, but it is explicitly stated in mathematically complete characterizations of the theory. It is the clock hypothesis, and one can find it stated in [12, p. 52]:

> If $\alpha : [a,b] \rightarrow M$ is a timelike worldline in M, then $L(\alpha)$ [my τ_{AB} above] is interpreted as the time lapse between the events α (a) and α (b) as measured by an ideal standard clock carried along by the particle whose world line is represented by α.

These standard clocks are ideal in that they continue to register proper time even though accelerated and hence subject to forces.[11] But these clocks are idealizations of the clocks around us and, even more so, of ourselves, since we are crude clocks (as any of us who have traveled across more than two time zones have been made aware). It is through clocks, and as clocks, that we experience time. Timelike curves are connected to ideal clocks via the clock hypothesis, and ideal clocks are an idealization of real clocks. These connections show why timelike lines are indeed *time*like and why the succession of nows along a timelike curve should count as an idealized version of the passage of *time*.

There is no similar spatial phenomenon (as the objector noted above) and so no connection from spacelike curves to it through any parallel or analogous hypothesis.

---

[11] Herein could lie a long and fascinating digression. See [21, Section 6.2.1].

360   S.F. Savitt

That is why there is no spatial becoming, no transient *here*, even though spacelike
curves can be parameterized too.

Another (fourth) problem that is often raised to accounts of transience is this: if
time is supposed to "pass", then it must make sense to ask how quickly it passes. Yet,
it is objected, there is no sensible answer to this question. Consider, for example,
Price's presentation [22, p. 13] of this "stock objection" to the so-called "block
universe" view of time:

> [T]he stock objection is that if it made sense to say that time flows then it would make sense
> to ask how fast it flows, which doesn't seem to be a sensible question. Some people reply
> that time flows at one second per second, but even if we could live with the lack of other
> possibilities, this answer misses the more basic aspect of the objection. A rate of seconds
> per second is not a rate at all in physical terms. It is a dimensionless quantity, rather than a
> rate of any sort. (We might just as well say that the ratio of the circumference of a circle to
> its diameter flows at π seconds per second!)

I think there are two ways of meeting this objection. The first is a "bite the bullet"
strategy. One of the things we learn from the four-vector formulation of special
relativity, as we saw above in Section 4, is that all massive objects move through
spacetime with speed 1. That is, the velocity four-vector **u** of all such objects has
length −1, and the absolute value of this length is a speed.[12] This result holds for
an object in its rest frame, when it is not moving through space at all. There is
only one dimension left in which it has this speed, then, the temporal dimension.
If these objects move through time with speed 1, however, then our anti-Dobsonian
argument ensures that we can equally well say that, for these objects, time passes
with speed 1.

Does this make sense? Maudlin [13] has a heroic argument that it does. Consider
by way of analogy, he says, exchange rates. The exchange rate for the US dollar
might be (say) $1.10 Canadian dollars per US dollar but only 0.7 Euros. What is the
exchange rate for the US dollar itself? One US dollar per US dollar, of course. Isn't
this limiting rate a valid exchange rate? The why isn't 1 s per s a valid "flow" rate?

In addition, one can point out that in many standard presentations of the special theory, time is given by ct, a distance. Speeds are dimensionless when time is
indicated this way, and this result has not been perceived as problematic.

There is another, and perhaps subtler, way to look at this matter. N. David
Mermin deduces from the invariance of the spacetime interval that for any massive
object[13]

$$(T_0/T)^2 + v^2 = 1, \tag{9}$$

where T_0 is proper time, T is coordinate time in a fixed frame, and v^2 is velocity
squared. Then he comments:

> Now a stationary clock moves through time at 1 nanosecond [of proper time] per nanosec-
> ond [of frame or coordinate time] and does not move through space at all. But if the

---

[12] See [11, Section 5.2].

[13] That is, the events for which Mermin considers the interval must be timelike separated, since
one clock can be present at each. Mermin has told me that his discussion was inspired by Greene
(1999) [23, p. 47–51].

clock moves through space—i.e. the larger v is—the slower it moves through time—i.e. the smaller T_0/T is—in such a way as to maintain the sum of the squares of the two at 1. It is as if the clock is always moving through a union of space and time—spacetime—at the speed of light. If the clock is stationary then the motion is entirely through time (at a speed of 1 nanosecond per nanosecond) [14, p. 86].

This second strategy for meeting the objection does provide a genuine ratio of one quantity to another in order to make a rate of transience or passage. In the classical case, this strategy is disastrous. Introducing a second temporal dimension re-raises the question of its "flow" and so leads to an infinite regress of temporal dimensions. The two kinds of time are already in the special theory, however. No new time dimension has to be added in order to solve the problem of transience, nor do both of the time dimensions have to "flow".

In fact, in my view transience is to be found in proper time, rather than coordinate time, and I would prefer to work with the reciprocal of the ratio in Eq. (9). In this case, one would have to think of (coordinate) time as speeding up as an object's spatial speed increased, and hence as moving clocks running faster. Mermin notes this [14, p. 163] but seems to think of coordinate time as that which passes.

The fifth and final reservation is one that is difficult to state and may prove the most difficult to overcome. Our cognitive life seems to be structured in a very deep way around classical Nows, around thinking in terms of the-world-at-an-instant. It is hard to have an intuitively satisfying picture of time, of local transience, of the sort sketched above.

We seem to have evolved to have a false picture of time for the Steinian reason given near the end of the previous section. I note that there are similar but in some odd way dual difficulties in understanding quantum mechanics, where features that we took to be local turn out not to be (instead of a global structure turning out to be, if I am right local). Perhaps these difficulties are in some deep way connected. Perhaps it is no coincidence that a similar picture seems to underlie the following remark of Lee Smolin, a prominent researcher in the field of quantum gravity [24, p. 55]:

> A causal universe is not a series of stills following on, one after the other. There is time, but there is not really a notion of a moment of time. There are only processes that follow one another by causal necessity.

A universe with local passing of time is unfamiliar to most of us, puzzling to common sense; but it looks as if this is the picture of time we must learn to live with if we are to understand our universe.

References

1. Carnap, R. (1963), "Intellectual autobiography" in *The Philosophy of Rudolf Carnap*, ed. by P. A. Schilpp. LaSalle: Open Court.
2. Weyl, H. (1949), *Philosophy of Mathematics and Natural Science*. (Revised and augmented English edition, based on a translation by O. Helmer.) Princeton: Princeton University Press.

362   S.F. Savitt

3. Davies, P. C. W. (1974, 1977), *The Physics of Time Asymmetry*. Berkeley and Los Angeles: University of California Press.
4. Rovelli, C. (1995), "Analysis of the distinct meanings of the notion of 'time' in different physical theories," *Il Nuovo Cimento* 110B(1): 81–93.
5. Shimony, A. (1993), "Reality, causality, and closing the circle" in *Search for a Naturalistic World View*, Volume I. Cambridge: Cambridge University Press.
6. Gödel, K. (1949), "A Remark about the relationship between relativity theory and idealistic philosophy" in *Albert Einstein: Philosopher-Scientist*, ed. by P. A. Schilpp. La Salle: Open Court.
7. Dieks, D. (ed.) (2006), "Becoming, relativity, and locality" in *The Ontology of Spacetime*, 1. Amsterdam, Elsevier.
8. Mermin, N. D. (2005), *It's About Time*. Princeton: Princeton University Press.
9. Dobson, H. A. (1905), *Collected Poems*. London: Kegan Paul, Trench, Trübner & Co. Ltd. Amsterdam, Elsevier.
10. Weyl, H. (1921, 1952), *Space-Time-Matter*. 4th edn. Tr. by Henry L. Brose. New York City: Dover Publications.
11. Hartle, J. B. (2003), Gravity: *An Introduction to Einstein's General Relativity*. San Francisco: Addison Wesley.
12. Naber, G. L. (1988), *Spacetime and Singularities*: An Introduction. Cambridge: Cambridge University Press.
13. Maudlin, T. (2002), "Remarks on the passing of time," *Proceedings of the Aristotelian Society* 102(3): 237–252.
14. Winnie, J. (1977), "The causal theory of space-time" in *Foundations of Space-Time Theories*. Minnesota Studies in the Philosophy of Science, Volume VIII, ed. by Earman et al. Minneapolis: University of Minnesota Press.
15. Dainton, B. (2001). *Time and Space*. Montreal and Kingston: McGill-Queen's University Press.
16. Broad, C. D. (1938) *Examination of McTaggart's Philosophy*, Volume II. Cambridge: Cambridge University Press.
17. Einstein, A. (1905), "Zur elektrodynamik bewegter Körper," *Annalen der Physik* 17: 891–921. (English translation by W. Perrett and G. B. Jeffrey in *The Principle of Relatvity: A Collection of Original Memoirs on the Special and General Theory of Relativity*. Dover Publications, Inc., 1952.)
18. Stein, H. (1991), "On relativity theory and the openness of the future," *Philosophy of Science* 58: 147–167.
19. Savitt, S. (2006), "Presentism and eternalism in perspective" in *The Ontology of Spacetime*, 1, ed. by D. Dieks, Amsterdam, Elsevier.
20. Savitt, S. (2002), "On absolute becoming and the myth of passage" in *Time, Reality, and Experience*, ed. by C. Callender. Cambridge: Cambridge University Press.
21. Brown, H. (2005), *Physical Relativity*. Oxford: Oxford University Press.
22. Price, H. (1996), *Time's Arrow and Archimedes' Point*. Oxford: Oxford University Press.
23. Greene, B. (1999), *The Elegant Universe*. New York: W. W. Norton & Company.
24. Smolin, L. (2001), *Three Roads to Quantum Gravity*. New York: Basic Books.

Quantum in Gravity?

Michael Horne

Abstract After a review of the elementary relativistic wave function that makes quantum fringes move, contract and adjust their probability correctly in all inertial frames, I transform the wave function to a non-inertial frame. The non-inertial wave function is approximate.

1 Introduction

Quantum mechanics in its original wave mechanical form gave new physics to an old object—the particle. The new physics rests on the wave function, the complex probability amplitude extending over space–time regions, possibly large. For example, the wave function of a particle of definite momentum–energy is a plane wave spread over the whole space–time. Since the plane wave probability density is a constant over the whole space–time, it possesses no features, and hence cannot exhibit motion. Only the particle has motion here. However, a superposition of two plane waves has a density that is a new kind of object—the fringe. Fringes are pure quantum mechanical objects with observable motions and extended features, spatial and temporal, in every frame, inertial and non-inertial. To prepare the moving fringes, I'm imagining here the ubiquitous double-slit experiment with the detection screen sliding parallel to itself and perpendicular to the fringes. Observation of the fringes at high speed requires both high particle count rates and an area detector with fine spatial and temporal resolution, since each fringe contracts and takes less time to pass. Theoretical description of the fringes requires relativistic quantum mechanics, special relativity (SR) in the inertial frames and general relativity (GR) in the non-inertial. This paper reports my attempt to set up these wave mechanical descriptions.

M. Horne
Physics and Astronomy Department, Stonehill College, North Easton, MA, USA
e-mail: mhorne@stonehill.edu

W.C. Myrvold and J. Christian (eds.), *Quantum Reality, Relativistic Causality, and Closing the Epistemic Circle,* The Western Ontario Series in Philosophy of Science 73,
© Springer Science+Business Media B.V. 2009

364   M. Horne

The correct fringe contraction and motion and even that the fringes move at all
must originate in the relativistic wave mechanics itself. In the first section of the
paper I review the rigorous details of how all that happens in the inertial frames.
In the second section I try to extend the wave-mechanical description of the fringe
features and motions to an accelerating frame. This takes the elements of wave mechanics to the edge of general relativity. Note that since I never leave flat space-time,
I never consider any "real" gravity. Thus a lengthier but more accurate title would
be Relativistic Motions of Quantum Interference Fringes in Inertial and Non-Inertial
Frames.

2 Inertial Frames

To introduce the fringe state and to show my methods, I first review some elementary
states and their Lorentz transforms. The fringe state will be a superposition of two
of these. Consider the state, phase, and density for a particle of rest mass m at rest,

$$\psi = e^{i\phi} \tag{1a}$$

$$\phi = -mc^2 t/\hbar \tag{1b}$$

$$\rho = \psi^* \psi = 1 \tag{1c}$$

In another frame where the same particle has speed u in the x' direction, these are

$$\psi' = \gamma^{\frac{1}{2}}(u)e^{i\phi'} \tag{2a}$$

$$\phi' = \gamma(u)m\left(ux' - c^2 t'\right)/\hbar \tag{2b}$$

$$\rho' = \gamma(u) \tag{2c}$$

The phase ϕ' follows from the fact that $kx - \omega t$ transforms to $k'x' - \omega' t'$. Then, I've
just inserted for k' and ω' their values as functions of the transform speed. Note,
then, that the space–time transforms are not directly employed in writing ϕ'; but they
entered indirectly via the proof of phase invariance. The magnitude enhancement
$\gamma^{\frac{1}{2}}(u)$ on ψ' is a consequence of the equation

$$\rho = \psi^* \psi \tag{3}$$

for density and a requirement of interframe agreement on probability,

$$\rho dx = \rho' dx' \tag{4}$$

How can the symmetric Lorentz transform introduce an asymmetry between the
wave functions, primed and unprimed? Physically, asymmetry is allowed because
ψ is the particle at rest while ψ' is the same particle with speed u. Since proper
spatial interval dx in the particle rest frame contracts to $dx' = dx/\gamma(u)$ in the primed
frame, both the primed density and wave function must be enhanced to compensate
for the smaller interval. In short, the probability of finding the particle in a region
must be frame independent.

Quantum in Gravity? 365

Now consider another particle whose state in some inertial frame (primed) is a superposition of two states (2a), one with $+u$ and the other with $-u$,

$$\psi' = \frac{1}{\sqrt{2}} \gamma^{\frac{1}{2}}(u) \left(e^{i\phi'_+} + e^{i\phi'_-} \right) \tag{5a}$$

$$\phi'_{\pm} = \gamma(u)m(\pm ux' - c^2 t')/\hbar \tag{5b}$$

$$\rho' = \gamma(u) \left[1 + \cos\left(\phi'_+ - \phi'_- \right) \right]$$
$$= \gamma(u)[1 + \cos\left(2\gamma(u)mux'/\hbar \right)] \tag{5c}$$

This density shows the stationary fringes of interest. These are the fringes of perfect visibility seen on the screen in an ideal double-slit experiment. The fringe width is

$$w' = \pi\hbar/\gamma(u)mu \tag{5d}$$

Finally, transform (5a) to another inertial frame (unprimed) moving at speed $+v$ along the x' axis, i.e. sliding right along the usual Young's screen. The state, phases, density, and fringe width are

$$\psi = \frac{1}{\sqrt{2}} \gamma^{\frac{1}{2}}(v) \gamma^{\frac{1}{2}}(u) \left(e^{i\phi_+} + e^{i\phi_-} \right) \tag{6a}$$

$$\phi_{\pm} = \gamma(v)\gamma(u)m \left[(\pm u - v)x - \left(c^2 \mp vu \right) t \right]/\hbar \tag{6b}$$

$$\rho = \gamma(v)\gamma(u) \left[1 + \cos(\phi_+ - \phi_-) \right]$$
$$= \gamma(v)\gamma(u) \left[1 + \cos\left(2\gamma(u)mu \left\{ \gamma(v) \left[x + vt \right] \right\}/\hbar \right) \right] \tag{6c}$$

$$w = \pi\hbar/\gamma(v)\gamma(u)mu \tag{6d}$$

The wave function ψ picks up an additional $\gamma^{\frac{1}{2}}(v)$ by the same heuristic argument as before, except, notably, here the resting object in the original frame (here primed) is the fringe. The phases $\phi_{\pm}$ are found by the same method as the ϕ in (2b), i.e. evaluate each phase function $kx - \omega t$ by inserting the correct momentum and energy values for the new frame; again the new coordinates come along for free without direct use of the space-time transforms. Note, however, that I have used the identities

$$\gamma\left(\overline{\alpha + \beta} \right) \left(\overline{\alpha + \beta} \right) = \gamma(\alpha)\gamma(\beta)(\alpha + \beta) \tag{7a}$$

$$\gamma(\overline{\alpha + \beta}) = \gamma(\alpha)\gamma(\beta)(1 + \alpha\beta) \tag{7b}$$

to clear the momenta and energies of relativistic velocity addition, denoted by the bar. These identities are derived from relativistic addition and hence from the space-time transforms. The density ρ shows that the fringes move, contract and enhance correctly. Since the curly bracket is the Lorentz transform of the x' coordinate in the fringe rest frame, the ρ could be derived directly as the transform of ρ', independently of the quantum origin of the fringes. The states in this section are the relativistic wave mechanics behind the fringes and their motion.

Standard treatments of relativistic wave mechanics do not enhance the magnitudes of the wave function when transforming from the rest frame of an object (particle or fringe), but assume that $\psi' = \psi$ for all transforms. The necessary enhancement of the density is achieved by replacing (3) with

$$\rho = \frac{i\hbar}{2mc^2} \left(\psi^* \frac{\partial \psi}{\partial t} - \psi \frac{\partial \psi^*}{\partial t} \right). \tag{8}$$

If any of the above states is inserted in (8), but with enhancements omitted, the identical densities are recovered. In short, density (3) with the enhancements is exactly equal to (8) without the enhancements.

3 A Non-inertial Frame

Equations (6) are the fully relativistic wave mechanics of fringes as seen from an inertial frame moving through the fringes at speed v, with Lorentz transform of (5) as foundation. Now imagine a frame moving through the fringes at constant proper acceleration g, with the following transforms of (5) as foundation,

$$x' = (1+x)\cosh\tau - 1 \tag{9}$$
$$t' = (1+x)\sinh\tau \tag{10}$$

Here the primed coordinates are those of (5), except here dimensionless instead of conventional,

$$x' = \frac{g}{c^2}x'_{conv.}, \quad t' = \frac{g}{c}t'_{conv.} \tag{11}$$

and x and τ are the tentative new coordinates, also dimensionless. At the new origin, $x = 0$, these equations are rigorous SR, with τ the proper time carried by a clock riding the accelerating origin. With non-zero x, these are an attempt to hang an extended spatial frame on the accelerating origin point and spread τ into a global time—Fermi-Walker coordinates. Note that at $\tau = 0$ when the new frame is at rest everywhere, all clocks in both frames are zeroed and spatial coordinates in the two frames match.

Some important consequences of these space-time transforms can be determined by simply paralleling the standard SR derivation of velocity addition. First define (dimensionless) speed in each frame,

$$u' \equiv \frac{dx'}{dt'} \tag{12}$$

$$u \equiv \frac{dx}{d\tau} \tag{13}$$

Quantum in Gravity? 367

then take differentials of (9) and (10), employ (12) and (13), and thereby derive the velocity relation

$$\frac{u}{1+x} = \frac{u'-v}{1-u'v} \tag{14}$$

Here

$$v \equiv \frac{dx'}{dt'}\bigg|_{x=const.} = \frac{t'}{1+x'} = \tanh \tau \tag{15}$$

is the instantaneous speed of the point in the accelerating frame coincident with the inertial frame event x',t'. The $(1+x)$ denominator in (14) has three disappointing effects. First it spoils what would otherwise be standard velocity addition, with a space-time dependent speed v of the accelerating frame. Second it spoils equality of relative speeds of the frames at coincident space-time points. To see this, insert $u' = 0$ into (14), and thereby get $-(1+x)v$, instead of $-v$, for u. Third, the valuable identities (7), that are direct consequences of velocity addition, are not valid.

Fortunately all three are fixed if an accelerating frame clock at x keeps a new time t, not the τ time of the origin clock, where

$$dt = (1+x)d\tau \tag{16}$$

Then the troublesome $(1+x)$ denominator in (14) gets absorbed into a corrected speed. That is, instead of (13), the correct speed u of an object in the accelerating frame is

$$u = \frac{dx}{dt} = \frac{dx}{(1+x)d\tau} \tag{17}$$

which fixes all three breakdowns. Equation (16) is the general relativistic (GR) effect on clocks in an accelerated frame.

With restoration of velocity addition, equality of relative speed of the frames, and the identities (7), the inertial frame wave function (6) can be employed here in the accelerating frame if the space-time dependent speed v of Eq. (15) is inserted for the constant speed v of the Lorentz transform. Integrate the GR differential relation (16) to obtain

$$\tau(x,t) = \frac{t}{1+x}(?) \tag{18}$$

The question mark concerning this integral will be discussed below. From (15) and (18), the relative speed of the frames, in the non-inertial coordinates, is

$$v = c\tanh\left[\frac{\frac{gt}{c}}{1+\frac{gx}{c^2}}\right] \tag{19}$$

where the c is inserted to convert dimensionless speed to conventional speed and the space–time coordinates have also been returned to conventional. Then the accelerating frame wave function, phases, density and fringe width are, respectively, Eqs. (6a), (6b), (6c), and (6d) with the substitution (19).

4 Approximations

There is evidence, independently of the QM, that the space–time of GR has not been adequately incorporated and that the integral (18) is the prime suspect. When the integrated $\tau(x,t)$ is inserted in the original transforms (9) and (10) and the differentials are run again, now directly in terms of x and t, the exact relativistic velocity addition is not recovered. As expected, the old troublesome $(1+x)$ denominator on the left side of (14) does disappear, but now new small-departure-from-one factors appear against the v and the $u'v$ terms on the right side. This inconsistency is direct evidence that (18) is unsound and thus that (6) with (19) inserted for v are approximations.

There is also direct quantum mechanical evidence that the wave function (6a) with v from (19) is approximate. When the hand-inserted magnitude enhancements are omitted and the wave function is inserted in the density (8), the density (6c), with (19) substituted for v, is not exactly recovered. As expected, the correct enhancements do come down, but another troublesome small-departure-from-one factor comes down with them. This factor arises because of the time dependence in the momenta and energies. If the time derivative operators in (8) were "blind" to these time dependencies, the correct (6c) is recovered. I take this to be additional evidence that the accelerated wave function is approximate.

A Proposed Test of the Local Causality of Spacetime

Adrian Kent

Abstract A theory governing the metric and matter fields in spacetime is *locally causal* if the probability distribution for the fields in any region is determined solely by physical data in the region's past, i.e. it is independent of events at space-like separated points. General relativity is manifestly locally causal, since the fields in a region are completely determined by physical data in its past. It is natural to ask whether other possible theories in which the fundamental description of space-time is classical and geometric—for instance, hypothetical theories which stochastically couple a classical spacetime geometry to a quantum field theory of matter—might also be locally causal.

A quantum theory of gravity, on the other hand, should allow the creation of spacetimes which violate local causality at the macroscopic level. This paper describes an experiment to test the local causality of spacetime, and hence to test whether or not gravity behaves as quantum theories of gravity suggest, in this respect. The experiment will either produce direct evidence that the gravitational field is not locally causal, and thus weak confirmation of quantum gravity, or else identify a definite limit to the domain of validity of quantum theory.

1 Introduction

Abner Shimony's many profound contributions to theoretical physics have greatly deepened our understanding of the nature of physical reality. This paper is devoted to subjects on which Abner's work is particularly celebrated, namely the theoretical

---

A. Kent
Centre for Quantum Computation, DAMTP, Centre for Mathematical Sciences,
University of Cambridge, Wilberforce Road, Cambridge CB3 0WA, UK
Perimeter Institute for Theoretical Physics, 31 Caroline Street North, Waterloo, Ontario,
Canada N2L 2Y5
e-mail: a.p.a.kent@damtp.cam.ac.uk

W.C. Myrvold and J. Christian (eds.), *Quantum Reality, Relativistic Causality, and Closing the Epistemic Circle,* The Western Ontario Series in Philosophy of Science 73,
© Springer Science+Business Media B.V. 2009

definition and understanding of locality and local causality and the ways in which these properties can be experimentally tested in Nature.

General relativity and quantum theory are both impressively confirmed within their domains of validity, but are, of course, mutually inconsistent. Despite decades of research, there are still deep conceptual problems in formulating and interpreting quantum gravity theories: we don't have a fully consistent quantum theory of gravity, nor do we know precisely how we would make sense of one if we did.

One initially natural-seeming possibility is to combine general relativity and quantum theory in a semi-classical theory that couples the metric to the expectation of the stress-energy tensor via the Einstein equations [1–3]. However, the problems with this suggestion are well-known. In particular, if the unitary quantum evolution of the matter fields is universal, then it would imply that the complete state of the matter fields in the current cosmological era ought to be a superposition of many (in fact, presumably an infinite continuum of) macroscopically distinct cosmologies. A semi-classical theory of gravity coupled to these matter fields would imply, inter alia, that the gravitational fields in our solar system and galaxy correspond to the weighted average over all possible matter distributions, rather than the actual distribution we observe. This would be grossly inconsistent with the observed data. It is also contradicted by terrestrial experiment [4].

One might try to rescue the hypothesis by supposing, instead, that unitary quantum evolution is not universal and that the metric couples to the expectation of the stress tensor of non-unitarily evolving matter fields. Obviously, this requires some explicit alternative to unitary quantum theory, such as a dynamical collapse model [5]. It is not presently known whether such a theory can be combined with a metric theory of gravity in a generally covariant way. An interesting related possibility is that a classical metric might be coupled to quantum matter via stochastic equations [6, 7]: however, no consistent and generally covariant theory of this type has yet been developed either.

I take here a possibly controversial stance. It seems to me that, because we haven't made any really certain progress in understanding how general relativity and quantum theory are unified, we should take more seriously the possibility that the answer might take a rather different form from anything we've yet considered. On this view, even apparently rather basic and solid intuitions are worth questioning: if an intuition can be tested experimentally, and we can unearth a sliver of motivation for speculating that it might possibly fail, we should test it.

2 Gravity, Local Causality and Reality

2.1 Sketch of Experiment

Before getting into technicalities, let me summarise the proposed experiment.

We start with a standard Bell experiment, carried out on an entangled pair of elementary particles, in which the measurement choices and measurement outcomes on both wings are spacelike separated.

A Proposed Test of the Local Causality of Spacetime   371

The choices and outcomes are then amplified to produce distinct local gravitational fields, on both wings. This amplification can be carried out by any practical means, for example by recording the choice and outcome on each wing in an electronic signal, and feeding this signal into a circuit connected to a device that moves a macroscopic quantity of matter to one of four possible macroscopically distinct configurations. Note that this amplification need *not* necessarily maintain quantum coherence.

These gravitational fields produced are then directly measured, by observing their influence on small masses in the relevant region, for example by Cavendish experiments. This is done quickly enough that the region A_2, in which the amplified gravitational field on wing A is measured, is spacelike separated from the region B_1 in which the Bell measurement choice on wing B was made, and similarly A_1 is spacelike separated from B_2. The results of these measurements are recorded and compared, to check whether they display the correlations which quantum theory predicts for the relevant Bell experiment.

2.2 Standard Expectations and Why They Should Be Tested

Almost all theoretical physicists would, I think, fairly confidently predict that any experiment of this type will indeed produce exactly the same non-local correlations as those observed in standard Bell experiments. What I want to argue is that there are some coherent—although of course speculative—theoretical ideas which would imply a different outcome, and that these provide scientific motivation enough to justify doing the experiment. To justify this, one needn't argue that the standard expectation is likely to be wrong (indeed, I think it's very likely right). One need only argue that there are some alternative lines of thought which have some non-negligible probability of being closer to the truth.[1]

2.2.1 One Possible Motivation

One view of quantum theory, advocated by Bell and taken seriously by many, is that the theory is incomplete without some mathematical account of "beables" or "elements of reality" or "real events"—the quantities which, ultimately, define the sample space for quantum probabilities, i.e. which are the things which quantum probabilities are probabilities of. Most attempts to resolve this problem postulate that the beables are at least approximately localised in space-time.

Now, a standard Bell experiment ensures that the particles in the two wings enter detectors at space-like separated points, in a sense which we can justify intuitively

---

[1] Obviously, there's no precise way to quantify how likely a surprising outcome must be to make an experiment worth doing. But to give a rough illustration, a probability of 10^{-5} of a surprising answer here would seem to me more than sufficient justification for carrying out an experiment that requires relatively modest resources.

within the quantum path integral formalism (and more precisely in some interpretations of quantum theory). But this does not ensure that any beables or real events associated with the measurements are necessarily space-like separated. For instance, if the beables or real events are associated with the collapse of the wavefunction, and if this collapse takes place only when a measurement result is amplified to macroscopic degrees of freedom, then the relevant question is whether these amplification processes on the two wings take place in space-like separated regions.

Consider now:

Assumption I Bell experiments appear to produce non-local correlations, consistent with the predictions of quantum theory, when the relevant beables are time-like separated (i.e. when there is time for information about the first relevant real event to propagate to the second), but *not* when they are space-like separated.

Assumption II in all Bell experiments to date, the relevant beables have indeed been time-like separated.

If both assumptions were correct, the apparent demonstration of non-locality in Bell experiments to date would be an artefact. The assumptions may, however, at first sight seem purely conspiratorial and completely lacking in theoretical motivation. Surprisingly, though, it *is* possible to sketch an alternative version of quantum theory which appears to be internally consistent, is not evidently refuted by the data, and implies both **I** and **II** [8].

Now, let us extend this speculation further. It is sometimes suggested that the solution to the quantum measurement problem is tied up with the link between quantum theory and gravity. Consider

Assumption III to ensure that a real event (selecting one outcome and one of the possible fields) takes place requires a measurement event whose different possible outcomes create measurably distinct gravitational fields.

If **(I–III)** were all true, the gravitational Bell experiment described above would indeed produce a different outcome from standard Bell experiments. To be sure, taking this possibility seriously requires one to take seriously three non-standard hypotheses. From the perspective of a firm believer in the universality of unitary quantum evolution and in quantum gravity, each of these hypotheses might be seen as quite implausible. It is worth stressing, though, that none of these hypotheses is an ad hoc invention, produced specifically for the purposes of the present discussion. Each of them has an independent motivation:

(I) results from a nonstandard but interesting way of trying to reconcile beable quantum theory and special relativity.

(II) becomes quite plausible if one takes seriously the idea of wave function collapse as a real physical process defined by explicit equations. Models, such as those defined by Ghirardi–Rimini–Weber–Pearle [5], which have this feature and which are consistent with other experiments tend to imply that collapse only takes place quickly (on a scale of μs) as the measurement result becomes amplified to a macroscopic number of particles (of order 10^{17}). In other words, according to these models, collapse need not take place at all quickly

A Proposed Test of the Local Causality of Spacetime 373

in the photo-detectors or electronic circuits used in standard Bell experiments. Hence, it need not necessarily be the case that there are spacelike separated collapses in the two wings of such experiments: as far as I am aware, in all Bell experiments to date, reasonable choices of the GRWP collapse parameters would imply that no significant collapse occurs until later, after the data have been brought together and stored.

(III) is a widely considered, if non-standard, intuition about the possible form of a theory unifying quantum theory and gravity. It is also related to another motivation for the proposed experiment, to which we now turn.

2.2.2 A Second Possible Motivation

Perhaps quantum theory and general relativity are unified, not via a quantum theory of gravity, but by some theory which somehow combines a classical description of a space-time manifold with a metric together with a quantum description of matter fields. Any such theory would presumably have to have a probabilistic law for the metric, since it seems essentially impossible to reconcile a deterministic metric evolution law with quantum indeterminism. That is, a fundamental law of nature selects a four-geometry drawn from a probability distribution defined by some set of principles, which also define the evolution of matter. Also, to be consistent with observation to date, these principles must tend to produce spacetimes approximately described by the Einstein equations on large scales.

Granted, we don't even know whether there *is* a consistent generally covariant theory of this form. Before dismissing the entire line of thought as thus presently unworthy of attention, though, one should remember that we don't know if there's a consistent quantum theory of gravity either. The idea of a stochastic hybrid theory, with a classical manifold coupled to quantum matter, has some attraction, despite its difficulties, as it suggests a possible way around some of the conceptual problems that arise when trying to make sense of a quantum theory of spacetime.

Suppose then that we agree to take this idea as serious enough to be worth contemplating exploring a little. Given the central role of causality in general relativity, it seems reasonably natural to consider the class of metric theories whose axioms require the metric encode some version of Einstein causality. Such theories would preclude the gravitational field exhibiting the type of non-local correlations that quantum theory predicts for matter fields—and so would have surprising and counter-intuitive features. Once again, it needs to be stressed that we neither want nor need to argue that this is the likeliest possibility, only that it has *some* theoretical motivation and has testable consequences. In the next section we define a local causality principle adapted to non-deterministic metric theories, and examine its consequences.

374   A. Kent

3 Local Causality for Metric Theories: Technicalities

One key feature on which various theories and proto-theories of gravity differ is the
causal structure of the classical or quasi-classical space-time which emerges. Bell's
definition of local causality [9] applies to physical operations taking place in a fixed
Minkowski space-time. As Bell famously showed, quantum theory is not locally
causal. The possibility of adapting the definition to apply to theories with a variable
space-time geometry (or a variable structure of some sort from which space-time
geometry is intended to emerge) has been considered by Rideout and Sorkin [10]
and Henson [11], among others. The following definition is a modified version of
one suggested by Dowker [12].

Define a *past region* in a metric spacetime to be a region which contains its own
causal past, and the *domain of dependence* of a region R in a spacetime S to be the
set of points p such that every endless past causal curve through p intersects R.

Suppose that we have identified a specified past region of spacetime Λ, with
specified metric and matter fields, and let κ be any fixed region with specified metric
and matter fields.

Let Λ' be another past region, again with specified metric and matter fields. (In
the cases we are most interested in, $\Lambda \cap \Lambda'$ will be non-empty, and thus necessarily
also a past region.)

Define

$$\mathrm{Prob}(\kappa | \Lambda \perp \Lambda')$$

to be the probability that the domain of dependence of Λ will be isometric to κ,
given that $\Lambda \cup \Lambda'$ form part of space-time, and given that the domains of dependence
of Λ and Λ' are space-like separated regions.

Let κ' be another fixed region of spacetime with specified metric and matter
fields.

Define

$$\mathrm{Prob}(\kappa | \Lambda \perp \Lambda'; \kappa')$$

to be the probability that the domain of dependence of Λ will be isometric to κ,
given that $\Lambda \cup \Lambda'$ form part of space-time, that the domain of dependence of Λ'
is isometric to κ', and that the domains of dependence of Λ and Λ' are space-like
separated.

We say a metric theory of space-time is *locally causal* if for all such $\Lambda, \Lambda', \kappa$ and
κ' the relevant conditional probabilities are defined by the theory and satisfy

$$\mathrm{Prob}(\kappa | \Lambda \perp \Lambda') = \mathrm{Prob}(\kappa | \Lambda \perp \Lambda'; \kappa').$$

4 Testing Local Causality of Metric Theories

By definition, general relativity is locally causal, since the metric and matter fields
in the domain of dependence κ of Λ are completely determined by those in Λ via
the Einstein equations and the equations of motion. If we neglect (or believe we

A Proposed Test of the Local Causality of Spacetime   375

can somehow circumvent) the fact that quantum theory is not locally causal (in Bell's original sense), it would also seem a natural hypothesis that any fundamental stochastic theory of space-time, or any fundamental stochastic theory coupling a classical metric to quantum matter, should be locally causal. One reason for considering this possibility is that, while it admittedly seems hard to see how to frame closed form generally covariant equations for any theory of this type, it seems particularly hard to see how to frame such equations for a non-locally causal theory. If we allow the evolution of the metric, and hence the causal structure, at any given point to depend on events at space-like separated points, it seems difficult to maintain any notion of causality, or to find any other ordering principle which ensures that equations have a consistent solution.

However, we should *not* expect a quasiclassical space-time emerging from a quantum theory of gravity to be locally causal, for the following reason. Consider a standard Bell experiment carried out on two photons in a polarization singlet state. For definiteness, let us say that the two possible choices of measurement on either wing are made by local quantum random number generators, and are chosen to produce a maximal violation of the CHSH inequality [13].

We suppose that the two wings of the experiment, A and B, are fairly widely separated. Now suppose that the measurement choices and outcomes obtained by the detectors in each wing mechanically determine one of four macroscopically distinct configurations. To be definite, let us suppose that the Bell experiment is coupled to local Cavendish experiments on each wing, in such a way that each of the two settings and two possible measurement outcomes on any given wing causes one of four different configurations of lead spheres—configurations which we know would, if the experiment were performed in isolation, produce one of four macroscopically and testably distinct local gravitational fields. Suppose also that the Cavendish experiments are arranged so that the local gravitational fields are quickly tested, using small masses on a torsional balance in the usual way. The separation of the two wings is such that the gravitational field test on either wing can be completed in a region space-like separated from the region in which the photon on the other wing is detected.

A quantum theory of gravity should predict that the superposition of quantum states in the singlet couples to the detectors in either wing to produce entangled superpositions of detector states, and thence entangled superpositions that include the states of the Cavendish experiments, and finally entangled superpositions of states that include the states of the local gravitational field. Extrapolating any of the standard interpretations of quantum theory to this situation, we should expect to see precisely the same joint probabilities for the possible values of the gravitational fields in each wing's experiments as we should for the corresponding outcomes in the original Bell experiment. As Bell [14] and Clauser et al. [13] showed, provided we make the standard and natural (although not logically necessary) assumption that the measurement choices in each wing are effectively independent from the variables determining the outcome in the other wing, these joint probabilities violate local causality in Bell's original sense.

We now make the further natural assumption that when, as in our proposed experiment, the measurement choices are made by the outputs of the local quantum random number generators, the choices made on each wing are independent of the metric and matter fields in the past of the measurement region on the other wing. Then, if κ is the region immediately surrounding the measurement choice and outcome in one wing of the experiment, κ' the corresponding region for the other wing, Λ the past of κ, and Λ' the past of κ', we have

$$\text{Prob}(\kappa|\Lambda \perp \Lambda') \neq \text{Prob}(\kappa|\Lambda \perp \Lambda'; \kappa').$$

Does such an experiment even need to be performed, given the impressive experimental confirmation of quantum theory in Bell experiments to date? In my view, it does.

Taking the Bell experiments to date at face value—that is, neglecting any remaining possible loopholes in their interpretation—they confirm predictions of quantum theory *as a theory of matter fields when gravity is negligible*. Specifically, they confirm predictions of quantum theory for experiments involving matter states when those states do not produce significant superpositions of macroscopically distinct gravitational fields.

The question at issue here is precisely how far quantum theory's domain of validity extends. When it comes to predicting whether or not the metric is locally causal, there is a genuine tension between intuitions extrapolated from quantum theory and those which one might extrapolate from general relativity. Examining and testing this question seems a very natural development of the line of questioning begun by Einstein, Podolsky and Rosen [15] and continued by Bell [14].

Standard Bell experiments test the conflicting predictions implied by quantum theory and by EPR's intuitions about the properties of elements of physical reality. EPR's intuitions can be motivated by a combination of classical mechanics (which suggests that the notion of an element of physical reality is a sensible one) and special relativity (which suggests the hypothesis that an element of physical reality has the locality properties ascribed to it by EPR). In the experiment considered here, we again have a tension between intuitions drawn from two successful theories—in this case quantum theory and general relativity.

5 Possible Counterarguments

But isn't this a crazy line of thought? How could the correlations obtained from Bell experiments *possibly* be altered by coupling classical devices to the detector outputs? Is the Bell experiment supposed to know that the classical devices are waiting for the data, and change its result because of that? Or, even more weirdly, is the gravitational field in each wing supposed to know that the classical lumps of matter are being moved around as the result of a Bell experiment, and change *its* behaviour—violating the predictions of Newtonian gravity as well as general relativity within a local region—because of *that*?

A Proposed Test of the Local Causality of Spacetime   377

I find it hard to accept the full rhetorical force of such objections, natural though they are. Nature has a capacity to surprise, and surprising experimental results sometimes have theoretical explanations which occurred to nobody beforehand. The "common sense" view just expressed implicitly assumes, among other things, first, that the outcomes of detector measurements in Bell experiments constitute local, macroscopic events that in some physically meaningful sense are definite and irreversible once they occur, and second, that the local gravitational fields respond instantly to these events in the same way as they would if they resulted from isolated experiments on unentangled states. These plausible propositions may very well be given precise meaning and completely justified by some deeper understanding of quantum theory and gravity than we currently have. Even if they don't turn out to have a precise and literal justification—for instance, because the fundamental theory contains no definition of definite local events—it seems very plausible that we nonetheless reach the right conclusion about Bell experiments and gravity by reasoning as though they were true. However, none of this is completely beyond reasonable doubt in the light of our current knowledge.

As we've already noted, there's some independent motivation for exploring variants of quantum theory in which definite local events *are* defined but in which photo-detector measurement outcomes aren't, so to speak, macroscopic enough to constitute such events.

There's also some motivation for exploring theories of quantum theory and gravity in which a probabilistic law defines a locally causal classical gravitational field. Standard reductionist reasoning would break down in such a theory—as it does, though in a different way, in quantum theory—and the behaviour of the gravitational field in one wing of a Bell experiment would indeed depend on the configurations of both wings of the experiment.

What, then, are the conceivable experimental outcomes, and what would they imply? One is that the violations of local causality predicted by quantum theory, and to be expected if some quantum theory of gravity holds true, are indeed observed. This would demonstrate that space-time is indeed not locally causal, as predicted by quantum theories of gravity, but not necessarily by other hypotheses about the unification of quantum theory and gravity. It would thus provide at least some slight experimental evidence in favour of the quantization of the gravitational field. It might be argued, pace Page and Geilker [4], that this would be the first such experimental evidence, since, as noted above, Page and Geilker's experiment tested a version of semi-classical gravity already excluded by astronomical and cosmological observation.

A second logical possibility is that the violations of local causality predicted by quantum theory fail to be observed at all in this particular extension of the Bell experiment: i.e., that the measurement results obtained from the detectors fail to violate the CHSH inequality. This would imply that quantum theory fails to describe correctly the results of the Bell experiment embedded within this particular experimental configuration, and so would imply a definite limit to the domain of validity of quantum theory.

378 A. Kent

A third logical possibility is that the Bell experiment correlations follow the predictions of quantum theory, but that the Cavendish experiments show gravitational fields which do not correspond to the test mass configurations in the expected way (or at least do not do so until a signal has had time to travel from one wing to the other). This would suggest the coexistence of a quantum theory of matter with some classical theory of gravity which respects local causality, but which has the surprising property that classical gravitational fields do not always couple to macroscopic matter in the way suggested by general relativity.

In summary: although our present understanding of physics leads us to expect the first outcome, the point at issue seems sufficiently fundamental, and our present understanding of gravity sufficiently limited, that it would be very interesting and worthwhile to carry out experiments capable of discriminating between some (and of course, ideally, all) of the possible outcomes outlined above.

Acknowledgments I am particularly indebted to Fay Dowker for very helpful comments on an earlier draft, for inspiring the definition of local causality used in the present version of the paper, and for many thoughtful criticisms. I am also very grateful to Nicolas Gisin, Valerio Scarani, Christoph Simon and Gregor Weihs for valuable discussions on various criteria for gravitationally induced collapse and experimental tests. Warm thanks too to Harvey Brown, Jeremy Butterfield, Robert Helling, Graeme Mitchison, Roger Penrose and Rainer Plaga for some very helpful comments. Last but by no means least, I would very much like to thank Abner for his characteristically kind and warm encouragement to pursue this idea.

References

1. C. Møller, in *Les Theories Relativistes de la Gravitation*, A. Lichnerowicz and M. Tonnelat (eds.) (CNRS, Paris, 1962).
2. L. Rosenfeld, Nucl. Phys. **40** 353–356 (1963).
3. T. Kibble, in *Quantum Gravity 2: A Second Oxford Symposium*, C. Isham, R. Penrose and D. Sciama (eds.) (Oxford University Press, Oxford, 1981).
4. D. Page and C. Geilker, *Phys. Rev. Lett.* **47**, 979 (1981).
5. G. Ghirardi, A. Rimini and T. Weber, *Phys. Rev. D* **34**, 470 (1986); G. Ghirardi, P. Pearle and A. Rimini, *Phys. Rev. A* **42**, 78 (1990).
6. R. Penrose, *The Emperor's New Mind* (Oxford University Press, Oxford, 1999) and refs therein.
7. E.g. L. Diosi, *Phys. Rev. A* **40**, 1165 (1989).
8. A. Kent, *Phys. Rev. A* **72**, 012107 (2005).
9. J.S. Bell, "The Theory of Local Beables," *Epistemological Letters*, **9**, March 1976; reprinted in *Dialectica* **39** 85–96 (1985) and in [16]; J.S. Bell, "Free Variables and Local Causality," *Epistemological Letters*, **15**, February 1977; reprinted in *Dialectica* **39** 103–106 (1985) and in [16].
10. D. Rideout and R. Sorkin, *Phys. Rev. D* **61**, 024002 (1999).
11. J. Henson, "Comparing Causality Principles," quant-ph/0410051.
12. F. Dowker, private communication (2005).
13. J. Clauser, M. Horne, A. Shimony and R. Holt, *Phys. Rev. Lett.* **23**, 880 (1969).
14. J. S. Bell, *Rev. Mod. Phys.* **38**, 447 (1966); J. S. Bell, *Physics* **1**, 195 (1964). Reprinted in [16].
15. A. Einstein, B. Podolsky and N. Rosen, *Phys. Rev.* **47**, 777 (1935).
16. J.S. Bell, *Speakable and Unspeakable in Quantum Mechanics* (Cambridge University Press, 1987).

Quantum Gravity Computers: On the Theory of Computation with Indefinite Causal Structure

Lucien Hardy

Abstract A quantum gravity computer is one for which the particular effects of quantum gravity are relevant. In general relativity, causal structure is non-fixed. In quantum theory non-fixed quantities are subject to quantum uncertainty. It is therefore likely that, in a theory of quantum gravity, we will have indefinite causal structure. This means that there will be no matter of fact as to whether a particular interval is time-like or not. We study the implications of this for the theory of computation. Classical and quantum computations consist in evolving the state of the computer through a sequence of time steps. This will, most likely, not be possible for a quantum gravity computer because the notion of a time step makes no sense if we have indefinite causal structure. We show that it is possible to set up a model for computation even in the absence of definite causal structure by using a certain framework (the causaloid formalism) that was developed for the purpose of correlating data taken in this type of situation. Corresponding to a physical theory is a causaloid, Λ (this is a mathematical object containing information about the causal connections between different spacetime regions). A computer is given by the pair $\{\Lambda, S\}$ where S is a set of gates. Working within the causaloid formalism, we explore the question of whether universal quantum gravity computers are possible. We also examine whether a quantum gravity computer might be more powerful than a quantum (or classical) computer. In particular, we ask whether indefinite causal structure can be used as a computational resource.

1 Introduction

A computation, as usually understood, consists of operating on the state of some system (or collection of systems) in a sequence of steps. Turing's universal computer consists of a sequence of operations on a tape. A classical computation is

---

L. Hardy
Perimeter Institute, Waterloo, Ontario N2L 2Y5, Canada
e-mail: lhardy@perimeterinstitute.ca

W.C. Myrvold and J. Christian (eds.), *Quantum Reality, Relativistic Causality, and Closing the Epistemic Circle,* The Western Ontario Series in Philosophy of Science 73,
© Springer Science+Business Media B.V. 2009

often implemented by having a sequence of operations on a collection of bits and a quantum computation by a sequence of operations on a collection of qubits. Such computations can be built up of gates where each gate acts on a small number of bits or qubits. These gates are defined in terms of how they cause an input state to be evolved. A physical computer may have some spatial extension and so gates may be acting at many different places at once. Nevertheless, we can always foliate spacetime such that we can regard the computer as acting on a state at some time t and updating it to a new state at time $t + 1$, and so on, till the computation is finished. Parallel computation fits into this paradigm since the different parts of the parallel computation are updated at the same time. The notion that computation proceeds by a sequence of time steps appears to be a fairly pervasive and deep rooted aspect of our understanding of what a computation is. In anticipation of more general computation, we will call computers that implement computation in this way *step computers* (SC). This includes Turing machines and parallel computers, and it includes classical computers and quantum computers.

Turing developed the theory of computation as a formalization of mathematical calculation (with pencil, paper, and eraser for example) [1]. Deutsch later emphasized that any computation must be implemented physically [2]. Consequently, we must pay attention to physical theories to understand computation. Currently, there are basically two fundamental physical theories, quantum theory (QT) and Einstein's theory of general relativity (GR) for gravity. However, we really need a physical theory which is more fundamental—a theory of quantum gravity (QG). A correct theory of QG will reduce to QT and GR in appropriate situations (including, at least, those situations where those physical theories have been experimentally verified). We do not currently have a theory of quantum gravity. However, we can hope to gain some insight into what kind of theory this will be by looking at QT and GR. Causal structure in GR is not fixed in advance. Whether two events are time-like or not depends on the metric and the metric depends on the distribution of matter. In quantum theory a property that is subject to variation is also subject to quantum uncertainty—we can be in a situation where there is no matter of fact as to the value of that quantity. For example, a quantum particle can be in a superposition of being in two places at once. It seems likely that this will happen with causal structure. Hence, in a theory of QG we expect that we will have indefinite causal structure.

Indefinite causal structure is when there is, in general, no matter of fact as to whether the separation between two events is time-like or not.

If this is, indeed, the case then we cannot regard the behaviour of a physical system (or collection of systems) as evolving in time through a sequence of states defined on a sequence of space-like hypersurfaces. This is likely to have implications for computer science. In particular, it is likely that a quantum gravity computer cannot be understood as an instance of a SC. In this paper we will explore the consequences of having indefinite causal structure for the theory of computation. In particular, we will look at how the causaloid framework (developed in [3]) can be applied to provide a definite model for computation when we have indefinite causal structure.

Quantum Gravity Computers   381

Although there are compelling reasons for believing that the correct theory of QG
will have indefinite causal structure, it is possible that this will not be the case. Nevertheless, in this paper we will assume that QG will have this property. There may
be other features of a theory of QG which would be interesting for the study of computation but, in this paper, we will restrict ourselves to indefinite causal structure.

2 General Ideas

2.1 What Counts as a Computer?

The idea of a computer comes from attempting to formalize mathematical calculation. A limited notion of computation would entail that it is nothing more than a process by which a sequence of symbols is updated in a deterministic fashion—such as
with a Turing machine. However, with the advent of quantum computation, this notion is no longer sufficient. David Deutsch was able to establish a theory of quantum
computation which bares much resemblance to the theory of classical computation.
Given that quantum computers can be imagined (and may even be built one day) we
need a richer notion of computation. However, a quantum computer still proceeds
by means of a sequence of time steps. It is a SC. The possibility of considering time
steps at a fundamental level will, we expect, be undermined in a theory of quantum
gravity for the reasons given above.

   This raises the question of whether or not we want to regard the behaviour of
a physical machine for which the particular effects of QG are important and lead
to indefinite causal structure as constituting a computer. We could certainly build a
machine of this nature (at least in principle). Furthermore, somebody who knows the
laws by which this machine operates could use it to address mathematical issues (at
the very least they could solve efficiently the mathematical problem of generating
numbers which would be produced by a simulation of this machine in accordance
with the known laws). Hence, it is reasonable to regard this machine as a computer—
a quantum gravity computer.

   At this point it is worth taking a step back to ask, in the light of these considerations, what we mean by a the notion of a computer in general? One answer is
that

> (1) **A computer** is a physical device that can give correct answers to well formulated ques-
> tions.

For this to constitute a complete definition we would need to say what the terms
in this definition mean. However, whatever a "well formulated question" means, it
must be presented to the computer in the form of some physical input (or program).
Likewise, whatever an "answer" is, it must be given by the computer in the form
of some physical output. It is not clear what the notion of "correctness" means.
However, from the point of view of the physical computer it must mean that the

device operates according to sufficiently well known rules. Hence, a more physical definition is that

(2) **A computer** is a physical device has an output that depends on an input (or program) according to sufficiently well known rules.

This still leaves the meaning of the word "sufficiently" unclear. It is not necessary that we know all the physics that governs a computer. For example, in a classical computer we do not need to have a detailed understanding of the physics inside a gate, we only need an understanding of how the gate acts on an input to produce an output. There remain interesting philosophical questions about how we understand the translation from the terms in definition (1) to those in definition (2) but these go beyond the scope of this paper.

These definitions are useful. In particular they do not require that the computational process proceed by a sequence of steps. We will see how we can meaningfully talk about computation in the absence of any spacelike foliation into timelike steps in the sense of definition (2) of a computer.

It is likely that, in going to QG computers, we will leave behind many of the more intuitive notions of computation we usually take for granted. This already happened in the transition from classical to quantum computation—but the the likely failure of the step computation model for a QG computer may cause the transition from quantum to quantum gravity computation to be even more radical.

2.2 The Church-Turing-Deutsch Principle

Consider the following

The Church-Turing-Deutsch principle: Every physical process can be simulated by a universal model computing device.

Deutsch [2] was motivated to state this principle by work of Church [4] and Turing [1] (actually he gives a stronger and more carefully formulated version). Deutsch's statement emphasizes the physical aspect of computation whereas Church and Turing were more interested in mathematical issues (note that, in his acknowledgements, Deutsch thanks "C. H. Bennett for pointing out to me that the Church-Turing hypothesis has physical significance"). We can take the widespread successful simulation of any number of physical processes (such as of cars in a wind tunnel, or of bridges prior to their being built) on a modern classical computer, as evidence of the truth of this principle. A principle like this would seem to be important since it provides a mechanism for verifying physical theories. The physical theory tells us how to model physical processes. To verify the physical theory there needs to be some way of using the theory to simulate the given physical process. However, there is a deeper reason that this principle is interesting. This is that it might lead us to say that the universe is, itself, a computer. Of course, the CTD principle does not actually imply that. Even though we might be able to simulate a physical process on

Quantum Gravity Computers 383

a computer, it does not follow that the computation is an accurate reflection of what is happening during that physical process. This suggests a stronger principle

The computational reflection principle: The behaviour of any physical process is accurately reflected by the behaviour of an appropriately programmed universal model computing device.

A proper understanding of this principle requires a definition of what is meant by "accurately reflected" (note that a dictionary definition of the relevant meaning of the word *reflect* is to "embody or represent in a faithful or appropriate way" [5]). We will not attempt to provide a precise definition but rather will illustrated our discussion with examples. Nevertheless, "accurate reflection" would entail that not only is there the same mapping between inputs and outputs for the physical process and the computation, but also that there is a mapping between the internal structure of the physical process and the computation. This relates to ideas of functional equivalence as discussed by philosophers.

We may think of a universal computer in the Turing model where the program is included in the tape. But we may also use the circuit model where the program is represented by a prespecified way of choosing the gates.

It is possible to simulate any quantum system with a finite dimensional Hilbert space (including quantum computers) to arbitrary accuracy on a classical computer. In fact, we can even simulate a quantum computer with polynomial space on a classical computer but, in general, this requires exponential time [6]. We might claim, then, that the CTD principle holds (though, since this is not exact simulation, we may prefer to withhold judgment). However, we would be more reluctant to claim that the CR principle holds since the classical simulation has properties that the quantum process does not: (i) It is possible to measure the state of the classical computer without effecting its subsequent evolution; (ii) the exponential time classical computer is much more powerful than a polynomial time quantum computer; and (iii) the detailed structure of the classical computation will look quite different to that of the quantum process.

2.3 Physics Without State Evolution

The idea of a state which evolves is deeply ingrained in our way of thinking about the world. But is it a necessary feature of any physical theory? This depends what a physical theory must accomplish. At the very least, *a physical theory must correlate recorded data*. Data is correlated in the evolving state picture in the following way. Data corresponding to a given time is correlated by applying the mathematical machinery of the theory to the state at that given time. And data corresponding to more than one time is correlated by evolving the state through those given times, and then applying the mathematical machinery of the theory to the collection of states so generated. However, there is no reason to suppose that this is the only way of correlating data taken in different spacetime regions. In fact, we have already other

384   L. Hardy

pictures. In GR we solve local field equations. A solution must simply satisfy the Einstein field equations and be consistent with the boundary conditions. We do not need the notion of an evolving state here—though there are canonical formulations of GR which have a state across space evolving in time. In classical mechanics we can extremize an action. In this case we consider possible solutions over all time and find the one that extremizes the action. Again, we do not need to use the notion of an evolving state. In quantum theory we can use Feynman's sum over histories approach which is equivalent to an evolving state picture but enables us to proceed without such a picture. In [3] the causaloid formalism was developed as a candidate framework for a theory of QG (though QT can be formulated in this framework). This enables one to calculate directly whether (i) there is a well defined correlation between data taken from two different spacetime regions and, if there is, (ii) what that correlation is equal to. Since this calculation is direct, there is no need to consider a state evolving between the two regions. The causaloid formalism is, in particular, suited to dealing with the situation where there is no matter of fact to whether an interval is time-like or not.

2.4 What is a Quantum Gravity Computer?

A quantum gravity computer is a computer for which the particular effects of QG are important. In this paper we are interested in the case where we have indefinite causal structure (and, of course, we are assuming that QG will allow this property).

As we discussed in Section 2.1, a computer can be understood to be a physical device having an output that depends on an input (or program) according to sufficiently well known rules. The computer occupies a certain region of spacetime. The input can consist of a number of inputs into the computer distributed across this region, and likewise, the output can consist of a number of outputs from the computer distributed across the region. Typically the inputs are selected (by us) in accordance with some program corresponding to the question we wish to use the computer to find an answer to. Usually we imagine setting the computer in some initial state (typically, in quantum computing, this consists of putting all the qubits in the zero state). However, physically this is accomplished by an appropriate choice of inputs prior to this initial time (for example, we might have a quantum circuit which initializes the state). Hence, the picture in which we have inputs and outputs distributed across the given region of spacetime is sufficient. We do not need to also imagine that we separately initialize the computer. This characterization of a computer is useful for specifying a QG computer since we must be careful using a notion like "initial state" when we cannot rely on having a definite notion of a single time hypersurface in the absence of definite causal structure. The QG computer itself must be sensitive to QG effects (as opposed to purely quantum or purely general relativistic effects). To actually build a QG computer we need a theory of quantum gravity because (i) this is the only way to be sure we are seeing quantum gravity effects and (ii) we need to have known physical laws to use the device as a computer.

Quantum Gravity Computers 385

In the absence of a theory of QG it is difficult to give an example of a device which will function as a QG computer. Nevertheless we will give a possible candidate example for the purposes of discussion. We hope that the essential features of this example would be present in any actual QG computer. We imagine, for this example, that our quantum gravity computer consists of a number of mesoscopic probes of Planck mass (about 20 µg) immersed in a controlled environment of smaller quantum objects (such as photons). There must be the possibility of having inputs and outputs. The inputs and outputs are distributed across the region of spacetime in which the QG computer operates. We take this region of spacetime to be fuzzy in the sense that we cannot say whether a particular interval in it is time-like or space-like. However, we can still expect to be able to set up physical coordinates to label where a particular input or output is "located" in some appropriate abstract space. For example, imagine that a GPS system is set up by positioning four satellites around the region. Each satellite emits a signal carrying the time of its internal clock. We imagine that the mesoscopic probes can detect these four times thus providing a position $x \equiv (t_1, t_2, t_3, t_4)$. Each satellite clock will tick and so x is a discrete variable. A given probe will experience a number of different values of x. Assume that each probe can be set to give out a light pulse or not (denote this by $s = 1$ or $s = 0$ respectively), and has a detector which may detect a photon or not (denote this by $a = 1$ or $a = 0$ respectively) during some given short time interval. Further, allow the value of s to depend on x. Thus,

$$s = F(x, n) \tag{1}$$

where n labels the probe. We imagine that we can choose the function F as we like. This constitutes the program. Thus, the inputs are given by the s's and the outputs by the a's. We record many instances of the data (x, n, s, a). We might like to have more complicated programs where F is allowed to depend on the values of previous outputs from other probes. However, we cannot assume that there is fixed causal structure, and so we cannot say, in advance, what will constitute previous data. Thus, any program of this nature must "physicalize" the previous data by allowing the probe to emit it as a physical signal, r. If this signal is detected at a probe along with x then it can form part of the input into F. Thus, we would have

$$s = F(x, n, r) \tag{2}$$

At the end of a run of the QG computer, we would have many instances of (x, n, r, s, a).

This is just a possible example of a possible QG computer. We might have the property of indefinite causal structure in this example since the mesoscopic probes are (possibly) sufficiently small to allow quantum effects and sufficiently massive to allow gravitational effects. Penrose's cat [7] consists of exploring the possible gravity induced breakdown of quantum theory for a Planck mass mirror recoiling (or not) from a photon in a quantum superposition.

Regardless of whether this is a good example, we will assume that any such computer will collect data of the form (x, n, s, a) (or (x, n, r, s, a)), and that a program can

386   L. Hardy

be specified by a function $F(x,n)$ (or $F(x,n,r)$). Whilst we can imagine more complicated examples, it would seem that they add nothing extra and could, anyway, be accommodated by the foregoing analysis. Importantly, although we have the coordinate x, we do not assume any causal structure on x. In particular, there is no need to assume that some function of x will provide a time coordinate—this need not be a SC.

3 The Causaloid Formalism

3.1 Analyzing Data

We will now given an abbreviated presentation of the causaloid formalism which is designed for analyzing data collected in this way and does not require a time coordinate. This formalism was first presented in [3] (see also [8, 9] for more accessible accounts). Assume that each piece of data $((x,n,s,a)$ or $(x,n,r,s,a))$ once collected is written on a card. At the end of the computation we will have a stack of cards. We will seek to find a way to calculate probabilistic correlations between the data collected on these cards. The order in which the cards end up in the stack does not, in itself, constitute recorded data and consequently will play no role in this analysis. Since we are interested in probabilities we will imagine running the computation many times so that we can calculate probabilities as relative frequencies (though, this may not be necessary for all applications of the computer). Now we will provide a number of basic definitions in terms of the cards.

The full pack, V, is the set of all logically possible cards.

The program, F, is the set of all cards from V consistent with a given program $F(x,n,s,a)$ (or $F(x,n,r,s,a)$). Note that the set F and the function F convey the same information so we use the same notation, the meaning being clear from the context.

A stack, Y, is the set of cards collected during a particular run of the computer.

An elementary region, R_x, is the the set of all cards from V having a particular x written on them.

Note that
$$Y \subseteq F \subseteq V \tag{3}$$

We will now give a few more definitions in terms of these basic definitions.

Regions. We define a composite spacetime region by

$$R_{\mathcal{O}_1} = \bigcup_{x \in \mathcal{O}_1} R_x \tag{4}$$

We will often denote this by R_1 for shorthand.

Quantum Gravity Computers 387

The outcome set in region R_1 is given by

$$Y_{R_1} \equiv Y \cap R_1 \tag{5}$$

This set contains the results seen in the region R_1. It constitutes the raw output data from the computation. We will often denote this set by Y_1.

The program in region R_1 is given by

$$F_{R_1} \equiv F \cap R_1 \tag{6}$$

This set contains the program instructions in region R_1. We will often denote it by F_1.

3.2 Objective of the Causaloid Formalism

We will consider probabilities of the form

$$\text{Prob}(Y_2|Y_1, F_2, F_1) \tag{7}$$

This is the probability that we see outcome set Y_2 in R_2 given that we have procedure F_2 in that region and that we have outcome set Y_1 and program F_1 in region R_1. Our physical theory must (i) determine whether the probability is *well defined*, and if so (ii) determine its value. The first step is crucial. Most conditional probabilities we might consider are not going to be well defined. For example if R_1 and R_2 are far apart (in so much as such a notion makes sense) then there will be other influences (besides those in R_1) which determine the probabilities of outcomes in R_2, and if these are not take into account we cannot do a calculation for this probability. To illustrate this imagine an adversary. Whatever probability we write down, he can alter these extraneous influences so that the probability is wrong. Conventionally we determine whether a probability is well defined by simply looking at the causal structure. However, since we do not have definite causal structure here we have to be more careful.

To begin we will make an assumption. Let the region R be big (consisting of most of V).

Assumption 1: We assume that there is some condition C on F_{V-R} and Y_{V-R} such that all probabilities of the form

$$\text{Prob}(Y_R|F_R, C) \tag{8}$$

are well defined.

We can regard condition C as corresponding to the setting up and maintenance of the computer. We will consider only cases where C is true (when it is not, the computer is broken or malfunctioning). We will regard region R as the region in which the

388   L. Hardy

computation is performed. Since we will always be assuming C is true, we will drop
it from our notation. Thus, we assume that the probabilities $\text{Prob}(Y_R|F_R)$ are well
defined.

The probabilities $\text{Prob}(Y_R|F_R)$ pertain to the global region R. However, we nor-
mally like to do physics by building up a picture of the big from the small. We will
show how this can be done. We will apply three levels of *physical compression*.
The first applies to single regions (such as R_1). The second applies to composite
regions such as $R_1 \cup R_2$ (the second level of physical compression also applies to
composite regions made from three or more component regions). The first and second levels of physical compression result in certain matrices. In the third level of
physical compression we use the fact that these matrices are related to implement
further compression.

3.3 First Level Physical Compression

First we implement *first level physical compression*. We label each possible pair
(Y_{R_1}, F_{R_1}) in R_1 with α_1. We will think of these pairs as describing measurement
outcomes in R_1 (Y_{R_1} denotes the outcome of the measurement and F_{R_1} denotes the
choice of measurement). Then we write

$$p_{\alpha_1} \equiv \text{Prob}(Y_{R_1}^{\alpha_1} \cup Y_{R-R_1} | F_{R_1}^{\alpha_1} \cup F_{R-R_1}) \qquad (9)$$

By Assumption 1, these probabilities are all well defined. We can think of what
happens in region $R - R_1$ as constituting a generalized preparation of a state in region
R_1. We define the state to be that thing represented by any mathematical object
which can be used to calculate p_{α_1} for all α_1. Now, given a generalized preparation,
the p_{α_1}'s are likely to be related by the physical theory that governs the system. In
fact we can just look at linear relationships. This means that we can find a minimal
set Ω_1 such that

$$p_{\alpha_1} = \mathbf{r}_{\alpha_1}(R_1) \cdot \mathbf{p}(R_1) \qquad (10)$$

where the state $\mathbf{p}(R_1)$ in R_1 is given by

$$\mathbf{p}(R_1) = \begin{pmatrix} \vdots \\ p_{l_1} \\ \vdots \end{pmatrix} \quad l_1 \in \Omega_1 \qquad (11)$$

We will call Ω_1 the fiducial set (of measurement outcomes). Note that the probabilities p_{l_1} need not add up to 1 since the l_1's may correspond to outcomes of incompatible measurements. In the case that there are no linear relationships relating the
p_{α_1}'s we set Ω_1 equal to the full set of α_1's and then $\mathbf{r}_{\alpha_1}$ consists of a 1 in position
α_1 and 0's elsewhere. Hence, we can always write (10). One justification for using
linear compression is that probabilities add in a linear way when we take mixtures.

Quantum Gravity Computers 389

It is for this reason that linear compression in quantum theory (for general mixed states) is the most efficient. The set Ω_1 will not, in general, be unique. Since the set is minimal, there must exist a set of $|\Omega_1|$ linearly independent states $\mathbf{p}$ (otherwise further linear compression would be possible). First level physical compression for region R_1 is fully encoded in the matrix

$$\Lambda_{\alpha_1}^{l_1} \equiv r_{l_1}^{\alpha_1} \tag{12}$$

where $r_{l_1}^{\alpha_1}$ is the l_1 component of $\mathbf{r}_{\alpha_1}$. The more physical compression there is the more rectangular (rather than square) this matrix will be.

3.4 Second Level Physical Compression

Next we will implement second level physical compression. Consider two regions R_1 and R_2. Then the state for region $R_1 \cup R_2$ is clearly of the form

$$\mathbf{p}(R_1 \cup R_2) = \begin{pmatrix} \vdots \\ p_{k_1 k_2} \\ \vdots \end{pmatrix} \quad k_1 k_2 \in \Omega_{12} \tag{13}$$

We can show that it is always possible to choose Ω_{12} such that

$$\Omega_{12} \subseteq \Omega_1 \times \Omega_2 \tag{14}$$

where $\times$ denotes the cartesian product. This result is central to the causaloid formalism. To prove (14) note that we can write $p_{\alpha_1 \alpha_2}$ as

$$
\begin{aligned}
\mathrm{prob}(Y_{R_1}^{\alpha_1} \cup Y_{R_2}^{\alpha_2} \cup Y_{R-R_1-R_2} & | F_{R_1}^{\alpha_1} \cup F_{R_2}^{\alpha_2} \cup F_{R-R_1-R_2}) \\
&= \mathbf{r}_{\alpha_1}(R_1) \cdot \mathbf{p}_{\alpha_2}(R_1) \\
&= \sum_{l_1 \in \Omega_1} r_{l_1}^{\alpha_1}(R_1) p_{l_1}^{\alpha_2}(R_1) \\
&= \sum_{l_1 \in \Omega_1} r_{l_1}^{\alpha_1}(R_1) \mathbf{r}_{\alpha_2}(R_2) \cdot \mathbf{p}_{l_1}(R_2) \\
&= \sum_{l_1 l_2 \in \Omega_1 \times \Omega_2} r_{l_1}^{\alpha_1} r_{l_2}^{\alpha_2} p_{l_1 l_2} \tag{15}
\end{aligned}
$$

where $\mathbf{p}_{\alpha_2}(R_1)$ is the state in R_1 given the generalized preparation $(Y_{R_2}^{\alpha_2} \cup Y_{R-R_1-R_2}, F_{R_2}^{\alpha_2} \cup F_{R-R_1-R_2})$ in region $R - R_1$, and $\mathbf{p}_{l_1}(R_2)$ is the state in R_2 given the generalized preparation $(Y_{R_1}^{l_1} \cup Y_{R-R_1-R_2}, F_{R_1}^{l_1} \cup F_{R-R_1-R_2})$ in region $R - R_2$, and where

$$p_{l_1 l_2} = \mathrm{prob}(Y_{R_1}^{l_1} \cup Y_{R_2}^{l_2} \cup Y_{R-R_1-R_2} | F_{R_1}^{l_1} \cup F_{R_2}^{l_2} \cup F_{R-R_1-R_2}) \tag{16}$$

390   L. Hardy

Now we note from (15) that $p_{\alpha_1 \alpha_2}$ is given by a linear sum over the probabilities $p_{l_1 l_2}$ where $l_1 l_2 \in \Omega_1 \times \Omega_2$. It may even be the case that we do not need all of these probabilities. Hence, it follows that $\Omega_{12} \subseteq \Omega_1 \times \Omega_2$ as required.

Using (15) we have

$$
\begin{aligned}
p_{\alpha_1 \alpha_2} &= \mathbf{r}_{\alpha_1 \alpha_2}(R_1 \cup R_2) \cdot \mathbf{p}(R_1 \cup R_2) \\
&= \sum_{l_1 l_2} r_{l_1}^{\alpha_1} r_{l_2}^{\alpha_2} p_{l_1 l_2} \\
&= \sum_{l_1 l_2} r_{l_1}^{\alpha_1} r_{l_2}^{\alpha_2} \mathbf{r}_{l_1 l_2} \cdot \mathbf{p}(R_1 \cup R_2)
\end{aligned}
$$

We must have

$$
\mathbf{r}_{\alpha_1 \alpha_2}(R_1 \cup R_2) = \sum_{l_1 l_2} r_{l_1}^{\alpha_1} r_{l_2}^{\alpha_2} \mathbf{r}_{l_1 l_2}(R_1 \cup R_2) \tag{17}
$$

since we can find a spanning set of linearly independent states $\mathbf{p}(R_1 \cup R_2)$. We define

$$
\Lambda_{l_1 l_2}^{k_1 k_2} \equiv r_{k_1 k_2}^{l_1 l_2} \tag{18}
$$

where $r_{k_1 k_2}^{l_1 l_2}$ is the $k_1 k_2$ component of $\mathbf{r}_{l_1 l_2}$. Hence,

$$
r_{k_1 k_2}^{\alpha_1 \alpha_2} = \sum_{l_1 l_2} r_{l_1}^{\alpha_1} r_{l_2}^{\alpha_2} \Lambda_{l_1 l_2}^{k_1 k_2} \tag{19}
$$

This equation tells us that if we know $\Lambda_{l_1 l_2}^{k_1 k_2}$ then we can calculate $\mathbf{r}_{\alpha_1 \alpha_2}(R_1 \cup R_2)$ for the composite region $R_1 \cup R_2$ from the corresponding vectors $\mathbf{r}_{\alpha_1}(R_1)$ and $\mathbf{r}_{\alpha_2}(R_2)$ for the component regions R_1 and R_2. Hence the matrix $\Lambda_{l_1 l_2}^{k_1 k_2}$ encodes the second level physical compression (the physical compression over and above the first level physical compression of the component regions). We can use $\Lambda_{l_1 l_2}^{k_1 k_2}$ to define a new type of product—the causaloid product—denoted by $\otimes^\Lambda$.

$$
\mathbf{r}_{\alpha_1 \alpha_2}(R_1 \cup R_2) = \mathbf{r}_{\alpha_1}(R_1) \otimes^\Lambda \mathbf{r}_{\alpha_2}(R_2) \tag{20}
$$

where the components are given by (19).

We can apply second level physical compression to more than two regions. For three regions we have the matrices

$$
\Lambda_{l_1 l_2 l_3}^{k_1 k_2 k_3} \tag{21}
$$

and so on.

3.5 Third Level Physical Compression

Finally, we come to third level physical compression. Consider all the compression matrices we pick up for elementary regions R_x during first and second level compression. We have

Quantum Gravity Computers

$$
\begin{pmatrix}
\Lambda^{l_x}_{\alpha_x} & \text{for all } x \in \mathcal{O}_R \\[2ex]
\Lambda^{k_x k_{x'}}_{l_x l_{x'}} & \text{for all } x, x' \in \mathcal{O}_R \\[2ex]
\Lambda^{k_x k_{x'} k_{x''}}_{l_x l_{x'} l_{x''}} & \text{for all } x, x', x'' \in \mathcal{O}_R \\[2ex]
\vdots & \vdots
\end{pmatrix}
\tag{22}
$$

where $\mathcal{O}_R$ is the set of x in region R. Now, these matrices themselves are likely to be related by the physical theory. Consequently, rather than specifying all of them separately, we should be able to specify a subset along with some rules for calculating the others

$$
\Lambda \equiv (\text{subset of } \Lambda's; \text{RULES})
\tag{23}
$$

We call this mathematical object the *causaloid*. This third level of physical compression is accomplished by identities relating the higher order Λ matrices (those with more indices) to the lower order ones. Here are some examples from two families of such identities. The first family uses the property that when Ω sets multiply so do Λ matrices.

$$
\Lambda^{k_x \cdots k_{x'} k_{x''} \cdots k_{x'''}}_{l_x \cdots l_{x'} l_{x''} \cdots l_{x'''}} = \Lambda^{k_x \cdots k_{x'}}_{l_x \cdots l_{x'}} \Lambda^{k_{x''} \cdots k_{x'''}}_{l_{x''} \cdots l_{x'''}} \quad \text{if} \quad \Omega_{x \cdots x' x'' \cdots x'''} = \Omega_{x \cdots x'} \times \Omega_{x'' \cdots x'''}
\tag{24}
$$

The second family consists of identities from which Λ matrices for composite regions can be calculated from some pairwise matrices (given certain conditions on the Ω sets). The first identity in this family is

$$
\Lambda^{k_1 k_2 k_3}_{l_1 l_2 l_3} = \sum_{k_2' \in \Omega_{2\bar{3}}} \Lambda^{k_1 k_2}_{l_1 k_2'} \Lambda^{k_2' k_3}_{l_2 l_3} \quad \text{if} \quad \Omega_{123} = \Omega_{12} \times \Omega_{\bar{2}3} \quad \text{and} \quad \Omega_{23} = \Omega_{2\bar{3}} \times \Omega_{\bar{2}3}
\tag{25}
$$

where the notation $\Omega_{\bar{2}3}$ means that we form the set of all k_3 for which there exists $k_2 k_3 \in \Omega_{23}$. The second identity in this family is

$$
\Lambda^{k_1 k_2 k_3 k_4}_{l_1 l_2 l_3 l_4} = \sum_{k_2' \in \Omega_{2\bar{3}}, k_3' \in \Omega_{3\bar{4}}} \Lambda^{k_1 k_2}_{l_1 k_2'} \Lambda^{k_2' k_3}_{l_2 k_3'} \Lambda^{k_3' k_4}_{l_3 l_4} \quad \text{if} \quad \begin{array}{l} \Omega_{1234} = \Omega_{12} \times \Omega_{\bar{2}3} \times \Omega_{\bar{3}4} \\ \Omega_{23} = \Omega_{2\bar{3}} \times \Omega_{\bar{2}3} \\ \Omega_{34} = \Omega_{3\bar{4}} \times \Omega_{\bar{3}4} \end{array}
\tag{26}
$$

and so on. These identities are sufficient to implement third level physical compression for classical and quantum computers. However, we will probably need other identities to implement third level physical compression for a QG computer. The task of fully characterizing all such identities, and therefore of fully characterizing third level physical compression, remains to be completed.

3.6 Classical and Quantum Computers in the Causaloid Formalism

Since third level compression has been worked out for classical and quantum computers we should say a little about this here (see [3, 8] for more details). Consider a classical (quantum) computer which consists of pairwise interacting (qu)bits. This is sufficient to implement universal classical (quantum) computation. This situation is shown in Fig. 1. Each (qu)bit is labeled by $i, j, \ldots$ and is shown by a thin line. The nodes where the (qu)bits meet are labeled by x. Adjacent nodes (between which a (qu)bit passes) have a link. We call this diagram a *causaloid diagram*. At each node we have a choice, s, of what gate to implement. And then there may be some output, a, registered at the gate itself (in quantum terms this is both a transformation and a measurement). We record (x, s, a) on a card. The program is specified by some function $s = F(x)$. We can use our previous notation. Associated with each (x, s, a) at each gate is some $\mathbf{r}_{\alpha_x}$. It turns out that there exists a choice of fiducial measurement outcomes at each node x which break down into separate measurement outcomes for each of the two (qu)bits passing through that node. For these measurements we can write $l_x \equiv l_{xi} l_{xj}$ where l_{xi} labels the fiducial measurements on (qu)bit i and l_{xj} labels the fiducial measurements on the other (qu)bit j. All Ω sets involving different (qu)bits factorize as do all Ω sets involving non-sequential clumps of nodes on the same (qu)bit and so identity (24) applies in these cases. For a set of sequential nodes the Ω sets satisfy the conditions for (25, 26) and related identities to hold. This means that it is possible to specify the causaloid for a classical (quantum) computer of pairwise interacting (qu)bits by

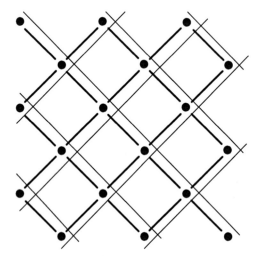

Fig. 1 This figure shows a number of pairwise interacting (qu)bits. The (qu)bits travel along the paths indicated by the thin lines and interact at the nodes. At each node we can choose a gate

Quantum Gravity Computers 393

$$\Lambda = \left(\{\Lambda_{\alpha_x}^{l_{xi}l_{xj}} \; \forall \; x\}, \{\Lambda_{l_{xi}l'_{x'i}}^{k_{xi}k_{x'i}} \; \forall \; \text{adjacent} \; x, x'\}; \left\{ \begin{array}{c} \text{clumping method} \\ \text{causaloid diagram} \end{array} \right\} \right) \quad (27)$$

where the "clumping method" is the appropriate use of the identities (24, 25, 26) and related identities to calculate general Λ matrices. The causaloid diagram is also necessary so we know how the nodes are linked up and how the (qu)bits move. There is quite substantial third level compression. The total number of possible Λ matrices is exponential in the number of nodes but the number of matrices required to specify the causaloid is only linear in this number. There may be simple symmetries which relate the matrices living on each node and each link. In this case there will be even further compression.

3.7 Using the Causaloid Formalism to Make Predictions

We can use the causaloid to calculate any $\mathbf{r}$ vector for any region in R. Using these we can calculate whether any probability of the form (7) is well defined, and if so, what it is equal to. To see this note that, using Bayes rule,

$$p \equiv \text{Prob}(Y_1^{\alpha_1} | Y_2^{\alpha_2}, F_1^{\alpha_1}, F_2^{\alpha_2}) = \frac{\mathbf{r}_{\alpha_1 \alpha_2}(R_1 \cup R_2) \cdot \mathbf{p}(R_1 \cup R_2)}{\sum_{\beta_1} \mathbf{r}_{\beta_1 \alpha_2}(R_1 \cup R_2) \cdot \mathbf{p}(R_1 \cup R_2)} \quad (28)$$

where β_1 runs over all (Y_1, F_1) consistent with $F_1 = F_1^{\alpha_1}$ (i.e. all outcomes consistent with the program in region R_1). For this probability to be well defined it must be independent of what happens outside $R_1 \cup R_2$. That is, it must be independent of the state $\mathbf{p}(R_1 \cup R_2)$. Since there exists a spanning set of linearly independent such states, this is true if and only if

$$\mathbf{r}_{\alpha_1 \alpha_2}(R_1 \cup R_2) \; \text{is parallel to} \; \sum_{\beta_1} \mathbf{r}_{\beta_1 \alpha_2}(R_1 \cup R_2) \quad (29)$$

This, then, is the condition for the probability to be well defined. In the case that this condition is satisfied then the probability is given by the ratio of the lengths of these two vectors. That is by

$$\mathbf{r}_{\alpha_1 \alpha_2}(R_1 \cup R_2) = p \sum_{\beta_1} \mathbf{r}_{\beta_1 \alpha_2}(R_1 \cup R_2) \quad (30)$$

It might quite often turn out that these two vectors are not exactly parallel. So long as they are still quite parallel we can place limits on p. Set

$$\mathbf{v} \equiv \mathbf{r}_{\alpha_1 \alpha_2}(R_1 \cup R_2) \quad \text{and} \quad \mathbf{u} \equiv \sum_{\beta_1} \mathbf{r}_{\beta_1 \alpha_2}(R_1 \cup R_2) \quad (31)$$

Define $\mathbf{v}^{\|}$ and $\mathbf{v}^{\perp}$ as the components of $\mathbf{v}$ parallel and perpendicular to $\mathbf{u}$ respectively. Then it is easy to show that

$$\frac{|\mathbf{v}^{\|}|}{|\mathbf{u}|} - \frac{|\mathbf{v}^{\perp}|}{|\mathbf{v}|\cos\phi} \leq p \leq \frac{|\mathbf{v}^{\|}|}{|\mathbf{u}|} + \frac{|\mathbf{v}^{\perp}|}{|\mathbf{v}|\cos\phi} \tag{32}$$

where ϕ is the angle between $\mathbf{v}$ and $\mathbf{v}^{\perp}$ (we get these bounds using $|\mathbf{v} \cdot \mathbf{p}| \leq |\mathbf{u} \cdot \mathbf{p}|$).

3.8 The Notion of State Evolution in the Causaloid Formalism

In setting up the causaloid formalism we have not had to assume that we can have a state which evolves with respect to time. As we will see, it is possible to reconstruct an evolving state even though this is looks rather unnatural from point of view of the causaloid formalism. However, this reconstruction depends on Assumption 1 of Section 3.2 being true. It is consistent to apply the causaloid formalism even if Assumption 1 does not hold. In this case we cannot reconstruct an evolving state.

We choose a nested set of spacetime regions R_t where $t = 0$ to T for which

$$R_0 \supset R_1 \supset R_2 \cdots \supset R_T \tag{33}$$

where $R_0 = R$ and R_T is the null set. We can think of t as a "time" parameter and the region R_t as corresponding to all of R that happens "after" time t. For each region R_t we can calculate the state, $\mathbf{p}(t) \equiv \mathbf{p}(R_t)$, given some generalized preparation up to time t (that is in the region $R - R_t$). We regard $\mathbf{p}(t)$ as the state at time t. It can be used to calculate any probability after time t (corresponding to the region R_t) and can therefore be used to calculate probabilities corresponding to the region R_{t+1} since this is nested inside R_t. Using this fact it is easy to show that the state is subject to linear evolution so that

$$\mathbf{p}(t+1) = Z_{t,t+1}\mathbf{p}(t) \tag{34}$$

where $Z_{t,t+1}$ depends on $Y_{R_t - R_{t+1}}$ and $F_{R_t - R_{t+1}}$.

Thus, it would appear that, although we did not use the idea of an evolving state in setting up the causaloid formalism, we can reconstruct a state that, in some sense, evolves. We can do this for *any* such nested set of regions. There is no need for the partitioning to be generated by a foliation into spacelike hypersurfaces and, indeed, such a foliation will not exist if the causal structure is indefinite. This evolving state is rather artificial—it need not correspond to any physically motivated "time".

There is a further reason to be suspicious of an evolving state in the causaloid formalism. To set up this formalism it was necessary to make Assumption 1 (in Section 3.2). It is likely that this assumption will not be strictly valid in a theory of QG. However, we can regard this assumption as providing scaffolding to get us to a mathematical framework. It is perfectly consistent to suppose that this mathematical framework continues to be applicable even if Assumption 1 is dropped. Thus it is possible that we can define a causaloid and then use the causaloid product and

Quantum Gravity Computers 395

(29, 30) to calculate whether probabilities are well defined and, if so, what these probabilities are equal to. In so doing we need make no reference to the concept of state. In particular, since we cannot suppose that all the probabilities $\text{prob}(Y_R|F_R,C)$ are well defined, we will not be able to force an evolving state picture. The causaloid formalism provides us with a way of correlating inputs and outputs across a region of space time even in the absence of the possibility of an evolving state picture.

4 Computation in the Light of the Causaloid Formalism

4.1 Gates

In the standard circuit model a computer is constructed out of gates selected from a small set of possible gates. The gates are distributed throughout a spacetime region in the form of a circuit. Hence we have a number of spacetime locations (label them by x) at which we may place a gate. At each such location we have a choice of which gate to select. The gates are connected by "wires" along which (qu)bits travel. This wiring represents the causal structure of the circuit. Since the wiring is well defined, causal structure cannot be said to be indefinite. In fact in classical and quantum computers we can work with a fixed wiring and vary only the choice of gates. The wires can form a diamond grid like that shown in Fig. 1. Where the wires cross two (qu)bits can pass through a gate. As long as we have a sufficient number of appropriate gates we can perform universal computation. In Section 3.6 we outlined how to put this situation into the causaloid formalism.

In the causaloid model we have spacetime locations labeled by x. At each x we have a choice of setting s. This choice of setting can be regarded as constituting the choice of a gate. Since we may have indefinite causal structure we will not be able to think in terms of "wiring" as such. However information about the causal connections between what happens at different x's is given by the Λ matrices which can be calculated from the causaloid. For example the matrix $\Lambda_{l_x l_{x'}}^{k_x k_{x'}}$ tells us about the causal connection between x and x'' by quantifying second level compression. Thus, the matrices associated with second level compression (which can be deduced from the causaloid, Λ) play the role of wiring. Since we do not have wires we cannot necessarily think in terms of (qu)bits moving between gates. Rather, we must think of the gates as being immersed in an amorphous interconnected sea quantitatively described by the causaloid. In the special case of a classical or quantum computer we will have wiring and this can be deduced from Λ.

Typically, in computers, we restrict the set of gates we employ. Thus, assume that we restrict to $s \in \{s_1, s_2, \ldots, s_N\} \equiv S$ where S is a subset of the set, S_I, of all possible s. Then a computer is defined by the pair

$$\{\Lambda, S\} \text{ where } S \subset S_I \tag{35}$$

The program for this computer is given by some function like $s = F(x,n)$ (or $s = F(x,n,r)$) from Section 2.4 where $s \in S$. This is a very general model for computation. Both classical and quantum computers can be described in this way as well as computers with indefinite causal structure.

4.2 Universal Computation

Imagine we have a class of computers. A universal computer for this class is a member of the class which can be used to simulate any computer in the class if it is supplied with an appropriate program. For example, a universal Turing machine can be used to simulate an arbitrary Turing machine. This is done by writing the program for the Turing machine to be simulated into the first part of the tape that is fed into the universal Turing machine. It follows from their definition that universal computers can simulate each other.

Given a causaloid, Λ and some integer M we can generate an interesting class of computers—namely the class C_Λ^M defined as the class of computers $\{\Lambda, S\}$ for all $S \subset S_I$ such that $|S| \leq M$. We will typically be interested in the case that M is a fairly small number (less than 10 say). The reason for wanting M to be small is that usually we imagine computations being constructed out of a small set of basic operations.

We can then ask whether there exist any universal computers in this class. We will say that the computer $\{\Lambda, S_U\}$ with $|S_U| \leq M$ is universal for the class C_Λ^M if we can use it to simulate an arbitrary computer in this class. This means that there must exist a simple map from inputs and outputs of the universal computer to inputs and outputs (respectively) of the computer being simulated such that the probabilities are equal (or equal to within some specified accuracy). We will then refer to S_U as a universal set of gates.

If we choose the causaloid Λ of classical or quantum theory discussed in Section 3.6 then it is well established that there exist universal computers for small M. This is especially striking in the quantum case since there exist a infinite number of gates which cannot be simulated by probabilistic mixtures of other gates. One way to understand how this is possible in the classical and quantum cases is the following. Imagine that we want to simulate $\{\Lambda, S\}$ with $\{\Lambda, S_U\}$. We can show that any gate in the set S can be simulated to arbitrary accuracy with some number of gates from the set S_U. Then we can coarse-grain on the diamond grid to larger diamonds which can have sufficient gates from S_U to simulate an arbitrary gate in S. In coarse-graining in this way we do not change in any significant way the nature of the causal structure. Thus we can still link these coarse-grained diamonds to each other in such a way that we can simulate $\{\Lambda, S\}$. This works because, in classical and quantum theory, we have definite causal structure which has a certain scale invariance property as we coarse-grain.

Quantum Gravity Computers 397

However, if we start with a class of computers C_Λ^M generated by a causaloid, Λ, for which there is indefinite causal structure, then we do not expect this scale invariance property under coarse-graining. In particular, we would expect, as we go to larger diamonds, that the causal structure will become more definite. Hence we may not be able to arrange the same kind of causal connection between the simulated versions of the gates in S as between the original versions of these gates. Hence, we cannot expect that the procedure just described for simulation in the classical and quantum case will work in the case of a general causaloid.

This suggests that the concept of universal computation is may not be applicable in QG. However the situation is a little more subtle. The classical physics that is required to set up classical computation should be a limiting case of any theory of QG. If a given causaloid, Λ, corresponds to QG then we expect that it is possible to use this to simulate a universal classical computer if we coarse-grain to a classical scale. We can also build random number generator since we have probabilistic processes (since QT is also a limiting case). This suggests a way to simulate (in some sense of the word) a general QG computer in the class corresponding to Λ. We can use the classical computer to calculate whether probabilities are well defined and, if so, what they are equal to arbitrary accuracy from the causaloid by programming in the equations of the causaloid formalism. We can then use the random number generator to provide outputs with the given probabilities thus simulating what we would see with a genuine QG computer. We might question whether this is genuine simulation since there will not necessarily be a simple correspondence between the spacetime locations of these outputs in the simulation and the outputs in the actual QG computation. In addition, in simulating the classical computer from the quantum gravitational Λ, we may need a gate set S with very large M. Nevertheless, one might claim that the Church Turing Deutsch principle is still true. However, it seems that the computational reflection principle is under considerable strain. In particular, the classical simulation would have definite causal structure unlike the QG computer. But also the detailed causal structure of the classical simulation would look quite different from that of the QG computer it simulates. There may also be computational complexity issues. With such issues in mind we might prefer to use the QG causaloid to simulate a universal quantum computer (instead of a universal classical computer) and then use this to model the equations of the causaloid formalism to simulate the original causaloid. This may be quicker than a classical computer. However, the computational power of a QG computer may go significantly beyond that of a quantum computer (see Section 4.3).

If the computational reflection principle is undermined for QG processes then we may not be able to think that the world is, itself, a computational process. Even if we widen our understanding of what we mean by computation, it is possible that we will not be able to define a useful notion of a universal computer that is capable of simulating all fundamental quantum gravitational processes in a way that accurately reflects what is happening in the world. This would have an impact on any research program to model fundamental physics as computation (such as that of Lloyd [10]) as well as having wider philosophical implications.

4.3 Will Quantum Gravity Computers have Greater Computational Power than Quantum Computers?

Whether or not we can define a useful notion of universal QG computation, it is still possible that a QG computer will have greater computational power than a quantum computer (and, therefore, a classical computer). Are there any reasons for believing this?

Typically we are interested in how computational resources scale with the input size for a class of problems. For example we might want to factorize a number. Then the input size is equal to the number of bits required to represent this number. To talk about computational power we need to a way of measuring resources. Computer scientists typically make much use of SPACE and TIME as separate resources. TIME is equal to the number of steps required to complete the calculation and SPACE is equal to the maximum number of (qu)bits required. Many complexity classes have been defined. However, of most interest is the class P of decision problems for which TIME is a polynomial function of the size of the input on a classical computer (specifically, a Turing machine). Most simple things like addition, multiplication, and division, are in P. However factorization is believed not to be. Problems in P are regarded as being easy and those which are not in P are regarded as being hard. Motivated by the classical case, BQP is the class of decision problems which can be solved with bounded error on a quantum computer in polynomial time. Bounded error means that the error must be, at most, $1/3$. We need to allow errors since we are dealing with probabilistic machines. However, by repeating the computation many times we can increase our certainty whilst still only requiring only polynomial time.

In QG computation with indefinite causal structure we cannot talk about SPACE and TIME as separate resources. We can only talk of the SPACETIME resources required to complete a calculation. The best measure of the spacetime resources is the number of locations x (where gates are chosen) that are used in the computation. Thus, if we have $x \in \mathcal{O}$ for a computation then $\text{SPACETIME} = |\mathcal{O}|$.

In standard computation, the SPACE used by a computer with polynomial TIME is, itself, only going to be at most polynomial in the input size (since, in the computational models used by computer scientists, SPACE can only increase as a polynomial function of the number of steps). Hence, if a problem is in P then SPACETIME will be a polynomial function of the input size also. Hence, we can usefully work with SPACETIME rather than TIME as our basic resource.

We define the class of problems $BP_{\{\Lambda,S\}}$ which can be solved with bounded error on the computer $\{\Lambda, S\}$ in polynomial SPACETIME. The interesting question, then, is whether there are problems which are in $BP_{\{\Lambda,S\}}$ but not in BQP for some appropriate choice of computer $\{\Lambda, S\}$. The important property that a QG computer will have that is not possessed by a quantum (or classical) computer is that we do not have fixed causal structure. This means that, with respect to any attempted foliation into spacelike hypersurfaces, there will be backward in time influences. This suggests that a QG computer will have some *insight* into its future state (of course, the terminology is awkward here since we do not really have a meaningful notion of "future"). It is possible that this will help from a computational point of view.

Quantum Gravity Computers   399

A different way of thinking about this question is to ask whether a QG computer
will be hard to simulate on a quantum computer. Assuming, for the sake of argument, that Assumption 1 is true then, as seen in Section 3.8, we can force an evolving
state point of view (however unnatural this may be). In this case we can simulate
the QG computer by simulating the evolution of $\mathbf{p}(t)$ with respect to t. However,
this is likely to be much harder when there is not the kind of causal structure with
respect to t which we would normally have if t was a physically meaningful time
coordinate. In the classical and quantum cases we can determine the state at time
t by making measurements at time t (or at least in a very short interval about this
time). Hence, to specify the state, $\mathbf{p}(t)$, we need only list probabilities pertaining to
the time-slice $R_t - R_{t+1}$ rather than all of R_t. The number of probabilities required to
specify $\mathbf{p}(t)$ (i.e. the number of entries in this vector) is therefore much smaller than
it might be if we needed to specify probabilities pertaining to more of the region
R_t. If, however, we have indefinite causal structure, then we cannot expect to have
this property. Hence the state at time t may require many more probabilities for its
specification. This is not surprising since the coordinate t has no natural meaning in
this case. Hence, it is likely that we will require much greater computational power
to simulate the evolution of $\mathbf{p}(t)$ simply because we will have to store more probabilities at each stage of the evolution. Hence we can expect that it will be difficult
to simulate a QG computer on a quantum computer. However, an explicit model is
required before we can make a strong claim on this point.

5 Conclusions

It is likely that a theory of quantum gravity will have indefinite causal structure. If
this is the case it will have an impact on the theory of computation since, when all
is said and done, computers are physical machines. We might want to use such QG
effects to implement computation. However, if there is no definite causal structure
we must depart from the usual notion of a computation as corresponding to taking
a physical machine through a time ordered sequence of steps—a QG computer will
likely not be a step computer. We have shown how, using the causaloid formalism,
we can set up a mathematical framework for computers that may not be step computers. In this framework we can represent a computer by the pair $\{\mathbf{\Lambda}, S\}$. Classical
and quantum computers can be represented in this way.

We saw that the notion of universal computation may be undermined since the
nature of the causal structure is unlikely to be invariant under scaling (the fuzzyness
of the indefinite causal structure is likely to go away at large enough scales). If
this is true then it will be difficult to make the case that the universe is actually a
computational process.

It is possible that the indefinite causal structure will manifest itself as a computational resource allowing quantum gravity computers to beat quantum computers for
some tasks.

An interesting subject is whether general relativity computers will have greater computational powers. There has been some limited investigation of the consequences of GR for computation for static spacetimes (see [11–13]). General relativity has not been put into the causaloid framework. To explore the computational power of GR we would need to put it into an operational framework of this nature.

The theory of quantum gravity computation is interesting in its own right. Thinking about quantum gravity from a computational point of view may shed new light on quantum gravity itself—not least because thinking in this way forces operational clarity about what we mean by inputs and outputs. Thinking about computation in the light of indefinite causal structure may shed significant light on computer science—in particular it may force us to loosen our conception of what constitutes a computer even further than that already forced on us by quantum computation. Given the extreme difficulty of carrying out quantum gravitational experiments, however, it is unlikely that we will see quantum gravity computers any time soon.

We have investigated the issue of QG computers in the context of the causaloid framework. This is a candidate framework for possible theories of QG within which we can use the language of inputs and outputs and can model indefinite causal structure (a likely property of QG). The main approaches to QG include String Theory [14], Loop Quantum Gravity [15–17], Causal Sets [18], and Dynamical Triangulations [19]. These are not formulated in a way that it is clear what would constitute inputs and outputs as understood by computer scientists. Aaronson provides an interesting discussion of some of these approaches and the issue of quantum gravity computation [20]. He concludes that it is exactly this lack of conceptual clarity about what would constitute inputs and outputs that prohibits the development of a theory of quantum gravity computation. Whilst the causaloid formalism does not suffer from this problem, it does not yet constitute an actual physical theory. It is abstract and lacks physical constants, dimensionalful quantities, and all the usual hallmarks of physics that enable actual prediction of the values of measurable quantities.

Issues of computation in the context of quantum gravity have been raised by Penrose [21, 22]. He has suggested that quantum gravitational processes may be non-computable and that this may help to explain human intelligence. In this paper we have chosen to regard quantum gravitational processes as allowing us to define a new class of computers which may have greater computational powers because they may be able to harness the indefinite causal structure as a computational resource. It is likely that QG computers, as understood in this paper, can be simulated by both classical and quantum computers so they will not be able to do anything that is non-computable from the point of view of classical and quantum computation. However, it may require incredible classical or quantum resources to simulate a basic QG computational process. Further the internal structure of a QG computation will most likely be very different to that of any classical or quantum simulation. Hence, the "thought process" on a QG computer may be very different to that of a classical or quantum computer in solving the same problem and so, in spirit if not in detail, the conclusions of this paper may add support to Penrose's position. Of course, QG computation can only be relevant to the human brain if it can be shown that the particular effects of QG can be resident there [22, 23].

Dedication

It is a great honour to dedicate this paper to Abner Shimony whose ideas permeate the field of the foundations of quantum theory. Abner has taught us the importance of metaphysics in physics. I hope that not only can metaphysics drive experiments (Abner's "experimental metaphysics") but that it can also drive theory construction.

References

1. A. Turing, "On Computable Numbers, with an Application to the Entscheidungsproblem", *Proc. of the London Math. Soc.*, Series 2, Vol. 42 (1936) 230–265.
2. D. Deutsch, "Quantum Theory, the Church-Turing Principle and the Universal Quantum Computer", *Proc. of the Roy. Soc. of London, Series A, Math. Phys. Sci.*, Vol. 400, No. 1818, pp. 97–117 (1985).
3. L. Hardy, "Probability Theories with Dynamic Causal Structure: A New Framework for Quantum Gravity", gr-qc/0509120 (2005).
4. A. Church, "An Unsolvable Problem of Elementary Number Theory", *Am. J. Math.*, 58, 345–363 (1936).
5. C. Soans and A. Stevenson (eds.) *Concise Oxford English Dictionary*, 11th edition (OUP, 2004) Oxford.
6. E. Bernstein and U. Vazirani, "Quamtum Complexity Theory", *SIAM J. Comput.*, 26(5), 1411–1473 (1997).
7. R. Penrose, "Gravity and State Vector Reduction", in R. Penrose and C. J. Isham (eds.), *Quantum Concepts in Space and Time*, pp 129–146 (Clarendon, Oxford, 1986).
8. L. Hardy, "Towards Quantum Gravity: A Framework for Probabilistic Theories with Non-Fixed Causal Structure", *J. Phys. A: Math. Theor.* **40**, 3081–3099 (2007).
9. L. Hardy, "Causality in Quantum Theory and Beyond: Towards a Theory of Quantum Gravity", PIRSA#:06070045 (pirsa.org, 2006).
10. S. Lloyd, "The Computational Universe: Quantum Gravity from Quantum Computation", quant-ph/0501135 (2005).
11. M. L. Hogarth, "Does General Relativity Allow an Observer to View an Eternity in a Finite Time?", *Found. Phys. Lett.*, 5, 173–181 (1992).
12. M. L. Hogarth, "Non-Turing Computers and Non-Turing Computability", *PSA: Proc. of the Biennial Meeting of the Philos. of Sci. Assoc., Vol. 1994, Volume One: Contributed Papers*, pp. 126–138 (1994).
13. I. Pitowsky, "The Physical Church Thesis and Physical Computational Complexity", *Iyyun* 39, 81–99 (1990).
14. J. Polchinski, *String Theory*, Vols. 1 and 2 (Cambridge University Press 1998) Cambridge.
15. C. Rovelli, *Quantum Gravity* (Cambridge University Press, 2004) Cambridge.
16. T. Thiemann, "Lectures on Loop Quantum Gravity", *Lecture Notes in Physics* 541 (2003).
17. L. Smolin, "An Invitation to Loop Quantum Gravity", hep-th/0408048 (2004).
18. R. D. Sorkin "Causal Sets: Discrete Gravity", qr-qc/0309009 (2003).
19. J. Ambjorn, Z. Burda, J. Jurkiewicz, and C. F. Kristjansen, "Quantum gravity represented as dynamical triangulations", *Acta Phys. Polon. B* **23**, 991 (1992).
20. S. Aaronson, "NP-complete Problems and Physical Reality", quant-ph/0502072 (2005).
21. R. Penrose, Emperors New Mind (OUP, 1989) Oxford.
22. R. Penrose, Shadows of the Mind (OUP, 1994) Oxford.
23. M. Tegmark, "Importance of quantum decoherence in brain processes", *Phys. Rev. E* 61, 4194–4206 (2000).

"Definability," "Conventionality," and Simultaneity in Einstein–Minkowski Space-Time

Howard Stein

For Abner Shimony—in gratitude for the light and warmth I have received from a lifelong friendship.

Abstract In this article, I attempt to clarify certain misunderstandings that have contributed to continuing controversy over the status of the concept of relative simultaneity in the special theory of relativity. I also correct a number of technical errors in the literature of the subject, and present several new technical results that may further serve to clarify matters.

Controversy over the status of the concept of relative simultaneity in the special theory of relativity has proved remarkably durable. Very recently (within two days of first writing these words), as a result of ruminations on a recent paper of Adolf Grünbaum's, I have come to believe that an important contributing factor to the persistence of the dispute is the use of certain key words (or phrases) in quite different senses by some of the disputants. One central aim of this paper, therefore, is to (try to) clarify these misunderstandings, and thereby both to reduce the number of the points of disagreement, and for the remaining points—for one can hardly expect all disagreement to be thus dispelled—at least to help clarify what the disagreements really *are*. A second aim is to correct some technical errors in the literature of the subject, and to state and prove some new technical results that may help contribute to clarity in the matter.

1 David Malament's Contribution: (a) Remarks on Some Technical Objections; (b) Refinements of the Theorem

In a recent "revisiting" of David Malament's well-known discussion of this subject [1], Mark Hogarth [2, p. 492] writes as follows, quoting a remark of mine from [3]:

> Just how decisive is Malament's result for the issue of conventionality of simultaneity? Howard Stein echoes a common sentiment:

H. Stein
Professor of Philosophy Emeritus, The University of Chicago

W.C. Myrvold and J. Christian (eds.), *Quantum Reality, Relativistic Causality, and Closing the Epistemic Circle*, The Western Ontario Series in Philosophy of Science 73, © Springer Science+Business Media B.V. 2009

404   H. Stein

> The issue ... has been dealt with—in my opinion, conclusively—by David Malament, who
> pointed out that the Einstein–Minkowski conception of relative simultaneity is not only
> characterizable in a direct geometrical way within the framework of Minkowski's geom-
> etry ... but is the only possible such conception that satisfies certain very weak 'natural'
> constraints.

Hogarth then notes that Malament has made use of the assumption that change of
scale is an automorphism of the structure of space-time; objects that physics itself
is *not* invariant under change of scale; concludes that Malament's result is after
all inconclusive; and offers an argument of his own to show that invariance under
change of scale can be replaced by another requirement, and the uniqueness of the
standard conception of simultaneity thereby rescued.[1]

In correspondence before his paper was published, I remarked to Hogarth that
there is a footnote to the statement he quotes from me—see [3, p. 153, n. 1]—
in view of which the statement cannot reasonably be interpreted to mean that—in
my view—Malament "had, as it were, dotted all the i's and crossed all the t's of
the subject"; that, rather, what I had meant was "that Malament had redirected the
discussion, away from the consideration of alternative ways of introducing 'a time-
coordinate,' to the consideration of what purely geometric notions are available in
Minkowski geometry." The footnote in question reads:

> There is one slightly delicate point to be noted: Malament's discussion, which is concerned
> with certain views of Grünbaum, follows the latter in treating space-time without a dis-
> tinguished time-orientation. To obtain Malament's conclusion for the (stronger) structure
> of space-time with a time-orientation, one has to strengthen the constraints he imposes on
> the relation of simultaneity: it suffices, for instance, to make that relation (as in the text
> above) relative to a *state of motion* (i.e., a time-like *direction*), rather than—as in Mala-
> ment's paper—to an *inertial observer* (i.e., a time-like *line*).

I did not think it necessary to *demonstrate* the stated fact, taking it for granted that
anyone who cared to would easily see how a proof would go.

Some years after [3] appeared, a paper was published by Sahotra Sarkar and
John Stachel [4]—under the title, "Did Malament Prove the Non-Conventionality
of Simultaneity in the Special Theory of Relativity?" These authors criticize Malament's argument on the very grounds mentioned in the footnote; but they elaborate
upon these grounds in a way that in my opinion is very defective, and is in serious

---

[1] Malament, requiring invariance under change of scale, adds, besides the condition of invariance
under automorphisms of space-time, only the assumption that simultaneity relative to O (a) is an
equivalence relation, and (b) holds for *at least one* pair of points (p,q) with p on O and q not on
O, but does not hold for *every* pair of points.—Hogarth appeals to quantum field theory for the
fact that physics is not invariant under change of scale (this is indeed already clear in classical
physics, since the fundamental classical physical constants allow us, in more than one way, to *determine a unit of length*; moreover, at the very beginning of the modern science of physics, Day 1
of Galileo's *Two New Sciences* opens with this paradox: a scale model of, for instance, a ship, may
be perfectly stable, but the ship built from this model may collapse under its own weight—and
Galileo's spokesman Salviati says that, although geometry is invariant under change of scale, this
non- scale-invariance of "machines" can be *explained* by geometry).

Hogarth, renouncing appeal to scale-invariance, strengthens Malament's assumption (b) by requiring that, for any "inertial observer world-line" O and any point p in space-time, there is one and
only one point q on O to which p is simultaneous relative to O. (See also Supplementary Note 1.)

"Definability," "Conventionality," and Simultaneity in Einstein–Minkowski Space-Time   405

need of clarification. They also offer a proof of a result akin to that indicated in the footnote; but their proof is fallacious, and the result therefore *does* demand the explicit proof that I thought superfluous. Since Hogarth's theorem is subject to the same objection raised by Sarkar and Stachel—Hogarth, too, makes crucial use of reflection-invariance—it is clearly desirable to establish a conclusion that depends *neither* upon invariance under scale-change *nor* upon invariance under reflections.[2]

But first I wish to make it clear that Malament's theorem itself—exactly as he has formulated it—is entirely correct. This seems important to emphasize, because Sarkar and Stachel have challenged the entire correctness of this theorem.[3] A discussion of the point is especially desirable because it concerns the question, just what is meant when one says that a notion is "definable *from*" something or other— I shall say, definable from the elements of *a given kind of structure*, or from the basic notions of *a given mathematical theory* (understood as being "the theory of that kind of structure"); and it turns out that a misunderstanding on this score is also relevant to points raised by Adolf Grünbaum, which we shall consider later.

Malament's theorem explicitly refers to a relation "definable from κ and O," where κ is the binary relation, on the set of points of Minkowski space-time, "p and q are such that one of them may causally influence the other," and where O is a *given* straight time-like world-line (what I shall henceforth refer to—and have already referred to, in Supplementary Note 1—as an "observer-line"). (So strictly speaking, if the phrasing of the end of the preceding paragraph is taken pedantically, the "theory" concerned is "the theory of the relation κ in a Minkowski space-time with a particular observer-line O singled out.") Malament then says [1, p. 297],"If an n-place relation is definable from κ and O, in any sense of 'definable' no matter how weak, then it will certainly be preserved under all O causal automorphisms [that is: mappings that preserve both O and κ]." It is this statement that Sarkar and Stachel challenge: they acknowledge the correctness of Malament's theorem *if "causal definability"* (as they put it) *is construed in Malament's way*; but they deny that this is an appropriate way to construe such definability—and they give an alleged counterexample.

Now, the issue this raises is simply one of *logic*. And the logical situation should be altogether clear: for the logician—or the mathematician—to *define* a notion in terms of certain basic concepts is in effect to introduce an abbreviated mode of expression; any statement phrased using the defined notion may be rephrased using only those basic concepts: the "new" notion is simply eliminable. But that a statement using only the basic notions of a theory is unaffected by automorphisms of the

---

[2] Sarkar and Stachel mention in passing [4, p. 215 n. 11, and p. 217] that they do not use scale-invariance in their proof; but, as remarked—and as will be shown below—the proof is invalid, so this fact is irrelevant.

[3] If this formulation seems odd, it should—it is designedly so; for although Sarkar and Stachel say, "Clearly, something is amiss with Malament's theorem," they add, "A *correct* mathematical result [emphasis added] seems to be contradicted by patently good counterexamples"; what they challenge is, not the soundness of Malament's argument, but "the interpretation of one of the conditions that Malament imposes on simultaneity relations" [4, p. 214]. So the result is "correct"—but not *entirely* so.

object treated of by the theory is an immediate consequence of the very definition of "automorphism." So "in principle," it is hard to see how a controversy can arise here.

"In practice," however, one sees how confusion *did* occur. Sarkar and Stachel point to the facts that "the distinction between the [two, oppositely directed,][4] half null cones of [the full null cone of an event e] can be made using causal definability alone" [4, p. 213], and that this distinction can be made "coherently" for all point-events of the Minkowski space-time concerned. They conclude that the following two relations, satisfying Malament's other criteria, can be defined from the null-cone structure of the space-time, and thus from the relation κ alone (O plays no role): (a) p lies on the mantle of the backwards null-cone at q; (b) p lies on the mantle of the forwards null-cone at q; and conclude therefore that both (a) and (b), applied to point-pairs (p,q) with q on O, constitute relations definable from κ and O that satisfy all Malament's other conditions on a simultaneity relation. These examples, they accordingly maintain, show that his theorem is incorrect when "definable from κ and O" is rightly construed.

The error here turns on an ambiguity in the notion of "distinguishing" two things—that is, "telling the difference between" them. A Minkowski space-time has two possible "time-orientations." This statement certainly implies that one can "tell them apart"—enough, at least, to count them. And yet, one can't "tell which is which"—the two time-orientations are like two bosons in this respect.—Well, to continue with amusing word-games in the vernacular (even the vernacular of physics), would soon grow tiresome, and might amplify confusion; that is why logical pedantry has its legitimate place: One can define from κ and O a set of relations, each of which satisfies Malament's other requirements. The set I mean[5] contains *two* relations—the two described by Sarkar and Stachel. But one *cannot* "distinguish"— "single out"—*either one* of those two relations in distinction from the other; that is, one cannot do this *in terms of κ and O alone*. If only one could *once* "simply point to" one of the two half null-cones at *one* space-time point, the structure of κ would allow this choice to be "spread around" all through space-time, and we should have one of Sarkar and Stachel's examples; and why, indeed, should one *not* be able to do so?—But that isn't the question: the question is, What is *definable* from κ and O? And Malament's answer stands.

There remains the point that this may not be "the right question"; and I think this may be argued from two different points of view. It *could*—it seems to me—be claimed quite plausibly by an adherent of a "causal theory of time," Grünbaum for instance, that such a view of time is not restricted to what can be characterized in terms of the *symmetric* relation κ—that, rather, the *non-symmetric* relation "causal influence may be propagated from p to q'' is part of the basic apparatus of that

---

[4] Sarkar and Stachel say "backwards and forwards"; but they emphasize—rightly—that these adjectives are mere labels, not to be taken as denoting the "past" and the "future"; so I have preferred to substitute a neutral characterization, lest the reader suppose that they have overstepped at this point.

[5] This—a set that bears upon the Sarkar-Stachel examples—is by no means the *only* set of relations (besides the one relation of Malament's theorem) that can be defined from κ and O—there are infinitely many others; but there is no need to consider them here.

"Definability," "Conventionality," and Simultaneity in Einstein–Minkowski Space-Time   407

theory. (I am not sure whether Grünbaum himself would take this position; he has not, so far as I know, actually done so; and we shall see presently that his real disagreement with Malament has an entirely different basis.) And it could—*can*—also be based, as in my remark in the cited footnote and in Sarkar and Stachel's further discussion [4, pp. 214 ff.], on the consideration that time-reversal (as well as reversal of spatial orientation, and change of scale) cannot be considered as clearly demanded by physics.[6]

Turning now to the theorem Sarkar and Stachel claim to prove that covers the vulnerable points in that of Malament, their formulation reads as follows [4, p. 216]:

> Standard simultaneity is the only non-vacuous simultaneity relation causally' definable from κ and O that depends only on an inertial frame, and not on the particular world line O initially chosen to define that inertial frame.

By a "non-vacuous simultaneity relation" is here meant a relation S that satisfies the following three conditions, borrowed from Malament: (1) S is an equivalence relation; (2) there exists a point p on O and a point q not on O such that $S(p,q)$; (3) S does not hold for *every* pair of space-time points. By "the inertial frame defined by a (straight, time-like) world-line O" is meant (of course) the family of all lines parallel to O (I shall also use the term "inertial system" for such a family). And by "causally' definable" is meant: invariant under all automorphisms of the Minkowski space-time that preserve the inertial frame and that are continuously connectible to the identity ("causal' automorphisms," as Sarkar and Stachel call them). This class of mappings excludes time-reversals and "spatial" reflections—it is precisely the class of all those automorphisms (in the standard sense) that preserve both time-orientation and the orientation of the whole manifold. In particular, *it includes changes of scale.*

The last point deserves to be emphasized. *With* it, the theorem is true. But Sarkar and Stachel say—rightly—that they actually do not make any use of scale-invariance in their proof; and *without* the assumption of scale-invariance, the theorem becomes *false*. There are infinitely many counterexamples, of which the simplest is this: Let $S(p,q)$ be the relation: "p and q belong to hyperplanes orthogonal to the direction of O, having, for some integer n, the orthogonal (time-like) Minkowski distance $n\tau$, where τ is a particular distance (given once for all)".[7] This relation clearly satisfies the conditions on a "non-vacuous simultaneity relation,"

---

[6] I should not wish to be taken as endorsing every aspect of the critical discussion of this point in [4], but there is no need to argue the matter here: the general point, as I have here stated it, suffices for our purposes.—Let me add that the fact that invariance under change of scale might be challenged was likewise suggested in [3]—see p. 149, near the top—and that, accordingly (see p. 250), an alternative proof was indicated for the theorem demonstrated on p. 249, avoiding appeal to scale-invariance. Having mentioned this I must add to the criticisms I have made of passages by others, one directed to a passage of my own: the argument [3. p. 149] leading up to the theorem mentioned is sound, but the formulation of the theorem itself does *not* state correctly what the argument has proved: the theorem as there stated is *false*!—On this, see Supplementary Note 2 below.

[7] The other counterexamples are fairly obvious variants of this one, taking the coefficients of τ to be, for instance, rational numbers, or elements of a given real algebraic number-field, or of any given proper subfield of the real numbers, or indeed of any given additive proper subgroup of the real numbers.

408   H. Stein

and as clearly is invariant under all automorphisms (reflections too!) that preserve
the inertial frame to which O belongs, *except* scale-invariance.

If one wants to avoid reliance on scale-invariance, then, Malament's conditions
on the relation S must be strengthened. (This is quite acceptable: Malament has
simply chosen a very weak set of conditions, not for any "ideological" reason, but
simply to show how restricted the class of "causally definable" relations in his sense
is.) A suitable condition is this: that, for a given point p, $S(p,q)$ holds for exactly
one point q on each line of the given inertial system. But a far weaker condition suffices: namely, that (a) for at least one observer-line O of the given inertial system, no
two distinct points p, q on O satisfy $S(p,q)$, and (b) at least one pair of distinct points
(p,q) in space-time does satisfy $S(p,q)$. (It is obvious, just from invariance under
translations—which *ipso facto* take the inertial system to itself—that condition (a)
entails that no two distinct points on *any* line of the system satisfy S.) To forestall any
possible ambiguity of terminology, I shall henceforth use the word "automorphism"
to refer to the full class of mappings normally considered by mathematicians to be
automorphisms of a Minkowski space-time (including, therefore, all reflections and
changes of scale); these coincide with the "causal automorphisms" of Malament;
I shall use the phrase "proper automorphisms" for those called "causal' automorphisms" by Sarkar and Stachel (these still include changes of scale, but do not include reflections of any sort, "spatial" or "temporal"); and I shall use the phrase
"strict automorphisms" for proper automorphisms that preserve the scale.

Let us proceed to the proof of two theorems: the one stated by Sarkar and Stachel
for a "non-vacuous" simultaneity relation and requiring invariance under change of
scale, and one using our strengthened condition on the relation but not requiring
scale-change invariance (both, however, relativize simultaneity to an inertial system, not just to a "single observer"). For convenience, the two theorems will be
formulated as a single one with two "cases":

Theorem 1. *In a Minkowski space-time of three or more dimensions let there be
singled out an inertial system* **I**, *and let S be an equivalence relation on the space-
time points satisfying one of the following two sets of conditions:*[8]

(a)(1) *S is invariant under all strict automorphisms of the space-time that preserve*
       I;
   (2) *S(p,q) does not hold if p and q are distinct points on a single world-line of* **I**;
   (3) *S(p,q) holds for at least one pair of distinct points;*
(b)(1) *S is invariant under all proper automorphisms of the space-time that preserve*
       I;
   (2) *S(p,q) does not hold for every pair of points;*
   (3) *S(p,q) holds for at least one pair of points not on the same world-line of* **I**;

*—then for any distinct points p, q, S(p,q) holds if and only if the line pq is orthogonal
to the lines of* **I**.

---

[8] $S(p,q)$ will also be expressed by saying "p and q are simultaneous for **I**," or "p and q are
I-simultaneous"; and that a line, or vector, or direction, is perpendicular to the lines of **I** will
also be expressed by saying that it is "perpendicular—or orthogonal—to **I**."

"Definability," "Conventionality," and Simultaneity in Einstein–Minkowski Space-Time   409

Proof. Let p, p', be any points on different lines of **I**—O and O' respectively—that are simultaneous for **I** (such points exist in both cases, (a) and (b)); let s be the orthogonal (space-like) distance of p' from O; and let M be a three-dimensional subspace of our space-time that contains the plane spanned by the parallel time-like lines O and O'. Under all rotations of M about O as axis, the images of the point p' are all the points of the circle through p', in the (space-like) plane of M orthogonal to **I**, having its center at the intersection of that plane with O. Since each such rotation can be extended to a strict automorphism of the entire space-time preserving **I** and leaving p fixed (just allow it to act as the identity on any orthogonal complement to M in the space-time), all the points of that circle are simultaneous with p, and therefore also with one another; and the vectors $p'p''$ from p' to the other points of that circle (i) are all orthogonal to **I** and (ii) have lengths that fill the half-open interval of real numbers $(0, 2s]$. We therefore have established the following.

Lemma 1. *If p' has space-like orthogonal distance s from the **I**-line through a point p with which it is **I**-simultaneous, then for every real number s'' with $0 < s'' \le 2s$ there is a point p'', **I**-simultaneous with p', such that (i) $p'p''$ is orthogonal to **I** and (ii) the space-like length of $p'p''$ is s''. (This holds, be it also noted, for both cases, (a) and (b).)*

   But this has the following as an almost immediate corollary: *Any two distinct space-time points q, q' such that the vector qq' is orthogonal to **I** are **I**-simultaneous.*—Indeed, for our special point p' above, the lemma shows that the property possessed by the real number s is also possessed by (among others) $2s$ (for if p_1 is the point diametrically opposite p' in the circle, the space-like orthogonal distance of p' from the **I**-line through p_1 is just the space-like distance of p' from p_1 itself, namely $2s$); therefore, by successive doubling, that property is possessed by arbitrarily large real numbers;[9] and then, by the full conclusion of the lemma, it is possessed by every positive real number smaller than some "arbitrarily large" one—i.e., by all real numbers. Now let r be the space-like distance between our points q *and* q'. There is a point p'', by what has already been shown, **I**-simultaneous with p', such that $p'p''$ is orthogonal to **I** and has space-like length r. The translation taking p' to q is a strict automorphism of the space-time that preserves **I**, so the image q'' of p'' by that translation is **I**-simultaneous with q, and the vector qq'' is orthogonal to **I** and has space-like length r. Moreover, there is a rotation that leaves q fixed, preserves **I**, and takes q'' to q'; therefore q' is **I**-simultaneous with q.
   We now know that any two points satisfying the "standard" condition for simultaneity relative to **I** are **I**-simultaneous. But to establish the converse is trivial: For case (a), we have only to remark that if Σ is a hyperplane orthogonal to **I**, p any point of Σ, and q any point **I**-simultaneous with p, the line of **I** through q meets Σ in a point q' that is **I**-simultaneous with q by the preceding result, and therefore is also simultaneous with q. So in case (a), q' must coincide with q—i.e., q must belong to Σ—since we cannot have distinct **I**-simultaneous points on a time-like line. As for case (b), the argument just given shows that if Σ is not the complete class of points

---

[9] Formally, an argument by mathematical induction is of course required.

I-simultaneous with p, there must be two distinct points q, q', that belong to the same line of **I** and that are **I**-simultaneous with p and therefore with one another. By changes of scale, the vectors qq' and $q'q$ can be transformed to vectors of arbitrary time-like length, and of both senses, all pointing in the direction of **I**. But such vectors, starting from all the points of Σ, reach all the points of the space-time; and since their end-points are simultaneous, we shall have that every point of the space-time is **I**-simultaneous with a point of Σ; so, since all the points of Σ are themselves **I**-simultaneous, all the points of space-time will be **I**-simultaneous, contradicting (b)(2): the proof is complete.[10]

The main point of these results is, as I have intimated, rather obvious; if the proofs are a bit lengthy and a little intricate, that is the result of making the conditions posited very Spartan. Thinking of the matter more broadly, it is easy to see—and to prove—that a relation, satisfying conditions one would surely ask of simultaneity, that is invariant under (a) every rotation about a line of the inertial system **I** and (b) every translation of space-time can have no other equivalence-classes than the hyperplanes orthogonal to the lines of **I**. Relativizing simultaneity to an inertial system is of course in consonance with the original procedure of Einstein, who envisaged, as coordinating their spatio-temporal observations, a "community of observers," in a shared "inertial state." But it has occurred to me to ask whether one can reach any sort of result from a weaker assumption—one that does not require *ab initio* that the "observers" of this community be at rest relative to one another; and indeed—in a certain sense—one can. I do not think that the results I shall now present are of great *philosophical* interest (this I shall discuss later); but I think they are of some—although again I should not say of "great"—*mathematical* interest; they are *not* "obvious."

Let us, then, make the following assumptions, for a Minkowski space-time of at least three dimensions (as we shall see, the two-dimensional case is notably different):

With each observer-line O there is associated an equivalence relation S_O on the whole space-time, in such a way that:

$(S1)$ any strict automorphism that maps O to O' transforms S_O to $S_{O'}$;
$(S2)$ if Σ is an equivalence-class of S_O then:

    (a) every observer-line O' has one and only one point in common with Σ, and
    (b) for every point p' of Σ there is an observer-line O' containing p' such that Σ is an equivalence-class of $S_{O'}$ (as well as of S_O).

—From these assumptions it does *not* follow that S_O is the standard Einstein–Minkowski[11] relation; let us examine what does follow.

---

[10] The counterexamples already given to the Sarkar-Stachel theorem with scale-changes excluded make it plain that the inference from the assumption that an **I**-line contains two simultaneous points to the conclusion that all its points are simultaneous could not be made without the condition of scale-change invariance. (For further discussion, locating the particular fallacy in the Sarkar–Stachel proof, see Supplementary Note 3.)

[11] I continue to use this designation, rather than "Poincaré–Einstein" as in [4], because I think it historically far more justified. Poincaré's treatment of what we now call "the special theory of

"Definability," "Conventionality," and Simultaneity in Einstein–Minkowski Space-Time   411

Let Σ be the equivalence-class of the relation S_O (for a given O), let p be the point of O in Σ, and let p' be any other point of Σ (note that p, as well as p', is an *arbitrary* point of Σ—for, by $(S2)(b)$, *every* point of Σ is the point, in Σ, of *some* observer-line having Σ as an equivalence-class). A key lemma in our discussion will be this: *If O' is the observer-line containing p' for which Σ is an equivalence-class, then O' and O are coplanar* (that is, if we call an observer-line having Σ as an equivalence-class an "axis" of Σ, then *every pair of axes of Σ is a pair of coplanar world-lines*).

In any event, O and O' can be embedded in a three-dimensional Minkowskian subspace M of our space-time; for if they are not coplanar, then there is a unique three-dimensional affine subspace M containing them (and this is necessarily Minkowskian, since it contains time-like lines), whereas if they are coplanar the two-dimensional subspace that they span can be extended to a three-dimensional subspace M. Any strict automorphism of M can be extended to the entire space-time (e.g., by making it act trivially on an orthogonal complement of M); therefore any such automorphism maps the intersection of Σ and M to itself (this follows obviously from $(S1)$). Consider, first, the family of all rotations of M about O as fixed axis. Under these rotations, p' generates a circle C, whose plane is orthogonal to O, and all of whose points are in Σ (since for any such point p'' we have $S_O(p'', p)$—p itself being fixed under all these rotations). Let l be the tangent-line to C at p'; I claim that O' must be orthogonal to l. Suppose it is not so. Then the plane through p' orthogonal to O' does not contain l. Now, that plane is space-like, and it separates M into two connected components—call them A and B. Since l meets, but does not lie entirely in, this plane, it contains—arbitrarily close to the point p' in which it meets the plane—points of a and points of B. The same must then be true of the circle C, to which l is tangent: in fact, any arc of C which contains p' in its interior and which does not extend as far as the point diametrically opposite to p' is divided by p' into a part that lies in A and one that lies in B; and each of these contains points arbitrarily close to p, and therefore points whose perpendicular distances from O' are arbitrarily small. In particular, C *contains a pair of points, say q in A and q' in B, that are of equal orthogonal distance from O'.* Under rotation of M about O', therefore, q and q' describe circles, in planes orthogonal to O', one in A and one in B, lying on a single cylinder having O' as axis; and all the points one *both these circles, together*, are in Σ, since q and q'—as points of C—lie in Σ, and Σ is invariant under rotations about O'. And this leads to a contradiction of $(S2)(a)$: for the generating lines of our cylinder, which are parallel to O' and thus time-like,

---

relativity" is quite wonderful; but (a) although he had previously discussed the lack of clarity of the notion of simultaneity for distant events, there is not a single word about distant simultaneity in his great essay [5], *except* for what is implicit in the spatio-temporal transformation equations (represented as changes of coordinates)—and equations, which Poincaré attributes to Lorentz, are indeed those introduced by Lorentz [6]. Further, Poincaré expresses in the introduction to this essay deep reservations about the theory he is presenting, as one that seems artificial and might perhaps some day be simplified by a critical consideration of measurement (so far is he from *offering* such a consideration here!). So (a) there is something to be said for "*Lorentz*-Einstein," rather than "Poincaré-Einstein," but (b) on the other hand, it was Einstein who made a clarification of simultaneity a central theme, and it was Minkowski who geometrized that clarification; hence my preference.

each meet *both* the circle generated by q and that generated by q', whereas these two circles do *not* meet (since one lies in A and the other in B). So it is established that O' is orthogonal to l, as claimed.

But this result means that O' and O are coplanar—that is, it establishes our lemma: for l, as the tangent-line to C, lies in the plane of C—i.e., the plane orthogonal to O; so any plane orthogonal to l is *parallel* to O; but the plane orthogonal to l at the point p', since it contains the center of C, *meets* O, and so must *contain* O; and since this plane, as we have just seen, also contains O', O and O' are indeed coplanar.

There are now two possibilities: O and O' intersect, or they are parallel.[12] Let O'' be any axis of the same equivalence-class Σ that does not lie in the same plane as O and O' (there are such, of course: take an axis at any point of Σ outside that plane—that such points exist is guaranteed, in the light of our assumption that the space-time is at least three-dimensional, by condition $(S2)(a)$). If O and O' are parallel, then since O'' cannot meet both of them it must be parallel to one—but, then, to both; any other axis is non-coplanar with either O and O' or O and O'', and is therefore parallel to O: in other words, in this case *all the axes of Σ are parallel*; and by the invariance of S_O under translations in the direction of O, the same must be true of *all* the equivalence-classes of S_O; and then, since there is a strict automorphism that takes O to any other observer-line, for all the equivalence-classes of any observer at all. But this puts us in the situation of Theorem 1 above (either (a) or (b)—the conditions of either are satisfied); so for this case it *does* follow that S_O is Einstein–Minkowski simultaneity: the equivalence-classes of simultaneity are hyperplanes orthogonal to the axes of simultaneity.

Suppose, then, that O and O' intersect. Then O'' (chosen as above) cannot be parallel to either—say, to O—by the immediately preceding result (putting O'' for O' and O' for O''). So we have three non-coplanar lines that intersect in pairs; from which it follows that all three intersect in *one common point* p_0. It follows by an obvious argument[13] (appealing to the fact that if, of three non-coplanar lines, every two intersect, then all three have a point in common) that *all* axes of Σ meet in p_0. (Note that at this point we have established—we have not assumed—that through each point q of Σ there passes *exactly one* axis of Σ: in the previous case, the line parallel to O; in this case, the line p_0q [that p_0 is *distinct* from q, so that there is a determinate line p_0q, is clear, since p_0 is not in Σ].)

To complete the analysis of this case, we must note that any strict automorphism that maps Σ onto itself must also take the set of axes of Σ one-to-one onto itself. For let ϕ be such an automorphism, let p be any point of Σ, let the axis of Σ at p be O, let $\phi(p)$ be p', and let $\phi(O)$ be O'. By the invariance assumption $(S1)$, ϕ transforms S_O to $S_{O'}$; so, since Σ is invariant under ϕ, Σ is an equivalence-class of $S_{O'}$; in other words, O' is an axis of Σ—and therefore, since there is only one axis of Σ at each point of Σ, it is the axis of Σ at p'; and this shows that ϕ does indeed map the set of axes of Σ one-to-one onto itself.

---

[12] If we adopt the point of view of projective geometry, introducing the "projective completion" of our space-time, these alternatives merge into one.

[13] From the projective point of view, the same argument as in the former case.

"Definability," "Conventionality," and Simultaneity in Einstein–Minkowski Space-Time   413

But from this it follows that any strict automorphism that maps Σ onto itself *leaves fixed* the point p_0 in which all the axes of Σ meet. Now, if p and p' are any two points of Σ, there is a strict automorphism taking p to p' and taking the axis of Σ at p to that at p' (for, given *any* points p and p', and *any* time-like lines l and l' containing p and p' respectively, there is a strict automorphism of the space-time taking p to p' and l to l'). That automorphism maps Σ to itself, and therefore leaves p_0 fixed. Since a strict automorphism preserves the length and the time-orientation of a time-like vector, the vectors $p_0 p$ and $p_0 p'$ have the same length and the same time-orientation. Let us call any class of all the time-like vectors that agree in length and time-orientation with a given one v the *temporally oriented radius* $[v]$ determined by v. Then what we have just established is that, *for any point p on an observer-line O, and any point p', if $S_O(p, p')$ holds then the vector $p_0 p'$ is time-like and the temporally oriented radius $[p_0 p']$ is the same as $[p_0 p]$*. We may describe this situation by saying that the equivalence-class Σ of S_O that contains p is contained in the "Minkowski hemisphere" with temporally oriented radius $[p_0 p]$; but it must then be the entire hemisphere, since otherwise it will not be the case that every observer-line meets Σ (i.e., assumption $(S2)$(a) will be violated).—However, this formulation is elliptical: it does not identify the point p_0. The answer to that objection is that what must be given (once for all) to determine the function S that assigns to every observer-line its "simultaneity-relation" S_O is *a temporally oriented radius r*. Then, given p, and O containing p, the point p_0 in the foregoing statement—the "center" of the Minkowski hemisphere containing p and having O as an axis—is the unique point *on O* such that $[p_0 p] = r$.—It is of course in singling out a particular (time-like) distance and time-orientation that we make essential use of the fact that we have not postulated invariance under change of time-orientation or change of scale.

To sum up, we have established the following:

Theorem 2. *In a Minkowski space-time of three or more dimensions let there be given an assignment, to each observer-line O, of an equivalence relation S_O on the whole space-time, in such a way that:*

(S1) any strict automorphism that maps O to O' transforms S_O to $S_{O'}$;
(S2) if Σ is an equivalence-class of S_O then:

  (a) every observer-line O' has one and only one point in common with Σ, and
  (b) for every point p' of Σ there is an observer-line O' containing p' such that Σ is an equivalence-class of $S_{O'}$ (as well as of S_O);

—then EITHER *(1) for every O, p, and q, we have that $S_O(p,q)$ holds if and only if the line pq is orthogonal to O,* OR *(2) there is a temporally oriented radius r such that for every O, every p on O, and every point q, we have that $S_O(p,q)$ holds if and only if $[p_0 q] = r$, where p_0 is the (unique) point on O for which $[p_0 p] = r$.* *(Equivalently we may say, in case (2), that for every O, p, and q, $S_O(p,q)$ holds if and only if, for some p_0 on O, $[p_0 p] = [p_0 q] = r$.)*

Two ways suggest themselves to strengthen the hypotheses of this theorem so as to eliminate the alternative (2) and leave standard Einstein–Minkowski simultaneity as the only possibility: we may replace $(S2)$(b) either by:

414   H. Stein

(S2) (b′) for every pair of space-time points p, p′, and every observer-line O containing p, there is an observer-line O′ containing p′ such that $S_{O'}$ coincides with S_O;

or by:

(S2) (b″) for every observer-line O, every equivalence-class Σ of S_O and every space-time point p, there is an observer-line O′ containing p such that Σ is an equivalence-class of $S_{O'}$ (as well as of S_O).

—The first of these alternative assumptions obviously rules out the alternative conclusion (2), since in the latter the *families* of equivalence-classes of any two distinct observer-lines have, *ipso facto*, their "moving centers" on different lines (the centers coincide only where the lines intersect). The second rules out conclusion (2) because for Σ to be an equivalence-class of $S_{O'}$ it is necessary that O' pass through the center p_0 of the "Minkowski hemisphere" Σ,[14] and of course not every observer-line does so.

Let us now consider a two-dimensional Minkowski space-time. This case differs, as has been intimated, in important ways from that of any higher dimension, and I think the difference is worth noting. (One point is obvious from the outset: namely, that the lemma we have exploited in the proof of Theorem 2 holds trivially in two dimensions—all the axes of a simultaneity-class are coplanar, because *everything* is coplanar; but this fact, in just this case, is of no help at all, and a quite different line of attack is required.) I shall begin by reviewing a few basic facts about the geometry of a Minkowski space-time of two dimensions:

(1) In two dimensions, there is complete symmetry in the geometry as between space and time, since the fundamental quadratic form has, in diagonal form, one positive and one negative coefficient.[15] It would, indeed, be possible in this case to extend the notion of an "automorphism" of the space so as to include an interchange of "space" and "time."[16] One aspect of this (as it were)

---

[14] This condition is also sufficient.

[15] There is in the literature some variation in the choice of signs: in the older convention, introduced by Minkowski (the "time-coordinate" as an imaginary number), the negative sign is assigned to the one temporal dimension, the positive sign to the *n* (for physics, 3) spatial ones; but the reverse choice is often made. This *is*, in the clearest sense, a pure matter of "convention": which of them one adopts makes no real difference. Accordingly, in the above discussion, I have never indicated a preference on this point: I have referred to vectors and subspaces as "time-like" or "space-like," without assigning an algebraic sign to the one or the other. I shall continue to do this in what follows.

[16] As we know, the wider sense of "automorphism" usual in mathematics for Minkowski space-times includes change of scale; this is tantamount to regarding, as characterizing the geometry, not a given non-degenerate quadratic form of appropriate signature, but a *class* of such forms, arising from one another through multiplication by arbitrary positive real factors. Nothing prevents one from admitting instead multiplication by arbitrary *nonzero* real factors. This, in the general case, of $n+1$ dimensions with $n > 1$, would make no difference at all to the theory; but in the case $n = 1$, it would automatically allow as automorphisms maps that preserve the linear (more exactly, affine) structure, preserve the relation of orthogonality, but take "time-like" vectors to "space-like" ones and *vice versa*.

"Definability," "Conventionality," and Simultaneity in Einstein–Minkowski Space-Time   415

"time-likeness of space" in a two-dimensional space-time is that, just as it is possible to divide time-like vectors into two classes that are topologically separated from each other, so this is also possible for space-like vectors. In fact, under "proper automorphisms," the non-zero vectors of a two-dimensional Minkowski space-time fall into eight distinct classes: two of time-like vectors (we may call them "future-pointing" and "past-pointing"), two of space-like vectors ("right-pointing" and "left-pointing"), and four classes of null vectors ("right-and-future pointing," "left-and-future pointing," "left-and-past point-ing," "right-and-past pointing"): one non-zero vector can be taken to another by a proper automorphism if and only if the two belong to the same class.[17] I shall call two non-zero vectors—or two lines—"like," or "of the same charac-ter," if they are both time-like, both space-like, or both null; and shall call (as above) two like vectors "of the same class" if they are similarly oriented—i.e., if they are "equivalent" under proper automorphisms.—Two vectors are like if and only if their inner products with themselves have like signs (here zero is to be counted as a sign in its own right, distinct from plus and minus); two non-null like vectors are of the same class if and only if their inner product has the same sign as the inner product of each with itself.[18]—The fact that proper automorphisms—and, *a fortiori*, strict automorphisms—preserve spatial as well as temporal orientation *of individual vectors* will prove to be important in the following.

(2) We shall have occasion to make use of the following facts about triangles— equivalently, about three vectors of which one is the sum of the other two—in a Minkowski plane:

(a) Let two vectorial sides—AB, BC—of a triangle be space-like and of the same class; then the side AC (their vector-sum) is also space-like and of the same class with them.

—This is perhaps obvious; to prove it, we have—representing the inner product by angle brackets—that, since $AC = AB + BC$, $< AC, AC >= < AB, AB > + 2 < AB, BC > + < BC, BC >$. By the criterion stated in (1) above, all three terms have the same sign; therefore the inner product of AC with itself has the same sign as those of AB and of BC with themselves: AC is space-like. And $< AC, AB >=< AB, AB > + < BC, AB >$—again a sum of terms of like sign—so it, too, has the same sign as the inner product of AB with itself, which shows that AC and AB are of the same class.

(b) If A, B, C are non-collinear points and D is a point on the space-like line BC such that AD is orthogonal to BC, then:

---

[17] To avoid a possible misunderstanding: "automorphism" is *not* here used in the extended sense mentioned in the previous note; and indeed, even if it were, since "proper" automorphisms are those that belong to the connected component of the identity in the Lie group of all automorphisms, the ones "interchanging time and space" would be excluded.

[18] Although it will be of no importance in what follows, it perhaps ought to be noted explicitly that this criterion fails for null vectors: two null-vectors directed along the same line have inner product zero whether their "senses" are the same or opposite.

(1) the vectors AB and AC are of like character and equal in length if and only if D is the midpoint of the segment BC (i.e., if and only if A is on the perpendicular bisector of BC);

(2) if this is indeed the case—i.e., if D is the midpoint of BC—then:

    (i) if AB and AC are space-like, then BA, AC, and BC are all of the same class, and the lengths of BA and of AC are *less than half* that of BC;

    (ii) if AB and AC are time-like, then AB, AD, and AC are all of the same class.

—For, setting for convenience $x = AD$, $y = BD$, $z = DC$ (all as vectors), and noting that $AB = x - y$ and $AC = x + z$, the necessary and sufficient condition for AB and AC to be of like character and equal in length is that $< x - y,\ x - y > = < x + z,\ x + z >$; i.e., in view of the orthogonality of x to y and to z, that $< x,\ x > + < y,\ y > = < x,\ x > + < z,\ z >$; i.e., that $< y,\ y > = < z,\ z >$; and since y and z are in the same line, this means that $y = \pm z$. But $y = -z$ is impossible, because that would mean $BD + DC = 0$, i.e., $C = B$, whereas our assumption that A, B, and C are non-collinear implies that these are three distinct points. So the condition for AB and AC to be of like character reduces to $y = z$, i.e., $BD = DC$, which is to say: D is the midpoint of BC; so (1) is proved.—Proceeding to (2)(i), we are now to assume that $z = y$ and, further, that AB and AC are space-like, which is to say that $< x - y,\ x - y >$ and $< x + y,\ x + y >$ (which of course are both equal to $< x,\ x > + < y,\ y >$) have the same sign as $< y,\ y >$ (this in view of the fact that y is space-like). What has to be proved is that $y - x$, $x + z$ (i.e., $x + y$), and $y + z$ (i.e., $2y$) are of the same class. It suffices to show that each of the first two is of the same class with the third; which is to say, that the inner products $< y - x,\ 2y >$ and $< y + x,\ 2y >$ have each the same sign as that of the inner product of a space-like vector with itself. But—again, since x and y are orthogonal—these are both equal to $2 < y,\ y >$ and, y being space-like, the point is established. The claim about the lengths of BA and of AC—that is, of $y - x$ and of $x + y$—follows almost immediately from the expression for both of them, $< y,\ y > + < x,\ x >$: since the terms of this sum are of opposite sign and the sum, by hypothesis, has the sign of $< y,\ y >$, it is smaller in absolute value than $< y,\ y >$; so the (common) length of $y - x$ and of $y + x$ is less than that of y—i.e., than half the length of BC.—Finally, as for (2)(ii), we now have to suppose that $< x,\ x > + < y,\ y >$ has the same sign as $< x,\ x >$, and to show that $x - y$, x, and $x + y$ are of the same class. Analogously to the case of (i), it suffices to show that the inner product with x of each of the other two has the same sign as that of x with itself; but this is even more obvious than in the other case: each of these inner products is *equal* to $< x,\ x >$. The proof of (b) is complete.

(c) In an isosceles triangle ABC with the two vectorial sides AB, AC space-like, equal in length, and of the same class, the "base" BC is time-like.

—This is most easily seen by considering the sum $AB + AC$, which by (a) is space-like. Its inner product with BC—since the latter is $AC - AB$ [for $AB +$

"Definability," "Conventionality," and Simultaneity in Einstein–Minkowski Space-Time   417

$BC = AC]$—is $< AB, AC > - < AB, AB > + < AC, AC > - < AC, AB >$.
The first and last terms of this sum cancel, by the commutativity of the inner product, and so do the middle two terms, since AB and AC are like and of equal length; so the whole is zero—i.e., BC is orthogonal to the space-like vector $AB + AC$.

(3) Finally, some facts about the strict automorphisms of a Minkowski plane (or the connected component of the identity in its Poincaré group):

(a) There are two kinds of automorphisms besides the identity: the translations, which have no fixed points and which leave invariant one complete family of parallel lines; and the "Minkowski rotations" ("Lorentz transformations," "boosts"), each of which has a unique fixed-point and leaves invariant only the two null-lines through the fixed-point. If p and p' are any two points, there is a unique translation that maps p to p'. If p and p' are any two points, and if l and l' are any two non-null lines of the same character through p and p' respectively, there is a unique strict automorphism that takes p to p' and l to l'.[19] For there is a translation that takes p to p', and this may be followed by a "boost" that leaves p' fixed and takes the image of l under the translation to l'. (That the resulting automorphism is unique follows from the fact that the identity is the only automorphism that leaves a point fixed and a non-null line through that point invariant; this—which is not hard to prove—I here simply take for granted.)

(b) If one considers all the strict automorphisms having a given fixed-point (here including the identity), they constitute a one-parameter subgroup of the Poincaré group. Unlike the case of Euclidean rotations, a one-parameter group of "Minkowski rotations" is non-compact—as the parameter varies from $-\infty$ to $+\infty$, the mapping from parameter values to group elements is one-to-one. In spite of this fact, one can—as in the Euclidean case—make a "natural" choice of parameter, which in the Euclidean case is the *angle* of rotation (the "natural" choice—there is not a unique one!—may be the radian-measure, or the measure by fractions of a full rotation).[20] In any event, the correspondence of the parameters to the rotations they parametrize is of such a kind that multiplying the parameter by, say, a positive integer n corresponds to "composing a rotation with itself" $n - 1$ times (the "minus one" comes from the fact that, e.g., "composing an operation with itself" *once* means "performing that operation" *twice*). Multiplication

---

[19] Although, once again, it is of no importance for us, let it be remarked that this is not the case if L and L' are both null: in this case, these lines must also be, let me say, "similarly inclined" (either they must both go from past and left to future and right, or both from past and right to future and left, for there to be any automorphism that takes one to the other; and then there will be infinitely many automorphisms that take p to p' and L to L').

[20] It is striking that in spite of the fact that in the Minkowski case there is no such thing as a "full rotation," there is nevertheless a "natural" analogue to the angle; but this is just one more manifestation of the marvelous interconnection of the trigonometric functions and the exponential function. (I trust the reader will forgive this gratuitous advertisement of the splendors of [even the rather elementary part of] mathematics.)

of the parameter by -1 amounts to taking the *inverse* of the given rotation (from which follows the interpretation of multiplying by any negative integer). Analogously—but this actually gives us something new—multiplying the parameter corresponding to a given rotation ϕ by the *reciprocal* of an integer n, we obtain a rotation which, "performed n times," results in the original rotation ϕ. Since the nth iterate is indicated by n as an exponent, the procedure just described yields what one might call the "nth root" of ϕ; but I shall call this instead—bearing in mind the parameter (the "quasi-angle") of the rotation—the result of *dividing* the rotation by n; and by an obvious extension, we obtain the interpretation of multiplication by any rational number.—Multiplication by *irrational* real numbers is an *entirely* new generalization of the simple idea of "composition of an operator with itself." (As will be seen below, what we shall have to consider is primarily the case $n = 2$ or a power of 2, and then multiplication by any rational number whose denominator is a power of 2; and I shall speak of "halving" the rotation, or of its "successive halving," etc., rather than of "taking square roots," or "taking to a rational power," etc.)

(c) The analogous situation for a translation is simpler: a translation (different from the identity) is represented by a (non-zero) vector, and when translations are considered separately, their composition is represented as *addition* of vectors (but when they are considered together with the other elements of the Poincaré group their composition with any group-element—translations themselves included—is represented multiplicatively). So, treating a translation and its compositions with *itself* (and iterations thereof), these amount to multiplications by a positive integer; the extension to positive and negative rational (or, for that matter, irrational) numbers remains in the domain familiar from the ordinary treatment of vectors, and the structure of the one-parameter group "generated by" a given non-zero translation is obvious. (Let it be noted that every non-zero multiple of a Minkowski rotation is a rotation and every non-zero multiple of a translation is a translation.)

In treating the higher-dimensional case, we excluded from the start (via condition $(S2)(a)$) the possibility that a simultaneity equivalence-class contains two distinct points with time-like separation, but we did not exclude in advance the possibility that such a class contains two distinct points with null separation—this emerged as a *consequence* of our hypotheses. In the two-dimensional case, if we wish to avoid this possibility in the end, we have to strengthen the assumptions (this will emerge clearly when the analysis is complete). Therefore, for the following discussion, I wish to strengthen condition $(S2)$, replacing clause (a) by the following: (a$'$)(i) *every observer-line O' meets Σ*; (ii) *any two distinct points of Σ have space-like separation*. The remaining hypotheses of Theorem 2 remain unaltered—except, of course, that we are now to consider a Minkowski space-time of two dimensions.

Let Σ again be an equivalence-class, containing the point p, of the "simultaneity" relation S_O, where p belongs to the axis O. It will be useful in the present case to begin by proving that there is no other axis of Σ at p. Suppose there were an axis O', distinct from O, containing p. Let q be a point of Σ with space-like separation

"Definability," "Conventionality," and Simultaneity in Einstein–Minkowski Space-Time   419

from p (that there are such points follows immediately from condition $(S2)(a')(i)$ and (ii), and from the fact that there are observer-lines that do not pass through p). Consider, now, the Minkowski rotation about p that takes O to O'. The image q' of q under this rotation belongs to Σ: for q, as a point of Σ, satisfies $S_O(p,q)$; therefore, by the invariance condition $(S1)$, q' satisfies $S_{O'}(p,q)$—and since O' is (assumed to be) an axis of Σ, q' belongs to Σ. On the other hand, the rotation about p that took q to q' took the vector pq to pq'; and since a rotation preserves the character, the length, and the class of a vector, pq' is space-like, equal in length to pq, and of the same character as pq. Therefore, by (2)(c) in the foregoing discussion, *if* there is at p an axis O' distinct from O—which implies that q' is distinct from q—there is a time-like line qq' containing two distinct points of Σ; and since this violates condition $(S2)(a')(ii)$, there cannot be an axis at p distinct from O.

Next we shall see that any strict automorphism that maps Σ *into* itself must (a) take the set of *axes* of Σ into itself, and must (b) take both the set of points, and the set of axes, of Σ, one-to-one *onto* themselves. For let ϕ be such an automorphism, let p be a point of Σ, let O be the axis of Σ at p, let p' be $\phi(p)$ and let O' be the image $\phi(O)$ of O under ϕ. We must first show that O' is an axis of Σ; so let q be any point such that $S_{O'}(p',q)$ holds. Since ϕ^{-1}, the inverse of ϕ, is (of course) also a strict automorphism, we have by invariance that $S_O(p,\phi^{-1}(q))$ holds; therefore $\phi^{-1}(q)$ belongs to Σ, so since ϕ maps Σ into itself $\phi(\phi^{-1}(q))$—which is to say, q—belongs to Σ; and this means that O' is indeed an axis of Σ. Now let q be any point of Σ, and let ϕ, p, O, and O' be as in the discussion of clause (a). We have just seen that O' is an axis of Σ, so $S_{O'}(p',q)$ holds (as was *assumed* of q in that discussion); therefore it follows again (or still!) that $\phi^{-1}(q)$ belongs to Σ, and accordingly that q is the image under ϕ of a point of Σ, which shows that ϕ maps Σ *onto* itself. Further, ϕ (as we already know) takes axes of Σ to axes of Σ; therefore it takes the axis of Σ at $\phi^{-1}(q)$ to that at q; and since q was an arbitrary point of Σ—and so the axis of Σ at q is an arbitrary axis of Σ—every axis of Σ is the image under ϕ of some axis of Σ, so the mapping by ϕ of the set of all such axes is *onto* that same set. Finally, as an automorphism of the space-time, ϕ is automatically one-to-one on *any* set of space-time points; and although an automorphism is *not* necessarily one-to-one on the set of time-like lines ("observer-lines"), an automorphism of Σ is one-to-one on its axes because the axes correspond one-to-one with their points of intersection with Σ, and the automorphism is one-to-one on that set of points. Thus clause (b) too has been fully demonstrated.

With all these perhaps tedious but comparatively trivial points now established, the main conclusion of the present an analysis could be rather quickly reached, on the basis of the discussion of one-parameter groups under (3)(b) and (c) above. But since just that part of the preliminary discussion contented itself with a vague indication of the proofs of its claims—by means, namely, in (3)(b), of the notion of "quasi-angles" (which is to say, the "hyperbolic trigonometry" of the Minkowskian plane), it seems preferable to base the results instead on more elementary geometric constructions.

Let, then, Σ still be an equivalence-class of a "simultaneity-relation," let p and p' be points of Σ, let O and O' be its axes at p and p' respectively, and let ϕ be the strict automorphism that takes p to p' and O to O'. Then ϕ maps Σ into itself (and therefore, by the preceding result, onto itself), because if q is in Σ if $q' = \phi(q)$, we have $S_O(p,q)$; therefore $S_{O'}(p',q')$; therefore q' is in Σ. We mean now to "halve" ϕ. To this end, consider the perpendicular bisector of the space-like line-segment pp'. This is a time-like line (cf. n. 21 above), and therefore contains a (unique) point p'' of Σ; let the axis of Σ at p'' be O'', and let the strict automorphism that takes p to p'' and O to O'' be ψ.[21] I maintain that ψ "*halves*" ϕ.—Proof: First, if $q = \psi p'')$ then q is a point of Σ, and $p''q$, as the image under ψ of pp'', is of the same character and length as the latter; but therefore, by (2)(b) of the preliminary discussion, of the same character and length as $p''p'$ as well. From this in turn, by (2)(c),[22] it follows that unless $q = p'$, qp' will be a time-like line containing two distinct points of Σ. Since this last is impossible, we must have $q = p'$; but then also the image under ψ of O'' must be O'; so since ψ takes p to p'' and p'' to p', and takes O to O'' and O'' to O', its "double" ψ^2 takes p to p' and O to O'; but these are the defining characteristics of ϕ, so the point is established.

Now observe that if we allow ϕ to "act iteratively on p," a countable set of points will be generated—all on Σ, equally spaced—proceeding from p on the one side; that ϕ^{-1} will similarly generate such a set proceeding from p on the opposite side; and that allowing ψ and its inverse to act similarly, we shall obtain another such pair of sets, which include all the points of the first pair, with new points interpolated between every two adjacent points of the first sets. The (equal) distances between successive points of the "refined" system will be *no more than half* that between successive points of the first system: namely, if three successive points such as p, p'', p' above form a triangle, the distances $|pp''|$ and $|p''p'|$ are *less than half* $|pp'|$ (see (2)(b) of the preliminary discussion), whereas if these points are on a single straight line then p'' is the midpoint of pp', so the former distances are *exactly* half the latter one.

We next conceive this process of "bisection" of the automorphism and the generation of new points on Σ to be iterated without bound. The result is a system of points "densely" distributed on Σ, in the sense that any one point has, on each "spatial side" of itself, others whose (space-like) distances from it are arbitrarily small.[23] We must now consider the geometric nature of the locus of these points.

---

[21] Note, by the way, that in light of what we now know we could characterize ψ equivalently as the strict automorphism that takes p to p'' *and maps Σ into* (or: *onto*) *itself.* By the same token, ϕ can be characterized as the strict automorphism that takes p to p' and maps Σ into (or: onto) itself.

[22] Here the reader may (should!) have a sense of *déjà vu*.

[23] Caution!—We cannot immediately conclude that these points are "dense *in the topology induced on Σ by that of the Minkowski plane*. For the space-like distance is not a metric in the standard sense of the theory of metric spaces—this because the triangle inequality is, as we have seen, "*the wrong one.*" Indeed, it is quite possible for a sequence of points to be (as it were) a "Cauchy sequence in the space-like distance," but not to converge; or for the distances of those points from a given point p to converge to zero, but the points *not* to converge to p. (This last is especially easy to see: let the "Cauchy sequence" converge to a point q, distinct from p, with *null separation* from p; then the "distances" from p will go to zero, but the sequence will not converge to p.) We shall have to (and we shall be able to) circumvent this difficulty eventually.

"Definability," "Conventionality," and Simultaneity in Einstein–Minkowski Space-Time   421

With Σ, p, p', p'', O, O', O'', ϕ, and ψ as above, suppose first that p, p'', and p' are collinear—so that p'' is the midpoint of pp'. Then the vectors pp'' and $p''p'$ are *the same*; so ϕ, mapping p to p'' and p'' to p', and therefore taking the vector pp'' to $p''p'$, leaves that vector *fixed*.[24] But a strict automorphism that leaves a non-null vector fixed also (a) leaves fixed all its multiples by scalars, and (b) leaves fixed any vector orthogonal to the given one (since *strict* automorphisms preserve the relation of orthogonality and preserve the length, character, *and class* of any vector).[25] But given a non-null vector, every vector can be represented as a sum of a multiple of that vector and a vector orthogonal to it; and automorphisms preserve sums; so we conclude that a strict automorphism that leaves a non-nulll vector fixed leaves every vector fixed. However, it is easy to see that an automorphism that leaves every vector fixed is a translation. So—in the case in which p, p', and p'' are collinear— the automorphism ψ is a translation—and, in consequence, all its powers, and also its inverse and all the powers of its inverse are translations.

Before we draw the (almost obvious) conclusion from this about the geometric character of the set Σ, it will prove best to consider the other case—that, namely, in which the points in question are *not* collinear, so that the segments—and the vectors (we have been using the same notation for both)—pp', pp'', and $p''p'$ are distinct, and the segments are the (space-like) sides of a (n isosceles) triangle. Consider, then, the perpendicular bisectors of the equal sides pp'' and $p''p'$. These meet (since the lines pp'' and $p''p'$ are not parallel, lines orthogonal to them are also not parallel) in a point p_0 that is equidistant from the end-points of both—i.e., from p, p'', and p'—and that therefore lies on the perpendicular bisector of pp' (see (2)(b) of the preliminary discussion). Since that perpendicular bisector is time-like, the vector p_0p'' is time-like; I claim that the vectors p_0p and p_0p' must then also be time-like, and of the same class as p_0p''. Indeed, we know already, by (2)(b) of the preliminary discussion applied to the triangle whose vertices are p_0, p, and p'', that p_0p and p_0p'' are of the same character, since p_0 lies on the perpendicular bisector of the space-like pp''; so p_0p must be time-like, because p_0p'' is so; and analogously for p_0p'. But then we know —again by (2)(b), now applied to the triangle whose vertices are p_0, p, *and* p', and in which p'' is the midpoint of pp'—that p_0p, p_0p', and p_0p'' are indeed of the same class.

Now, the strict automorphism ψ takes p to p'' and p'' to p'. It therefore takes the perpendicular bisector of pp'' to that of $p''p'$. Let r be the temporally oriented

---

[24] Lest there be confusion: an automorphism of a Minkowski space is to be regarded as acting *both* on the space of points *and* on the associated vector space (thus, in a more precise sense, as a *pair* of maps [which we nonetheless designate by the same symbol])—the connection between the two actions (or the two maps) being that if the *points* A and B are taken to A' and B' respectively, then the *vector* AB is taken to $A'B'$.

[25] Two points of clarification: (1) strictness is required only for the preservation of "class": without that, vectors orthogonal to a fixed one need not themselves be fixed, they may be "reversed"—i.e., multiplied by -1; (2) "non-null" is important here because a vector orthogonal to a null-vector in a Minkowski plane is a multiple of that null-vector: it is not true that every vector is a linear combination of the first vector and the second. (It *does* remain true for a non-zero null-vector— in two dimensions—that a strict automorphism that leaves one fixed leaves all vectors fixed; but another argument would be required to prove this, and we have no need of the fact.)

radius $[p_0 p]$; then the point p_0 may be characterized as the unique point q on the perpendicular bisector of pp'' for which $[qp] = r$; it may also be characterized as the unique point q on that perpendicular bisector for which $[qp''] = r$. And by the same token, p_0 may be characterized as the unique point q' on the perpendicular bisector of $p''p'$ for which $[q'p''] = r$. But ψ takes any point q on the perpendicular bisector of pp'' for which $[qp] = r$ to a point q' on the perpendicular bisector of $p''p'$ for which $[q'p''] = r$; that is, ψ takes p_0 to p_0 : p_0 *is a fixed-point of* ψ (and, as we know, there cannot be more than one such: p_0 is *the* fixed-point of ψ (which, incidentally, by this very argument *possesses* a fixed point: i.e., ψ *is a Minkowski rotation*.

This conclusion leads to a far-reaching consequence for our *other* case—that in which p, p', and p'' are collinear. We saw that in that case ψ is a translation. We can now infer that the result of *halving* ψ is *again* a translation. For we saw, in our previous analysis, that if, starting from p, p', and the strict automorphism ϕ that takes p to p' and maps Σ into itself (cf. n. 22 above), we construct the point p'' and the automorphism ψ that halves ϕ, then if p, p', p'' are collinear, ψ is a translation, and has no fixed-point; we have now seen that if those points are *not* collinear, ψ is a rotation, and *does* have a fixed-point. Applying this to the halving of ψ (with p'' playing the role that p' did previously, and with a new "third point" for p''), we see that unless the new triad of points is collinear, the result of halving ψ will be a rotation—a transformation that has a fixed-point. But this is impossible; for if a transformation has a fixed-point, so does the result of "doubling" it. Therefore, as claimed, the result of halving ψ will in that first case be again a translation; *and so on ad infinitum.* And it follows immediately that not only the "backwards and forwards *sequences* of points" that we constructed from ϕ and refined using ψ, but all the points of the subsequently constructed "dense" system in Σ lie on *a single straight line.*

The conclusion for the second case—in which the first three points of the construction, p, p', p are *non*-collinear—is obviously analogous: If ψ has a fixed-point p_0, all the subsequent results of halving must likewise have, not only a fixed-point, but *the same* fixed-point p_0—the result of doubling a rotation has the same fixed-point—or "center"—as the original rotation, and therefore the result of halving a rotation must have the same fixed point as the original rotation. It follows that the points of the "dense" array in Σ in this case *all lie on a "Minkowski semicircle" having the temporally oriented radius* $[p_0 p]$.

Two matters remain to be treated. First, we are obviously led to think that Σ itself just *is* the straight line or "semicircle" concerned; but this has to be proved. And then, we have so far determined nothing about the *axes* of Σ: in the higher-dimensional case, the fact that these axes (in the "hemispherical" case) have all a single point of intersection was crucial to our whole argument; but in the two-dimensional case, we have made very limited use of arguments involving the axes, and have so far draw *no conclusion whatever* about their geometrical configuration.

In order to deal with the first problem, there is one obstruction we have to remove. We have indeed seen that the range of points that lie on, in the one case a straight line, in the other a circle, extends "infinitely" in both directions,

"Definability," "Conventionality," and Simultaneity in Einstein–Minkowski Space-Time   423

and has "everywhere" points as close together as one wants; but "infinitely" and "everywhere" here are in an important respect misleading: we have shown that there are *infinitely many* points, in each direction, in our array; we have *not* shown—now taking advantage of the possibility of using the geometric loci, the straight line or the "semicircle," as a sort of standard—that our range of points extends from the initial point p, in both directions, *past any given point of the line or "semicircle" that we care to name.*

In the case of the straight line, this is trivial to deal with: at the very first stage, when we iterate the translation ϕ or ψ, we *do* obtain points arbitrarily far from p along the line in either direction; and of course this continues to be true at every stage of the subdivision. In the case of the "semicircle," the situation is quite analogous: the iteration of a rotation leads to points, in both directions from p along the curve, that extend (in point of the natural ordering of points on an "open" curve) *past* any given point. To prove this is not hard, but going into the details of a proof would not afford any new insight into the matter—any reader who is unfamiliar with the fact stated and who cares to have a proof should be able to find one—so I shall now take this for granted. More precisely, I shall make use of the fact that if one is given any positive real number a, and any point q of the straight line or "semicircle," we can find a sequence of points of our array, starting from p and continuing past q, such that the space-like separation between any two successive points of the sequence is less than a. Then, from this sequence, one can select *two* points, say q' and q'', *one on each side of q*, and with space-like separation less than a.

If this *is* granted, the question of the full geometric locus can be settled at once. Let q be any point on our line or "semicircle" that contains our "generated" array (q is not assumed to belong to the array); it is to be shown that q belongs to Σ. To this end, let l be any time-like line through q. By condition $(S2)(a')(i)$, l contains a point of Σ. By the fact stated just above, there are points q', q'' of our constructed point-array—points belonging both to Σ and to the line or "semicircle"—*on both sides of* the point of l in Σ and "arbitrarily close" to one another. Both q' and q'' have space-like separation from the point of l in Σ (since *all* points of Σ have space-like separation, by $(2)(a')(ii)$). But it is obvious that a point x that has space-like separation from points on the line or "semicircle" lying arbitrarily close to one another and on both spatial sides of x can only be a point that itself lies on the line or "semicircle". Therefore the point of l in Σ must be the point q in which l meets that geometric locus—as was to be shown.

Now that we know that every point of our line or "semicircle" belongs to Σ, we can easily show that these are the *only* points of Σ—i.e., that Σ *is* the line or "semicircle": for every point q of Σ belongs to some time-like world-line; and every time-like world-line meets the line or "semicircle" in some point q'; that point of intersection q' belongs to Σ (as we have just seen); but l has only *one* point in common with Σ. So the point q of Σ through which L passes is the same as the point q' in which l meets the line or "semicircle": the analysis of the geometrical nature of Σ is complete.

As to the axes of Σ—and this is the only point on which we shall find a difference *in the end result* from the higher dimensions—one can choose the axis at a given point q of Σ *arbitrarily*; that is, it can be any time-like line l through p. For the choice of such a line as axis can be specified in the following "objectively geometrical" way: Choose—"in advance" and once for all—a spatially oriented radius s, of absolute length less than unity. (It is obvious how "spatially oriented radius" is to be defined, except for the new proviso that here we allow the vector 0 to count as "space-like" and "oriented"; it therefore constitutes a class by itself—the spatially oriented radius 0. By the "absolute length" of a spatially or temporally oriented radius r we of course mean the absolute value of the length of any vector in the class r). Then let u be the time-like, "future-pointing," unit vector normal to the surface Σ at the point p; let v be the unique vector orthogonal to u such that $[v] = s$ (of course, if $s = 0$, v is the zero vector); and let l be the line through p in the direction of the vector $u + v$ (any time-like line through p—and only such lines—can be described in this way). Now, it can be seen without great difficulty (I forgo details here) that if we start with one particular locus Σ of the kind already determined; if we then, having chosen the spatially oriented radius s once for all, assign to every point p of Σ the corresponding line l as "axis of simultaneity" for Σ at p; and if, finally, we apply to this configuration all possible strict automorphisms of the two-dimensional Minkowski space-time; then the resulting configuration determines an assignment to every time-like line l of a relation S_l satisfying all our conditions.[26]

If one equivalence-class of our system is a straight (necessarily space-like) world-line, then they all are; all those that belong to some one axis are parallel; and all the axes of any one equivalence-class are parallel. The axes of different systems of equivalence-classes—that is, the observer-lines O with different associated relations S_O—are "inclined at the same angle" to the lines normal to their equivalence-classes: this can be taken to mean simply that there is a strict automorphism taking the one normal line to the other, and at the same time taking the one axis to the other.

If one of the equivalence-classes is a "Minkowski semicircle," then they all are—and, moreover, they all have the same temporally oriented radius r. No two equivalence-classes have the same set of axes: as in the higher-dimensional case, each observer-line has its own family of equivalence-classes. And just as in the case of the straight-line equivalence-classes, each axis of an equivalence-class is "inclined at a given angle"—the same for all axes and for all equivalence-classes—to the line normal to the equivalence-class at its point of intersection with the axis.

It is worth noting that whereas the possibility of "Minkowski-semicircular" equivalence-classes is tied to the fact that we are not requiring scale-invariance—so that we are free to choose a temporally oriented radius $r \neq 0$—the possibility, in the

---

[26] More precisely, for any observer-line l and any point p, there will be one and only one set Σ' that is an image of (the original) Σ under a strict automorphism such that p belongs to Σ', l is an axis of Σ' at its point of intersection with Σ', and for any points q, $S_l(p,q)$ holds if and only if q belongs to Σ'.—There are of course many details to check to justify the statement that this S satisfies all our conditions; the only matter that may seem doubtful is whether, for any given time-like line l and associated equivalence-class Σ with l oblique to the normal to Σ' at their point of intersection, the system of all translates of Σ' in the direction of l constitutes a foliation of the space—i.e., whether every point belongs to one and only one such translate. This can be made transparently so, in the usual Euclidean picture of the Minkowskian plane, through a judicious transformation of coordinates.

"Definability," "Conventionality," and Simultaneity in Einstein–Minkowski Space-Time   425

case of straight-line equivalence-classes, of "inclined axes," does *not* depend upon that weakening of the invariance requirement (although it *does* depend on giving up invariance under spatial or temporal reflections); for the construction of the axis described above may be modified as follows: instead of taking for u the *unit* future-pointing normal vector, take for it *any* future-pointing normal vector; and then take (as a preliminary step) a vector u' orthogonal to u, having as inner product with itself the negative of the inner product of u with itself and pointing "right," and then, having chosen (once for all) a non-zero (signed) real number s, let $v = su'$.

Either of the strengthened conditions $(S2)(b')$ or $(S2)(b'')$ will again restrict the equivalence-classes to the "straight" ones only—i.e., to the same classes given by the Einstein–Minkowski relative simultaneity relations. *But the relations of simultaneity relative to an observer are **not** necessarily those of Einstein–Minkowski*: for the *axes* retain their degree of arbitrariness: they can be "inclined at any given angle" (given once for all, that is) to the lines normal to the equivalence-classes.

One further fact seems worth pointing out, regarding the contrast with the higher-dimensional case: There, when the simultaneity set was a "hemisphere," all its axes of simultaneity (which of course were just the diameters of the corresponding "sphere") had a common point of intersection: the center of the "sphere." But in two dimensions, for an s different from zero, the axes of simultaneity of a "semicircle" do *not* all meet in a single point.—What particular geometrical configuration the set of axes form for r different from zero is a question that may here be left for the entertainment of Platonic philosophers.[27]

Summing up, our results for the case of a two-dimensional Minkowski space-time are as follows:

Theorem 3. *In a two-dimensional Minkowski space let there be given an assignment, to each observer-line O, of an equivalence relation S_O on the whole space-time, in such a way that:*

(S1) any strict automorphism that maps O to O' transforms S_O to $S_{O'}$;
(S2) if Σ is an equivalence-class of S_O then:

> (a') (i) *every observer-line O' meets Σ;*
>   (ii) *any two distinct points of Σ have space-like separation, and*
> (b) *for every point p' of Σ there is an observer-line O' containing p' such that Σ is an equivalence-class of $S_{O'}$ (as well as of S_O);*

—then (1) the system of equivalence-classes for one observer-line is either (a) a family of parallel space-like straight lines, or (b) a family of "Minkowski semicircles" of given temporally oriented radius r; (2) these alternatives hold "uniformly" for all observer-lines—that is, either (a) holds for all, or (b) holds for all—and then with the same time-oriented radius r for all the observers; and (3) in either case, there is a fixed spatially oriented radius s of absolute length less than unity (it may be zero), such that for any point p of any equivalence-class Σ, if u is the unit normal vector to Σ at p, and if v is a spatially oriented vector orthogonal to u such that the

---

[27] That is, lovers of geometry. (I have stated the result in Supplementary Note 4.)

426   H. Stein

oriented radius [v] is s, the axis of Σ *at p is the line through p in the direction of the vector* $u + v$.—*If condition* (S2)(b) *is replaced by either* $(S2)(b')$ *or* $(S2)(b'')$—(for which see the discussion following Theorem 2 above)—*then the alternative* (b) *is excluded: all the equivalence-classes are straight lines.*

Addendum*: In the case of straight-line equivalence classes, invariance holds under the wider class of all "proper" automorphisms; for this wider class, modify clause* (3) *as follows:* $(3')$ *there is a fixed non-zero (signed) real number s such that for any point p of any equivalence-class* Σ, *if u is any future-pointing normal vector and* u' *is a right-pointing vector orthogonal to u, the absolute value of whose inner product with itself is equal to the absolute vale of the inner product of u with itself, the axis of* Σ *at p is the line through p in the direction of the vector* $u + v$.[28]

2 Remarks on the Controversy

It is certainly no new observation that philosophical controversy is often vitiated by the fact that the disputants argue at cross-purposes; in particular, that they use the same words with different meanings. Locke was far from the first, Wittgenstein, Carnap, and Quine far from the last to see in a lack of clarity in linguistic use a prime source of the apparent intransigence of philosophical problems.[29] Nor, considering that the pointing out of this (in principle after all fairly obvious) fact has not so far notably lessened the evil, is it at all likely—to compare a minor writer with major ones—that I shall be the last one to do so either. Nonetheless, as I have indicated in the opening paragraph, I have *some* hope of helping a *little* to clarify this particular issue.

The most crucial notion that cries out here for clarification is that of the "conventional." Poincaré, whose emphasis upon this notion is the beginning of its latter-day philosophical prominence, argued that the great organizing principles of geometry and mechanics—and in part, of pure mathematics—are "conventions, or *definitions in disguise*." Now, it is clear that definitions *are* "conventions": they are stipulations—or agreements, since one assumes that the stipulation will be accepted at least within a given discussion by all the discussants—concerning how a word or phrase is to be used. But this does not help us, because Adolf Grünbaum has always been quite explicit that his claims concerning conventionality are *not* about the "trivial semantic conception" of conventionality (and of course, if they were, the claims themselves would be trivial, and there would be no need for discussion).

Now, I do not know how to characterize "conventional" and its contrary in a general sense that particularizes to the one that Grünbaum has in mind in this context, in a clear way.[30] It does not follow that it is impossible to give such a characterization;

---

[28] Of course this slightly modified construction could have been used in Theorem 3 itself.

[29] Or: the intransigence of apparent philosophical problems.

[30] The notion of a distinction between a "merely conventional" definition and one that is not so— i.e., that *is* "conventional" *only* in the "trivial semantical sense"—seems closely related to the distinction, in traditional philosophy, between (merely) "nominal," and "real definitions"—ones that

"Definability," "Conventionality," and Simultaneity in Einstein–Minkowski Space-Time   427

but since I am unable to give one, I shall content myself—a little later—with suggesting what I think is a helpful explanation of what, *just in this particular context*, would count for Grünbaum (and, I suppose, for his supporters) as *not* conventional (in the non-trivial sense of "convention"). I have some hope that my explanation will be acceptable to Grünbaum, because (I say a little shamefacedly) I shall there merely copy, or paraphrase, part of what I have read in his recent paper. If I deserve any credit for this, it is only that of having at last realized that Grünbaum's claim has all along been misunderstood by those—among whom I count myself—who have objected that the Einstein–Minkowski concept of simultaneity is *not* just a "convention."

And I continue to hold this latter position; but—I now hope—in a sense that does not conflict with Grünbaum's main contention, because I am using the word "convention" with a different meaning from his.

Before I offer my new understanding of what Grünbaum has contended for all along, I shall suggest a few corrections of detail to some of his adversarial remarks, and shall try to explain the sense in which I—and, I believe, Malament and others who have disagreed with Grünbaum—have understood the issue about simultaneity.

I pointed out in [3, p. 153] that the question of "conventionality" is a different one for the procedure of Einstein in 1905 from what it is for that of Minkowski in 1907–8: Einstein was *seeking a theory* that should satisfy certain requirements—a theory that did not yet exist; whereas Minkowski was seeking the *most cogent and instructive formulation* of a theory already in existence. We, of course, are not at all in the situation of Einstein;[31] but it seems worthwhile to discuss briefly how the issue looks from the perspective of that situation—all the more, in view of the fact that Grünbaum has cited some words of Einstein in support of the conventionality thesis.

It is indeed true, as Grünbaum remarks, that Einstein [7, p. 279][32] characterizes his conception of simultaneity as the result of a *Festsetzung*, or "stipulation" by means of a definition: a definition according to which, for an observer at rest in "a coordinate system in which the Newtonian mechanical equations are valid" (*ibid.*, p. 277), the time that light takes to get from A to B is the same as the time it takes to get from B to A. This is very clearly a convention, then—but it is "clearly" so *only* in the "trivial semantical sense": it is a definition. Should one conclude that it is a convention in a nontrivial sense? It seems to me difficult to draw this conclusion simply from Einstein's own words in this passage.

Grünbaum takes Michael Friedman to task for saying that the theory that resulted from Einstein's investigation "postulates" metrical relations that include the notion of "relative simultaneity" for distant events; Grünbaum's comment [8, p. 14] is:

---

in some sense define the "essence" of something; it is well known that many traditional philosophers have rejected such a notion entirely: this is the "nominalist" position.

[31] Not in that of Minkowski either, since Minkowski has performed his task; nonetheless our situation more nearly resembles Minkowski's, in that we are concerned with a critical discussion of the theory.

[32] The page reference is to the *Collected Papers*.

But, as we saw, Einstein stated emphatically that assertions of metrical simultaneity in the STR are *not* "hypotheses" which are "postulated" in Friedman's sense, ontologically on a par with, say, the postulate that light is the fastest causal chain. Why then does Friedman feel entitled to gloss over that important ontological difference by using the same term "postulate" for both?

But this ignores something else that Einstein says, *before* the passage about "stipulating" equal speeds of light in opposite directions: in the second paragraph of the introductory section of the paper, after speaking of empirical evidence that leads to the "conjecture"—*Vermutung*—that not only in mechanics but also in electrodynamics "no properties of the phenomena correspond to the concept of absolute rest, but that rather [—*vielmehr*—] for all coordinate systems for which the mechanical equations are valid the same electromagnetic and optical laws are also valid," Einstein has written (and I quote his words in German first, to make sure that no distortion is introduced by translation):

Wir wollen diese Vermutung (deren Inhalt im folgenden "Prinzip der Relativität" genannt werden wird) zur Voraussetzung erheben und außerdem die mit ihm nur scheinbar unverträgliche Voraussetzung einführen, daß sich das Licht im leeren Raume stets mit einer bestimmten, vom Bewegungszustände des emittierenden Körpers unabhängigen Geschwindigkeit V fortpflanze.

We intend to elevate this conjecture (whose content in the following will be called "Principle of Relativity") to a presupposition, and, besides, to introduce the presupposition—only in appearance incompatible with [the former one]—that light in empty space is always propagated with a determinate speed V, which is independent of the state of motion of the emitting body.

It would seem, then, that we have Einstein's authority after all for characterizing as a "postulate" (or "presupposition" or "hypothesis") the principle that the "speed of light is the same" in one direction as in the other. This of course does not decide the issue as to whether these postulates themselves should be regarded as "conventions" (as Poincaré *did* regard the axioms of geometry); it bears only on the particular appeal to Einstein's statements made by Grünbaum.

The situation, then, for Einstein's investigation was this: he *did* have reasons to want a theory that satisfies the two "presuppositions" he formulated. The urgent desirability of such a theory had been emphasized in 1900 by Poincaré [9]. When Poincaré himself *solved* this problem in 1905, he regarded the solution as a mere *tour de force* (and in subsequent writings, after the publication of his great paper [5] [see, e.g., Part 3 of *Science and Method*, which reproduces a review article of 1908], he never referred to his own work but only to the not quite satisfactory "new dynamics" of Lorentz). Whether or not he had read the 1900 report by Poincaré. Einstein was motivated by the same considerations as Poincaré; and if, both having found essentially the same theory, Einstein's *view* of it was radically different from Poincaré's, this rests to no small degree upon the fact that Einstein had subjected the concept of time to a much deeper criticism than had Poincaré, for whom the "transformed time-coordinate t''' was no more than a *mathematical trick* to make the theory work. (I believe that this is not something that Grünbaum will disagree with.)

So: the *problem* lay precisely in the "apparent contradiction" referred to by Einstein between his two "presuppositions." The *resolution* he found of this appar-

"Definability," "Conventionality," and Simultaneity in Einstein–Minkowski Space-Time   429

ent contradiction consisted precisely in his realization—and it evidently cost him considerable intellectual struggle (cf. [9])—that *the discrepancy in synchronization* between the time t and the transformed time t' in the Lorentz transformation—a discrepancy that had led Lorentz to call t' the "local time"—corresponds to a *conceptual gap* concerning the concept of "simultaneity" or "synchronization": that, as Einstein says [11, p. 61], in a passage also cited by Grünbaum [8, p. 14], "There is no such thing as simultaneity of distant events."[33]—This recognition opened the way to the introduction of a suitable definition to close that conceptual gap; and it is not at all surprising, therefore, that Einstein was concerned to emphasize to his readers that *a definition was here (a) needed and (b) (therefore) legitimate.*

In the light of all this—considering the fact that it was essential for the success of Einstein's project to find a systematization of spatiotemporal relations and measures that would satisfy two requirements: (1) that for investigators who use these measures, the laws of classical mechanics and of classical electrodynamics (including optics) hold (at least to high approximation) for the results of measurement, and (2) that this should be true for a system of teams of investigators, the investigators of each single team being mutually at rest, the investigators of different teams in arbitrary states of uniform motion relative to one another; the fact that the classical laws presuppose, for any single such team of observers, a standard of synchronization; and the fact that any standard of synchronization that meets these requirements *must* agree, in application, with the criterion proposed by Einstein's *Festsetzung*—it seems somewhat misleading to call the latter a "convention" in a deeper sense than the one applicable to all matters of linguistic usage.

Just one further turn regarding this aspect of the matter—i.e., Einstein's own procedure: Einstein could perfectly well have contented himself with the *Voraussetzungen* he formulates in his introductory section, and *instead* of "defining" simultaneity, have *deduced* from these two assumptions that light that travels back and forth (or *vice versa!*) between two "inertial" observers A and B at relative rest must take equal times both ways, and therefore can serve as a signal to synchronize clocks in precisely the way the "definition" prescribes.—I do not think this alternate expository mode changes anything essential: an upholder of the view of "conventionality" could just say that the *Voraussetzungen* themselves have to be counted as "conventions," rather as Poincaré did with respect to the axioms of geometry and mechanics. And I remind my readers (and myself!) that I do not here claim to "settle" the issue of conventionality—on the contrary, I have already said that for me the very notion (in general) of what is or is not a "convention" is distressingly unclear; I should really prefer to say the sort of things I have already said about the *role played* by

---

[33] Grünbaum urges the fact that, according to Einstein, this is one of the "insights of definitive character that physics owes to special relativity," as showing again that Einstein is on Grünbaum's side on the question of the "conventionality" of simultaneity. But what Einstein says in the passage cited, in the very next clause, is: "thus there is no unmediated distance action in the sense of Newtonian mechanics." The point—the *"definitive insight"*—is that there is no such thing in nature as simultaneity *schlechthin* of distant events: no *absolute*, but only (at most) *relative* simultaneity. At any rate, simultaneity "relative to" either an observer or an inertial system *is not mentioned by Einstein in this passage at all!*

Einstein's concept of simultaneity, and leave the word "convention" out of the discussion entirely.

Let us then proceed to the other point of view: that of the finished theory. The theorems of the first part of this paper—together, of course, with that of Malament—are for me the main text for this point of view; but they do require some comment.

First, then, I have already said that the generalization contained in Theorem 2 does not seem to me of much philosophical interest. The reason is that the "communities" of observers who share a simultaneity equivalence-class, in the "new" case—where the classes are "Minkowski hemispheres"—are, as noted, constantly changing; it is therefore impossible to refer any sort of stable measurements to them: the *purpose* for which Einstein's systems of inertial observers were introduced has been entirely lost.

But some remarks about that purpose also need to be made. These systems of inertial observers play, for the theory, the role of a kind of Platonic myth. In actual fact there are *no* such observers: first because the real world is *not* characterized by a Minkowski space-time, and second because "even if it were," it would be very hard to see how even *one* such system of inertial observers could be created.—What on earth can the meaning of "even if it were" be here?—Well, I should say (in the spirit still of "Platonic myth-making"), we might just envisage the possibility that a theory of gravitation (such as Poincaré and Minkowski independently did attempt to formulate) could be envisaged within the Einstein–Minkowski framework. If for a moment we consider that as a possibility, the first observation to make concerning our problem is that *we* certainly are *not* (would not be) inertial observers, since we live on a body that is not in a state of inertial motion. In order to produce an inertial environment, we should have to embark upon a program of space-travel expressly designed for that purpose: that is, to devise space vehicles whose navigational systems were designed to compensate exactly for gravitational forces.—I shall not continue with this fantasy; I am not good enough at science fiction. But I hope the point will be clear—how far from "reality" these envisaged inertial observers are. They are none the less useful, however, as vivid embodiments of the relationships expressed by the "congruence transformations" of Minkowskian geometry—i.e., of the Lorentz transformations (both homogeneous and inhomogeneous—in other words, the Poincaré group). But there are two corollaries of these elementary remarks: (1) that Einstein's "definition" of simultaneity, and analogous considerations, are best thought of, not as *quasi*-"operational" definitions, but as depicting" something like "thought experiments" to make vivid the situation in this theory; (2) that it is indeed the situation "in this theory" that is concerned—*not* that "in the real world": any insight provided into "the real world" comes only through the fact that the theory can claim to give "partial," or "approximate," information—or, perhaps, information of an "infinitesimal" kind—about the real world. To put it simply: any conclusions we are inclined to draw about such things as "conventionality" or the opposite should in principle refer, in the first instance, to how things stand, conceptually, *within the theory*.

With this understood, I have one remark to make that may be surprising: it is that unlike Theorem 2, I think there *is* a point of (mild) "philosophical interest" in Theorem 3—which deals with a space-time of two dimensions, hence a space

"Definability," "Conventionality," and Simultaneity in Einstein–Minkowski Space-Time   431

of only one dimension, which is extremely far from the "real world." This interest attaches to the easier and more familiar branch of the theorem—the case where the simultaneity equivalence-classes are straight lines. The point is that just here we find a "possible system of inertial observers" with a *deviant* concept of "relative simultaneity"; I hope it will be a little instructive to examine this case.

But just what *is* this deviant concept (or what *are these* deviant *concepts*)?— The answer is that they are very close to the ε-relations of Reichenbach, for whose viability Grünbaum has long contended; and they *coincide*—but under the drastic dimensional restriction—with a conception put forward by Allen Janis (cited by Grünbaum [7, pp. 9–10]). For if we choose a system of relative simultaneity relations in the way described in Theorem 3, the axes determined by the spatially oriented radius s (represented by a real number of absolute value less than 1, positive if "to the right," negative if "to the left"), then our "observers" are taking the ratio of "the speed of light to the right" to "the speed of light to the left" to be $(1-s)/(1+s)$; and this amounts, for a Reichenbachian observer, to choosing $\varepsilon = (1+s)/2$ for light sent, from his position, to the right, and $\varepsilon = (1-s)/2$ for light sent, from his position, to the left. (This differs from Reichenbach's principal example [12, p. 127] in that Reichenbach supposes one "central" observer who uses the same value of ε for all directions;[34] but he also gives an example [12, p. 162] in which ε depends on the direction, so our present situation does fall under the class of those he at least implicitly envisages.)

There are two reasons why this possibility *does* emerge in two dimensions but (from the point of view under discussion) does *not* in higher dimensions. One simple point that rules this out in higher dimensions is that such a choice of the relation violates the relativistic invariance principle. It would do so in two dimensions also, if (as in Malament's theorem) we required invariance under *reflections*. In higher dimensions it violates even the narrower invariance requirement, because any spatial direction can be transformed to any other by rotation (whereas when there is only one dimension of space, there is no room to turn around: we can distinguish left and right, and make the axes lean, or the speeds differ). The second reason is in a way more interesting: in higher dimensions, *there is no "intrinsic," or "objective," way to* INSTRUCT *an observer as to how to make the choice of a preferred direction.*[35]

This does not mean that such a choice could not be *made*; it only means that it could not be made *according to a "universal" rule*: a "team" of inertial (imaginary!—science-fictional!) observers, at rest with respect to one another, who wished to carry out systematic measurements and to determine (for instance) velocities and accelerations using a "Janis-simultaneity relation" would have to come to a special agreement with one another as to how to determine the direction of the

---

[34] This is not really clear from Reichenbach's text at this point; at least, it has not been clear to *me*: I had until recently always supposed that Reichenbach wanted the speed of light to be constant in a given direction—and this would obviously necessitate, for $\varepsilon \neq 1/2$, that ε be different in different directions.

[35] It might be argued that this "more interesting" point is really the same one as the first point; this bears on the question of the connection of "invariance with respect to" and "definable *from*"—on which there will be a little more below.

world-line that is to be their axis of simultaneity; and this agreement would have to use some special features of the "geography"—or, rather, the "cosmography" of their universe. This would in general not be an easy thing to do; its possibility would depend on the existence of recognizable, and *stable*, features of their cosmos to serve as (the analogue of) landmarks for determining and *redetermining* directions that "are the same" for all the observers and "remain the same" over time. In the alternative choice of a simultaneity relation described by Janis [8], the person who chooses this alternative notion does so "by specifying a set of three parameters," and thereby singling out a time-like direction that is inclined to the investigator's own. The three parameters required are the coordinates—in the projective space associated with space-time—of a time-like direction: three are needed because the projective space of the directions in a two-dimensional affine space is three-dimensional. *But* one can single out a direction in that way only if *coordinates have been laid down* for that space of directions. Since this is itself a task at least as complicated as that of determining a relation of simultaneity, to speak so cursorily of "specifying three parameters" partially masks the problem.—Note that the problem lies, not in the need to specify *the values of the parameters*, which, as real numbers, are available as "individual concepts" belonging to the logico-mathematical apparatus, but in the need to choose a *way of relating parameters to directions* (i.e., a "coordinatization" of the projective space). That is why the problem does not arise for a two-dimensional Minkowski space-time: the "space of directions" of such a space-time is one-dimensional, which means that there is an "intrinsic" association of directions with real numbers.

An analogy may help to make the main point clear. When *temperature* was first introduced into physics, and first *measured*, the quantity so named was not a single one at all—there were as many such "quantities" as there were types of thermometer, and the choice among them was *nothing but* "conventional." Indeed, for various investigators—or the same investigator in various experiments—who were concerned with temperature, whether the quantities they referred had simple relations to one another depended entirely upon stability in this respect; the quantities could "in principle" vary with the thermometric material, with the material in which (if the thermometric material was a gas or liquid) that material was contained, in the proportions of the containing vessel and of its cavity, etc. The situation became radically different after the development of the second law of thermodynamics: now a *theoretical definition* of temperature was available, that determined the quantity "temperature" precisely, leaving open only the choice of a unit for the temperature scale: the notion of *ratios of temperatures*—which, initially, would have seemed *least* likely to have any significance *at all* (since the zero-point of a temperature scale was at first entirely arbitrary)—had now received an "absolute" theoretical meaning.

It was perhaps—it *is* perhaps—still open to a disputant to argue that the so-called "absolute temperature" itself remains a matter of convention, chosen in the interest of "merely descriptive simplicity." I should not care to debate that point. But I *do* maintain, and think it important to recognize, that the difference, within the special theory of relativity, between the "simultaneity relative to a state of inertial motion"

"Definability," "Conventionality," and Simultaneity in Einstein–Minkowski Space-Time   433

of Einstein and Minkowski, and the simultaneity relations described by Janis or by Reichenbach, resembles in an important way the difference between the "material" conceptions of temperature as a quantity, and the "absolute" conception offered by developed thermodynamics.

Before coming specifically to Grünbaum's claims and the misunderstandings—on both sides—that I now believe to have muddied the issue, I want to mention one interesting point raised by Sarkar and Stachel. They speak of the possibility of "formulat[ing] the basic structure of the special theory of relativity without the use of any simultaneity convention" [4, p. 219]. This *is* certainly possible—there is more than one way to interpret their words; but in the strongest way of all, namely taking them to mean "without the use of any conception whatever of "distant simultaneity," it is surely possible. Indeed, the *general* theory of relativity, in its "most general case," altogether lacks any such notion as distant simultaneity; and this does not prevent the theory from being formulated. But: (1) A formulation that dispenses with any use notion of relative simultaneity must also dispense with any notion of relative velocity, and with the notion of acceleration in its usual form. Therefore, (2) such a formulation is not adapted to the comparison of the theory with Newtonian mechanics. On the other hand, (3) such a comparison is instructive; and since, as we know, it can be made, it must follow that (4) the formulation of special relativity that does not *use* a notion of relative simultaneity must nonetheless include the means of formulating such a notion whenever it is *desired* to do so. The reason a notion of distant simultaneity is not needed to formulate physical laws is that physical interactions, in this theory, are "infinitesimally near-by" interactions, governed entirely by partial differential equations. There *is* an "infinitesimal counterpart" of "simultaneity equivalence-class," relative, *not* to an "inertial observer," but to any state of (smooth) motion, inertial or not: it is the space-like hyperplane, in the tangent space to space-time, that is orthogonal to the tangent-line of the world-line of that motion. This infinitesimal (or "differential") notion is quite indispensable, both in the special and in the general theory. In the special theory, for the special case of "inertial motions that constitute, together, a state of relative rest," this "infinitesimal" notion is (uniquely) *integrable*; and *that* is the description of the Einstein–Minkowski relative simultaneity concept that, in my own view, presents the best case for its "true standing"[36] in the theory.

Turning, then (at last!) to Grünbaum's views as I now understand them, I have first to complement my earlier discussion of the notion "definable from" or "definable in terms of"; for this is one of the phrases in which it seems clear that people on each side of the debate have misunderstood their opponents' statements. For Grünbaum's part, he has certainly used that phrase in a sense very different from that of its customary use in logic and mathematics. This is not an intellectual crime—but it *is* a misfortune for all parties, when one of them uses a term in an unusual way without taking pains to explain this fact (of course, such an occurrence is not deliberate: it results from the fact that the party in question does not realize what the customary usage is). I was mystified to read that Grünbaum [7, p. 3] *rejected*

---

[36] Do I mean its "non-conventional" standing?—I have said that I should really prefer to express my opinions without using the word "conventional" at all.

Malament's condition, on a relation "definable in terms of the relation of causal connectibility," that it be invariant under all "causal automorphisms"—that is, one-to-one mappings of space-time to itself that *preserve* the relation κ of causal connectibility. I have explained above, in discussing the paper of Sarkar and Stachel, the grounds for this condition (although not long ago I should have thought this something too well known to require "explanation"). But *part* of the clarification of Grünbaum's use of the term appears when he attributes to Bas van Fraassen the remark that Malament defines the notion he is defending "in terms of κ *alone*" [8, p. 11, emphasis in the original]. I call this "part of" the clarification, because it shows that Grünbaum does not mean the same thing by "definable from..." as by "definable from... alone"; this still leaves us with the question, which for some time seemed to me hard to answer, what he *does* mean by the former expression. If, for example, one asks a geometer whether the concept of "the vertical direction" can be defined "from" the basic concepts of Euclidean geometry, the geometer would surely say no: "vertical" is, first of all, a concept of physics, not of geometry; and second, it is only defined for points near the earth's surface—or, in a more sophisticated view, for points at which there is a non-vanishing gravitational field. But it seems that in Grünbaum's usage the answer to that question must be yes: "vertical" means "in the direction of the gravitational field," and the notion of "direction" used here can be—ordinarily *is*—that based upon Euclidean geometry.—I may be wrong here in my interpretation of Grünbaum; if so, I am willing to be corrected. This of course still does not answer the question that I have said remains open; I shall try—again, under correction—to give at least a partial answer presently.

At any rate, Grünbaum has surely misread Malament when he writes of the latter as follows: "But before giving [[the]] proof, [[Malament]] declared: 'To be sure, there are other two-place relations [of relative simultaneity] which are definable from κ and O [i.e., relative simultaneity relations corresponding to non-standard synchrony, for example, some fixed $\varepsilon \neq 1/2$]. But all these are ruled out if minimal seeming innocuous conditions are imposed.'" The words in double brackets are substitutes by me for Grünbaum's words—substitutions made only to adapt the passage from its context in Grünbaum to the context here; the passages in single brackets are bracketed in Grünbaum's own text, and are interpretations offered by him of Malament's words. They are serious misinterpretations: (1) The other two-place relations Malament means are ones that are quite unsuited to serve as relative simultaneity relations—they are the relations that are ruled out by his conditions that a simultaneity relation S relative to an observer O be an equivalence relation; that it hold between some point on the world-line of O and some point not on that line; and that it not be the universal relation. (2) In particular, not only (as stated in (1)) are the excluded relations not relative simultaneity relations at all, but the Reichenbachian $\varepsilon \neq 1/2$ relations are *not* examples of relations "which are definable from κ and O." This strange misconception must have arisen from the fact that *after* discussing the matter of "definability from κ and O," Malament lists invariance as the first of his conditions on S: Grünbaum has failed to realize that Malament has done so for the sake of the exact mathematical formulation of his result, *not* because invariance is a special condition *added* to "definability from κ and O"; on the contrary, it is rela-

"Definability," "Conventionality," and Simultaneity in Einstein–Minkowski Space-Time   435

tions that satisfy this invariance condition, and *only* these, that Malament has called "definable from κ and O."

All this is pedantry; necessary, I think, but not edifying. Let us now consider what, in my present opinion, is the sound core of Grünbaum's view. In discussing this, I am going to have to dissent (still) from some of Grünbaum's particular *expressions* of that view; and to begin with, from this one, which initially puzzled me as much or more as did the one concerning definability: Grünbaum asks whether—and clearly means to deny that—"the facts of causal connectibility and non-connectibility *mandate* (dictate)" the standard (Einstein–Minkowski) relation of simultaneity relative to an inertial frame [8, p. 2; cf. pp. 3 and 5 for the explicit denial that this relation is "mandated" by those facts]. I was, and remain, still *more* puzzled by his claim that the standard relation of relative simultaneity, unlike the relation of causal connectibility, lacks a *fundamentum in re*: a "foundation in the thing," or "in nature" (pp. 12–13); and that "there is no fact to the matter" in ascriptions of this relation (pp. 1, 9, 12, 13)—that they lack "facticity"[37] (pp. 2, 3, 12). What is very strange here, in point of the "foundation in nature" especially, is this: we *know* that the whole metric structure of Minkowski space-time (without a distinguished spatio-temporal unit) is definable from the one basic relation of causal connectibility (the symmetric one if a time-orientation is not presupposed, the asymmetric one if such an orientation is presupposed). Therefore this structure—and in particular, the relation of orthogonality, which gives the Einstein–Minkowski relative-simultaneity relation—has a "foundation" in the relation κ of causal connectibility (again: symmetric or asymmetric). So if this relation has a "foundation in the thing," the standard relation of simultaneity relative to an inertial state has one also—assuming that the relation "A has a foundation in B" is transitive; which seems hard to deny. By the same token, it would seem that the ascription of the relation of relative simultaneity has "factual content." And as to being "mandated": a relation *"founded on" κ* would seem to derive whatever is meant by a "mandate" from that fact itself, as long as κ is regarded as "founded in things."

Of these puzzles, the one about having a foundation seems to me irresoluble: I may be wrong, but I think that Grünbaum has simply overlooked what I just referred to as the transitivity of "foundedness" (probably because he has not quite seen the importance, and the *strength*, of "definable from" in what I have called the usual sense). But the other two puzzles seem to me to have a solution. It is easiest to see with the question of factual content. Suppose I say that the space-time vector pq is orthogonal to the time-like straight world-line O. I maintain—since I have argued that the relation of orthogonality (and also the properties of linearity and time-like-ness, since they belong to the geometry that is derivable from the relation κ) is (are) "founded in things"—that that statement conveys factual content (always, of coursed, assuming, contrary to fact, that the world *is* Einstein–Minkowskian). I hope that I have said enough to persuade Grünbaum that this is so. What I think he will *deny* is that this admitted factual content is *about simultaneity*; so if I say that p and q are simultaneous relative to O, Grünbaum will deny that *this* statement

---

[37] Linguistic point: this word seems like the nominalization of the adjective "factitious," whose meaning is opposite to the one desired; I should suggest "factuality."

has factual content—even though I myself have *defined* "simultaneity relative to O" to be just the relation expressed by the former statement. In other words, what— according to Grünbaum—what the theory, or the facts of causal connectedness, does not "mandate" is *how the word simultaneity is to be used*.

Now, if this is all there is to it, it looks as if Grünbaum is after all defending nothing more than the "trivial semantic conventionality" of the use of the words "simultaneous" and "simultaneity." But I think that is not quite all there is to it. There is a certain traditional baggage carried by the word "simultaneous"; and I think Grünbaum is *rightly* maintaining that that baggage has no place in the special theory of relativity. But this statement is crudely metaphorical; is it possible to say clearly what this "baggage" is?—Probably not; but I think it was a mistake on the part of Wittgenstein when he produced his celebrated aphorism that whatever can be said at all can be said clearly: my own motto is that whatever one thinks is worth saying, one should try to say as clearly as one can. An *exact* statement *is* possible, and I shall make it; it is the one about which I expressed "some hope" that Grünbaum will accept it (as far as it goes). This exact statement, however, will not express what the "baggage" is, but only clarify something about the source of the latter. I shall then try to indicate—but only by indirection and example—something about what baggage the relativistic notion of simultaneity *does not* carry.

The exact statement is based upon Grünbaum's own very clear depiction of the state of affairs in pre-relativistic theory and in the special theory of relativity—that is, in the Newtonian world and that of Einstein [7] and Minkowski. I paraphrase what he has said thus: In each of these theories of physics, there is an "objective" division of the world, viewed from any point p, into the class of points that are *past* for p, those that are *future* for p, and those that are neither.[38] An admissible time-function[39] is a function τ on the entire space-time of the theory such that for any pair of points p, q, if p is in the past of q then $\tau(p) < \tau(q)$. If τ is a time-function, then the relation between p and q expressed by $\tau(p) = \tau(q)$ is an admissible simultaneity-relation. In the Newtonian theory there is a great abundance of admissible time-functions, but there is a *unique* admissible simultaneity-relation; in the Einstein–Minkowski theory, this is very far from true. In so far as any admissible time-function, and correspondingly any admissible simultaneity-relation, is compatible with the structure of the Einstein–Minkowski theory, the choice among them is a matter of "convention" (or, for that matter, "convenience"); in Grünbaum's terminology, all these functions, and all these relations, are "definable in terms of" the structure of Einstein–Minkowski space-time. An example of a space-time structure

---

[38] It is crucial, in Grünbaum's opinion, that this classification is grounded in facts about "causal connection." I am a skeptic in this matter (not a disbeliever, an agnostic). I agree that the notions of "causal past" and "causal future" are deeply important in the theories—perhaps especially so in the theory of relativity; but, on the other hand, (1) I think the notion of "cause" itself is in some degree problematic, so that what [if anything!] we "know" about this notion is derived from the knowledge we have gained in physics, and is not the "foundation" of the latter; (2) in assessing the knowledge we have from physics we cannot ignore quantum physics; and (3) how quantum physics ultimately affects the framework notions of relativity theory seems to remain a problem.

[39] My own term—this is intended as my own *paraphrase*, or formulation of what I believe *follows* from what Grünbaum has said.

"Definability," "Conventionality," and Simultaneity in Einstein–Minkowski Space-Time   437

"in terms of which" *there is no "definable" simultaneity-relation* is that of Gödel: in the original Gödel "rotating universe," there is in fact no admissible time-function whatever.[40]

But why should we accept this very liberal notion of what is "admissible," not alongside of, but *instead of*, the narrower—stronger!—notion of what is (in the usual sense) "definable from" the structure of Newtonian space-time on the one hand, Einstein–Minkowski space-time on the other?—Indeed, I do not think we *should* "accept [the former]... instead of [the latter]"; that is what most of this paper has been about. I think we should acknowledge, side by side, the points made by Grünbaum and the points made by Malament et al.

As to the matter of "baggage": I have already stated my objection to Grünbaum's claim that the Einstein–Minkowski notion lacks a *fundamentum in re*; but the baggage I have referred to is "metaphysical baggage." I suspect—I confess that I *hope*—that at least part of what Grünbaum means in rejecting *any* "metaphysical foundation" for relative simultaneity is what I myself meant, long ago, when I wrote the following, in criticizing metaphysical arguments of C. W. Rietdijk and of Hilary Putnam:

> [W]hat Einstein's arguments showed was that *a certain procedure of measurement singles out a time axis and gives numerical time differences dependent upon that distinguished axis*; not that an observer's state of motion imposes upon him a special view of the world's structure. This illegitimate metaphysical interpretation of the time-coordinate appears perhaps most plainly in Rietdijk's phrase describing C and A, when at rest with respect to one another, as "experiencing the same 'present'"; there is of course no such "experience": the fact that there is no experience of the presentness of remote events was one of Einstein's basic starting points [13, p. 16, n. 15].

The "baggage," then, can be said to be the carrying around of a special relation of simultaneity, as it were "in one's head." I believe that A. N. Whitehead thought something like this, when he contrasted, among "actual entities," the relation of "causal efficacy" and that of "presentational immediacy": in the latter, the mode of *perceptual space*, what we perceive is the entire present simultaneity slice "relative to us," *as if* it were characterized by the perceptual qualities that we experience. Perhaps I am wrong about Whitehead; at any rate, it is an impossible conception. If there were no other trouble with it, what are we to say about an observer who is *not* in a state of uniform motion? For such an observer, it is entirely possible—indeed, it is certain!—that "his or her" simultaneity-slice at one moment will contain "events" that are in the *future* of "his or her" simultaneity slice at a *later* moment; a perfect muddle! It is, then, certainly *not* the case that the special theory of relativity "mandates" that a sentient being carry such a relation around through that being's career. A paradigm of what I think of as the *poignancy* of the "old" notion of simultaneity that is quite lost to the new one is the old sentimental song-line, "I wonder who's kissing her now!"—In fact, for events that are within normal human spatio-temporal range of one another, the special (or the general) theory of relativity

---

[40] This, then, clarifies, at least to some extent, what Grünbaum's notion of "definable in terms of" can *exclude*.

438   H. Stein

provides a perfectly intelligible notion of "now" to *carry* that kind of poignancy; and it is *not* the *geometrical* notion of the "instantaneous now" relative to a state of inertial motion.[41]

The moral that I draw, then, is that although Grünbaum is (in my opinion) wrong to believe that, in so far as the causal theory of time has real—or factual—or fundamental content, the Einstein–Minkowski notion of relative simultaneity does *not* have such content, he is right to deny that this notion has content entirely comparable to that of the old Newtonian relation of absolute simultaneity. I do not believe that Malament, for one, would differ with Grünbaum on this point any more than I do.

3 Supplementary Notes

1. Hogarth's proof is far more complicated than the theorem requires; here is a simpler one: What is to be proved is:

   If to every "inertial observer world-line" (more briefly: "observer-line") O there is assigned an equivalence relation between space-time points, "p and q are simultaneous for O," (a) invariant under all maps of Minkowski space onto itself that preserve the Minkowski quadratic form, and such that (b) for every point p there is a unique point q on O that is simultaneous with p for O, then points p, q, are simultaneous for O if and only if they lie in a hyperplane orthogonal to O.

   Proof. First, let p and q be simultaneous for O. Let p_0 be the point on O that is simultaneous with p—and so also with q—for O, and let h be the hyperplane through p orthogonal to O. Reflection of space-time in the (space-like) hyperplane h is a map satisfying the conditions laid down; under it, p is fixed and O is mapped to itself; so, by the invariance condition (a), the image p_0' of p_0 is simultaneous with p for O. Unless p_0' and p_0 coincide—i.e., unless p_0 is in the hyperplane h—this implies that two distinct points, p_0 and p_0', both on O, are simultaneous with p for O. Since this violates condition (b), p_0 must lie in h. But q satisfies the same conditions as p, *vis-à-vis* O and p_0; p_0 therefore lies also in the hyperplane through q orthogonal to O. Since p_0 lies in only only one hyperplane orthogonal to O, and this is h, q too must lie in h. This establishes the "only if" clause of the theorem. Second, if h is a hyperplane orthogonal to O, and if p and q lie in h, let p_0 be the point of O simultaneous, for O, with p; then by what we have already seen, p and p_0 lie in a hyperplane orthogonal to O—and this can only be h. By the same token, q must be simultaneous, for O, with a point of O that lies in h—and this can only be p_0. It follows that p and q, since each is simultaneous for O with p_0, are themselves simultaneous for O; and this completes the proof.

2. As stated above (end of n. 6), the formulation of the theorem on p. 149 of [3] not only fails to state accurately what the argument preceding it has established, but is simply false. A correct statement is:

---

[41] What it *is*, is discussed, implicitly, in Stein [3, p. 159].

"Definability," "Conventionality," and Simultaneity in Einstein–Minkowski Space-Time   439

> If R is a reflexive, transitive relation on a Minkowski space (of any number of dimensions—
> of course at least two), invariant under automorphisms that preserve the time-orientation,
> and if Rxy does not hold for every pair of points (x, y) of the space, but does hold whenever
> xy is a past-pointing (time-like or null) nonzero vector, then Rxy holds if and only if xy is a
> past-pointing vector.

If the dimension is greater than two, the automorphisms considered may be restricted further to such as preserve the *spatial* orientation as well the time-orientation [equivalently: that preserve the orientation of the whole manifold as well as the time-orientation], and also preserve the scale.

3. In the proof—or proof sketch—they give for their Theorem 1 (which is essentially the same as case (b) of Theorem 1 above), the exposition of Sarkar and Stachel is not at all points quite clear: for instance, they refer, near the beginning of part (ii) of their argument [4, p. 217], to "the family of hypersurfaces of simultaneity," although they have not given any reason to suppose that the equivalence-classes of the simultaneity relation *are* hypersurfaces (the counterexamples given above show that in the absence of a requirement of scale-invariance this need not be the case), or even that the equivalence-classes *contain* hypersurfaces; so one cannot be entirely sure exactly what they may be assuming tacitly. Nevertheless, there *is* in their proof one passage containing a clearly identifiable and crucial paralogism. They have (almost) correctly remarked, at the beginning of (ii), that "[a]ccording to our definition, any simultaneity relation causally' definable from κ and O must be invariant under any transformation belong to the group of O causal' automorphisms. This implies that it must take the family of hypersurfaces of simultaneity onto itself under any such automorphism" (*sic*; but read, of course, "that any such automorphism must take [etc.]"). Some lines below this, however, they say of translations orthogonal to the world-lines of the inertial system, "If they are not to affect the simultaneity relation (which amounts to our assumption that the simultaneity relation is independent of the initially-chosen world line O), these translations must take each simultaneity hypersurface onto itself." This simply does *not* follow: what does, is just that each translation must take each equivalence-class to *some* equivalence-class, not necessarily to *itself*. The point is crucial, because it is only from the premise that each equivalence-class (a) is a hypersurface, and (b) is mapped *to itself* by translations orthogonal to the inertial system, that they conclude that the classes are hyperplanes orthogonal to the inertial system. Indeed, the mere assumption of *translation*-invariance, if one did not also postulate invariance under *rotations* that take the inertial system to itself, would allow the possibility of simultaneity hyperplanes "*inclined at a fixed angle*" to the world-lines of the inertial system; the system of such hyperplanes would, then, be invariant under translations orthogonal to the inertial-system's world-lines; but the *individual* hyperplanes would *not* be invariant under these translations. And Sarkar and Stachel—as they seem not to have noticed!—not only make no appeal to scale-change-invariance, they likewise make no appeal to rotation-invariance in their proof-sketch.

I have not discussed Theorem 2 of Sarkar and Stachel. It is fairly clear what this theorem is intended to say, and that what it is intended to say is true. But its

440   H. Stein

formulation [4, p. 218] is, when one looks at it closely, very obscure; and the proof
given for it there is garbled. As to the obscurity: in this theorem, the authors impose
the condition—condition (ii)—that "no event is simultaneous with one in its causal
future (past)"; the parenthesis is intended to imply an *alternative*: one of the two
simultaneity relations they arrive at has as its equivalence-classes the "backwards"
mantle of a null-cone, the other the "forwards" mantle; the former contains "no
event in the causal future" of the vertex of the cone, the latter "no event in its causal
past." But the condition as formulated certainly does not do what it is meant to:
simultaneity is—not just usually, but explicitly for Sarkar and Stachel—an equivalence relation. If event e is simultaneous with e', and if e' is on the backwards
mantle of the cone of e and thus "not in the causal future of e," then *ipso facto* e' is
simultaneous with e; but e is in the causal future of e'; so the condition as the authors have stated it *rules out* the mantles of the null-cone (forwards *or* backwards)
as simultaneity-classes. As to the proof: The theorem requires that simultaneity be
relativized to an inertial observer-line O, and that it be invariant, for every point e of
O, under "boosts" at O; this is condition (i) of the theorem. The proof sketch begins:
"Let p be any event not on O that is simultaneous with e. Consider the vector ep.
By condition (i) of the theorem, under boosts at e, the length of this vector must
remain invariant."—This, I say, is garbled: the conclusion has nothing to do with
condition (i) of the theorem; "boosts," which are among the transformations in the
Poincaré group, *ipso facto* preserve the lengths of vectors.—The proof continues:
"Thus, the locus of p under all such boosts is either the forward or backward null
cone or a time-like hyperboloid within the null cone."[42]—Again, this is just a fact
about the geometry of boosts; so far, none of the conditions of the theorem has been
actually used. The remainder of the proof is: "Now, e does not belong to any such
hyperboloid. Therefore, if such a hyperboloid were used to define the simultaneity
relation, e would not be simultaneous with itself violating the reflexivity condition
of an equivalence relation. Thus, only the two half null cones remain as potential
hypersurfaces of simultaneity. Condition (ii) restricts us to one of the two."

Well, we have already seen that condition (ii) cannot be helpful as it stands.
But as to the connection with condition (i): the conclusion that the locus of the
point p under boosts is either a half-cone or a lobe of a "hyperboloid," as already
remarked, is independent of condition (i); what that condition does now imply is
that this locus consists entirely of points simultaneous with e. However, what we
need for the conclusion drawn by Sarkar and Stachel is that these are the *only* points
simultaneous with e; and condition (i) does *not* imply this, without *some* further
stipulation.

Perhaps it was overstating the matter to say that it is "fairly clear" what the theorem is intended to say. Here is an attempt at it: If we require of a simultaneity relation
relative to an observer-line O (a) that for every point e on O, simultaneity is invariant
under boosts at O and (b) that no observer-line meet any of the equivalence-classes
of this relation in more than one point, then the only possibilities are the two Sarkar

---

[42] "Hyperboloid," of course, in the usual Euclidean model of a Minkowskian space-time. "Time-
like" is perhaps misleading: this is a hyperboloid of two branches; vectors from one to another
point of one branch are space-like; it is the separation between the two branches that is time-like.

"Definability," "Conventionality," and Simultaneity in Einstein–Minkowski Space-Time   441

and Stachel describe.—*This* is true; and it follows by a straightened-out redaction of their argument (as shown just above, one concludes that the class of events simultaneous with a given e on O is either one mantle of the null-cone at O, or a "branch of a hyperboloid" [the full cone or hyperboloid is ruled out by the fact that there are observer-lines that meet them in two points]; but this class must contain e—by reflexivity—whereas O meets the hyperboloid-branches in points other than e, and it must also contain e; but meeting an equivalence-class in more than one point has been excluded).

Second thoughts (or afterthought)—an alternative reconstruction of the intent of Theorem 2: it may be that condition (ii) was intended to mean that no event *on O* is simultaneous with an event in its causal future (alternatively: its causal past). This would do the trick (although the part of the proof invoking this condition would need a little rewriting).

4. In the two-dimensional case, if the equivalence-classes are "Minkowski semicircles" and if the axes are not normal to these curves, the configuration of a given equivalence-class Σ and its axes may be described as follows: The equivalence-class, as we know, has a given temporally oriented radius r. Let us represent this simply by a real number (positive or negative, denoting "future-pointing" or "past-pointing"—zero is not a possibility). In the usual Euclidean model of a Minkowski plane, Σ is a connected branch of a hyperbola whose principal semi-axis is r—understanding the sign of r to mean (taking the time-axis to be vertical) the "upper branch" if r is positive, the "lower branch" if r is negative. We may, as usual, take the center of the hyperbola—which represents the center of the "Minkowski semicircle"—to be at the origin of the system of coordinates. Now let there be given also a spatially oriented non-zero radius d (*also* representable as a real number—with an analogous convention about the sign: e.g., positive "to the right," negative "to the left". Consider a second "Minkowski semicircle," or hyperbolic branch, Ω, having the same center as Σ, but with the space-like oriented radius d; and take the axis of the equivalence-class Σ, at any given point p, to be *the line through p that is tangent to Ω*; so the family of all axes of Σ is just the family of all lines tangent to Ω: it is this that takes the place of the family of all lines through the center (which may indeed be considered as the limiting—degenerate—case of our "hyperbolic" construction when the spatial radius d goes to zero).—The radius d of the auxiliary hyperbola Ω is *not* the same as the spatially oriented radius s used in the construction described in Theorem 3 of the text above; r being given, the connection between the s and d (this is the one point that does necessitate a little calculation to determine), if we represent s, d, and r by real numbers, is given by the pair of equations, inverse to one another: $d = -sr/\sqrt{(1-s^2)}$, $s = -d/\sqrt{(d^2+r^2)}$. (Note that s must, as previously specified, be chosen with absolute value less than 1; this is guaranteed by the second equation. On the other hand, d is entirely arbitrary.—Note too that these equations also hold in the "degenerate" case (or perhaps, rather, the *normal* case!) $s = d = 0$.

References

1. Malament, David (1977). "Causal Theories of Time and the Conventionality of Simultaneity." *Noûs* **11**, 293–300.
2. Hogarth, Mark (2005). "Conventionality of Simultaneity: Malament's Result Revisited." *Foundations of Physics Letters* **18**, 491–497.
3. Stein, Howard (1991). "On Relativity Theory and Openness of the Future." *Philosophy of Science* **58**, 147–167.
4. Sarkar, Sahotra, and Stachel, John (1999). "Did Malament Prove the Non-Conventionality of Simultaneity in the Special Theory of Relativity?" *Philosophy of Science* **66**, 208–220.
5. Poincaré, Henri (1906). "La dynamique de l'électron." *Rendiconti del Circolo matematico di Palermo* **21**, 129–176; reprinted in *Oeuvres de Henri Poincaré*, vol. 9 (Paris: Gauthier-Villars, 1954), pp. 494–550. A brief summary of the main results had previously been given in the *Comptes rendus* of the Académie des Sciences, Paris, **140**, 1504–8 (5 June 1905); Poincaré, *Oeuvres*, vol. 9, pp. 489–93.
6. Lorentz, H.A. (1904). "Electromagnetic Phenomena in a System Moving with Any Velocity Less Than That of Light." *Proceedings of the Academy of Sciences of Amsterdam* **6**; reprinted in H.A. Lorentz, A. Einstein, H. Minkowski and H. Weyl, *The Principle of Relativity* (Methuen, 1923; Dover, 1952).
7. Einstein, Albert (1905). "Zur Elektrodynamik bewegter Körper." *Annalen der Physik*, series 4, **17**, 891–921; reproduced in *The Collected Papers of Albert Einstein*, vol. 2, ed. John Stachel (Princeton, 1989), pp. 276–306.
8. Grünbaum, Adolf (2001). "David Malament and the Conventionality of Simultaneity: A Reply." http://philsci-archive.pitt.edu/archive/00000184. (This is a prepublished essay and will appear as a chapter in Adolf Grünbaum, *Philosophy of Science in Action*, vol. 1, to be published by Oxford University Press, New York—Information obtained from online "Bibliography for Adolf Grünbaum.")
9. Poincaré, Henri (1900). "Les théories de la physique moderne." *Rapports du Congrés de Physique de 1900* vol. 1; reprinted as Ch. X of *La Science et l'Hypothese*. An English translation by G. B. Halsted can be found in the collection *The Foundations of Science* (Lancaster, PA: The Science Press, 1913); this translation is superior to that found in the Dover edition of *Science and Hypothesis*.
10. Pais, Abraham (1982). '*Subtle is the Lord...*'; *The Science and the Life of Albert Einstein.* Oxford University Press, New York.
11. Einstein, Albert (1949). "Autobiographical Notes" in *Albert Einstein: Philosopher-Scientist*, ed. P.A. Schilpp, Open Court Press, Evanston, IL.
12. Reichenbach, Hans (1958). *The Philosophy of Space and Time.* Dover, New York.
13. Stein, Howard (1968). "On Einstein–Minkowski Space-Time." *The Journal of Philosophy* **65**, 5–23.

Part VI
Concluding Words

Bistro Banter
A Dialogue with Abner Shimony and Lee Smolin

Abstract This is a transcript of a dialogue that took place between Abner Shimony, Lee Smolin, and members of the audience, on July 21, 2006, in the Black Hole Bistro at the Perimeter Institute for Theoretical Physics. A video of the discussion can be found online at http://www.pirsa.org/06070049.

Smolin

Welcome, everybody. My name is Lee Smolin. I'm one of the faculty here at Perimeter Institute and, first, it's been said before but all of our guests and friends here attending this conference are welcomed, and thank you for coming such a long way for many of you. The main reason for this evening, as well as for the conference, is to celebrate Abner Shimony; I think he needs no introduction here, so he won't get one.

The aim of this evening is for all of us to have a discussion with Abner. I will start it off by asking him some questions and, as we go along in the evening, anyone who wants also to pose some questions to Abner, please feel free. Here at Perimeter we're used to being very informal and also as part of the atmosphere here—I feel like in the old days when I used to play in bars—so, drink up, drinks are available all evening and I think food is still available for another fifteen minutes for those who want to order. The idea is to have an informal atmosphere here, to have fun as well as try to understand better Abner's thinking.

Here's a starting point. In preparing this I've been very intimidated, very daunted because I'm a theoretical physicist. Like many of us here I've been very influenced by Abner and was as a student. He played a crucial role in my educational development at a crucial stage for myself as well, I think, for many people here, and I know something of Abner's work related to foundations of quantum mechanics, I know even something of his work related to statistical physics, statistical mechanics, but his work has a much larger scope than this and we see this here in the

W.C. Myrvold and J. Christian (eds.), *Quantum Reality, Relativistic Causality, and Closing the Epistemic Circle*, The Western Ontario Series in Philosophy of Science 73,
© Springer Science+Business Media B.V. 2009

conference. There are some people who are physicists interested in quantum mechanics and other things, but there are also here professional philosophers with a range of interests and they reflect the fact that Abner is a philosopher. This is a little bit intimidating because he is also, as we see by the conference, an influential philosopher and a philosopher—I would say like the best, at least to my understanding, people who work in philosophy of science or philosophy of physics—his motivation is not just what some of our motivations are, that quantum mechanics may not make sense, or we'd like to go a little bit further, solve some problems and understand gravity or space and time or something a little bit better. He starts with the ancient and deep philosophical questions about ontology, epistemology, their relationship; and he has an agenda, a longstanding agenda, a longstanding program in philosophy and I think to understand Abner's thought we have to start there, with the basic questions in philosophy and his program. So that's where I'd like to start, Abner. So, if you don't mind just jumping in, can you - and remember some of us know this very well and some of us do not - can you tell us something about your larger philosophical program, what is the main ambition, what are the main ideas behind it?

Shimony

Well, curiosity is a wonderful human trait and I'm delighted that I'm endowed with it. It also can take one in odd directions. You can be curious about very general questions, about why human beings are here in the world, what their aims are, was there a purpose in having them here, what are the basic principles on which the universe is run—but you can also be curious about specific phenomena. My introduction to science, the earliest I can remember, was going down—this was during prohibition, by the way—going down to the basement, and my father would take a jug down and fill the jug from a barrel of wine and he would suck up the wine through a tube and then hold it up and the wine would go from the tube into the jug. How could it do *that*? So curiosity can be very specific about a phenomenon. So I became a scientist, and my father is partly responsible for it. That's how curiosity works. Well, what happens? You do try to systematize. Partly as a matter of efficiency, partly because any one problem you're curious about is going to somehow spill over into another. In about sophomore year in college I was attracted to become a philosophy major for reasons which were partly wrong. That, is I thought that philosophy would give me as much of a surrogate for omniscience as any branch of learning could. Well, there is no surrogate for omniscience; that was a misguided child's illusion. However, it does help organize one's thoughts and there have been very smart people who have tried to give systematic views of human knowledge, human values and the structure of the universe and how the various subdisciplines mesh. I think I'm grateful for my error as a sophomore because, even though I later decided that the standard approach to philosophy is too schematic, too lacking in detail, and I went back to my initial love of physics, and tried to get an education that would supplement the kind of things that are done in standard philosophy courses.

Bistro Banter

However, if it weren't for the philosophy program I wouldn't have made this attempt to somehow systematize, to the extent that I was capable of doing, my view of the world.

From audience

When you decided to go to graduate school in physics, was it with the idea of becoming a physicist rather than a philosopher?

Shimony

First, first, and as time went on—the thing is I was curious about everything. How to satisfy all those curiosities? Somehow, philosophy seemed to be the best way of doing it. It isn't. It isn't by itself. It has to be done with something else, but I didn't realize that right away. But, yes, my initial choice of a prospective major was physics. In fact, one week after I came to Yale I was assigned to the cyclotron and I wrote to my parents, "I'm just a freshman and I'm going to be working on the cyclotron" It turned out there was a strike of the handymen, and I was used as a scab; that already was a beginning of correction of illusions.
   [Remark from audience, inaudible]

Smolin

Yes, thank you for saying that. I was going to say it was not the last time Yale employed that strategy.

Shimony

Anyway, I did sweep the cyclotron room.

Smolin

Abner, a phrase that is associated with your philosophy is "closing the circle." Can you tell us what you mean by that?

Shimony

Well, let me go back a bit in the history of philosophy. There are great philosophers whom I revere. When I'm critical of them that doesn't mean that I don't revere them and that I haven't learned from them. I'll take two particular ones, Descartes, who was a rationalist, and Hume, who was an empiricist. Both of them shared a kind of architectural view of knowledge. That is, that there should be a solid foundation, and on that solid foundation a first floor, a second floor and so on, until the whole skyscraper of knowledge is constructed. It didn't work, because: How is the foundation to be laid? Is there a set of principles that are in no need of correction? I believe our experience over some four or five hundred years is that there may be reasonable starting points, but none of them that aren't in need of reexamination and reassessment in the light of what structure has been built upon that foundation. So, I think the alternative to this architectural model is "closing the circle." That is, one has to start somewhere, one starts with one's native endowment and with one's psychology, but also with one's culture and also with one's reading and education, and you start, and go where you can, and then in the course of learning more about the world, starting from your tentative first principles, you go back and reassess the first principles. It seems to me that the best strategy, the one that is most promising for human beings to have a reliable view of the world, is to follow this pattern of trying to close this circle of epistemology and ontology. That is, you start with tentative principles of methodology and assessment of beliefs, you build up, you use those to explore the world and then, in the course of exploring the world, you learn something about your own faculties. We've learned some psychology, we've learned where human beings fit in the Earth that they live on. We learn about human beings as products of evolution. That learning enables one to reassess the beginnings, and I think the best chance of having a reliable view of the world is to keep repairing the foundations in the light of what one has learnt. One of the Vienna circle, Neurath, has the wonderful simile that we are like sailors at sea whose ship needs repair. We can't go to port to repair it, we repair the ship while we're at sea, and I think—this may not work, the ship may sink—but if anything will work this will work. And if this doesn't work, our quest for knowledge is futile. I am a fallibilist concerning the quest for knowledge. That's Peirce's phrase, and I'm a very devoted follower of Peirce. One can be a fallibilist and think that any proposition that we're committed to may be wrong, may be subject to correction, but one can be an *optimistic* fallibilist. I'm a *highly* optimistic one. This is something of an aside but not entirely, it's an historical remark: the twentieth century was a time when the best-established physical theories were overturned. This has led sociologists of science and some historians of science to be very skeptical about any claim of approaching the truth about the world. My feeling is that's an entirely wrong misreading of the scientific revolutions of the twentieth century, because the theories that were overthrown—like Newtonian mechanics, Newtonian spacetime—the theories were retired, like presidents who have completed their terms honorably. They were not impeached. And of course Bohr's expression of a correspondence principle applies. The new theories do not discredit the old ones; they give the old theories their place as approximations. And this program that I've sketched of closing the circle of epistemology and ontology embraces the idea of approximation. Approximation fits in very naturally in such a view of the world.

Smolin

Now, I'm going to going to show my philosophical naïveté. A word that you didn't use is "realism."

Shimony

I'll use it now. REALISM.

But, Lee, I'm a fallibilistic realist. That is, any claim I make to a certain principle—it may be well-established in the textbooks as being a good approximation to the truth—is subject to revision and to reassessment, but that doesn't mean that one is skeptical in a wholesale way. Our experience in twentieth-century science has led us to be both skeptical of excessive claims of being right at the truth and also skeptical of claims about the hopelessness of obtaining knowledge of reality.

From audience

It seems to me that you can also use a simile of a "spiral" instead of a "circle."

Shimony

Yes. There's also a nice word that goes back to Plato, which is "dialectic." The dialectic is a procedure in which one starts where one starts, with whatever equipment one has, and learns as much as one can, including reassessment of the starting point. My optimism comes because the spiral is generally upward. Not always, but when you look, I think, in an unbiased way at the history of science, how can one fail to be impressed by the magnificent progress that has been made? It can't be just an accident that we have electromagnetic principles which enable us to make electromagnetic devices that work. Is that just a social conspiracy? How can one believe that?

Smolin

Abner, you're preaching to the converted here. Certainly there are big obstacles to closing the circle are there not, in your view?

Shimony

Yes. And we do not know in advance how bad the obstacles are.

Smolin

What are the obstacles you worry about?

Shimony

Well, for one, there are practical obstacles. We may annihilate ourselves.

Smolin

You're in Canada!

Shimony

But that's a political question, not an epistemological question. We really do not know how different—cosmologists of course talk about this, and you do—how different the early universe or the pre-universe is from the one we live in. If our knowledge of the great world in some way is an extrapolation from what we learn about locally, but extrapolated to other galaxies, fifteen billion years ago, back to the Big Bang and so on, how do we know that that process of extrapolation will remain as reliable as it has been? Look, it seems to me a miracle—just overwhelming— that we could learn about the chemical constitution of the stars, that it was possible to extrapolate knowledge of the chemical composition of matter nearby to far off galaxies because—well, what is our explanation?—matter is essentially the same in the far off galaxies as here, and the optical laws, the optical phenomena which carry information from the stars to us and allow us to do spectroscopy to learn the chemical composition, hold generally. The universe, from a logical point of view, need not have been made in such a way that we could proceed stepwise. How do we know in advance that this stepwise progress will continue indefinitely? There may come a time when there are singularities, discontinuities, and so this type of extrapolation that made us capable of learning the chemical composition of stars simply won't work. That's one such barrier.

Smolin

Surely. And I was just at a meeting to celebrate the birthday of another very influential person, Gerard 't Hooft, and a theme of the meeting was that essentially, since 1976, with the establishment of the standard model of particle physics, there's been no further progress in our understanding of the fundamental laws of nature, no definitive progress. So this question can always be asked of the reach of knowledge. But my understanding is that for you "closing the circle" particularly refers to the relationship between ontology and epistemology and somehow situating ourselves as the subject who knows and who learns, inside the picture, the story about the natural world that we describe. Am I right?

Shimony

Absolutely.

Smolin

And, how are we on that project? How are we doing?

Shimony

Well, in some ways we're doing very well. In some ways I'm not so optimistic. It seems to me the one great philosophical problem which I believe may not be solvable at all—and I certainly don't think it's solvable quickly—is the mind-body problem. There is a real problem for closing the circle of your epistemological beginning and your ontological structure. There's nothing we know better than that we have conscious experience. There's nothing that we know *much* better than that the matter that the world is made of is inanimate. There's nothing that we know much better than that somehow there was pre-biotic evolution, then biological evolution, and then here we are. Put those together; you don't have a solution, you have a puzzle, a terrible puzzle. How can it be that from inanimate beginnings something endowed with our type of conscious experience can emerge? Now, there are people who use computer models, who use neurological models. I believe they are deluded, because they are denying the immediate experience of their own phenomenology, that a felt pain or a felt desire is something that is different from a neural impulse. Some of you heard the little debate that Malin and I had on Whitehead. Now, I'm very sympathetic with Whitehead because Whitehead does give an answer to this by postulating a primitive universe which is not entirely inanimate; he calls his philosophy the "philosophy of organism." That is as promising as anything I know for a solution to the mind-body problem but it leaves out the details terribly.

Smolin

Now, my understanding is that for some time of your career you did embrace Whitehead and then you came to reject him. Can you say more about why you rejected Whitehead and where the problem stands now?

Shimony

Well, I gave it in the debate—some of you have heard it already.

From audience

You didn't give enough of it!

Shimony

I read a passage in Lovejoy's *The Revolt Against Dualism*. He has only about a page on Whitehead but it's a very, very good page. He says Whitehead tries to overcome the dualism of mentality and materiality by essentially postulating a primitive world which is mentalistic in character. That is, an electron is conceived of as a temporal string of *experiencing* but not conscious occasions, experience below the level of what we call conscious experience. He [Lovejoy] said that's a verbal trick. What do we know of the experience of an electron that allows us to use the word "experience" and apply it to ourselves and the electron without absolute equivocation? We know something—I believe we know something—about the experience of babies,

Bistro Banter 453

we know something about the experience of dogs, we can extrapolate down to simpler animals. When we get down to worms and crayfish and so on I'm not sure we understand what "experience" means applied to them, though we're tender-hearted creatures and we don't like to squash them—I don't like to squash them, anyway. What in the world are we talking about when we apply the term "experience" to an electron? It's simply an equivocation and we cannot take it literally. We do not have a solution to the mind-body problem. I think that's right. That seems to me a very powerful argument and the only way one can solve it is by having tremendously more information of a kind that I don't even understand what would constitute it. How are we going to learn what the experience of the lowest animals is like? What kind of information would give it to us? I don't see that the tools that are available now to biologists and neurologists and so on are likely to give us the information that Lovejoy is demanding.

Now, at this point I may be falling into just the kind of habit that I was deriding before, that is, excessive skepticism. Maybe what's needed is to be a little bit less rigid about what one can demand of explanations. And there are certain phrases that have been used about the primitive world. That there are "brute facts" concerning the constitution of this world, and the very word "brute" already is a gross word. It somehow underestimates the richness and the complexity of this primitive world. Maybe I'm being excessively skeptical in underestimating the richness with which the world began, but I don't know what to do! So I'm sympathetic with Whitehead's general program, but I do not see how to implement it. I don't see how to fill out the details. I don't see the tools in the making for filling out the details. So that's the answer to the question.

From audience: Geoff Hellman

I'd like to connect this with another major question that science confronted at the turn of the twentieth century, the notion of life was quite mysterious and very much up for debate and it was not unrespectable to be a vitalist, to believe that there were some special properties of living matter, some special forces at work that simply could not be brought within the framework of the physical sciences, including chemistry. Now, that's a dead idea. It was reasonable to take it seriously up until a point, but most people in the scientific community would agree that today it's not a viable idea and it's not needed. And so this would be some grounds for some more optimism that we might gain the tools—I think you're right to emphasize that we don't have them now, we don't seem to have them now and I don't at all mean to suggest that there aren't special problems about consciousness and experience that make it harder and different from the problem of life, but, what do you think about that problem? Isn't that some grounds for optimism that the biological sciences now have grappled with this, and essentially replaced the question with...

Shimony

I wish I could go along with you, but I think it's a mistake to conflate the problem of the nature of life with the problem of the nature of conscious experience.

Hellman

I'm not doing that.

Shimony

But I am! That is, I'm saying that's the sticking point for me. That is, one can analyze—the biologists have learnt how to analyze life (setting aside conscious experience) with extreme subtlety and with physical tools, that is, chemical cybernetics, completely formulated in terms of atomic and molecular physics. It's the basis of explanation of cell behavior, replication, metabolism, anything that you want about living creatures, provided you leave out the conscious experience. I don't see that we have the conscious experience in this way other than something that is correlated with neural behavior. But a correlation is not an identity. There's an enormous philosophical literature about the idea of identity here. My feeling is that it's very unconvincing. I have one principle, which I didn't say earlier, I call it the "phenomenological principle." The phenomenological principle is that anything that appears must be accepted as real *sui generis*. And what does that mean? It means we have to understand in terms of our ontology where and how that aspect of appearance has entered into our consciousness. If we haven't done justice to the phenomena *as they appear* something is still missing in our structure of knowledge. Now, I think that's a crude argument, but it's enough to recognize that there are some things which reductionist philosophers have fallaciously characterized as derivative. One of them is conscious experience; as *phenomenon* it's not the same as neurological impulses and transitions and so on. The sense of transience, absolutely central to a complete theory of time, it's not part of current spacetime theory but that may be the fault of current spacetime theory. Transience still has an ontological status.

Now, I'll make a little deviation. There are people who say that the sense of transience is entirely psychological.

Smolin

Can I interrupt you, because, before we—don't worry, we're going to talk about time and transience...

Shimony

I just want to say one thing. That is, because it's not really about time but about what happens when you attribute a phenomenon to consciousness. That's possible. It's possible and it may be so that this phenomenon, like color sense, like appreciation of music, is entirely psychological, though there are physical correlates, but when you do that you don't impoverish the ontology, you say the ontology must include a mentalistic component. The mentalistic component may be inseparable from the physicalistic as in Whitehead's philosophy of organism, or it may be a separate type of thing as in a dualistic ontology, Descartes' mind and body as two kinds of substance. I don't want to take a stand on that, I will just say that anybody who takes seriously certain appearances, like consciousness, like transience, but tries to give as the locus of these things the mental aspect of the world, has a coherent point of view but it doesn't mean that these things somehow are derivative. It means that they have a status different from the status of molecules.

I don't need to talk any more about transience.

Smolin

We will come back because many of us are interested in that. There was a third thing you were connecting with. You started of saying that there were three things that were connected. The issue of consciousness, transience, but there was a third.

Shimony

The very small. The experimentalists here can tell you how much more energy and how much more effort and how much more control is necessary in order to probe into smaller and smaller spatial intervals. So what is going to be necessary to probe the ten to the minus thirty-third of a centimeter? And if we can't probe by high enough frequencies of light how are we going to find out about the structure of spacetime in the very small? So that may be a barrier just because of the relation between wavelength and frequency and energy.

Smolin

I'm not going to take that bait but I have two questions for you. I could, but I'm not going to; we can later, if you want. You suggested that somebody who believes that consciousness is part of the ontology of the world could also incorporate, if I understood right, a belief in transience of the present moment as an aspect of that ontology.

Shimony

Indeed, that's a possibility.

Smolin

Now, have you noticed (because I have) that people who take an artificial intelligence or an identity theory view of the brain and the mind—that is, there is no issue of consciousness—often are people who are happy, at least in my experience, with timeless formulations of physics in which transience disappears? I wonder if the opposite of that is true amongst some of the people in our community.

Shimony

I'll tell you my crude reaction. If you can believe that you can believe anything. A timeless formulation? This is a really timeless formulation, not just denying transience? Time itself is not an independent parameter?

Smolin

Something that is quite commonly said in the field of quantum gravity, quantum cosmology...

Shimony

Barbour.

Smolin

Julian Barbour and Steven Hawking, Jim Hartle...

Shimony

He smuggles in time. Everybody who tries to get rid of time and then explains the derivation smuggles it in. He does it by that model of cards, putting data on cards, and then he arranges the cards. Tell me about arrangement without time.

Smolin

It was the Wheeler-de Witt equation that arranged the cards, allegedly. But just to say, because this is important for those of us, a number of us here in this building and at this conference who work on quantum gravity, and it's quite a respectable and may even be considered the mainstream position in quantum gravity, or a dominant position, that fundamentally nature is timeless and time "emerges" in a semi-classical description, and it sounds like you're not convinced by that.

Shimony

No, I'm sorry, you didn't quite understand my position. What I was really trying to do was open up what I called the phenomenological principle. I would say if one ascribes transience or ascribes consciousness to a type of reality which is not physical, that doesn't make it unreal, that just means that your ontology is richer than a materialistic ontology. It's certainly—I'm just talking at the level of sketching a metaphysics—it's certainly possible to have a coherent materialistic ontology, and we have a number of examples like Hobbes', or a materialistic or a mentalistic ontology like Berkeley's or a dualistic one like Descartes', or something of a synthesis. Whitehead's is an example of a synthesis, in which the entities are in a sense biased toward a mentalistic pole but still are endowed with certain things that we usually attribute to matter. Whitehead still tries to maintain a concept of energy and apply it to the actual occasion. Now, I don't want to judge among these various ontologies. They're all possibilities. I just want to say that when you take seriously an appearance which occurs as part of your phenomenology and then you say I want to find a niche for that in the great world, that doesn't make the phenomenon unreal; it just places the phenomenon in one part of the ontology. We need to go further to see whether it has a derivative status or an ultimate and primitive status in this ontology. And I would say—I won't use the 't' word again - but consciousness seems to me ineradicable. There's no way of getting rid of it.

Smolin

Okay, I'd like to query you about transience and about time. Here is a version of the physicists' argument that transience is unreal. We start with a belief in the configuration space in some Platonic sense. That is, there is, symbolized maybe by this blackboard, the space of all possible configurations of our physical system, let's say the universe or a subsystem of the universe. We then hypothesize equations of motion which, given a starting-point in the configuration space, some initial condition, give us a path in the configuration space. And then we say a history of the universe is that path in the configuration space and that is physical reality, and the present moment or transience has disappeared from the scene, we're just left with the curve frozen there in a Platonically existing space.

Shimony

It may be physical reality, but it's not *all* of reality. Let me quote one of my favorite philosophers, Max Beerbohm. Max Beerbohm has a wonderful play called *'Savonarola' Brown*. Savonarola, in that play, is in prison, and he is humiliated that he's in prison. And he knows that when he's released from prison people will point at him and say "That man hath done Time" and then he reflects: "but the worst of Time is not in having done it but in doing it." It's one thing to have done time and that's on your record, it's another thing sitting in the cell for year after year. *That* is left out in the scenario that you presented. The scenario is a partial scenario and it's a good one. It may be a good one. It may not be, that is if we find a physical correlate for transience, which I rather hope for, then it's not the whole story even physically, but even if it's the whole story physically it's not the whole story ontologically. That's why one of my favorite philosophers is Beerbohm.

From audience

For example, put consciousness in it.

Shimony

Well, fine, alright, that's a possibility.

From audience

Glad to hear it!

Bistro Banter

Shimony

That's a possible ontology.

Smolin

I'm not glad to hear it. It sounds like what you're saying is the way out of that dilemma is that physical reality is not all there is to ontology, but I would have hoped that the task of physics is to explain or model all of ontology.

From audience: Shimon Malin

The description is never the described. This is Krishnamurti. I think we get so attached to our descriptions, our concepts, our conceptual models, we forget, this is not the world, it's not the world at all. This is part of your point, is it not? We have a conceptual model and take it seriously; it's part of the ontology, but it's not all of the ontology.

Shimony

Look, a phenomenology is a very crude beginning to a view of the world but it is an important beginning. We are able to communicate among ourselves, we seem to be made similarly enough that we can talk about our phenomena of color, of tone, of pain and so on. That doesn't parcel out what parts of those phenomena have physical correlates and what parts are fantasies. That's a further job and a very subtle job. We know by now quite a lot about the physical correlates of our color sense. It's very subtle indeed, it's not just association of different frequencies with the different sensed colors because context makes a difference. There are the beautiful experiments of Land in which one can get color sense without the usual accompanying frequency, so it's a subtle matter. But, still, you're not going to expel color from the universe by saying that color is a combination of physical stimulus of certain types and contexts of certain types with mental capacities. That's sorting out and giving the details. There's a difference between the outline of the ontology and the details of the ontology.

From audience: Jonathan Hackett

You talk about different ontologies, and I want to comment on these possible ontologies. At a certain point, these ontologies involve consciousness. Where do you ever get to decide that one's better than the other or that one could be the actual ontology if these things all, in the end, end up being untestable—and I guess maybe I'm just showing my background as a physicist rather than a philosopher—my concern is whether or not, at some point if you've got all these options, if you can't choose between them through any logical process, then it doesn't matter that you have all these options.

Shimony

You need detailed knowledge. We've learned a lot from physiologists of vision, we've learned from neurologists who know something about visual centers in the brain, we know something about the evolution of color sense in lower animals, which ones have it and which ones don't. We have at least some reasonable explanations, given the lifestyle of, say, bees, why they should have color sense and why some other insects don't have color sense. There's no way to complete the details of the ontology without making use of our scientific knowledge about the world. The bare general principles of a mentalism versus a materialism or a dualism, those don't give you the details.

From audience: Philip Pearle

I just wanted to put in a partial answer to your [Hackett's] question.. There's a quotation from Feynman, I don't remember where it is, where it comes from, but it is that, the more ways you can look at something differently, the richer the phenomenon is. You don't need to make a choice. I think he might be talking about, for example, the path-integral formulation of quantum mechanics versus the formulation of Schrödinger—we don't need to choose between them and the world is richer to have these two different ways of looking at it, and I feel that way about what you call dualistic or multiple-istic ways of looking at things. You don't need to have just the physical explanation: the internal explanation of love, the physiological explanation of love and every other explanation you have of love is fine, the phenomenon is rich, that's wonderful.

Shimony

But one wants details, too! Look, here is a very important question which historians and sociologists of science have made really central in their investigations. To what extent are data theory-dependent? And the more skeptical of these have said they're very theory-dependent, so the experimentalist cannot help but see his experimental results in the light of his theoretical preconceptions and that becomes an argument for the fallibility of science. Now, let me tell you about one of my favorite experiments by Bruner and Postman, which shows how much more subtle the whole question is than that the data are independent of theory or that the data are dependent on theory. There are subtleties in the mind of shifting from one mode of perception to another. They did card experiments in which they exposed a card to a viewer very briefly or for longer intervals. The cards they used were doctored cards, like a black Ace of hearts: wrong color to match with the shape of the pip. They flashed the card for a brief time and almost always the subjects will say "Ace of hearts" or "Ace of spades," will make a guess. When they're exposed longer, a long time, then they say, "Oh, you're trying to fool me, that's a doctored card!" In other words, with enough time to reexamine them, the subject can overcome his theoretical knowledge of a normal deck of cards. The most interesting case is the right time, which of course is subject-dependent, not too long, not too short a time. The subject becomes very confused: "I don't even know what an Ace looks like, I don't know what a heart looks like, what are you doing to me?" They really are very confused. So, what is the moral of that kind of experiment? It does seem that our perceptual system is flexible. It is very good at making rapid decisions, which are important for a lot of activities in the world, and the cost of making rapid decisions is making mistakes. We're also good, when there's time enough to examine in a leisurely way, at correcting the errors. If you realize this flexibility in the multi-modal nature of the perceptual system then you have a very simple answer to the skeptics about scientific evidence being somehow dictated by the theory that the experimenter has initially. If they have time enough they can re-examine their data and correct their mistakes, and they take time.

From audience: Andre Mirabelli

Abner, I just want to understand you. Are you suggesting that we could come to deeply understand something like consciousness or transience and keep physics from coherently incorporating it into its worldview?

Shimony

I don't know the answer. Look, I'll give you various possibilities that people have played with. One is that the sense of transience is nothing but psychological. That doesn't make it unreal, it has its place in the universe, but it only comes in because we're creatures endowed with consciousness. Another is that there is a physical correlate of transience. One of my favorite physicists, Bialynicki-Birula, has an interesting paper in which he tries to relate transience to reduction of the wave packet. It's a possibility. Now, people who don't believe in reduction of the wave packet certainly wouldn't accept that, but those of us who think that that is also a physical phenomenon can be open-minded about saying, well, that may be the physical correlate of the sense of transience. It's not out of the question.

Mirabelli

That attaches transience to consciousness but, now what if they went further...

Shimony

No, no, I've given a physical correlate.

Mirabelli

Let me just stick to consciousness. Is it possible that we come to a deep understanding of consciousness and physics not coherently incorporate it into its worldview?

Shimony

My guess is you don't know what your words mean.

Smolin

I understood them!

Shimony

No, Andre, I'm not trying to tease you. I'm saying we're trapped by our words, we are trying for explanation and the price people pay for their avidness to have explanations is they start stretching words beyond their literal meanings. You start saying, "This is a physical explanation of conscious experience." What does the word "physical" mean anymore? My guess is, partly because I'm a residual Whiteheadian, is that "physical" no longer means physical in the sense of Lucretius's atoms;

Bistro Banter   463

it is something like physics but with the primitive entities modelled partly on little
minds. So, you're not using words literally any more.

Mirabelli

That's why I said *coherently* incorporated. Do you believe that it's really possible
that you'll end up with a complete dualism?

Shimony

Well, I would say a necessary condition for coherence, though not a sufficient one,
is to use words literally. If you don't use words literally you'll confuse yourself and
other people.

Smolin

May I try here, because this is related to something I wanted to take you back to.
I gave you this little syllogism against transience which I—I should say personally
I'm with you about transience—but in this syllogism one starts with this Platonic
configuration space, one posits the laws of motion, one writes down a solution and
of course that could be Hilbert space in the Schrödinger equation and so forth. If you
let me say those things then, somebody might argue, how could there be a physical
correlate of transience because you've already expressed the laws of physics entirely
in a framework which has none. So if there is a physical correlate of - do you agree, if
there is a physical correlate of transience there's something wrong, that is incorrect,
with that formulation of the laws of physics?

Shimony

No, what I'm saying is that accepting transience phenomenologically doesn't dictate a physical theory. It leaves open the possibility that your basic physical theory
has room for a correlate to sense-transience, it also leaves open the possibility that
you're now espousing that your complete physical theory does not have to pay attention to it at all. It can do physics, it can take into account all physical measurements,
all physical observations, without transience at all.

Smolin

But let me tell you what I think I argued. What I think I argued was that if I'm a Platonist about configuration space, that is it exists in some timeless sense, and if I'm deterministic about either the classical equations of motion or the Schrödinger equation then there is no place for a correlate of transience. Therefore, either I cannot be a Platonist about configuration space, that is I have to give up the notion that physics is done by first specifying the configuration space as some eternally existing space of possible configurations, and then proceeding. And if I take that point of view, which I have tried to, then it's a bit scary because how else are we to formulate the laws of physics?

Shimony

Okay. Look, a major reason why I'm sympathetic with Bialynicki-Birula's paper is that I think the problem of the reduction of the wave packet is unsolved. And I also believe it's not going to be solved without some fundamental change in physics. And one more thing, this is a very nice paradigm case of closing the circle of epistemology and ontology because, if you don't have physical data, you're not going to get a quantum mechanics, you're not going to get any physical theory at all. So, if you have a physical theory, as quantum mechanics is now constituted, the dynamics being a unitary dynamics which preserves superpositions, you don't get data, you get no experimental results. Therefore, you can't close the circle. One of the reasons why I spend lots of time on the problem of measurement, the problem of the reduction of superposition, is that it is a—I wouldn't say manageable case because it seems far from manageable—but it is a particularly rich case with a lot of information that's promising for future use and it is an example, almost a paradigmatic example, of the whole enterprise of closing a circle. So now, having said that, that's what I really care about in the quantum-mechanical problem. Then, as a secondary theme, because transience is not my main interest, as a secondary thing there may be a bonus. The bonus may be that whatever the modifications of quantum mechanics may be, sufficient to explain reduction of superpositions, may give us the gift of a physical correlate with phenomenological transience. I hope so, but I'm not obsessed with that question.

Smolin

What you seem to be suggesting, is that there's a cluster of beliefs here and there's another cluster of beliefs [there (gestured)]. So, this cluster of beliefs, that I think you're identifying yourself with, is that you deny the identity theory, that is, you say there is something irreducible about consciousness, you believe in transience, that there is something true and irreducible about transience, and you believe that there's a problem of measurement in quantum mechanics, that is, if you will, how the data come to exist. What I think is interesting to observe is that there is cluster of beliefs held by many of our friends which denies all three of those things. And I think it's interesting just to notice, if you like, phenomenologically, that the three things come together. And so I think if we're going to talk about - there are of course positions in between because academics can do anything

From audience

Especially philosophers!

Smolin

[to remark from audience] You don't hang out with people who do quantum cosmology; you have no idea what can happen. But having opened it up, and if we're going to talk about...

Shimony

I just want to make a little comment about the clustering of problems. It's sometimes very fruitful when you see problems which initially were posed in very different ways, because of different problematic situations, nevertheless can be solved by common means. We have a lot of medical examples like this. What we have learned about the immune system has carried over from prevention of smallpox to many, many diseases that do not seem to be contagious, that can be genetic, you see what I mean. But it's not always the case that one gains by consolidating problems. When the finance company says consolidate your debts, be skeptical!

Smolin

So, not to have too many philosophical debts, when we come to quantum mechanics I happen to agree with you, and this is where I know your views and I agree with you. So I don't want to say too much, I want to open up the floor to people who don't agree with you. Anton, for example. While Anton is gathering his thoughts, we have another question.

From audience

I wanted to direct this question more to Lee than to you. It sounds to me like you have an aversion to incorporating transience into the fundamental physics.

Smolin

My own view, this is not about my view but since you asked, my view is in agreement with Abner that transience is real and should be somehow, I hope it will be somehow in the future incorporated into fundamental physics. We had a very interesting small workshop here that some of us were in about this issue in which we phrased it in the language of "Do the laws of nature need to be eternal or could laws of nature evolve in time, could time in some sense go all the way down?" And a very interesting contributor to that discussion was a philosopher that I'm more and more interested in, Roberto Unger. I'm just sort of throwing out things to be provocative, he talks about the poisoned gift of mathematics to physics that allows us to replace description of causal relations with logical relations which perhaps live in some Platonic realm where they're eternally true and therefore are quite different from causal relations, and I'm convinced this is a problem. I'm also convinced that to get out of it is not trivial; it means finding–and this is the question I was posing–it means finding a way to do physics without the notion of a pre-existing configuration space. But, again, I'm happy to discuss my views elsewhere. Anton, have you collected your thoughts?

From audience: Anton Zeilinger

That's a good question! About your closing the circle: this is a very fertile idea. This is now related to the question does this pass? If you want to close the circle, you need to have some kind of principle. Your phenomenological principle is one of them. There must be other ideas, like a consistency requirement, or things like that. Now I want to know: which principles do we need which guide us along the circle?

Shimony

You have to begin with your natural faculties, however flawed they are. There's no doubt that our rational faculties and our various emotional faculties are very much entangled and it was a great cultural achievement to start sorting things out and to use rational analysis to examine values and motivations. But you have to start where you are and then you learn, and one of the things you learn is that with primitive

Bistro Banter

methodology one learns something about the regularities of the world. The Greeks were able to predict eclipses, the Egyptians could predict the time of the rising of the Nile, they had rather primitive scientific method but they had something. When one learns something about the regularities of the world, and it was certainly a cultural achievement to learn that some regularities of the world can be explained mathematically, then that went back and refined the methodology. I'm not sure there was an explicit formulation of scientific method in which the mathematical character of fruitful hypotheses was emphasized until probably about the time of Huygens. Huygens did talk about the hypothetico-deductive method and certainly talked about physical laws in mathematical form. Well, not long afterwards there was Thomas Bayes who did a mathematical theorem, probability treated with an equation relating posterior probabilities to priors and to likelihood. But this is partly a cultural achievement and it meant looking back at the primitive scientific method and realizing that there are various components in it. One component being the credibility of various starting points, various hypotheses. Another is the component of how well our various pieces of evidence, how well they are predicted by competing hypotheses. Those are the likelihoods. But it took quite a lot of intelligent analysis to pick apart the hypothetico-deductive method and put it into that quantitative form. Well, my feeling is the job isn't done yet. We still don't have a completely adequate Bayesian formulation of scientific method. The weak spot being the prior probabilities. I have a sketch of a Bayesian theory which I call "strategic Bayesianism." Strategic Bayesianism is in contrast to personal probability theory which says the probabilities are nothing more than assertions of your degree of belief. Then there are the logical probabilists who say that somehow assignments of probabilities are natural extensions of relations of implication. That was Keynes' and Carnap's and Johnson's view.

Strategic probability is more modest in a way. It's more modest than logical probability because it doesn't say that the probabilities are intrinsic in the relations among propositions. It does say that there is a human element. On the other hand it's less modest than subjective probabilities because it says there are good strategies and bad strategies. My favorite formulation of a good strategy is by Saki, who said "In baiting a mousetrap with cheese, be sure to leave room for the mouse." I think that's terrific strategy. You don't want to assign such low prior probability—even to a hypothesis that you don't like and is proposed by one of your rivals anyway—you don't want to don't want to give it such low prior probability that no envisageable amount of evidence in favor of that hypothesis will overcome its initial disabilities. You've got to leave room for the mouse. That's a sketch of strategic Bayesianism. My hope is that with some friends here we'll examine what Newton did in detail. I think Newton was a crypto-Bayesian and an extremely subtle one and we will learn how he did it. And if we can make articulate what he kept secret, that will revolutionize Bayesianism. That's a hope; we'll see.

Smolin

I don't know if I'm the mouse or the cheese, but Chris:

From audience: Chris Fuchs

When you said that you were a strategic Bayesian...

Shimony

I'm a strategic Bayesian. I invented the term; I'm entitled to be it!

Fuchs

The only thought that came to my mind was that you are a strategic Bayesian and I am a Bayesian strategist.

Shimony

"Bayesian strategist" is a wider term.

Smolin

It leaves more room.

Shimony

That's right, but you also have more room for error than I do!
    Listen, Bayesianism is still in its youth, after two hundred and fifty years. It really needs fine-tuning.

Smolin

Anton, are you satisfied?

Zeilinger

You start from a very fundamental, basic approach to the world. I think you still seem to subscribe to making a distinction between epistemology and ontology. What leads you to that? What leads you to make a distinction between epistemology and ontology?

Shimony

Well, they're different enterprises even though they may link. Epistemology is the discipline of assessing beliefs and refining assertions, refining propositions about the world. Ontology is the discipline of systematizing knowledge of how the world is constituted. You need the epistemology in order to build a structure of knowledge. That was recognized very well by Leibniz and by Descartes and by rationalists, that you need epistemology to make the whole structure of knowledge grow. What I think was not fully realized—Peirce is very good on this—is that you need some knowledge of the world in order to assess and refine the epistemology. I think you will find in the history of philosophy more philosophers who consider epistemology to be autonomous, and ontology a derivative or an application of epistemology than you find people who thought the two need to be mutually supportive, mutually corrective. But I'm in the latter line of thought.

If you look at the *Encyclopedia Britannica* around 1911 on philosophy you will see how strongly epistemological it was. You have more of that in the later twentieth century with analytic philosophy, but, well, I hope I'm part of a reaction to this.

[Shimony takes a break, leaves platform]

Smolin to audience

Let's think of good things to pose to Abner when he comes back. Meanwhile, I'm happy to be the cheese.

From audience to Smolin

I don't see how your comment before about evolving laws comes from transience. You can imagine that, if you come to an understanding of the evolving laws, and you make an overall law in the theory of everything, it has some, maybe local, but some time parameter, and then you have an eternal law, that has a time parameter in it, and still no transience exists in your world.

Smolin

That's one of the severe puzzles. I agree with you. I have two things to offer, neither of which is very helpful.

Well, here is Abner.

Shimony

May I make a philosophical comment?

Smolin

Please do so, yes.

Shimony

menschliches, allzumenschliches.

Smolin

For those of us who are badly educated, what does that mean?

Shimony

Human, all too human.

Bistro Banter   471

Smolin

So, the question which was posed was, and let's make this our last topic, it really comes back to the issue of time, and the question is the following. Since you're a great admirer of Peirce, let me set the question and then quote Peirce, which I think is related. Many of us went into physics with the belief that there are eternally true laws and, our job is to find them. A remarkable feature of the last twenty years in theoretical particle physics, in the efforts for unification, is that we have not found unique true laws. What has seemed to have happened is that—and this is all in the realm of theory, unconfirmed by experiment—as models appear to unify more, they have more freedom in the choice of laws, the choice of parameters, not less freedom. The old hope that more unification would lead to more uniqueness, larger symmetries, tighter structures, seems in fact to be going the other way. The more unification, the more symmetries, in fact, the more free parameters in the laws. And this is a...

Shimony

May I just add one thing? More broken symmetries, which makes things even more complicated.

Smolin

Indeed, and many of the parameters come not just from how the symmetries break but from the models required to model the process of broken symmetry. And this has led to, in some peoples' minds, a crisis, it goes under the name of "the landscape," to some extent it was anticipated twenty years ago by a few people but has now become, with Lenny Susskind, a general hue and a cry, without much wisdom coming from our philosopher friends to help straighten it out. And surely wisdom is needed here. One extreme point of view is to say maybe the mistake was in looking for ultimate explanation of physical theory in some notion of some Platonically, eternally true law and here I quote Peirce, who worried about this, as you know, and this seems to have been his response: "To suppose universal laws of nature..." (I'm sure you know the quote but for those of you who don't know it) "...capable of being apprehended by the mind and yet having no reason for their special forms, but standing inexplicable and irrational, is hardly a justifiable position. Uniformities are precisely the sort of facts that need to be accounted for. Law is *par excellence* the thing that wants a reason. Now, the only possible way of accounting for the laws of nature, and for uniformity in general, is to suppose them results of evolution."

Now, I'll let you respond, but the question that our friend here was asking is that if you buy something like that surely, whatever scenario that evolution is described within, will take place against a framework of yet another law which will be posited to remain true throughout the evolution and therefore there's an infinite regress. How do we escape from this?

Shimony

Okay, well I have lots of things to say to that. One is this. The experience of twenty years or so is such a short time. Really, if there's any wisdom to be learned from the history of science it's patience. Think of the time between 1913 and 1925. How confused people were on atomic structure. They knew they were onto something right. but things didn't fall into place. Then there were de Broglie, Heisenberg, Schrödinger and their followers and things did fall into place. But why should one despair if for twelve years or so there's confusion? So that's one reaction.

Another reaction. Peirce is right to cite, but you're with him. I wrote a paper called 'Can the fundamental laws of nature be considered products of evolution?' and there are four principle characters: Whitehead, Peirce, John Wheeler and Lee Smolin.

Smolin

Not Andrei Linde?

Shimony

So you're in good company. And of course these are people I sympathize with. I think that the idea of evolution as a meta-explanatory principle is *incredibly* powerful. It's not only attractive as a project, but we know how powerful it is in practice in biology and for that matter in, say, the history of languages. It's an incredibly powerful one. The question is, is it the whole story? That is, can the idea of the four of you, that essentially *all* laws of nature are products of evolution, can that be maintained?

From audience: Geoff Hellman

What does it mean? I've no idea what it means. What does it mean to say a law is a product of evolution? You don't mean human activity of thinking of the law...

Bistro Banter 473

Shimony

No, no, not the history of science but the history of the universe, that is...

Hellman

What does that mean?

Shimony

Well, alright, let's give an unequivocal example first and then we'll go on to more questionable cases. The notion of law in biology is not as rigorous and strict as the notion of law in physics but it means something to biologists. And it is, I suppose, a law of biology that the inheritance of genetic traits is governed by DNA organized in chromosomes. But that clearly is a product of evolution. We have evidence that DNA was not the first organization of the nucleotides, that RNA, which is a simpler, one strand structure rather than a double helix, almost certainly came first. But it wasn't as efficient for various purposes as DNA was. So there's an example of a biological law, something that biologists would consider law, having a history, having a ...

Hellman

It's a law-like phenomenon.

Shimony

Well, you see, what Peirce is saying is that all laws of nature are really law-like. Even, whatever you like, the second law of motion in Newtonian mechanics, or Maxwell's equations, they are all law-like in that they are the products of evolution. Various things have been tried out and these, in some way or another, are the more stable situations. I'm paraphrasing these people, Lee can...

Hellman

You mean, you treat it as a chaotic situation, and then order evolves.

Shimony

Yeah, but I think it's a truly wonderful idea and partly true. But can it be the whole truth? And I am really skeptical that it can be the whole truth. Let me speak for Peirce because I've studied him more than the others. Peirce slips at one point. He says that in the primitive universe, chaotic universe, in fact there wasn't a metrical structure of time yet, so you can't even talk about time order, it was so chaotic. But in the primitive universe anything could happen. If anything can happen, anything can unhappen because these primitive events didn't have staying-power. One thing that happened, by chance, was a—this is his phrase—'a germ of habit-formation'. A pretty phrase. And unlike the other things that happened by chance it had the proclivity to spread. It's habit-formation. So it spreads, and it spreads to its neighbors. Now you see where my skepticism sets in. What is he talking about when he talks about "its neighbors"? How do you have neighbors if you don't have a geometrical structure? What is this all about? So hasn't he smuggled some primitive structure in order for the law-like habits to grow? And my real question is simply one question to all four of you: can you dispense with some kind of primitive law which allows the more detailed laws to grow? And I don't know that any of you gave an answer to that.

Smolin

And I believe that's the question being asked.

Shimony

Lee, do you have anything to add to what you wrote?

Smolin

I struggle with this myself, not just in the context that you described but I think that those of us who think about quantum spacetime—after all, if you're right about quantum mechanics then maybe we don't have to do quantum spacetime. Maybe the theory that you hopefully soon will discover that supersedes quantum mechanics, with some of the other people here, will save us and will be more easily integrated..

Bistro Banter

Shimony

But wouldn't that then be the kind of primitive law that I'm talking about, that is needed in order to get the more detailed law?

Smolin

Yes. Like you, I am a—I forget your wonderful word, but I believe we make lots of mistakes and I'm hesitating to appear more foolish than I must necessarily be because...

Shimony

No, I don't accuse you of that. No, no, never ever.

Smolin

No, no, just all of *them* [indicating audience]! But these are the hard problems and the question, you know, it sounds—the little story there with Peirce, and the little kernel of habit-formation—this sounds like Andrei Linde again, or those of us who worry—the question of what does it mean to be a neighbor if you're below the level of classical spacetime is a question that people here think about and try to see if there's a crucial question. I think of those thinking about quantum gravity, Lucien, Fotini, Olaf, just to mention some of the people here, a question like that, what does it mean to be a neighbor? And is that structure there initially, in which case geometry was classical, was intrinsic to begin with and if it's not how does the notion of being a neighbor arise from some level where it doesn't exist? Olaf is nodding, I think he's spent years working on that.

[*Inaudible remark from Olaf Dreyer*]

Smolin continues

He's just the photographer, he says.

So there's nothing to do but to wonder at the clarity of the thought of somebody like Peirce who in his confusion can mark out confusions that a hundred and twenty years later are still with us at the forefront. I think this is why, from the point of view

476 Bistro Banter

of those of us who are scientists, this is why we value our friends and our mentors who are philosophers.

Shimony

Let me just ask a question about philosophical attitude in questions of this depth. My question is: is there anything wrong, in your opinion, *intrinsically* wrong that you would have to stay away from it, of having several different levels, and several different approaches appropriate to the different levels? One level being the rather primitive laws of great generality, and they allow the evolutionary process to proceed. And the other kind of laws are themselves products of evolution and they presuppose the first kind. Is there anything intrinsically wrong about that?

Smolin

No.

Shimony

Now, let me tell you why I prefer it—I would prefer not to have it but we may have to live with it. It sounds as if you then are adopting an attitude that really is very peculiar. You're saying that you are taking as the very basis of rational knowledge about the world something for which no reason can be given. Namely, that's the way the world is, these primitive...

Smolin

That's what Mr. Peirce is saying, yes.

Shimony

... these primitive laws are the way the world is and they are brute fact. Now, here's my reaction to my reaction: Stay away from the word that I use, "brute." "Brute" is already an insult. It may be that you have to say, yes, there is something that is there, it's the fundamental in the way the world is constituted. Why not use a more favorable simile, like the *richness of the womb of nature* instead of *brute fact*? But stay away from terms that are tendentious either for or against. Anyway, I would say that it's a possibility that we have to accept these fundamental laws, that's the way things are. And is it so bad?

Smolin

That sounds like wisdom to me and I think that we should thank Abner.
[*applause*]

Smolin

And the bar is still open, I believe.

From audience: Myrvold

And we should thank Lee.
[*more applause*]

Shimony

May I tell you something? What I want to tell you is that this was much more fun than I thought it would be!

Unfinished Work: A Bequest

Abner Shimony

Abstract The following is a list of projects on which some results have been achieved but are still incomplete. Participants in this Conference and their students and colleagues are invited to carry investigations further:

(1) A quantum mechanical limitation upon the possibility of exact measurement due to the existence of additive conserved quantities; (2) The apparent impossibility of achieving a quantum mechanical mixture of definite measurement outcomes by means of a measurement procedure that is reliable or even approximately reliable if the initial state of the object is a superposition of eigenstates with different eigenvalues; (3) The extension to a system of n particles, with n greater than two, of an established complementarity relation between one-particle and two-particle interferometric visibilities in a two-particle system; (4) The refinement and performance of a proposed experiment for testing the hypothesis that the validity of the Pauli Exclusion Principle is a time dependent phenomenon, holding with increasing accuracy with the aging of an ensemble of fermions; (5) The resolution of the conflict between the locality implied by the special theory of relativity and the non-locality exhibited by violations of Bell's Inequalities in entangled quantum mechanical systems.

1 Introduction

Like most elderly men I must sadly accept the fact that my life's work is fragmentary and incomplete. This Conference, however, is not only a great and surprising honor, but also an opportunity to make a bequest of some unfinished work. I shall summarize five projects in the foundations of quantum mechanics on which I have achieved some progress but have not been able to complete. They seem worthy of further effort, and I hope that some of my friends assembled here, or their colleagues and students, will be attracted to continue the work. If they obtain results while I am still alive I hope that they will tell me about them.

A. Shimony
Emeritus Professor of Philosophy and Physics, Boston University
e-mail: abner.shimony@gmail.com

W.C. Myrvold and J. Christian (eds.), *Quantum Reality, Relativistic Causality, and Closing the Epistemic Circle*, The Western Ontario Series in Philosophy of Science 73,
© Springer Science+Business Media B.V. 2009

480   A. Shimony

2 Limitations on Exact Measurement due to Additive Conserved Quantities

Consider an object associated with a Hilbert space Ω_1 and a self-adjoint operator M measured by an apparatus associated with Hilbert space Ω_2. Suppose that the measurement is non-distorting, in the sense that if the object is just before the measurement in a state represented by an eigenvector u of M with eigenvalue r, it is after the measurement still in a state represented by u even if r is degenerate. Wigner [1] and Araki and Yanase [2] demonstrated that such a measurement is possible only if M commutes with every operator L_1 which is the first term of a bounded linear operator of the form $L_1 \oplus L_2$, where L_1 and L_2 are operators on Ω_1 and Ω_2 respectively. Stein and Shimony [3] generalized this theorem by allowing $L_1 \oplus L_2$ to be unbounded and permitting the measuring procedure to be finitely distorting, that is, the eigenspace $E(r) \subseteq \Omega_1$ associated with an eigenvalue r of M is the direct sum of eigenspaces $E(r, j)$, each of which is finite dimensional and carried into itself under the measurement process. They were unable to prove the natural further generalization in which the limitation to finite dimensionality of the eigenspaces $E(r, j)$ is dropped, nor were they able to construct a counter-example to the theorem when the condition of finite dimensionality is dropped. Furthermore, they were not able to settle this uncertainty even in a simple concrete model [4] of the measurement process, nor has any one else succeeded in doing so in spite of some effort. This open problem is of limited physical importance, since less ideal but experimentally satisfying measuring procedures are possible even when M fails to commute with an additive conserved quantity [3, pp. 62–63], but we continue to be intrigued by the mathematical problem.

3 The Problem of Definite Measurement Outcomes

An idealized scheme of measurement, when applied to an initial state of an object that is a superposition of eigenstates of the measured quantity, leads to the conclusion that there is there is no definite measurement result. Let u_1 and u_2 be eigenvectors with distinct eigenvalues of the self-adjoint linear operator M being measured, hence orthogonal, and let v_0 be the initial (neutral) state of the measuring apparatus with the property that the linear dynamics of object plus apparatus implies

$$u_j \otimes v_0 \Rightarrow u_j \otimes v_j, \tag{1}$$

where v_i is an eigenvector of the operator Q of the apparatus with an eigenvalue indicative of the initial value of the measured operator M. Then

$$(c_1 u_1 + c_2 u_2) \otimes v_0 \Rightarrow c_1(u_1 \otimes v_1) + c_2(u_2 \otimes v_2). \tag{2}$$

Hence, if neither of the coefficients c_1 and c_2 is zero, then the apparatus does not exhibit a definite value. Thus the formalism of quantum mechanics seems to preclude the occurrence of definite measurement results, if the measurement procedure is treated in the foregoing idealized way.

It has often been maintained that the interaction of the apparatus with the environment requires a correction of this idealization, and that definite measurement results can be obtained if the initial state of the apparatus is represented by a statistical operator T influenced by the environment. There exists, however, a series of theorems throwing doubt upon this optimism. A pioneering theorem of this type by Wigner [5] assumes a statistical operator T as the initial (neutral) state of the apparatus but continues to idealize measurement by assuming that if the initial state of the object is represented by an eigenvector u_i of the measured operator M with eigenvalue a_i, then the statistical operator $W(t)$ of object plus apparatus at the final time t of the measurement process,

$$W(t) = U(t)[P(u_i) \otimes T])[U(t)]^{-1}, \tag{3}$$

has the property

$$W(t) = P(u_i) \otimes \Sigma_k c_k |\psi_{ik}> < \psi_{ik}|. \tag{4}$$

Here $P(u_1)$ is the projection onto the one-dimensional subspace (or ray) of Ω_1 spanned by eigenvector u_1, and ψ_{ik} is a vector in $\Omega_2/E(i)$ belonging to a subspace F_i of Ω_2 in which the indicative operator Q of the measuring apparatus exhibits the eigenvalue a_i of the measured operator M of the object; k is an index of degeneracy required because the initial state of the apparatus is a statistical operator The pioneering theorem of Wigner asserts that if the foregoing conditions are satisfied but the initial state of the object is represented by the superposition $u = (c_1 u_1 + c_2 u_2)$ as in Eq. (2), with non-zero coefficients c_i, then the final state of object plus apparatus cannot be expressed as a statistical operator of the form

$$W(t) = \Sigma_k c_k |\Phi_{ik}> < \Phi_{ik}|, \tag{5}$$

where Φ_{ik} is a vector in $\Omega_1 \otimes \Omega_2$ in which M has value a_i and Q has the indicative value exhibiting this eigenvalue of M, and k is an index of degeneracy; in other words the final state is not a mixture of product states of object plus apparatus, each member of which exhibits a definite outcome of the measurement of M. Thus, within the scope of the assumptions of Wigner's theorem the linear dynamics of quantum mechanics is incompatible with definite results of measurement.

Of course Wigner's theorem as it stands does not preclude a solution to the measurement problem within the framework of standard quantum mechanics, because its negative result could conceivably still be due to an idealized characterization of measurement in spite of its replacement of the pure initial state of the apparatus by a statistical operator. In particular, expression (4) above is essentially a requirement that the measurement of M on an object initially in a definite (though presumably unknown) eigenstate of M is accomplished without the possibility of error. Conceivably the barrier to obtaining definite measurement results can be removed by

482   A. Shimony

replacing Eq. (4) by the condition that if the initial state of object plus apparatus is
$P(u_i) \otimes T$, then the state at time t has a probability greater than one-half of lying in
the subspace $\{u_1\} \otimes E(i)$, where $E(i)$ is the subspace of Ω_2 in all of which Q has a
value indicative of the value a_i of the object operator M. This condition is expressed
more formally in Shimony [6]. In that reference it is then demonstrated that even
with the foregoing weakened condition on measurement a transition from the initial
superposition (2) to the desirable mixture (5) of eigenstates of the indexical operator
Q is impossible. The same theorem was proved in a different way by Stein [7].

A further weakening of conditions on the procedure of measurement is to allow
measurable quantities to be represented by positive operator valued measures [8]
instead of the traditional self-adjoint operators, which are projection-valued measures. It was shown by Busch and Shimony [9] that the same impossibility result
demonstrated in reference [5] holds when the procedure of measurement is thus
generalized. Hence a radical and apparently promising avenue towards a solution
within the standard formalism of quantum mechanics of the problem of obtaining
definite measurement outcomes (which is also called "the measurement problem"
and "the problem of the reduction of superpositions") would confront an impasse.

The open problem which I am bequeathing to the participants in this Conference
and their students and colleagues is the assessment of some optimistic statements of
Machida and Namiki [10], Araki [11], and of Namiki and Pascazio [12] for solving
the measurement problem.

> Our answer to the measurement problem is affirmative. In fact we have explicitly derived the
> wave-function collapse by measurement... by taking into account the statistical fluctuations
> in the measuring apparatus, in the limit of infinite number of degrees of freedom of the
> apparatus system.

> It is very important to remark that the exact wave-function collapse takes place only in
> the infinite limit of N (the number of degrees of freedom) and is to be regarded as an
> asymptotic process, like a phase transition. However, in practice, a finite but very large
> N suffices to produce the wave-function collapse, as was repeatedly discussed and was
> shown by numerical simulations. Of course, as long as we keep N finite, the present theory
> yields only an approximate wave-function collapse, even though the exact collapse can be
> approximated up to any desired accuracy be increasing N. Do not forget that the present
> theory describes the exact wave-function collapse as an asymptotic limit.

> For fixed and finite N, coherence among the branch waves engendered by the spectral de-
> composition is *partially* lost, and the measurement is not perfect. ... Up to what extent a
> measurement is imperfect depends on the details of the physical process taking place in the
> detector. [12, p. 405]

Several questions need to be investigated concerning this program. One is
whether the authors mentioned have succeeded in rigorously proving that in a system with infinitely many degrees of freedom the wave-function collapse occurs in
the measurement process without violating the principles of quantum mechanics—
that is, that the negative result established by Wigner and generalized by Shimony
and Busch concerning the final statistical operator $W(t)$ in Eq. (5) is rigorously
avoided by assuming that the number N of degrees of freedom is infinite. A second
question concerns the claim regarding finite but very large N that this suffices for
"an approximate wave-function collapse." One aspect of this question is to clarify

the meaning of "approximately". A possible meaning is that the eigenvectors ψ_k of the indexical operator N do not have exact real values of the indexical quantity but rather have support on very small intervals α_k which are non-overlapping for different values of the index k. If the result claimed by Machida, Namiki, et al. holds for this sense of "approximately" then the reasonable requirement of the epistemology of measurement would be satisfied, since sensory differentiation of the different intervals would be possible. An entirely different meaning of "approximately" would sacrifice the exact orthogonality of the ψ_k. With this sense, the claimed result would not be epistemologically satisfactory, because *there would be no definite results registered by the measuring apparatus, but only probabilities of results,* which are just "potentialities" in Heisenberg's terminology. And even if all but one of these probabilities is very small and only one is non-negligible, one still would not have an *actual* result. And therefore the conceptual problem of explaining the transition from potentiality to actuality, which is the heart of the problem of wave-function collapse, strictly construed, would remain unsolved. A third question is partly conceptual and partly historical, concerning the relation between the program under discussion and the cluster of theories called "consistent history" [13] and "decoherence" [14] interpretations of quantum mechanics: does the mathematics of Machida and Namiki et al. fill in gaps in the informal reasoning of those theories, and do those theories suffice to clear up the obscurities concerning "approximately" when N is finite but very large?

4 Complementarity in n-Particle Interferometery

Two-particle interferometry has been investigated extensively, both theoretically [15] and experimentally [16]. A typical arrangement is shown in Fig. 1.

S is a source from which pairs of particles are emitted within a short time interval, with particle 1 propagating to the left through two apertures into paths A and A' and particle 2 propagating to the right into paths B and B'. The quantum states determined by paths A and A' are assumed to constitute a complete set of states for the Hilbert space of particle 1, and analogously for B and B' concerning particle 2. A phase shifter ϕ_1 is inserted in path A, and a phase shifter ϕ_2 is inserted in path B. Paths A and A' impinge on a symmetrical beam-splitter H_1 from which

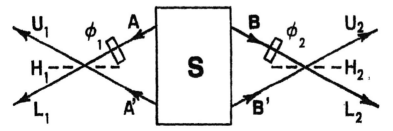

Fig. 1 Schematic two-particle four-beam inteferometer, using beam splitters H_1 and H_2

484   A. Shimony

there is probability $1/2$ to trigger a detector fed by path U_1 and probability $1/2$ to trigger a detector fed by path L_1. Analogously B and B' impinge on symmetric beam splitter H_2, each yielding probability $1/2$ for particle 2 to trigger the detector fed by U_2 and probability $1/2$ to trigger the detector fed by L_2. The most general state of the particle pair $1 + 2$ is

$$|\Psi> = \gamma_1|A> |B> + \gamma_2|A> |B'> + \gamma_3|A'> |B> + \gamma_4|A'> ||B'>, \qquad (6)$$

which by Schmidt's theorem [14, p. 85] can be expressed as

$$|\Psi> = \alpha|C> |D> + \beta|C'> |D'> . \qquad (7)$$

Here $|C>$ and $|C'>$ constitute a basis in the space apanned by $|A>$ and $|A'>$, while $|D>$ and $|D'>$ constitute a basis the space spanned by $|B>$ and $|B'>$, and the coefficients α and β (where $\alpha^2 + \beta^2 = 1$) can be chosen real by using the phase options of the vectors $|C>$. $|C'>$, $|D>$, $|D'>$. The most general unitary unimodular mapping relating the domain and counterdomain for particle 1 can be expressed in the $|C>$, $|C'>$ basis as

$$T_1|C> = a \exp(i\theta_1)|U_1> + b \exp(i\theta_1')|L_1> \qquad (8a)$$
$$T_1|C'> = -b \exp(-i\theta_1')|U_1> + a \exp(-i\theta_1)|L_1>, \qquad (8b)$$

And likewise the most general unitary unimodular mapping for particle 2 is

$$T_2|D> = c \exp(i\theta_2)|U_2> + d \exp(i\theta_2')|L_2> \qquad (9a)$$
$$T_2|D'> = -d \exp(-i\theta_2')|U_2> + c \exp(-i\theta_2)|L_2>, \qquad (9b)$$

where c and d are real numbers whose squares sum to unity.

The phases in Eqs. (8a, b) and (9a, b) are determined by the phase shifts ϕ_1 and ϕ_2 above and by the scalars in Eqs. (6), (8a), (8b), (9a), and (9b),

$T = T_1 \otimes T_2$ is the unitary unimodular mapping from the space initially associated with the pair $1 + 2$ into the space of output states. The probabilities of joint outcomes by ideal detectors placed in the output beams are

$$P(U_1U_2) = (\alpha ac)^2 + (\beta bd)^2 + \alpha\beta abcd \ \cos\Theta, \qquad (11)$$

where
$$\Theta = \theta_1 + \theta_2 + \theta_1' + \theta_2' \qquad (12)$$

Likewise,

$$P(U_1L_2) = (\alpha ad)^2 + (\beta bc)^2 - 2\alpha\beta abcd \ \cos\Theta, \qquad (13)$$
$$P(L_1U_2) = (\alpha bc)^2 + (\beta ad)^2 - 2\alpha\beta abcd \ \cos\Theta, \qquad (14)$$
$$P(L_1L_2) = (\alpha bd)^2 + (\beta ac)^2 + 2\alpha\beta abcd \ \cos\Theta, \qquad (15)$$

The probabilities of single outcomes are

$$P(U_1) = P(U_1U_2) + P(U_1L_2) = \beta^2 + a^2(\alpha^2 - \beta^2) \qquad (16)$$
$$P(U_2) = P(U_1U_2) + P(L_1U_2) = \beta^2 + a^2(\alpha^2 - \beta^2). \qquad (17)$$

Unfinished Work: A Bequest   485

The fringe visibilities for single outcomes are defined in the standard manner:

$$V_{single} = \{[P(U_i)]_{max} - [P(U_i)]_{min}\}/\{[P(U_i)]_{max} + [P(U_i)]_{min}\} = \alpha^2 - \beta^2 \; for \; i = 1 \; or \; 2$$
(18)

(Note: same evaluation if L is substituted for U.)

It is tempting to use the analogue of Eq. (18) to define the two-particle fringe visibility,

$$V_{pair} = \{[P(U_iU_2)]_{max} - [P(U_iU_2)]_{min}\}/\{[P(U_iU_2)]_{max} + [P(U_iU_2)]_{min}\}$$

This expression would yield a nonzero value even if $|\Psi >$ were a product state, for in that case $P(U_1U_2)$ would equal the product of $P(U_1)$ and $P(U_2)$, and these factors vary respectively with T_1 and T_2. We therefore "correct" the expressions for joint probability by subtracting the product $P(U_1)P(U_2)$ and adding a constant as a compensation against excessive subtraction:

$$P^*(U_1U_2) = P(U_1U_2) - P(U_1)P(U_2) + {}^1\!/_4.$$
(19)

Now the two-particle visibility is defined as

$$V_{pair} = [P^*(U_iU_2)]_{max} - [P^*(U_iU_2)]_{min}/[P^*(U_iU_2)]_{max} + [P^*(U_iU_2)]_{min}.$$
(20)

The foregong equations yield, after some calculation [18], a complementarity relation between V_{single} and V_{pair}:

$$(V_{single})^2 + (V_{pair})^2 = 1,$$
(21)

a relation which provides an alternative to Bohr's explanation of the complementarity between determining the position of a particle and exhibiting its contribution to an interference pattern interference pattern [18, p. 54].

An open problem is to establish a generalization of the complementarity relation Eq. (21) for entangled n-tuples of particles, with n greater than 2. Some results have been obtained for special cases of $n = 3$ by Horne [19], but no general expression. There is reason to suspect complications for n greater than 2, because then entanglement can be achieved in various ways: for example, a measurement performed on a single one of three entangled particles may yields a product state of the three individual particles, or it may yield a product state of two subsystems, one of which consists of a single particle and the other of two particles in an entangled state—a situation named "entangled entanglement" by Anton Zeilinger. Obviously, the taxonomy of types of entanglement becomes more complicated as n increases. It should also be noted that Horne does not introduce a "corrected" joint probability analogous to Eq. (19) in his treatment of $n = 3$.

Another open problem is to find a generalization of Eq. (20) when the initial state of the pair of particles $1 + 2$ is quantum mechanically described by a statistical operator rather than by a pure entangled state. I conjecture that the generalization is simply Eq. (21) with $=$ replaced by $\leq$.

5 Proposed Experiment to Test the Possible Time Dependence of the Onset of the Pauli Exclusion Principle (PEP)

Corinaldesi [20] has conjectured that the symmetry of integral spin particles under exchange and the anti-symmetry of half-integral spin particles under exchange are not kinematic principles but are the time-dependent consequences of interactions among the particles. Hence, a freshly constituted ensemble of electrons may exhibit violations of PEP, but as the ensemble ages the violations become much less frequent. I have proposed [21] an experiment to test Corinaldesi's conjecture, in which a beam of Ne^+ ions in a linear accelerator is crossed by a beam of electrons from an electron gun at variable positions along the flow direction of the ions (see Fig. 2). Some of the ions capture electrons, at a rate monitored by detectors sensitive to the photons emitted in the capture process. A PEP-violating electron can make a transition from the outermost level to the doubly occupied 1s level of a Ne atom, emitting a photon of approximately 1 keV. The rate of detection of such photons, which diminishes with the age of the ensemble and hence with the distance of the detector from the point of capture, permits in principle a calculation of the equilibration constant of Corinaldesi's conjecture. Reasonable assumptions about the parameters of the experimental arrangement indicate that if the conjecture is correct and the equilibration constant is not shorter than $10^{-15}\,s$, the proposed experiment can determine the value of this constant.

The main unfinished work of this project is finding a research team with a linear accelerator willing and able to perform the proposed experiments. When the characteristics of this linear accelerator are specified there will obviously be further problems of adapting and fine-tuning the proposed experiment.

If the experiment is performed and Corinaldesi's conjecture is vindicated, some important theoretical questions will be raised. What interactions among the electrons are responsible for the onset of an "equilibrium" regime in which PEP holds?

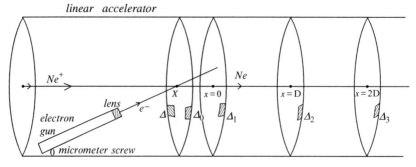

Fig. 2 Micrometer screw moves X in steps of 10^{-4} cm. Lens moves X in steps of 10^{-6} cm. Detectors Δ_0 and Δ' at X, both movable. Detectors $\Delta_1,\ldots,\Delta_{10}$ fixed, separated by 80 cm. Diameter of each detector: 2.5 cm. Δ' monitors the number of electrons captured by neon ions in a small interval around X. All other detectors Δ_i monitor anomalous electrons making transition to its level. Diagram not drawn to scale

In particular, is the Hamiltonian proposed by Corinaldesi himself in good agreement with the detailed experimental results? What explanation can be offered for the breakdown of Pauli's theorem on the connection of spin and statistics [22], which of course is incompatible with time dependence of PEP? One of the premises of Pauli's theorem is the validity of Lorentz Invariance in the small.

Evidence for the time-dependence of the onset of PEP would conceivably indicate a limitation on the validity of Lorentz Invariance.

6 Tension Between Relativistic Locality and the Non-locality Exhibited in Tests of Bell's Inequalities

Figure 3 presents schematically an experiment involving particle pairs $1+2$ emitted from a source S, with particle 1 subject to analysis by an analyzer with a controllable parameter a and particle 2 subject to analysis by an analyzer with controllable parameter b. The possible outcomes of analysis of particle 1 are s_m ($m = 1,2,\ldots$), and $-1 \leq s_m \leq 1$; the possible outcomes of the analysis of particle 2 are t_n ($n = 1\ldots$) and $-1 \leq t_m \leq 1$. The complete state (possibly the quantum state and possibly a more detailed state containing "hidden variables") of the pair $1+2$ at the time of emission from S will be denoted by k, and the space of states k will constitute a probability space $<K, \Sigma(K), \rho>$, where $\Sigma(K)$ is a sigma-algebra of subsets of K, and ρ is a probability measure on $\Sigma(K)$. Even when the complete state k of $1+2$ is specified, the outcomes of analysis of the two particles may be given only probabilistically:

$Pk(mn|ab)$ = probability that when the complete state is k and a is the parameter of the analyzer of 1 and b is the parameter of the analyzer of 2, then the outcomes of analysis are respectively s_m and t_n.

If the factorization condition

$$p_k(mn|ab) = p_k(m|a)p_k(n|b) \qquad (22)$$

is satisfied, then various inequalities can be derived regarding the probabilities determined by the complete states k, and by integrating these probabilities using the probability measure ρ one obtains inequalities governing the integrated probabilities

$$p(mn|ab) = \int_\rho p_k(mn|ab)d\rho, \qquad (23)$$

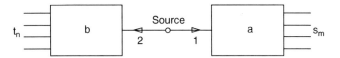

Fig. 3 An ensemble of particle pairs $1+2$ is emitted in a uniform manner from the source. Particle 1 enters an analyzer with a controllable parameter a, and the possible outcomes are s_m ($m = 1,2,\ldots$). Particle 2 enters an analyzer with controllable parameter b, and the possible outcomes are t_n ($n = 1,2,\ldots$).

or by calculating expectation values with these probabilities:

$$E(ab) = \Sigma_{mn} p(mn|ab) s_m t_n. \tag{24}$$

Quantities (23) and (24) are the theoretical counterparts of the frequencies and averages obtained with the experimental arrangement of Fig. 3. The collection of inequalities obtained in this way are collectively called "Bell's Inequalities", since the pioneering inequality of this class was derived by Bell [23] in 1964. An inequality experimentally investigated more frequently than Bell's pioneering inequality is that of Clauser, Horne, Shimony, and Holt (CHSH) [24], which will be stated explicitly:

$$-2 \leq E(ab) + E(ab') + E(a'b) - E(a'b') \leq 2. \tag{25}$$

The foregoing formalism is mainly a straightforward application of probability theory to the outcomes of analysis of the ensemble of pairs $1 + 2$. The factorization condition (Eq. (21), however) is a crucial physical condition, which Bell himself refers to as a "locality condition." This name is appropriate if the events of choosing the parameters a and b for analyzing particles 1 and 2 are spacelike separated in Einstein–Minkowski space, and the equivalence (established independently by Bell [25, pp. 56–57, 64–65] and Jarrett [26]) of Eq. (23) to the conjunction of the following two conditions (26) and (27) is applied:

Parameter Independence

$$p_k(m|ab) \text{ is independent of } b \text{ and can be written } p_k(m|a) \tag{26a}$$
$$p_k(n|ab) \text{ is independent of } b \text{ and can be written } p_k(n|b) \tag{26b}$$

Outcome Independence

$$p_k(m|abn) \text{ is independent of } n \text{ and can be written } p_k(m|ab) \tag{27a}$$
$$p_k(n|abm) \text{ is independent of } m \text{ and can be written } p_k(n|ab) \tag{27b}$$

Equations (26) and (27) are extensions to probabilistic connections of the usual limitation in the Special Theory of Relativity on direct causal connection, given the assumption above of the spacelike separation of the choice of parameters a and b.

Bell and his followers showed that the quantum mechanical predictions for analyses of certain pairs $1 + 2$ according to the scheme of Fig. 3 violates Ineq. (25) and others of Bell's Inequalities. More generally, Gisin [27] and Popescu and Rohrlich [28] have shown that *any* ensemble of pairs of particles whose complete state k is an entangled two-particle quantum state will predict a violation of Ineq. (25). Consequently, from the foregoing discussion, the quantum mechanics of such an ensemble predicts a violation of either Parameter Independence or Outcome Independence, and in this sense quantum mechanics is a non-local theory. Which one is violated, or are both violated? The violation of Outcome Independence is exhibited explicitly in quantum mechanical examples studied by Bell and his followers in the references cited above. That quantum mechanics does not violate Parameter Independence

Unfinished Work: A Bequest

has been demonstrated independently by Eberhard [29], Ghirardi et al. [30], and Page [31]. If Parameter Independence failed, then information could be encoded at the first locus by the choice of parameter a and would be probabilistically received at the locus of analysis of particle 2. The maintenance of Parameter Independence precludes the success of such encoding. On the other hand, no message can be conveyed by the failure of Outcome Independence, because the outcome m after parameter a has been chosen in the analysis of particle 1 is a matter of chance, not under the control of an experimenter at that locus.

The impossibility of capitalizing on quantum nonlocality for the purpose of sending superluminal messages has inspired hope that there is a kind of "peaceful coexistence" between this peculiarity of quantum mechanics and the special theory of relativity. For a long time I was attracted by this way of reconciling two of the fundamental branches of physics. With some regret I now find the following argument by Bell to be convincing:

> Do we then have to fall back on 'no signaling faster than light' as the expression of the fundamental structure of theoretical physics? This is hard for me to accept. For one thing we have lost the idea that correlations can be explained, or at least this idea awaits reformulations. More importantly, the 'no signaling...' notion rests on concepts that which are desperately vague, or vaguely applicable. The assertion that 'we cannot signal faster than light' immediately provokes the question:
>
> Who do we think *we* are?
>
> *We* who can make 'measurements', *we* who can manipulate 'external fields'. *we* who can 'signal' at all, even if not faster than light? Do *we* include chemists, or only physicists, plants, or only animals, pocket calculators, or only mainframe computers? [32, p. 254]

If Bell is right, that "peaceful coexistence" between relativistic locality and quantum nonlocality is not achieved by exhibiting the inability of the latter to permit superluminal signaling, then my bequest to you contains a very difficult problem. It is hard for me to believe that this is an isolated problem, because there are other difficulties in combining quantum mechanics with space-time theory, notably the problem of quantizing general relativity theory and the problem of maintaining the concept of a space-time continuum at the Planck level (around $10^{-33}\,cm$), where quantum uncertainties undermine the meaningfulness of the metric. One radical theory proposed for dealing with this tangle of problem is due to Heller [33], based on the non-commutative geometry of Connes [34].

The heuristic for this geometry is to approach ordinary differential geometry through algebra, and then to generalize the resulting algebraic formulation.

The information about a differentiable manifold is contained in the algebra of smooth functions on the manifold, by identifying a point in the manifold with the set of smooth functions vanishing at it. The algebra of smooth functions is commutative if the ordinary multiplication of function is taken to be the multiplication operation of the algebra. If, however, a non-commutative algebra, such as the algebra of bounded operators on a Hilbert space, is given instead of this commutative algebra, then in the corresponding space the local notions of point and neighborhood have no meaning. The distinction between singular and non-singular states breaks down, which looks very unpromising for extending the geometric concepts of general relativity, but a compensation may be a new way of dealing with the cosmological problem of the beginning of the universe. The breakdown of the concept of neighborhood seems to undermine the possibility of physical dynamics, which ordinarily requires the use of continuous functions; but the use of Leibniz's rule—$D(fg) = f(Dg) + (Df)g$—provides an algebraic surrogate for the ordinary operation of derivation, thereby making a generalized dynamics possible.

How seriously Heller's scheme should be taken depends upon success in several directions: first, the recovery of standard quantum mechanics and standard general relativity as "correspondence limits" at appropriate physical scales; second, the illumination of puzzling features of standard theories by considering them to be limiting cases of a physics with a pervasive global character—notably the feature of nonlocality in entangled quantum states, and the singularities and horizon problems of general relativity theory; and third, and most important, experimental evidence confirming predictions of the new physics at the Planck level. It is conceivable that some relevant experimental evidence will be provided if the test proposed above favors the time-dependence of the onset of the Pauli Exclusion Principle, and if, furthermore, the generalized dynamics at the Planck level predicts the observed time-dependence.

Of all the unfinished work that I am offering you as my bequest, this proposal about physics in the small is the most speculative and difficult. In the long run it probably will be accomplished, but I nevertheless must remind you of Pogo's warning, "We are surrounded by unsurmountable opportunities."

References

1. E.P. Wigner, "Die Messung quantenmechanischer Operatoren," *Zeitschrift f. Physik* **133**, 101–108 (1952).
2. H. Araki and M. Yanase, "Measurement of quantum mechanical operators," *Physical Review* **120**, 622–626 (1960).
3. H. Stein and A. Shimony, "Limitations on measurement," in B. d'Espagnat (ed.) *Foundations of Quantum Mechanics* (Academic, New York/London, 1971), 56–76.
4. A. Shimony and H. Stein, "A problem in Hilbert space theory arising from the quantum theory of measurement," *American Mathematical Monthly* **86**, 292–293 (1979).
5. E.P. Wigner, "The problem of measurement," *American Journal of Physics* **31**, 6–15 (1963).
6. A. Shimony, "Approximate measurement in quantum mechanics—II," *Physical Review D* **9**, 2321–2323 (1974). Reprinted in *Search for a Naturalistic World View*, vol. 2 (Cambridge University Press, Cambridge, UK, 1993), 41–47.
7. H. Stein, "Maximal extension of an impossibility theorem concerning quantum measurement," in R.S. Cohen, M. Horne, and J. Stachel (eds.) *Potentiality, Entanglement and Passion-at-a-Distance* (Kluwer Academic, Dordrecht/Boston/London, 1997), 231–243. Theoretical Physics **64**, 719 (1980).
8. P. Busch, P. Lahti, and P. Mittelstaedt, *The Quantum Theory of Measurement*, 2nd ed. (Springer, Berlin-Heidelberg, 1996), 6.
9. P. Busch and A. Shimony, "Insolubility of the quantum measurement problem for unsharp observables," *Studies in the History and Philosophy of Modern Physics* **27B**, 399 (1997).
10. S. Machida and M. Namiki, "Theory of measurement of quantum mechanics—mechanism of reduction of wave packet" I, II, *Progress in Theoretical Physics* **63**, 1457–1473, 1833–1843 (1980).

Unfinished Work: A Bequest 491

11. H. Araki, "A remark on Machida-Namiki theory of measurement," *Progress in Theoretical Physics* **64**, 719–730 (1980).
12. M. Namiki and S. Pascazio "Quantum theory of measurement based on the many-Hilbert-space approach," *Physics Reports* **232**, 302–411 (1993).
13. R.B. Griffiths, *Consistent Quantum Theory* (Cambridge University Press, Cambridge, UK, 2002).
14. R. Omnès, *Quantum Philosophy* (Princeton University Press, Princeton, NJ, 1999).
15. M.A. Horne, A. Shimony, and A. Zeilinger, "Two particle interferometry," *Physical Review Letters* **62**, 2209–2212 (1988).
16. A.F. Abouraddy, M.B. Nasr, B.E.A. Saleh, A.V. Sergienko, and M.C. Teich, "Demonstration of the complementarity of one- and two-photon interference," *Physical Review A* **63**, 8031–8036 (2001).
17. R.B. Griffiths, *Consistent Quantum Theory* (Cambridge University Press, Cambridge, UK, 2002), 85.
18. G. Jaeger, A. Shimony, and L. Vaidman, "Two interferometric complementarities," *Physical Review A* **51**, 63 (1995).
19. M. Horne, "Complementarity of fringe visibilities in three-particle quantum mechanics," in D. Greenberger, W. Reiter, and A. Zeilinger (eds.) *Epistemological and Experimental Perspectives on Quantum Physics* (Kluwer Academic, Dordrecht/Boston/London, 1999), 211–219.
20. E. Corinaldesi, "Model of a dynamical theory of the Pauli Principle," *Supplemento al Nuovo Cimento serie I* **5**, 937–943 (1967).
21. A. Shimony, "Proposed experiment to test the possible time dependence of the onset of the Pauli Exclusion Principle," *Quantum Information Processing* **5**, 277–286 (2006).
22. W. Pauli, "The connection between spin and statistics," *Physical Review* **58**, 716–722 (1940).
23. J.S. Bell, "On the Einstein-Podolsky-Rosen Paradox," *Physics* **1**, 195–200 1964. Reprinted in J.S. Bell, *Speakable and Unspeakable in Quantum Mechanics*, 2nd ed. (Cambridge University Press, Cambridge, UK, 2004), 14–21.
24. J. Clauser, M.A. Horne, A. Shimony, and R.A. Holt, Proposed experiment to test local hidden-variable theories," *Physical Review Letters* **23**, 880–884 (1969).
25. J.S. Bell, "The theory of local beables," in *Speakable and Unspeakable in Quantum Mechanics*, 2nd ed. (Cambridge University Press, Cambridge, UK, 2004), 56–57, 64–65.
26. J. Jarrett, "On the physical significance of the locality conditions in the Bell arguments," *Noûs* **18**, 569 (1984).
27. N. Gisin, "Bell's inequality holds [note: should read "is violated"] for all non-product states," *Physics Letters A* **154**, 201–202 (1991).
28. S. Popescu and D. Rohrlich, "Generic quantum nonlocality," *Physics Letters A* **166**, 293–297 (1992).
29. P. Eberhard, "Bell's theorem and the different conceptions of locality," *Nuovo Cimento B* **46**, 392–419 (1978).
30. G.-C. Ghirardi, A. Rimini, and T. Weber, "A general argument against super-luminal transmission through the quantum mechanical measurement process," *Lettere al Nuovo Cimento* **27**, 293–298 (1980).
31. D. Page, "The Einstein-Podolsky-Rosen physical reality is completely described by quantum mechanics," *Physics Letters A* **91**, 57–60 (1982).
32. J. S. Bell, "La Nouvelle Cuisine," in *Speakable and Unspeakable in Quantum Mechanics*, 2nd ed. (Cambridge University Press, Cambridge, UK, 2004).
33. M. Heller, "Cosmological singularity and the creation of the universe," *Zygon* **35**, 665–685 (2000); M. Heller, W. Sasin, and D. Lambert, "Groupoid approach to noncommutative quantization of gravity," *Journal of Mathematical Physics* **38**, 5840–5853 (1997).
34. A. Connes, *Noncommutative Geometry* (Academic, New York, 1994).

Bibliography of Abner Shimony

1947

1. "An Ontological Examination of Causation". *Review of Metaphysics* **1**, 52–68.

1948

2. "The Nature and Status of Essences". *Review of Metaphysics* **2**, 38–79.

1954

3. "Braithwaite on Scientific Method". *Review of Metaphysics* **7**, 444–460. An Italian translation appeared in 1978 in A. Meotti (ed.), *L'Induzione e l'Ordine dell'Universo*, pp. 125–131. Milano: Edizioni di Communità.

1955

4. "Coherence and the Axioms of Confirmation". *Journal of Symbolic Logic* **20**, 1–28. Reprinted in [110].
5. Theory of Information in the Discrete Communication System. Technical Memorandum 1690. Fort Monmouth, NJ: Signal Corps Engineering Laboratory.

1956

6. "Comments on David Harrah's 'Theses on Presupposition' ". *Review of Metaphysics* **9**, 121–122.

494   Bibliography of Abner Shimony

1962

7. (With Joel L. Lebowitz.) "Statistical Mechanics of Open Systems". *Physical Review* **128**, 1945–1959.

1963

8. "Role of the Observer in Quantum Theory". *American Journal of Physics* **31**, 755–773. Reprinted in [110].
9. "Stochastic Process of Keilson and Storer". *Physics of Fluids* **6**, 390–391.

1965

10. "Comments on the Papers of Prof. S. Schiller and Prof. A. Siegel". *Synthese* **14**, 189–195.
11. "Quantum Physics and the Philosophy of Whitehead". In M. Black (ed.), *Philosophy in America*, pp. 240–261. London: Allen and Unwin. Reprinted in 1965 in R.S. Cohen and M. Wartofsky (eds.), *Boston Studies in the Philosophy of Science* vol. 2 pp. 307–330. New York: Humanities Press. Reprinted in [110].

1966

12. Review of Alfred Landé, *New Foundations of Quantum Mechanics, Physics Today* **19**, (September), 85–90.

1967

13. "Amplifying Personal Probability Theory". *Philosophy of Science* **34**, 3–8.

1968

14. Review of Patrick Heelan, *Quantum Mechanics and Objectivity, Philosophical Review* **77**, 524–526.
15. Review of Max Jammer, *The Conceptual Development of Quantum Mechanics, Synthese* **18**, 118–120.
16. (With John Earman.) "A Note on Measurement". *Il Nuovo Cimento X*, **54B**, 332–334.

1969

17. (With John F. Clauser, Michael A. Horne, and Richard A. Holt.) "Proposed Experiment to Test Local Hidden-Variable Theories". *Physical Review Letters* **23**, 880–884. Reprinted in J.A. Wheeler and W.H. Zurek (eds.), *Quantum Theory and Measurement*, pp. 409–413. Princeton, NJ: Princeton University Press.

Bibliography of Abner Shimony

1970

18. "Resolution of the Paradox: a Philosophical Puppet Play". In W. Salmon (ed.), *Zeno's Paradoxes*, pp. 1–3. Indianapolis, IN: Bobbs-Merrill.

1971

19. "Homage to Rudolf Carnap". In R.C. Buck and R.S. Cohen (eds.), *PSA 1970, Boston Studies in the Philosophy of Science VIII*, pp. xxv–xxvii. Dordrecht: Reidel.
20. "Scientific Inference". In R. Colodny (ed.), *The Nature and Function of Scientific Theories*, pp. 79–172. Pittsburgh, PA: University of Pittsburgh Press. Reprinted in [110].
21. "Filters with Infinitely Many Components". *Foundations of Physics* 1, 325–328. Reprinted in [110].
22. (With Kenneth Friedman.) "Jaynes's Maximum Entropy Principle and Probability Theory". *Journal of Statistical Physics* 3, 381–384.
23. (With Howard Stein.) "Limitations on Measurement". In B. d'Espagnat (ed.), *Foundations of Quantum Mechanics (Proceedings of the International School of Physics "Enrico Fermi")*, pp. 56–76. New York: Academic.
24. "Babar et les Variables Cachées" (a poem), ibid., p. 170.
25. "Experimental Test of Local Hidden-Variable Theories". Ibid., pp. 182–194. Reprinted in [110].
26. "Philosophical Comments on Quantum Mechanics". Ibid., pp. 470–478.
27. "Perception from an Evolutionary Point of View". *Journal of Philosophy* 7, 571–583. Reprinted in [110].

1973

28. "Status of Hidden-Variable Theories". In P. Suppes, L. Henkin, A. Joja, and C.R. Moisil (eds.), *Logic, Methodology and Philosophy of Science*, vol. 4, pp. 593–601. Amsterdam: North-Holland.
29. "Comment on the Interpretation of Inductive Probabilities". *Journal of Statistical Physics* 9, 187–191.
30. (With Michael A. Horne.) "Hidden Variables and Quantum Uncertainty". *Epistemological Letters of the Institut de la Méthode* (Association Ferdinand Gonseth, Vienne, Switzerland), 1, 1–24.

1974

31. (With Mary H. Fehrs.) "Approximate Measurement in Quantum Mechanics — I". *Physical Review D* 9, 2317–2320. Reprinted in [110].
32. "Approximate Measurement in Quantum Mechanics — II". *Physical Review D* 9, 2321–2323. Reprinted in [110].

1975

33. "Carnap on Entropy". In J. Hintikka (ed.), *Rudolf Carnap; Logical Empiricist*, pp. 381–395. Dordrecht: Reidel. Reprinted in [48], pp. viii–xxii.

496   Bibliography of Abner Shimony

34. "Vindication: A Reply to Paul Teller". in G. Maxwell and R. Anderson (eds.), *Induction, Probability, and Confirmation*, pp. 204–211. Minneapolis, MN: University of Minnesota Press.

1976

35. "Reply to Dr. Lochak". *Epistemological Letters of the Institut de la Méthode* **8**, 1–5.
36. (With John F. Clauser and Michael A. Horne.) "Comment on Bell's Theory of Local Beables". Ibid., **13**, 1–8. Reprinted in [77,110].
37. (With Bror Hultgren.) "The Lattice of Verifiable Propositions of the Spin-1 System". *Journal of Mathematical Physics* **18**, 381–394.
38. "Comments on Two Epistemological Theses of Thomas Kuhn". In R.S. Cohen, P. Feyerabend, and M. Wartofsky (eds.), *Essays in Memory of Imre Lakatos (Boston Studies in Philosophy of Science*, vol. 39), pp. 569–588. Dordrecht: Reidel. Reprinted in [110].

1977

39. "Is Observation Theory-Laden? A Problem in Naturalistic Epistemology". In R. Colodny (ed.), *Logic, Laws, and Life*, pp. 185–208. Pittsburgh, PA: University of Pittsburgh Press. Reprinted in 1979 in D. Boyer, P. Grim, and J. Sanders, *The Philosopher's Annual, 1978*, pp. 116–145. Totowa, NJ: Rowman and Littlefield. Reprinted in [110].
40. "Some Comments on Philosophical Method". In Paul Weiss, *First Philosophy*, pp. 187–189. Carbondale, IL: Illinois University Press.
41. "Comments on Papers by Tisza and Brush". In F. Suppe and P. Asquith (eds.), *PSA 1976*, vol. 1, pp. 609–616. East Lansing, MI: Philosophy of Science Association.
42. (With Joseph Hall, Christopher Kim, and Brien McElroy.) "Wave-Packet Reduction as a Medium of Communication". *Foundations of Physics* **7**, 759–767. Reprinted in [110].
43. (With Michael A. Horne.) "Comment on Zeilinger's Talk". *Progress in Scientific Culture* **1/4**, 459–460.
44. "Reply to Bell". *Epistemological Letters of the Institut de la Méthode*, **18**, 1–4. Reprinted in [77,110].

1978

45. "Some Theses and Questions concerning the Foundations of Quantum Mechanics". *Epistemological Letters of the Institut de la Méthode* **19**, 23–24.
46. "Reply to Costa de Beauregard". *Epistemological Letters of the Institut de la Méthode* **20**, 33–36.
47. "Metaphysical Problems in the Foundations of Quantum Mechanics". *International Philosophical Quarterly* **18**, 8–17. Reprinted in R. Boyd, P. Gasper, and J.D. Trout (eds.), *The Philosophy of Science*, pp. 517–528. Cambridge, MA: MIT Press.
48. Edited Rudolf Carnap, *Two Essays on Entropy*, with an Introduction. Berkeley, CA: University of California Press.
49. (With John F. Clauser) "Bell's Theorem: Experiments and Implications". *Reports on Progress in Physics* **41**, 1881–1927. Reprinted with additional notes in S. Stenholm (ed.), *Lasers in Applied and Fundamental Research*, Bristol, UK: Adam Hilger.
50. Review of S.S. Chissick and W.C. Price (eds.), *Uncertainty Principle and Foundations of Quantum Mechanics*, *Science* **99**, 168–169.
51. Review of B. d'Espagnat, *Conceptual Foundations of Quantum Mechanics, American Journal of Physics* **46**, 590–591.

Bibliography of Abner Shimony 497

1979

52. Review of Thomas Kuhn, *Black-Body Theory and the Quantum Discontinuity, 1894–1912*, *Isis* **70**, 434–437.
53. "Quantum Mechanics". In *McGraw-Hill Yearbook of Science and Technology*, pp. 341–348. New York: McGraw-Hill.
54. "Proposed Neutron Interferometer Test of Some Nonlinear Variants of Wave Mechanics". *Physical Review A* **20**, 394–396. Reprinted in [110].
55. (With Howard Stein.) "A Problem in Hilbert Space Theory Arising from the Quantum Theory of Measurement", *American Mathematical Monthly* **86**, 292–293.

1980

56. "An Analysis of the Proposal of Garuccio and Selleri for Superluminal Signalling". *Epistemological Letters of the Institut de la Méthode* **25**, 55–57.
57. "The Point We Have Reached". *Epistemological Letters of the Institut de la Méthode* **26**, 1–7.
58. "Comments on Fine and Suppes". In A. Asquith and R. Giere (eds.), *PSA 1980*, vol. 2, pp. 572–582. East Lansing, MI: Philosophy of Science Association.

1981

59. "Integral Epistemology". In M.B. Brewer and B.E. Collins (eds.), *Scientific Inquiry and the Social Sciences*, pp. 98–123. San Francisco, CA: Jossey-Bass. Reprinted in [83] and [110].
60. (With Penha Dias.) "A Critique of Jaynes' Maximum Entropy Principle". *Advances in Applied Mathematics* **2**, 172–211.
61. (With T.K. Lo.) "Proposed Molecular Test of Local Hidden-Variables Theories". *Physical Review A* **23**, 3003–3012.
62. "Réflexions sur la Philosophie de Bohr, Heisenberg, et Schroedinger". *Journal de Physique* **42** C2, 81–93. Reprinted in English translation in [67,110].
63. "Meeting of Physics and Metaphysics" (a review of D. Bohm, *Wholeness and the Implicate Order*), *Nature* **291**, 435–446.
64. Review of Hans Primas, *Chemistry, Quantum Mechanics and Reductionism*, *American Journal of Physics* **51**, 1159–1160.

1982

65. "Laszlo Tisza's Contributions to Philosophy of Science". In A. Shimony and H. Feshbach (eds.), *Physics as Natural Philosophy*, pp. 412–428. Cambridge, MA: MIT Press.

1983

66. Review of E.G. Beltrametti and G. Cassinelli, *The Logic of Quantum Mechanics*, *Physics Today* **36** (December), 62–64.
67. "Reflections on the Philosophy of Bohr, Heisenberg, and Schroedinger" (translation of [62]). In R.S. Cohen and L. Laudan (eds.), *Physics, Philosophy and Psychoanalysis (Boston Studies in Philosophy of Science*, vol. 76), pp. 209–221. Dordrecht: Reidel. Reprinted in [110].
68. "The Status of the Principle of Maximum Entropy", *Synthese* **63**, 35–53. Reprinted in [110].

498   Bibliography of Abner Shimony

69. "Some Proposals Concerning Parts and Wholes". In P. Sällstrom (ed.), *Parts and Wholes*, vol. 1, pp. 115–122. Stockholm: Swedish Council for Planning and Coordination of Research.

1984

70. "Contextual Hidden Variables Theories and Bell's Inequalities". *British Journal for the Philosophy of Science* **35**, 25–45. Reprinted in [110].
71. "Controllable and Uncontrollable Non-locality". In S. Kamefuchi et al. (eds.), *Foundations of Quantum Mechanics in the Light of New Technology*, pp. 225–230. Tokyo: The Physical Society of Japan. Reprinted in [110].
72. (With Michael A. Horne.) "Comment on Angelidis". *Physical Review Letters* **53**, 1296.
73. Response to "Comment on 'Proposed Molecular Test of Local Hidden-Variables Theories'", *Physical Review A* **30**, 2130–2131.
74. "Further Comments on Parts and Wholes". In P. Sällstrom (ed.), *Parts and Wholes*, Vol. 2. Stockholm: Swedish Council for Planning and Coordination of Research. Reprinted in [110].

1985

75. Comment on "Consistent Inference of Probabilities for Reproducible Experiments". *Physical Review Letters* **55**, 1030.
76. Review of Henry Folse, *The Philosophy of Niels Bohr: The Framework of Complementarity, Physics Today* **38** no. 10, 108–109.
77. (With John S. Bell, J. Clauser, and Michael A. Horne.) "An Exchange on Local Beables". *Dialectica* **39**, 83–110. Includes reprints of [36, 44].

1986

78. "Bulletins Delivered from the Quantum World". *Nature* **320**, 578.
79. "Events and Processes in the Quantum World". In R. Penrose and C.J. Isham (eds.), *Quantum Concepts in Space and Time*, pp. 182–203. Oxford: Oxford University Press. Reprinted in [110].
80. "The significance of Jarrett's 'Completeness Condition'". In L.M. Roth and A. Inomata (eds.), *Fundamental Questions in Quantum Mechanics*, pp. 29–31. New York: Gordon & Breach.
81. "Summary of Conference". Ibid., pp. 331–332.
82. "Concluding Remarks". In D. Greenberger (ed.), *New Techniques and Ideas in Quantum Measurement Theory*, pp. 629–632. New York: The New York Academy of Sciences.

1987

83. *Naturalistic Epistemology: A Symposium of Two Decades*, A. Shimony and D. Nails (eds.), (*Boston Studies in the Philosophy of Science*, vol. 100). Dordrecht: Reidel. Includes the following by A. Shimony: Introduction, pp. 1–13, 'Integral Epistemology,' pp. 299–318 (reprint of [59], Comment on Capek, pp. 112–114, Comment on Reed, pp. 230–234, Comment on Levine, pp. 291–294, Comment on Sagal, pp. 333–336, Comment on Agassi, pp. 352–355. Comment on Wartofsky, pp. 375–377.

Bibliography of Abner Shimony   499

84. "The Methodology of Synthesis: Parts and Wholes in Low Energy Physics". In R. Kargon and P. Achinstein (eds.), *Kelvin's Baltimore Lectures and Modern Theoretical Physics*, pp. 399–424. Cambridge, MA: MIT Press. Reprinted in [110].

1988

85. "Evidence of the Quantum World". *Scientific American* **258** (January), 46–53, reply to a letter from O. Piccioni. Ibid., July 9
86. "Philosophical Problems in the Work of Louis de Broglie". In G. Lochak (ed.), *Louis de Broglie que Nous Avons Connu*, pp. 207–209. Paris: Fondation Louis de Broglie.
87. (With Elida de Obaldia and Frederick Wittel.) "Amplification of Belinfante's Demonstration of the Nonexistence of a Dispersion-Free State", *Foundations of Physics* **18**, 1013–1021.
88. "Conceptual Foundations of Quantum Mechanics". In P. Davies (ed.), *The New Physics*, pp. 373–395. Cambridge, UK: Cambridge University Press.
89. "Physical and Philosophical Issues in the Bohr–Einstein Debate". In H. Feshbach (ed.), *Niels Bohr: Physics and the World*, pp. 285–304. Chur: Harwood. Reprinted in [110].
90. Comment on Martin Eger, "A Tale of Two Controversies". *Zygon* **23**, 333–340. Reprinted in [110].
91. (With Michael A. Horne and Anton Zeilinger.) "Two-Particle Interferometry". *Physical Review Letters* **62**, 2209–2212.
92. "An Adamite Derivation of the Principles of the Calculus of Probability". In J.H. Fetzer (ed.), *Probability and Causality: Essays in Honor of Wesley Salmon*, pp. 79–89. Dordrecht: Reidel. Reprinted in [110].

1989

93. "Search for a World View that will Accommodate our Knowledge of Microphysics". In J. Cushing and E. McMullin (eds.), *Philosophical Consequences of Quantum Theory: Reflections on Bell's Theorem*, pp. 25–37. Notre Dame: Notre Dame University Press. Reprinted in [110].
94. "The Non-existence of a Principle of Natural Selection". *Biology and Philosophy* **4**, 255–273. Reprinted in [110].
95. Reply to Sober, Ibid., 280–286. Reprinted in [110].

1990

96. "The Theory of Natural Selection as a Null Theory". In R. Cooke and D. Costantini (eds.), *Statistics in Science (Boston Studies in the Philosophy of Science*, vol. 122), pp. 15–26. Dordrecht: Kluwer/Reidel.
97. "An Exposition of Bell's Theorem". In A. Miller (ed.), *Sixty-two Years of Uncertainty*, New York: Plenum, pp. 33–43. Reprinted in [110].
98. (With Michael A. Horne and A. Zeilinger.) "Introduction to Two-Particle Interferometry". Ibid., pp. 113–119.
99. Some Comments and Reflections, ibid., pp. 309–310.
100. (With Joy Christian.) "The Relation between the Aharonov-Anandan Phase and the Berry Phase in a Simple Model". In S. Kobayashi et al. (eds.), *Proceedings of the Third International Symposium Foundations of Quantum Mechanics in the Light of New Technology*, pp. 93–97. Tokyo: The Physical Society of Japan.

500   Bibliography of Abner Shimony

101. (With Joy Christian.) "Geometric Phase in a Simple Model". In J. Anandan (ed.), *Quantum Coherence*, pp. 121–124. Singapore: World Scientific.
102. (With Michael A. Horne and Anton Zeilinger.) "Down-Conversion Photons: A New Chapter in the History of Entanglement". Ibid., pp. 356–372.
103. (With Michael A. Horne and Anton Zeilinger.) "Quantum Optics: Two-Particle Interferometry". *Nature* **347**, 429–430.
104. (With Daniel Greenberger, Michael A. Horne, and Anton Zeilinger.) "Bell's Theorem without Inequalities". *American Journal of Physics* **58**, 1131–1143.
105. "Desiderata for Modified Quantum Dynamics". In A. Fine, M. Forbes, and L. Wessels (eds.), *PSA 1990*, pp. 49–59. East Lansing, MI: Philosophy of Science Association. Reprinted in [110].

1991

106. (With Martinus Veltman and Valentine Telegdi.) Obituary of John S. Bell, *Physics Today* **44** no. 8, 1991–1993, 82–86.
107. "Meditation on Time" (a poem). In A. Ashtekar and J. Stachel (eds.), *Conceptual Problems in Quantum Gravity*, p. 125. Basel: Birkhaeuser. Reprinted in [130].

1992

108. "That There Exists no Greatest Prime" (a poem). *Synthese* **92**, 313–314.
109. "On Carnap: Reflections of a Metaphysical Student". *Synthese* **93**, 261–274.

1993

110. *Search for a Naturalistic World View*, vols. I and II. Cambridge, UK: Cambridge University Press. Volume I: Acknowledgments; reprints of [4, 27, 36, 39, 59, 68, 74, 90, 92, 93]; and two previously unpublished essays, "Reality, Causality, and Closing the Circle". pp. 21–61, and "Reconsiderations on Inductive Inference", pp. 274–300, Volume II: Reprints of [8, 11, 20, 21, 25, 31, 32, 38, 42, 44, 54, 62, 67, 69, 70, 71, 84, 89, 94, 95, 97, 98, 105, 129]; and two previously unpublished essays: (with Michael A. Horne) "Proposed Neutron Interferometer Observation of the Sign Change of a Spinor Due to 2π Precession", pp. 72–74, and "Toward a Revision of the Protophysics of Time", pp. 255–270.
111. (With Daniel M. Greenberger, Herbert J. Bernstein, Michael A. Horne, and Anton Zeilinger.) "Proposed GHZ Experiments Using Cascades of Down-Conversions". In H. Ezawa and Y. Murayama (eds.), *Quantum Control and Measurement*, pp. 23–28. Amsterdam: Elsevier Science Publishers.
112. (With Joy Christian.) "Non-Cyclic Geometric Phases in a Proposed Two-Photon Interferometric Experiment". *Journal of Physics A* **26**, 5551–5567.
113. (With Gregg Jaeger and Michael A. Horne.) "Complementarity of One-Particle and Two-Particle Interference". *Physical Review A* **48**, 1023–1027.
114. "Dr. Hegels Geistesschwitzengymnasium". *Bostonia*, Fall, 88.

1994

115. (With Gregg Jaeger.) "Complementarity and Path Distinguishability: Some Recent Results Concerning Photon Pairs". In D. Han, Y.S. Kim, M.H. Rubin, Y. Shih, and W.W. Zachary

Bibliography of Abner Shimony 501

(eds.), *Third International Workshop on Squeezed States and Uncertainty Relations*, pp. 523–533. Greenbelt, MD: NASA Conference Publications.

116. "The Relation between Physics and Philosophy". Ibid., pp. 617–623.

117. "Objectivity, Evolution et cetera in Shimony's Naturalistic World View". *Physics Today* **47** no. 7, 9.

118. Review of Henry P. Stapp, *Mind, Matter, and Quantum Mechanics, American Journal of Physics* **62**, 956–957.

120. "Ten Philosophical Poems". In C.C. Gould and R.S. Cohen (eds.). *Artifacts, Representations and Social Practice* (*Boston Studies in the Philosophy of Science*, vol. 154), pp. 43–48. Dordrecht: Kluwer Academic.

121. Two plays, "The Confrontation" and "Monadology". In I.C. Jarvie and N. Laor (eds.), *Critical Rationalism, Metaphysics and Science* (*Boston Studies in the Philosophy of Science*, vol. 161), pp. 29–32. Dordrecht: Kluwer Academic.

1995

122. (With Gregg Jaeger and Lev Vaidman.) "Two Interferometric Complementarities". *Physical Review A* **51**, 54–67.

123. (With Gregg Jaeger.) "Optimal Distinction between Two Non-orthogonal States". *Physics Letters A* **197**, 83–87.

124. "Degree of Entanglement". In D. Greenberger and A. Zeilinger (eds.), *Fundamental Problems in Quantum Theory*, pp. 675–679. New York: New York Academy of Sciences.

125. (With Michael A. Horne.) "Multipath Interferometry of the Biphoton", Ibid., pp. 664–674.

126. "Cybernetics and Social Entities". In K. Gavroglu, J. Stachel, and M. Wartofsky (eds.), *Science, Politics and Social Practice* (*Boston Studies in the Philosophy of Science*, vol. 164), pp. 170–194. Dordrecht: Kluwer Academic.

127. "Empirical and Rational Components in Scientific Confirmation". In D. Hull, M. Forbes, and R.M. Burian (eds.), *PSA 1994*. vol. 2, pp. 146–155. East Lansing, MI: Philosophy of Science Association.

128. (With M. Freyberger, P.K. Aravind, and Michael A. Horne.) "Proposed Test of Bell's Inequality without a Detection Loophole by Using Entangled Rydberg Atoms". *Physical Review A* **53**, 1232–1244.

129. "The Transient Now". In L.E. Hahn (ed.), *The Philosophy of Paul Weiss*, pp. 331–348. Chicago, IL: Open Court.

130. "Philosophical Reflections on EPR". In A. Mann and M. Revzen (eds.), *The Dilemma of Einstein, Podolsky and Rosen—60 Years Later. An International Symposium in Honour of Nathan Rosen, Haifa, Israel*, Annals of Israel Physical Society **12**, 27–41.

131. "Measures of Entanglement". Ibid., pp. 163–176.

1996

132. "Three Poems on Time". In U. Ketvel et al. (eds.), *Vastakohtien Todellisuus*, p. 147. Helsinki: Helsinki University Press.

133. (With Paul Busch.) "Insolubility of the Quantum Measurement Problem for Unsharp Observables". *Studies in History and Philosophy of Modern Physics* **27**, 397–404.

134. (With Millard Baublitz.) "Tension in Bohm's Interpretation of Quantum Mechanics". In J. Cushing, A. Fine, and S. Goldstein (eds.), *Bohmian Mechanics and Quantum Theory: an Appraisal* (Boston Studies in Philosophy of Science, vol. 184), pp. 251–264, Dordrecht: Kluwer Academic.

135. "A Bayesian Examination of Time-Symmetry in the Process of Measurement". *Erkenntnis* **45**, 197–208. See Errata.

502 Bibliography of Abner Shimony

1997

136. Edited with annotations Vol. III, Part II of *The Collected Papers of Eugene P. Wigner, Foundations of Quantum Mechanics*, Berlin/Heidelberg: Springer.
137. "On Mentality, Quantum Mechanics and the Actualization of Potentialities". In Roger Penrose (ed.), *The Large, the Small and the Human Mind*, pp. 144–159. Cambridge, UK: Cambridge University Press.
138. (Illustrated by Jonathan Shimony) *Tibaldo and the Hole in the Calendar*, New York: Copernicus.
139. "Some Historical and Philosophical Reflections on Science and Enlightenment", *Philosophy of Science* **64**, S1–S14.

1998

140. "Implications of Transience for Spacetime Structure". In S.A. Huggett, L.J. Mason, K.P. Tod, S.T. Tsou, and N.M.J. Woodhouse (eds.), *The Geometric Universe: Science, Geometry, and the Work of Roger Penrose*, pp. 161–172. Oxford/New York: Oxford University Press.
141. "Comments on Leggett's 'Macroscopic Realism". In Richard A. Healey and Geoffrey Hellman (eds.), *Quantum Measurement: Beyond Paradox*, pp. 21–31. Minneapolis/London: University of Minnesota Press.
142. "On Ensembles that are both Pre- and Post-selected". *Fortschritte der Physik* **46**, 725–728. See Errata.

1999

143. "Can the Fundamental Laws of Nature be the Results of Evolution?" In J. Butterfield and C. Pagonis (eds.), *From Physics to Philosophy*, pp. 208–223. Cambridge UK: Cambridge University Press.
144. (With Gregg Jaeger.) "An Extremum Principle for a Neutron Diffraction Experiment". *Foundations of Physics* **28**, pp. 435–444.

2000

145. "Philosophical and Experimental Perspectives on Quantum Physics". In D. Greenberger, W.L. Reiter, and A. Zeilinger (eds.), *Epistemological and Experimental Perspectives on Quantum Physics: Vienna Circle Institute Yearbook 1999*, pp. 1–18. Dordrecht/Boston, Kluwer Academic.
146. Edited, with M. A. Horne and A. Zeilinger, a Festschrift for Daniel Greenberger. *Foundations of Physics* **29**, nos.3, 4.
147. "Holism and Quantum Mechanics". In H. Atmanspacher, A. Amann, and U. Müller-Herold (eds.), *On Quanta, Mind and Matter: Hans Primas in Context*, pp. 233–246. Dordrecht/Boston/London: Kluwer Academic.

2001

148. "The Logic of EPR". *Annales de la Fondation Louis de Broglie* **26**, n° spécial, 399–409.
149. (With Howard Stein.) "Comment on 'Nonlocal Character of Quantum Theory', by Henry P. Stapp". *American Journal of Physics* **69**, 848–853.

Bibliography of Abner Shimony   503

2002

150. Review of Shimon Malin, *Nature Loves to Hide*. *Americnn Journal of Physics* **70**, 848–853.
151. (With Roman Jackiw.) "The Depth and Breadth of John Bell's Physics". *Physics in Perspective*. **4**, 78–116.
152. "John S. Bell: Some Reminiscences and Reflections". In R.A. Bertlmann and A. Zeilinger (eds.), *Quantum [Un]speakables*, pp. 51–60. Berlin/Heidelberg/New York: Springer.
153. "The Character of Howard Stein's Work in Philosophy and History of Physics". In David B. Malament (ed.), *Reading Natural Philosophy*, pp. 1–8. Chicago/LaSalle, IL: Open Court.
154. "Some Intellectual Obligations of Epistemological Naturalism". Ibid., pp. 297–313.

2003

155. (With Howard Stein.) "On Quantum Non-locality, Special Relativity, and Counterfactual Reasoning". In J. Renn et al. (eds.), *Space-Time, Quantum Entanglement and Critical Epistemology: Essays in Honor of John Stachel*". Dordrecht/Boston, FL: Kluwer Academic.
156. "Four Poems Dedicated to David Mermin". *Foundations of Physics* **33**, 1699–1700.

2004

157. "Comment on Norsen's Defense of Einstein's 'Box Argument'". *American Journal of Physics* **72**, 1–2.
158. "Wigner's Contributions to the Quantum Theory of Measurement". *Acta Physica Hungarica* **B20**, Wigner Centennial Volume, 59–72.

2005

159. "An Analysis of Ensembles That Are Both Pre- and Post-Selected". *Foundations of Physics* **35**, 215–232.
160. Review of Michael Epperson's *Quantum Mechanics and the Philosophy of Alfred North Whitehead, Transactions of the Charles S. Peirce Society* **41**, 714–723.

2006

161. Review of GianCarlo Ghirardi, *Sneaking a Look at God's Cards: Unraveling the Mysteries of Quantum Mechanics, Physics in Perspective* **8**, 247–255.
162. "An Analysis of Stapp's 'A Bell-Type Theorem Without Hidden Variables'". *Foundations of Physics* **36**, 61–72.
163. "Dialogue: Shimony-Malin". *Quantum Information Processing* **5**, 261–276.
164. "Proposed Experiment to Test the Possible Time Dependence of the Onset of the Pauli Principle". Ibid., 277–286.
165. Edited and contributed an Introduction and part of a Dialogue to Martin Eger's *Science, Understanding, and Justice: The Philosophical Essays of Martin Eger*. Chicago, IL: Open Court.

504   Bibliography of Abner Shimony

2007

166. "Aspects of Nonlocality in Quantum Mechanics". In James Evans and Alan S. Thorndike
     (eds.), *Quantum Mechanics at the Crossroads*, pp. 107–123. Berlin/Heidelberg/New York:
     Springer. See Errata.

2008

(With Roman Jackiw.) "John Stewart Bell". In Noretta Koertge (ed.), *The New Dictionary of Scientific Biography*, Vol. 1, pp. 236–243. Detroit: Thomson Gale.

To Be Published

"Probability in Quantum Mechanics". In Brigitte Falkenburg, Dan Greenberger,
    Klaus Hentschel and Friedel Weinert (eds.), *A Compendium of Quantum Physics:
    Concepts, Experiments, History and Philosophy*, Berlin/Heidelberg: Springer.
"Noncontextual and Contextual Models". Ibid.
"Hao Wang on the Cultivation and Methodology of Philosophy"
A Birthday in Bologna—a play or filmscript based on *Tibaldo and the Hole in the
    Calendar*
Perfect Pottos (A children's book with illustrations by Amanda Mouselaer).
A Flight of Ideas: Plays, Fables, and Poems
Edited, with Introduction, Annemarie Shimony, *Iroquois Portraits*, Syracuse, New
    York: University of Syracuse Press.

Published works of Abner Shimony: Errata

129. "The Transient Now", reprinted in [110] *Search for a Naturalistic World View*, vol. II, paper
     18, Eq. 4: the radical in the denominator should be in the numerator. Also, in footnote 5,
     "Zelicovici" should be "Zeilicovici," and likewise for all other occurences of this name.
132. In the first poem, "The Present," line 3, "thrill" should be replaced by "trill".
135. "A Bayesian Examination of Time-Symmetry in the Process of Measurement": the summa-
     tions with respect to i in the numerator and with respect to j in the denominator should be
     outside the absolute value signs. Consequently, this paper is not reliable.
141. "On Ensembles That Are Both Pre- and Post-Selected": in Eq. (3) the W on the lhs should be
     W', and the rhs should be divided by $\mathrm{Tr}[WQ(c')]$. In Eq. (5) the summation should be with
     respect to i rather than c. The argument against time symmetry in the latter part of the paper
     should be carefully re-examined.
166. "Aspects of Nonlocality in Quantum Mechanics" has serious errors in Sections 6.3 and 6.4.
     These two sections should be replaced by the following.

The demonstration of a generalized Bell's Inequality is correct, but the demonstration that some of the calculation of quantum mechanical violation of this Inequality
is erroneous.

   The discrepancy between Bell's Inequality and some of the predictions of quantum mechanics will be demonstrated in the special case of a pair of photons, propagating in opposite senses along the z-direction, hence with polarizations in the

xy-plane. A possible basis for the space of polarization states of photon 1 consists of the two vectors $|x_+>$ and $|x_->$, the former representing a state of polarization in the x-direction, the latter a state of polarization in the y-direction, hence blocked by an analyzer that allows with certainty light polarized in the x direction to pass. The pair of vectors $|x'_+>$ and $|x'_->$ represent analogous polarization states for photon 2. The pair of photons is assumed to be quantum mechanically described by the entangled two-photon state

$$|\Psi> = (1/\sqrt{2})[|x_+>|x'_+> + |x_->|x'_->] \qquad (1)$$

Other polarization states can of course be expressed in terms of these basis vectors. The state of the first photon with polarization in the direction in the xy-plane making an angle a with the x-axis will be represented by $|a_+>$ and that with polarization in the perpendicular direction is $|a_->$; analogous states for the second photon are represented by $|a'_+>$ and $|a'_->$. The self-adjoint operator representing polarization in the prescribed direction a is A, with the property

$$A|a_+> = |a_+>, A|a_-> = -|a_-> \qquad (2)$$

Other capital roman letters will represent polarization in other directions for both first and second photons. In terms of the bases that have been chosen:

$$|a_+> = (1/\sqrt{2})[\cos a|x_+> + \sin a|x_->], \qquad (3a)$$
$$|a_-> = (1/\sqrt{2})[-\sin a|x_+> + \cos a|x_->], \qquad (3b)$$

and likewise for $|b'_+>$ and $|b'_->$ in terms of the basis vectors $|x'_+>$ and $|x'_->$.
In order to expedite the calculation of the expectation values

$$E_\Psi(AB) = < \Psi|AB|\Psi>, \qquad (4)$$

it is useful to write the action of the operators A, B etc. on the basis vectors:

$$A|x_+> = \cos 2a||x_+> + \sin 2a|x_->, \qquad (5a)$$
$$A|x_-> = \sin 2a||x_+> - \cos 2a|x_-> \qquad (5b)$$

It is then straightforward to show that

$$E_\Psi(AB) = \cos 2(a - b), \qquad (6)$$

where a and b are the angles made by the directions of polarization of photons 1 and 2 with the x and x' axes.

To exhibit a discrepancy between Bell's Inequality, Eq. (6.13) on p. 114 of "Quantum Mechanics at the Crossroads" and a quantum mechanical prediction it suffices to choose two angles $a = \pi/4$ and $f = 0$ for polarization measurements on photon 1 and two directions $b = \pi/8$ and $g = 3\pi/8$ for polarization measurements on photon 2.

Since

$$\cos 2(a-b) = \cos 2(a-g) = \cos 2(f-b) = -\cos 2(f-g) = -.707, \quad (7)$$

one obtains

$$E_\Psi(AB) + E_\Psi(AG) + E_\Psi(FB) - E_\Psi(FG) = 2.828, \quad (8)$$

in disagreement with Bell's Inequality (6.13) on p. 114.

(Note the correction of the order of letters BF and BG in Eq. (6.14) on p. 115.)

Index

A

Absence of orbital precession, 45
Adler, S., 274
Analytical Society Memories (Charles Babbage and John Herschel), 24
Anderson, R., 10, 15
An Essay Concerning Understanding (Locke), 26
Anomalous Brownian motion (ABM), 305
Anton Zeilinger's 60th birthday conference, 63
"Applications of the Quantum Theory to Atomic Problems in General" (Bohr), 73
Approximate ideality, 238–239
Approximate joint measurement, 247–248
Approximate repeatability, 237–238
Areal rate, 46
Aristotle, 25, 350
Arons, A. B., 69–70
Aspect, A., 125, 212

B

Baldwin, J. M., 4
Bassi, A., 279
Bayesian learning model, 44
Bechmann-Pasquinucci, H., 135
Bell correlations, 66
Bell experiment, 370
  assumptions in, 372–373
  counterarguments in, 376–378
Bell hidden variable proof, 150
Bell inequalities, 4, 126–127. *See also* Bell's concept of locality
  bipartite, 132–133
  of Clauser–Horne, 151–154
  diagonal, 134
  elegant, 135–137
  fundamental questions, 128–130

loopholes, 212
for nonlocal resources, 131–132
original, 149–151
questions relevant to experimental issues, 130–131
tension between relativistic locality and the non-locality, 487–490
tests for correlations, 131
tight, 127
usage of, 126
violations of, 127
  with d outcomes, 129
  n-party pure entangled states, 129
Bell, J., 262–263
Bell-like inequalities, 96
Bell's concept of locality, 142
  and classical (hidden variable) theory, 144–145
  "Parameter Independence" and "Outcome Independence," 142–144
  quantum mechanical probabilities, 146–148
Bell's theorem, GHZ application
  elements of reality, 90–91
  EPR postulates, 91–94
  functions, 90–91
  polarizations of the four particles, 91
  review, 89–90
  two-particle state, 90
Bell-type correlations, 111
Bennett, C. H., 382
Bessel function, 178
Binary collision equation, 318
Bipartite quantum state, 96
Blake depiction, of Newton, 57–58
"Block universe" view of time, 360
"Blurred" two-photon image, calculation of, 200

Bohemian mechanics, 218–219, 224–226
Bohr–Kramers–Slater theory, 76
Bohr–Rosenfeld paper, 80
Bohr's photon concept
  agreement with Einstein's views, 81
  Bohr–Kramers–Slater theory, 76
  complementary descriptions, 78–80
  Correspondence Principle, 72–75
  and Einstein's experiments, 77–78
  first theory of atom, 71
  starting point of, 70–71
  textbook discussions, 69–70
  use of "phenomenon," 80–81
Borel algebra of subsets, 231
Borel set, 237
Bose–Einstein statistics, 318
Bothe–Geiger experiments, 70, 76
Boulliau, 46–47
Brans, C., 56
Brans–Dicke theory, 55
Brent, J., 16–17
Broad, C. D., 313
Brown, E., 53–54
Brownian motion, 305
Brown's measurement, 53–54
Bucket experiment, of Newton, 55–56
Burgers, J. M., 63
Burks, A. W., 17
Busch, P., 229

C
Calhoun, R., 6
Campbell, D., 4
Campos, D., 35
Carlo, G., 263
Carnap, R., 7, 349, 426
Carnap's λ-principle, 311
Causaloid diagram, 392
Causaloid formalism
  classical and quantum computers, 392–393
  computations
    out of gates, 395–396
    universal, 396–397
  data analysis, 386–387
  first level physical compression, 388–389
  notion of state evolution in, 394–395
  objective of, 387–388
  second level physical compression, 389–390
  third level physical compression, 390–391
  use in making predictions, 393–394
Cavendish experiments, 371
Ct–distribution, 318
Cell entropy, 321
Cell occupation vector, 314

Centripetal accelerations, 45
Centripetal direction force, 45
Centripetal force, 45
Chapman–Kolmogorov equation, 317
CHSH inequality, 127–128, 130
Church, A., 382
Church-Turing-Deutsch principle, 382–383
Church-Turing hypothesis, 382
Classical (hidden variable) theory, 144–145
Classical theory language, 264
Clauser–Horne (CH) inequality, 102–103,
    151–154
Clauser, J., 7
Closing the circle, 350–351, 356. *See also*
    Shimony, A.
Cognitions, truth-tracking, 5
Coherence cells, notions of, 300–301
  of spherical solid body, 302–303
Coherence width, 301
Collapse problems and their resolutions,
    261–262
  conservation law problem, 275–279
  experimental problem, 272–275
  framework for, 258–261
  legitimization problem, 285–290
    cosmological creation of negative w-field
      energy, 287–289
    gravitational considerations, 286–287
    other cosmological considerations,
      289–290
  relativity problem
    in Gaussian noise, 281
    in quasi-relativistic collapse model,
      284–285
    in Tachyonic noise, 281–284
    in white noise field, 279–281
  tails problem
    qualified possessed value criterion,
      264–272
    smeared mass density (SMD) criterion,
      263–264
Collinear propagated signal–idler photon pair,
    171
Collinear SPDC, 176
Complementary descriptions, concept of,
    78–80
Completely positive operations, 232
"Completely Quantized Collapse" (CQC)
    model, 275–277
Completeness, concept of, 93, 108–111,
    115–123, 160
Complete positivity, 232
Compton–Simon experiments, 70
Compton wavelength, of neutron, 300

Computational reflection principle, 383
Computation, definition, 379
Conditional behavior, of particle, 201
Condition C, physical interpretation, 108–109, 115–123
Condition S, physical interpretation, 118–120
Conservation law
  problem
    "Completely Quantized Collapse" (CQC) model, 275–277
    conservation of energy, 277–279
    and quantum limitations of measurements, 239–241
Conservation of energy, 277–279, 322
Continuous spontaneous localization (CSL)
  center of mass wave packet, 269–270, 272–273
  density matrix constructed, 260
  dynamical equation, 259–261
  formalism of, 260–261
  ionization of Ge atom, 274
  lite, 258–260
  probability associated, 259
Cooke, E. F., 29–30
Copenhagen versions, of probability measures, 219
Correspondence principle, 72–75
Correspondence, state of physical system, 266
Costantini, D., 311
CQC Hamiltonian, 286
Creation condition, 315–316
CTD principle, 382–383
C-thermodynamics, of very small systems, 320–323
CW monochromatic laser pump, 165

D
Darwin, G., 73
Darwinian evolution, 4
Davies, P. C. W., 350
De Broglie, L., 77
Decay factor, 273
Delta wavefunction, 158
De Morgan, 27
De Motu (Newton), 50
Descartes, 19, 31, 328
Destruction condition, 315–316
Determinism, 106, 108, 111
Deutsch, D., 380, 382
Development of the Concepts of Physics (Arnold B. Arons), 69
Dewey, J., 4
Dicke, R. H., 56
Dieks, D., 352

Discontinuity, element of, 70–71
Discourse on the Cause of Gravity (Huygen), 51
Dispersion relation $k(\omega)$, 169
Distinct measurement-types, 107
Dobson, H. A., 353
Doppler effect, 337
Dowker, F., 374
Duhemian-Quinean "holism" of testing, 212
Dynamical collapse theory language, 267–268

E
Eger, M., 9
Ehrenfest's adiabatic principle, 75
Eigenstate-eigenvalue link, 265
Einstein–Minkowski space-time
  assumptions, 404
  automorphisms, 408
  causal influences, 406–407
  clarifications, 426–438
  criticism, 404–405
  impact of scale-invariance, 408
  logical situations, 405–406
  relation "definable from κ and O", 405–406
  and Sarkar–Stachel theorem, 407
    proofs, 408–426
    supplementary notes, 438–441
  time-orientation, 404, 406
Einstein–Podolsky–Rosen (EPR), 376
  completeness, 213
  element of reality, 87
  entangled signal–idler photon pair of SPDC, 175–176
  gedankenexperiment, 170, 195
    Bohm's version, 144
Einstein–Podolsky–Rosen (EPR)-Separability, 111, 113–117
Einstein's incompleteness, of quantum mechanics, 111
Einstein's light quantum hypothesis, 70, 81
Einstein's optical experiments, 77–78
Einstein's principle of separability of physical states, 212
Einstein's theory of relativity, 54–57
Ekert, A., 125
Elastic binary collision, 316
Elastic dipole, 320–321
Elektroweak quantum field theory, 296
Elementary particles statistics, characterization
  Abner comments, 312
  c-thermodynamics of very small systems, 320–323
  destruction and creation probabilities, 314–316

equilibrium distribution, 318–319
problem of indistinguishability, 312–314
result on random variables, 314
stochastic dynamics, 316–317
throwing two imaginary quantum coins, 313
Elementary region, 386
Element descriptions (complexions), 315
Element of reality, 92
Elements of Logic (Whatley), 23
Elements of the History of Mathematics
(Bourbaki), 33–34
Energy density operator, 279
Ensemble vector, 277–278
Entangled quantum mechanical systems, 3
EPR/Bell correlations, 66
EPR-Holism, 114
EPR-separable theory, 120–121
EPR system, 158–159
criteria, 159–160
Equilocs, 329
Equitemps, 329
Error bar width, 250–251
Event, definition, 329
Everett, H., 220
Everett interpretation, 220–224
Evolutionary epistemology, 4
Exchangeability, definition, 314
Expansion-contraction cycles, notions of,
304–305
Experimental metaphysics, 214
and Bohemian mechanics, 224–226
Experimental problem, 272–275

F
Faraday Lecture, 79
Fermi–Dirac statistics, 318
Fermion density, 283–284
Field operators, 202
Fields of acceleration, 45
Fisch, M., 17
Foot (ft), defined, 330
Fourier conjugate variables, 160–161
Fourier spectroscopy, 202
Fourier transforms, 160
of the spectrum amplitude function, 172
Free $W(x)$-field time-translation generator, 277
Frenkel, A., 293
Fresnel paraxial approximation, 172, 180
Fresnel phase factor, 180
Fresnel propagation-diffraction, 180, 205–207
Fresnel propagator, 180
Friedman, M., 427
Fry, E. S., 141
Fuchs, C., 468

Full pack, 386
Fundamental theorem, quantum measurement
theory, 233

G
Garibaldi, U., 311
Gaussian function, 173, 180
Gaussian thin lens equation, 177, 180, 190,
196
Generalization, of completeness, separability
and condition C, 121–123
Gentile, G. Jr., 314
Geometry's dual nature, 28–29
Geroch, R., 220–221
Ghirardi, G.C., 95
Ghirardi–Rimini–Weber–Pearle model, 372
"Ghost imaging" experiment
application of Gaussian functions, 192
calculation of Green's functions, 190–192,
194
calculation of transverse coordinates, 190,
193
EPR δ-functions, 189
Gaussian thin lens equation, 190, 192–193
measurement of the signal and the idler
subsystem, 189
object-aperture function, 193–194
recording of coincidence counts, 188
schematic setup of, 187
unfolded setup of, 189
Gibbs Inequality, 115
Gisin, N., 125
Glauber theory, 170
Global time—Fermi-Walker coordinates, 366
Glymour, C., 46
Gödel, K., 352
Goldenberg, H. M., 56
Goudge, T. A., 34
GPS system, 385
Greenberger, D.M., 9, 87
Greenberger–Horne–Zeilinger (GHZ)-type
argument, 87
Green's function, 171, 182, 206–207
theory, 296–297
Grünbaum, A., 426–427
Grundlagen (Cantor), 28

H
Haack, S., 29
Hackett, J., 459–460
Halley, E., 50
Hall's hypothesis, 53–54
Hamilton, Sir W., 27
Hardy, L., 57–58, 379

Index 511

Hardy's nonlocality conditions, 96–102
Harper, W., 43
Hartshorne, C., 17
Harvey, B., 257
Heisenberg's idea of potentiality, 19
Heisenberg's treatment, wave packet light, 77
Hellman, G., 211, 453
Henson, J., 374
Hermitian conjugate, 166
Hidden variable models, 98
Hilbert space, 96, 101–103, 113, 128–129,
     135–136, 215–217, 222, 225, 230–233,
     236, 246, 313, 383, 463, 480, 489
Hogarth, M., 403
Holst, H., 75
Holt, R., 7
Hookway, C., 35
Horne, M., 7, 87, 363
Howard, D., 3, 111
Hume, 31
Huygens-Fresnel principle, 205
Hypothetico-deductive (H-D) model of
     scientific method, 43–45

I
Ideal measurements, 236–237
Image plane, defined, 180
Indefinite causal structure, 380
Indeterminacies
     Einsteinian space-time, 295–296, 308–309
     in relative phases of quantum state, 298–299
     synchronization, moments of time, 296–297
     in vaccum state, 297–298
Indistinguishability, problem of, 312–314
"Interaction Hamiltonian" density, 280
Invariance, definition, 314
Ippoliti, E., 279
Isochron, 329

J
James, W., 4, 17
Jarrett, J.P., 8, 105
Jarrett "locality" condition, 8
Jarrett's theorem, 143
Joint measurability, 247
Joint probability functions, 106
Joswick, H., 34

K
Kalckar, J., 77
Károlyházy cycles, 305
Károlyházy, F., 294–295
Kent, A., 369
Kepler's area law, 45–46

Kepler's harmonic law, 45–46
Ketner, K. L., 17, 26, 33
Klein, M., 75
Klein, O., 72
Kramers, H., 75
Kuhn, T., 53, 55

L
Legitimization problem
     cosmological creation of negative w-field
          energy, 287–289
     gravitational considerations, 286–287
     other cosmological considerations, 289–290
Leibniz's rule, 490
Lightlike event, 335
Light rectangle, notion of, 328
Local determinism, 212
LOCALITY, parameter independence
     condition, 107–108, 142–144, 159
Local model, 96
Local polytope, 127
Locke, J., 426
Lorentz transforms, 364–366, 429
Lüders measurement, 237–238
Lüders theorem, 239

M
Macrostates, equilibrium probability
     distribution, 320–322
Malament, D., 403
Malin, S., 63, 459
Marginal probability functions, 107
Marinatto, L., 95
Markov chain distribution, 317–318
Markov transition matrix, 319
Mass density operator, 279
Mathematische Grundlagen der Quanten-
     mechanik (von Neumann), 233
Maudlin, T., 360
Maximum entropy principle, 321
Maxwell–Boltzmann statistics, 318
Maxwell, J.C., 29
Mean-distance, of Keplerian orbit, 47
Mercury precession problem, 53–57
Mermin contraption, 106
Mermin, N. D., 327, 360
Michelson interferometer, 202
Microstates, equilibrium probability
     distribution, 317–318
Minkowskian space-time, 296–297, 349, 355
Minkowskian zero value, 297
Minkowski's space-time diagrams, 328
Mirabelli, A., 461–463
MKB inequality, 130

Monthly Notices of the Royal Astronomical Society, 54
Mutual information, 117–118
Myrvold, W., 57, 477

N
Naturalistic epistemology, 4
Negative *w*-field energy, cosmological creation of, 287–289
Newton's laws of motion, 46
Newton's methodology
  bucket experiment, 55–56
  center of mass resolution, 50
  classic inferences from phenomena, 45–49
  comments on Newton, 57–59
  contrary hypotheses, 52
  empirical success of theories, 46
  vs. hypothetico-deductive (H-D) model of scientific method, 43–45
  Mercury precession problem, 53–57
  orbits of the primary planets, 48–49
  provisional acceptance of theoretical propositions, 46
  sixth corollary, 48
  Strong Ideal of Empirical Success, 52
  theory-mediated measurements, 46
Nonlocality, aspects of, 142–144
  Bell-like inequalities, 96
  Clauser-Horne inequality, 102–103
  Hardy's nonlocality without inequalities proof, 96–102
  Hilbert space, 101
  joint-probability distributions, 95
  set-theoretic inequality, 99
  theorem I, 100–102
  theorem II, 102–103
"Notch" function, 202–203
Null vectors, 354

O
Oseen, C.W., 71
Outcome independence, 8, 143, 213
Outcome set in region, 387
Ozawa's model, sharp position measurement, 234

P
Pais' biography, 54
Panpsychism, proto-mentality/mentalism, 20–21
Parameter independence, 8, 107, 142, 213
Parametrized Post Newtonian (PPN) Formalism, 56
Φ-particle propagator, 283

"Passion at a distance," 3
Passionate intensity, 3
Passionate sympathy, 3
Past region, of spacetimeΛ, 374
Pauli Exclusion Principle (PEP), 486–487
Pauli matrices, 146
Pearle, P., 257, 460
Peirce, B., 21, 36
Peirce, C.S.
  collection of papers
    style and ideas, 17
    usage of metaphors, 18
    use of quotations, 18
  life and thoughts, 16–20
  on mathematics, 20–26
    abstraction, 35
    application in study of nature, 26–37
    and Boole's work, 23–24
    characterizing logical inference, 33
    creation of hypotheses, 32–33
    early education, 21–22
    features of, 23
    generalization, 28
    hypothetical nature, 26–27
    meeting place of mathematics and nature, 35–36
    metaphor functions, 30–31, 37
    observation in, 24–25
    perceptual judgments, 25
    Platonism in, 32–33
    position of fallibilism, 29–30
    priori thinking, 24
    Pythagorean fusion, 37
    reasoning, 24–25
    relation of logic and mathematics, 22–23
    role in developing a philosophy, 22
    space imagination, 28
    "theorematic" reasoning, 34
    use in physics, 29
  role of mathematics in natural sciences, 26–37
  and scholarship, 16
  selected topics, 16
  themes
    characterization of methods of scientific inquiry, 20
    epistemology, 20
    fallibilism, 18–19
    proto-mentality or mentalism, 20–21
    theory of mind, 21
    universe as evolutionary, idea of, 20
    use of probability theory, 19
Penrose, R., 400
Penrose's cat, 385

Index   513

"Phenomenon "concept, of Bohr, 80–81
Photon trajectories, 332
Physical compression, 388–391
Pitowsky's "quantum gambling devices," 219
Planck length, 295
Planck mass, 303
Planck's constant, 320
Planck's "second theory" of radiation, 70–71
Planck time, 295
Plane geometry
    application of light rectangles, 341–342
    clock postulate, 328–329
    creation of plane diagram
        application of light rectangles, 341–342
        area Ω of the light rectangle determination,
            339–340
        concept of relativity, 332–334
        estimation of distance, 344–345
        estimation of squared interval, 340–341
        intersection of an equiloc and equitemp,
            329–330
        photon trajectories, 332, 344–345
        ratio of each scale factors, 338–339
        rectangle of photon lines, determination
            of, 335–337
        relating to spatial and temporal scales,
            330–332
        two different scale determination, 334–335
    interchange of space and time, 328
Poincaré, H., 426, 428
Poincaré sphere, 137
Point-spread function, of the imaging system,
    180
Polya distribution, 321
Polytopes, Bell inequalities, 127
Popper, K., 4, 44, 194
Popper's "propensity interpretation" of QM,
    218
Popper's thought experiment, 194–201
    measurements, 198–201
    modified version of, 197
Positive-operator-valued measure (POVM),
    129, 231–233, 234, 240, 243
Possessed value, 265–266, 268
Preliminary Discourse (Herschel), 27
Principia (Newton), 45–46, 48–51
Probability of subspace, $P(ij; kl)$, 151–152, 154
Probability reproducibility condition, 232
Probability rule equation, 258–259
"Problems of The Atomic Theory" (Bohr), 74
Process philosophy and QM. *See* Whitehead's
    philosophy
Program in region, 387
Projection operator, 147

Proper time, 355
Purcell, E., 327
Pure state, for a system, 164
Putnam, H., 17, 26

Q

Qualified possessed value, 268–270
Quantum entanglement, concept of, 158
    classical inequality *vs.* EPR inequality, 162
    entangled state of a two-particle system,
        164–170
        correlation measurement, 170–177
        of a pair of lower frequency signal-idler
            photons, 165–166, 168
        two-photon superposition, 176
    EPR system, 158–160
    "ghost imaging" experiment, 187–194
    Popper's experiment, 194–201
    practical engineering applications, 177–187
    realistic model, 162–163
    subsytems, 201–204
    violation of the uncertainty principle,
        160–161
Quantum gravity computer, 380, 384–386
    power, 398–399
Quantum imaging
    classical imaging, 179–180
    coherent, 181
    determination of Fourier transforms,
        184–185
    Green's function, 182–183
    important aspects of physics in, 186–187
    incoherent, 181
    features, 177
    image plane, 180
    lens for, 177
    perfect image, method for, 177–178
    point-spread function of, 180
    standard imaging setup, 178
    use of quantum entangled states, 179
Quantum limitations, of measurements
    additive conserved quantities, 480
    approximate ideality, 238–239
    approximate repeatability, 237–238
    complementarity in n-particle interferome-
        tery, 483–485
    complementarity principle, 242–244
    complementarity *vs.* uncertainty, 245–246
    Conservation laws, 239–241
    discrete observable, 239
    discrete sharp observable, 237
    effects, 231
    examples, 233–234
    formalism of, 234–241

inequality, observable marginals, 249–250
Lüders theorem, 239
measurement schemes, 231–233
measures of inaccuracy and disturbance, 246–253
no information gain without disturbance, 235
no measurement without entanglement, 235–236
observable marginals, 249
Ozawa's model, 234
problem of definite measurement outcomes, 480–483
quantum measurement theory, 230–234
repeatability and ideality, 236–237
repeatable observable measurement, 237
selfadjoint operators, 246
set of states, 230
sharp observable admits repeatable measurement, 240–241
trade-off relation, observable marginals, 251
uncertainty principle, 244–245
unitary mapping, 236
Von Neumann model of an unsharp position measurement, 233–234
Quantum measurement theory, fundamental theorem, 233
Quantum mechanical analogs, 147
Quantum mechanical states
in Bell's inequality, 146–148
completeness, concept of, 108–111, 115–120
condition C, 108–109, 115–120
condition S, 118–120
determinism, 106, 108, 111
distinct measurement-types, 107
Einstein, Podolsky, and Rosen (EPR)-separability, 111, 113–117
joint probability functions, 106
locality, 107–108
marginal probability functions, 107
Mermin contraption, 106
mutual information, 117–118
parameter independence, 107
relativity theory, 107
separability, 112–113, 115–120
strong locality, 110
theoretical predictions, 106
Quantum probability, interpretations of, 215–220
Bohmian mechanics, 224–226
"Everett interpretation," 220–224
two-by-two matrix of interpretative possibilities, 216–218

Quasi-Relativistic Collapse Model, 284–285
Quinean philosophy of science, 211
Quine. W. V. O., 426
Qu, X., 141

R
Reichenbach, H., 7
Reichenbach's conception of "screening off," 225
Relativistic "locality" constraint, 107
Relativistic quantum mechanics
approximations, 368
inertial frames, 364–366
non-inertial frames, 366–367
Relativity problem
in Gaussian noise, 281
in quasi-relativistic collapse model, 284–285
in Tachyonic noise, 281–284
in white noise field, 279–281
Relativity theory, 107
Repeatable measurements, 236–237
$(d,1-\varepsilon)$-repeatable measures, 238
Resolution width, 251
Rest energy phase factor, 298
Reverse-logicism, 23
The Revolt Against Dualism (Lovejoy), 63, 452
Rideout, D., 374
Rovelli, C., 350

S
Sarkar, S., 404
Sarkar–Stachel theorem, 407
proofs, 408–426
supplementary notes, 438–441
Satellite clock, 385
Saunders, S., 312–313
Savitt, S. F., 349
Schlick, M., 7
Schrödinger, E., 77
Schrödinger's cat, quantum entanglement, 157–158
Schrödinger's "Principle of objectivation," 64–65
Schrödinger wave function, 298
Scully, M. O., 141
Second-order correlation functions, 174
Separability, 112–113, 115–123
Separable theory, 120–121
Separation principle, 9
Sequential joint probability, 232
Shannon information, 112
Shapiro, I., 56
Shapiro's time delay, 56
Sharp observables, 231

Index 515

Shih, Y., 157
Shimony, A., 3, 107, 263, 327
  achievements, 9
  on Bell's theorem, 6–8
  contributions, 4
  dialogue with Lee Smolin
    about "connecting," 455
    Bayesianism, 467–468
    Bialynicki-Birula's paper, 464
    chemical constitution of the stars, 450–451
    closing the circle, 448–449, 451–452, 466–467
    clustering of problems, 464–465
    coherence, 463
    configuration space, 464
    consciousness, 455–456, 461–462
    dialectic, 449
    disagreements with Whitehead, 452–453
    epistemology *vs.* ontology, 468
    fundamental physics, 466
    notion of law in biology, 473–477
    notion of life, 453–454
    ontologies, 457, 459–460
    Peirce quotes, 471–472
    phenomenological principle, 457, 459
    philosophy of science *vs.* physics, 446–447
    quantum spacetime, 474–476
    realism, 449
    strategic probability, 467
    theoretical preconceptions, 460–461
    timeless formulation, 456–457
    transience and time, 454, 458, 461–463, 466
  elementary particles statistics, 312
  experimental metaphysics, 212
  intellectual passion, 4
  life history, 9–10
  lineage of Peirce's ideas, 19–20
  metaphysics, 6, 9
  naturalistic epistemology, 4, 9
    vs. W.V.O. Quine, 4–5
  "naturalist" perspective of brain, 21
  professional and personal experiences, 6–7
  quantum physics, 7
  reflections on Jon Jarrett's work, 8
  scientific inferences, 5–6
  tail state's squared amplitude, 271–272
  theory *h*, 52
Shimony, A.A., 10
Shimony, D. F., 9
Shimony, E., 10
Shimony, J., 10
Shimony, M. (Moshe), 9

Shimony, S. A., 9
Sikic, M., 10
Silliman Lecture, correspondence principle, 73–74
Smeared mass density (SMD), 263–264
Smolin, L., 445
Somb-function, 180–181, 183
Sorkin, R., 374
Spacelike event, 335
Spacelike vectors, 354
Spacetime, local causality of
  discussion, 376–378
  expectations from experiments, 371–373
  overview of experiment, 370–371
  technicalities, 374
  testing of local causality of metric theories, 374–376
Space-time transforms, 363–366
SPDC Hamiltonian, 166
Speed, defined, 366
Spontaneous parametric downconversion (SPDC) process, 165–167
Stachel, J., 9, 69, 309, 404–405
Stack, causaloid formalism, 386
Standard error, 248–249
Standard quantum theory language, 265–266
State, idea of, 383–384
Stein, H., 8–9, 45, 403
Step computers (SC), 380
Stern–Gerlach apparata (SGA), 144
Stewart's view, of mathematics, 33
Stochastic dynamics, 316–317
Stochastic hybrid theory, 373
Strauss, R., 328
Strong Locality, 110
Submacroscopic decay, 306–307
Sudbury Neutrino Observatory experiment, 274
Superposition principle break down, 296–297
  indetermnancies
    anomalous Brownian motion, 305
    Einsteinian space-time, structure, 295–296, 308–309
    localizing effect *vs.* expansion–contraction cycles, 306–307
    notions of coherence cells, 300–303
    notions of expansion-contraction cycles, 304–305
    observation-independent rule for, 304
    relative phases of a quantum state, 298–299
    vaccum state, 297–298
    submacroscopic decay, 306–307
    transition between regions, 303–304

Sylvester, J.J., 24, 29
System of the Worldi (Newton), 49
Szilard, L., 6

T
Tails problem
    qualified possessed value criterion, 264–272
    smeared mass density (SMD) criterion,
      263–264
Tempered personalism, 5
Theoretical predictions, 106
Theory of harmonic vortices, 51
Theory of mind, 21
Thompson prism, 187
Tibaldo and the Hole in the Calendar, 4, 312
Timaeus (Plato), 65
Timelike curve, 355
Timelike event, 335
Timelike vectors, 354
Tisza, L., 9
Trace distance, of mixed state, 96
Transience, 352
Transient *nows*
    Carnap's response, 350–351
    classical, 351–352
    Dieks' argument, 352
    reservations and rejoinders, 358–361
    specious/psychological, 356–358
    and theory of relativity, 354–356
    transience of time, 353–354
Transition values, microscopic *vs.* macroscopic
    behavior, 303–304
Trivial observable, 231
Two-by-two matrix, interpretative possibilities,
    216–218
Two-particle entangled state experiment
    elements of reality, analysis, 90–91
    EPR postulates, 91–94
    functions, 90–91
    polarizations of the four particles, 91
    review, 89–90
    schematic diagram, 90
Two-photon effective wavefunction, 172
Two-photon spatial correlation, 174–177
Two-photon state, of SPDC, 165–170
Two-photon temporal correlation, of SPDC,
    173–174

Two-photon wavefunction, 176
Type II SPDC, 169
Type-I SPDC, 169

U
Universal computer, 396
Universal gravitation, 48
Unsharpness, 251
Unsharp observables, 231
Unsharp reality, notion of, 238

V
Vacchini, B., 279
"Vacuum" state, 276
Van Fraassen, B.C., 53
"Volkswagen state" (VW state), 89
Von Neuman entropy, 203
Von Neumann model, 233–234

W
Walker, H. C., 10
Weber, T., 263
Weiss, P., 6, 17
Werner's joint measurement uncertainty
    relation, 249
Werner states, 128–129
Weyl, H., 349–350, 353–354
W-field energy-momentum, in CQC, 286
Whately's Logic, 23, 27
Wheeler, J.A., 6, 220
Whewell, W., 33
Whitehead's philosophy
    Abner's modifications, 63–64
    and QM, 64
      acceptance of Whitehead's statement,
        67–68
      correspondence of the collapse to an actual
        entity, 65
      EPR/Bell correlations, 66
      Lovejoy's objection, 67
Wick's theorem, 282
Wigner, E., 6, 9

Z
Zeilinger, A., 87, 466, 469
Zukowski, M., 87

CPSIA information can be obtained at www.ICGtesting.com
Printed in the USA
LVOW070304100412

276918LV00002B/65/P